Advances in Neurobiology

Volume 30

Series Editor
Arne Schousboe
Department of Drug Design & Pharmacology
University of Copenhagen
Copenhagen, Denmark

Advances in Neurobiology covers basic research in neurobiology and neurochemistry. It provides in-depth, book-length treatment of some of the most important topics in neuroscience including molecular and pharmacological aspects. The main audiences of the series are basic science researchers and graduate students as well as clinicians including neuroscientists (neurobiologists and neurochemists) and neurologists. *Advances in Neurobiology* is indexed in PubMed, Google Scholar, and the Thompson Reuters Book Citation Index.

Matthew Macaluso
Sheldon H. Preskorn • Richard C. Shelton
Editors

Drug Development in Psychiatry

Springer

Editors
Matthew Macaluso
Department of Psychiatry and Behavioral Neurobiology
University of Alabama at Birmingham Heersink School of Medicine
Birmingham, AL, USA

Sheldon H. Preskorn
Department of Psychiatry and Behavioral Science
Kansas University School of Medicine-Wichita campus
Wichita, KS, USA

Richard C. Shelton
Department of Psychiatry and Behavioral Neurobiology
University of Alabama at Birmingham Heersink School of Medicine
Birmingham, AL, USA

ISSN 2190-5215 ISSN 2190-5223 (electronic)
Advances in Neurobiology
ISBN 978-3-031-21053-2 ISBN 978-3-031-21054-9 (eBook)
https://doi.org/10.1007/978-3-031-21054-9

This Springer imprint is published by the registered company Springer Nature Switzerland AG
The registered company address is: Gewerbestrasse 11, 6330 Cham, Switzerland

Contents

About the Editors

Matthew Macaluso is the Bee McWane Reid Professor and Vice Chair for Clinical Affairs in the Department of Psychiatry and Behavioral Neurobiology as well as the Clinical Director of the UAB Depression and Suicide Center. His research focus is on clinical psychopharmacology and its translation to clinical practice with a focus on treatment-resistant major depression.

As clinical director, Dr. Macaluso oversees clinical trials of novel mechanism of action drugs, devices, and biologics to treat patients with severe forms of depression that are not responsive to currently marketed treatments. He is a highly experienced investigator in complex clinical trials and the translational neuroscience of mood disorders and has contributed most significantly in the area of novel treatment development.

Dr. Macaluso completed medical school at the University of Medicine and Dentistry of New Jersey in Stratford, NJ, and graduated from the psychiatry residency program at the University of Kansas School of Medicine, where he was on faculty for 11 years before joining UAB in 2020.

Sheldon H. Preskorn is generally considered one of the world's foremost experts in psychiatric drug development research having worked with over 145 pharmaceutical, biotechnology, devices, and diagnostic companies around the work and was a principal site investigator in all antidepressants and antipsychotics marketed in a 25-year period.

His overarching goal has been throughout his career to bring science to the practice of psychiatry. His research focus is on clinical psychopharmacology and its translation to clinical practice with a focus on otherwise treatment-resistant psychiatric illnesses.

In addition to being a consultant broadly to companies bringing products to the market, he has also worked with the FDA in many different capacities.

Dr. Preskorn did his basic medical training at the University of Kansas School of Medicine where he also completed a two-year residency in anatomical pathology with a focus on neuropathology. He did his psychiatric residency at Washington

University School of Medicine in St. Louis MO. During his residency, he did seminal work on the role of the locus coeruleus in the brain. He has continued related work throughout his more than 40-year career in academic medicine.

Richard C. Shelton is the Charles Byron Ireland Professor and Co-director of the UAB Depression and Suicide Center in the Department of Psychiatry and Behavioral Neurobiology in the Heersink School of Medicine at the University of Alabama at Birmingham. He is an internationally recognized researcher in the areas of translational neuroscience and clinical psychopharmacology of mood disorders.

Dr. Shelton founded and co-directs the UAB Depression and Suicide Center with Dr. Yogesh Dwivedi. The Center has a clinic that provides advanced treatment options for depression including esketamine, electroconvulsive therapy, and vagal nerve stimulation, and it also conducts research ranging from basic molecular neuroscience and genomics to clinical applied research in mood disorders and suicide. Dr. Shelton is a translational neuroscientist, and his work spans from molecules to clinical trials. Recent work has included research on genomic predictors of depression and suicide and advanced therapies for depression.

Dr. Shelton graduated from the University of Louisville School of Medicine and graduated from the psychiatry residency program at the Massachusetts Mental Health Center (now part of the Brigham and Women's residency program) of Harvard Medical School in Boston. He then was a research fellow in the intramural program of the National Institute of Mental Health. He was a professor in the Department of Psychiatry at Vanderbilt University School of Medicine for 26 years before joining UAB in 2012.

Chapter 1
Drug Development in Psychiatry: The Long and Winding Road from Chance Discovery to Rational Development

Sheldon H. Preskorn

Abstract Based extensively on tables and figures, this chapter reviews drug development in psychiatry with an emphasis on antidepressants from the 1950s to the present and then looks forward to the future. It begins with the chance discovery drugs and then moves to through their rational refinement using structure activity relationships to narrow the pharmacological actions of the drugs to those mediating their antidepressant effects and eliminating the effects on targets that mediate adverse effects. This approach yielded newer antidepressants which compared to older antidepressants are safer and better tolerated but nevertheless do still not treat the approximately 40% of patients with major depression (MD) which is unresponsive to biogenic amine mechanisms of action. This form of MD is commonly referred to as treatment resistant depression. Esketamine is an antidepressant which has a novel mechanism of action: blockade of the glutamate NMDA receptor. These studies coupled with earlier studies with other NMDA drugs suggest approximately 60% of patient with TRD are rapidly and robustly responsive to this mechanism of action. Thus, there appears to be three forms of MD based on pharmacological responsiveness: (a) 60% responsive to biogenic amine mechanisms of action, (b) 24% (i.e., 40 × 60%) responsive to NMDA but not to biogenic amine mechanisms of action, and (c) 16% (i.e., 40–24%) not responsive to either of these mechanisms of action. Scientific investigation of these three groups may yield important information about the pathophysiology and/or pathoetiology of these different forms of MD. This information coupled with studies into the neurobiology (e.g., imaging studies, connectomes to name a few approaches being used) and genetics of MD should provide the fundamental knowledge which will permit a rational search for and discovery of newer antidepressant drugs and other somatic and psychotherapeutic approaches to the treatment of patients with different forms of MD based on pathophysiology and

S. H. Preskorn (✉)
Department of Psychiatry and Behavioral Science, Kansas University School of Medicine-Wichita Campus, Wichita, KS, USA
e-mail: spreskorn@kumc.edu

M. Macaluso et al. (eds.), *Drug Development in Psychiatry*, Advances in Neurobiology 30,
https://doi.org/10.1007/978-3-031-21054-9_1

pathoetiology. Examples are given of how such discovery and development have occurred in other areas of medicine and even in central nervous system (CNS) space including six novel mechanisms of action CNS drugs which have been successfully developed and marketed over the last 25 years.

Keywords Antidepressants · Central nervous system biogenic amines · Drug development · Esketamine · Major depression · Mechanism(s) of action · Psychiatric diagnosis · Relative receptor binding · Structure-activity relationships

> [For] knowledge of mental diseases one must have: (a) knowledge of the physical changes in the cerebral cortex, and (b) [knowledge of] the mental symptoms associated with them.
>
> Until this is known, we cannot hope to understand the relationship between symptoms of disease and the physical processes underlying them.—Emil Kraepelin [1], Father of modern psychiatry
>
> Symptoms and behaviors are the output of brain function whereas syndromes are man-made constructions.—Sheldon Preskorn [8]

This Chapter, which was adapted with permission from the Springer Nature book, *Antidepressants: From Biogenic Amines to New Mechanisms of Action*, will discuss the history of antidepressant drug development and put it into the broader context of psychiatric drug development. This chapter will focus on the history of and current status of antidepressant drug development but will also incorporate other concepts relevant to future antidepressants and other central nervous system (CNS) drug development. It will be heavily dependent on the writings of the author on these topics over the last 30 years. The chapter will be primarily focused on illustrative figures and tables with the minimum amount of text needed to explain the figures and tables, put them in context, and then transition to the next topic. All the articles in which figures and tables originally appeared are cited in the reference list. The reader who wants additional text and references on a given topic can do so by referring to the specific cited article of interest.

1.1 Current Status of Psychiatric Diagnosis as a Rate-Limiting Step in Rational Psychiatric Drug Development

In all of medicine, there are four levels of increasing sophistication of diagnosis as illustrated in Fig. 1.1 [12].

The first level is symptomatic diagnosis which is generally the presenting complaint of the patient to the treatment provider. For patients suffering from major depressive disorder (MDD), that presenting complaint may be feeling tired, absence of enjoyment, insomnia, or even headache to name but a few.

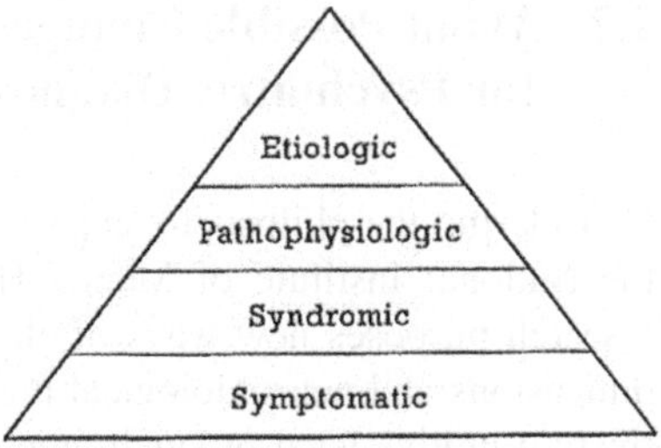

Fig. 1.1 Diagnostic criteria pyramid – the four levels of increasing diagnostic sophistication. (Reproduced with permission from Preskorn and Baker [12]. © Preskorn, 2002)

In general, the psychiatrist is then taught to advance to a second level of diagnostic sophistication which is the syndromic level. The result may be that the patient presenting with these initial complaints may meet criteria for major depressive disorder or perhaps acquired immunodeficiency disorder (AIDS) if the patient also has Kaposi's sarcoma, an opportunistic infection, and generalized wasting.

To reach the third level of diagnostic sophistication illustrated in Fig. 1.1 requires testing for pathophysiological findings. In the case of AIDS, that would be a lowering of the CD 4 count or a positive Western blot test or a high HIV titer. In the case of MDD, there is no generally established testing, but some practitioners might test for cortisol nonsuppression or REM latency which have both been proposed as biochemical test for "endogenous major depression."

To reach the fourth level of diagnostic sophistication illustrated in Fig. 1.1 requires the establishment of a test for the etiological agent or a neurobiological condition which is not established for most psychiatric disorders with the possible exception being testing for the presence of autoantibodies against the NMDA receptor for patients suffering from NMDA receptor-mediated neuroencephalitis. In the case of AIDS, it would be to test for the presence of the etiological agent, the HIV virus.

The above illustrates the basic problem with psychiatric drug development: The field is currently principally stuck at the syndromic diagnosis and has not been able – in general – to advance to the pathophysiological or to the even higher etiological level. However, that is not completely true. In the early 1900s, approximately 20% of admission to psychiatric hospitalization no longer exist. Those conditions were pellagra and general paresis of the insane. The former was due to vitamin D deficiency and the latter to tertiary syphilis. Once those etiological causes were identified and specific treatments identified, those conditions essentially no longer exist in the modern age and instead are consigned to being historical footnotes. In the future, the same will likely be true for major depressive disorder and other similar currently syndromic psychiatric diagnoses.

1.2 What Possible Changes Lie Ahead for Psychiatric Diagnoses?

Considering the philosophy expressed in my quote at the beginning of this paper, the National Institute of Mental Health (NIMH) in 2008 began to develop for research purposes new ways of classifying mental disorders based on behavioral dimensions and neurobiological measures. The goal being to move from the relatively primitive level of syndromic diagnoses to the next level pathophysiological diagnoses (Fig. 1.1).

The author proposed a similar approach in a paper published 34 years earlier and illustrated in Fig. 1.2 [3]. The concept expressed in this figure is that there may be both syndromes which have an underlying biology and dimensional aspects of traits such as impulsivity, IQ, and introversion to extroversion which are independently, biologically, and environmentally determined which can modify the expression of the syndromic cluster such as agitated versus psychomotor retard MDD. Treatments addressing the pathophysiology or even better – perhaps – the pathoetiology of the syndromic diagnosis (MDD) and the pathophysiology of the modifying dimension (e.g., impulsivity) might be the ideal way to approach a given patient.

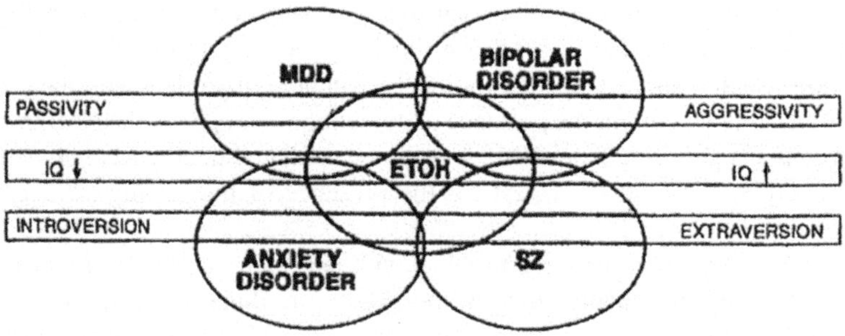

Fig. 1.2 Future of psychopharmacology. Interaction among syndromic diagnoses and between such diagnoses and dimensional aspects of personality. Space and the constraints of being a two-dimensional drawing of three-dimensional phenomena place limitations on this figure. In a three-dimensional figure, it would be clear that there is the potential for overlap between any two syndromic diagnoses and that the syndromic diagnoses are not on a personality trait continuum with respect to each other but rather that such traits are dimensionally present in all diagnoses and influence their expression. This figure also is not meant to imply that there are only three personality traits nor that the three depicted here are necessarily the most important (MDD major depressive disorders, ETOH alcoholism, SZ schizophrenia). (Reproduced with permission from Preskorn [3]. © Preskorn 1990)

1.3 The History of Current Psychiatric Drug Development: Chance Discovery and Rationale Refinement

The current treatment armamentarium for major depressive disorder (and psychotic disorders for that matter) owes their existence to two factors: first, chance discovery and then rationale refinement (Table 1.1) [4–6]. That is particularly true for the treatments aimed at the two of the most major syndromic diagnoses: affective and psychotic disorders.

Chlorpromazine can be viewed as the "Adam" or "Eve" (whichever the reader prefers) to both the family of modern antipsychotics and modern antidepressants as illustrated in Fig. 1.3 [4–6]. In the interest of space and because the themes are the same, this text will not cover the antipsychotic line of the family of drugs while acknowledging that the first widely used class of antidepressants [i.e., tricyclic antidepressants (TCAs)] resulted from a failed medicinal chemistry attempts to develop better antipsychotics. The interested reader can review the primary papers cited in the reference list for details on the antipsychotic lineage if they wish.

Briefly, chlorpromazine begat imipramine as a failed attempt by relatively blind medicinal chemistry to develop a better antipsychotic. The structural change leads to the loss of antipsychotic efficacy (i.e., no to weak D-2 receptor blockade) but the emergence of antidepressant efficacy (due to most likely the ability to inhibit the neuronal uptake of either norepinephrine or serotonin uptake).

About the same time, there was a failed attempt to develop better antitubercular drugs based on the structure of isoniazid produced effective antidepressants. These drugs are called monoamine oxidase inhibitors (i.e., MAOIs) because they presumably work via their ability to inhibit monoamine oxidase, the rate-limiting enzyme in the degradation of three biogenic amine neurotransmitters: dopamine (DA), epinephrine (E), norepinephrine (NE), and serotonin (SE). The antidepressant activity of the MAOIs coupled with the antidepressant efficacy of the TCAs reinforced the

Table 1.1 Early drugs that targeted the central nervous system

Drug	Class	Decade of discovery
Amphetamine	Stimulant	1880s
Cocaine	Analgesic/stimulant	1850s
Chlorpromazine	Antipsychotic	1950s
Diazepam	Anti-anxiety	1950s
Imipramine	Antidepressant	1950s
Isocarboxazid	Antidepressant	1950s
Lithium	Mood stabilizer	1940s
Morphine	Analgesic	2100 BC
Phenobarbital	Anticonvulsant	1930s
Reserpine	Antipsychotic	1950s

Reproduced with permission from Preskorn [4]. © Preskorn 2010

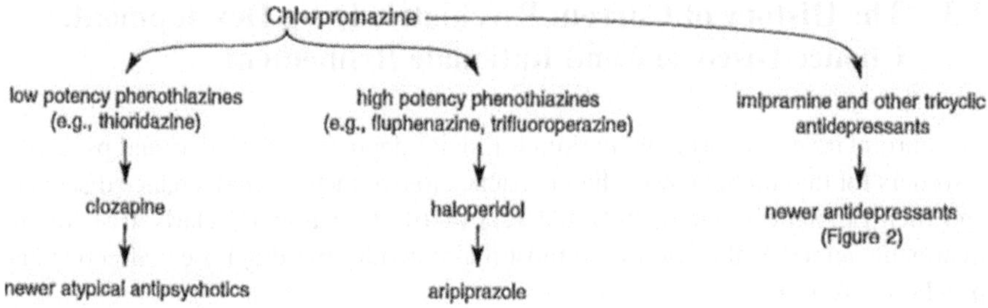

Fig. 1.3 Drug development based on chlorpromazine. (Reproduced with permission from Preskorn [5]. © Preskorn 2010)

idea that deficiency in either SE or NE neurotransmission was responsible for the depressive symptoms seen in patients with MDD.

Armed with the knowledge of the antidepressant activity of TCAs and MAOIs in the 1970s coupled with the ability to use structure-activity relationships and in vitro methods to examine in vitro receptor binding lead to the development via medicinal chemistry of new compounds which were capable of blocking either SE or NE transporters either selectively or in a sequential manner to develop molecules (i.e., 10 times more potent at one than the other or both sequentially over less than a ten-fold concentration range). The former were SE or NE selective reuptake inhibitors, whereas the latter were combined SE and NE reuptake inhibitors over their dosing range (i.e., generally capable of blocking SE reuptake at low concentrations and NE uptake inhibition at higher concentrations) (Fig. 1.4) [4–6]. In the case of bupropion, the goal was to develop a molecule capable of blocking NE and dopamine (DA) reuptake pumps, but the concept is otherwise the same.

The "pharmacological refinement approach" allowed the development of drugs capable of affecting the desirable target (e.g., the SE transporter) at concentrations low enough to not engage from other targets which produce undesirable effects (e.g., acetylcholine muscarinic receptors). Importantly, this approach meant that the new drug did not have a novel mechanism of action different from the earlier antidepressants but instead had a more limited range of pharmacologic actions making it more focused and with a more limited adverse effect profile by eliminating effects on targets capable of mediating adverse effects which were off target.

This strategy has led to the development of selective serotonin reuptake inhibitors (SSRIs) and serotonin-norepinephrine reuptake inhibitors (SNRIs) which are the latest, generally accepted antidepressants.

The consequence of this iterative step without knowledge of the fundamental biology underlying the disorder has led to a plethora of drugs capable of treating patient suffering from a form of the illness which is responsive to their mechanism of action. Table 1.2 shows the relative receptor binding of most currently marketed antidepressants relative to the receptors currently known to be clinically relevant in terms of either producing antidepressant efficacy or "off-target" adverse effects [11].

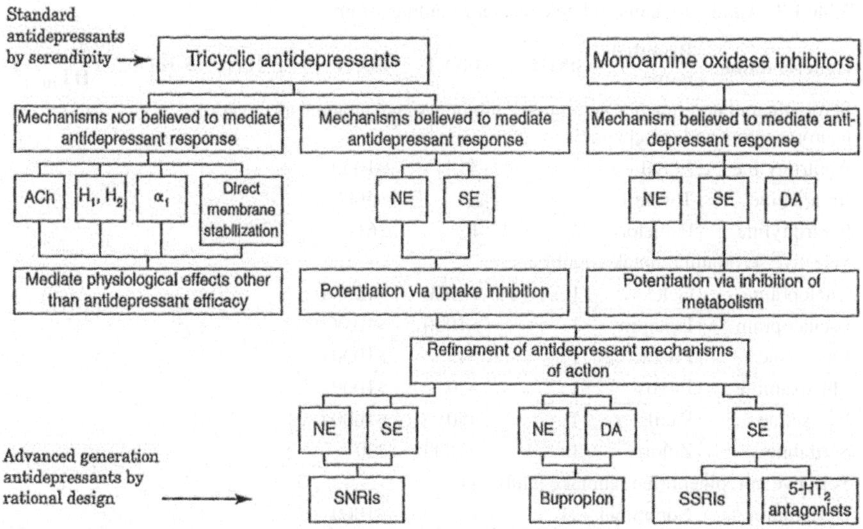

Fig. 1.4 Evolution of antidepressants. ACh acetylcholine, H histamine, α_1 alpha adrenergic, NE norepinephrine, SE serotonin, DA dopamine, SNRI serotonin-norepinephrine reuptake inhibitor, SSRI selective serotonin reuptake inhibitor. (Reproduced with permission from Preskorn [4]. © Preskorn 1996)

All the drugs shown in Table 1.3 [9] are essentially a "rehash" or a realignment of the mechanisms previously suggested to play a role in producing an antidepressant response. The question is: Do they offer anything which is meaning- fully new in terms of additional efficacy? In general, the answer is no based on the results of the largest sequential trial of currently marketed antidepressants ever funded by the NIMH, the Sequenced Treatment Alternatives to Relieve Depression (STAR*D). That study showed that perhaps 40% of patients with MDD have a form of the illness which is not responsive to multiple trials of antidepressants which work via effects on biogenic amine antidepressants (i.e., SE, NE, or DA).

That finding is the reason for the interest in antidepressants which work via non-biogenic amine antidepressants such as ketamine and related drugs.

1.4 The Future or Where to Go from Here?

On the downside, one could look at the last 50 years of psychiatric drug development particularly regarding antidepressants and antipsychotics as an era in which the same mechanisms were rehashed repeatedly. That is simply because these mechanisms were known to work, and not enough was known about the biology of the illness to take many chances on speculative targets. Admittedly, some development work was tried on speculative targets but failed which is the reason why it is

Table 1.2 Antidepressants' relative receptor binding affinity[a]

Generic name	Branded name	hSET	hNET	hDAT	5-HT$_{1A}$	5-HT$_{1B}$	5-HT$_{1D}$	5-HT$_{2A}$
Serotonin and norepinephrine reuptake inhibitors and antagonists at various neuroreceptors and ion channels								
Amitriptyline	Elavil	4	34	>1000				
Imipramine	Tofranil	1	26	>5000				
Nortriptyline	Pamelor	4	1	261				
Selective serotonin reuptake inhibitors								
Citalopram	Celexa	1	>1000	>10,000				
Escitalopram	Lexapro	1	>1000	>10,000				
Fluoxetine	Prozac	1	545	>1000				
Fluvoxamine	Luvox	1	620	>1000				
Paroxetine	Paxil	1	450	>1000				
Sertraline	Zoloft	1	>1000	220				
Selective norepinephrine reuptake inhibitors								
Desipramine[b]	Norpramin	21	1	>1000				
Reboxetine	Vestra	8	1	>1000				
Dual serotonin and norepinephrine (SE ≥ NE) reuptake inhibitors								
Desvenlafaxine	Pristiq	1	27	>1000				
Duloxetine	Cymbalta	1	7.5	504				
Levomilnacipran	Fetzima	1	8	>1000				
Milnacipran	Savella	1	8	>1000				
Venlafaxine	Effexor	1	16	>10,000				
5-HT$_{2A}$ antagonist and weak serotonin reuptake inhibitors								
Flibanserin	Addyi				1	>1000	>1000	49
Nefazodone	Serzone	9	18	17				1
Trazodone	Oleptro	21	>1000	929				1
Specific histamine, serotonin, and norepinephrine receptor antagonist								
Mirtazapine	Remeron	>100,000	>10,000	>100,000				
Dopamine and norepinephrine (weak) reuptake inhibitor								
Bupropion	Wellbutrin	17	95	1				
SSRIs + specific SE receptor activity								
Vilazodone	Viibryd	1	>500	370	21			
Vortioxetine	Brintellix	1	71	>1000	9	33	21	

Table 1.2 (continued)

Generic name	p5-HT$_{2C}$	5-HT$_3$	5-HT$_7$	h alpha1	hM$_1$	gpH$_1$	D3	D4
Serotonin and norepinephrine reuptake inhibitors and antagonists at various neuroreceptors and ion channels								
Amitriptyline	–			25	16	1		
Imipramine	–			65	65	8		
Nortriptyline	–			148	34	1		
Selective serotonin reuptake inhibitors								
Citalopram	>1000			757	894	179		
Escitalopram	>1000			>1000	>1000	257		
Fluoxetine	65			>1000	638	>1000		
Fluvoxamine	>1000			560	>5000	>5000		
Paroxetine	>10,000			>10,000	720	>100,000		
Sertraline	>10,000			>1000	>1000	>100,000		
Selective norepinephrine reuptake inhibitors								
Desipramine[b]	–			156	235	132		
Reboxetine	875			>1000	933	44		
Dual serotonin and norepinephrine (SE ≥ NE) reuptake inhibitors								
Desvenlafaxine	>1000			>1000	>1000	>1000		
Duloxetine	>1000			>1000	>1000	>1000		
Levomilnacipran								
Milnacipran	917			>1000	>1000	>1000		
Venlafaxine	>1000			>1000	>1000	>1000		
5-HT$_{2A}$ antagonist and weak serotonin reuptake inhibitors								
Flibanserin		>10,000	990				>100	>10,000
Nefazodone	–			1.2	522	1		
Trazodone	1			5	>1000	45		
Specific histamine, serotonin, and norepinephrine receptor antagonist								
Mirtazapine	–			>1000	>1000	1		
Dopamine and norepinephrine (weak) reuptake inhibitor								
Bupropion	–			10	95	10		
SSRIs + specific SE receptor activity								
Vilazodone								
Vortioxetine		2	12					

Reproduced with permission from Preskorn [10]. © Preskorn 2017

Table 1.2 (continued)

Key: *h* human, *SET* serotonin transporter, *NET* norepinephrine transporter, *DAT* dopamine transporter, *p* porcine, *5-HT* serotonin, *gp* guinea pig, *H* histamine, *M* muscarinic, *D* dopamine, *SE* serotonin, *NE* norepinephrine, *SSRIs* selective serotonin reuptake inhibitors

[a]Relative binding affinity (RRB) is the binding affinity of the drug for every receptor reported in the package insert in relationship to the drug's highest affinity site. To calculate the relative binding affinity for each drug, its *Ki* for its highest affinity site is divided by itself, yielding 1, and next the *Ki* for the highest affinity site (which is the smallest concentration of drug needed to bind to any site) is divided into all its Ki's for lower affinity sites (which is hence a higher concentration needed to bind to a lower affinity site); the result then is a number greater than 1. The larger that number, the higher the concentration needed to bind to the next potential target for the drug

[b]This drug is also a selective norepinephrine reuptake inhibitor

For each drug in this table, its highest affinity and its affinity expressed in nanomolar concentration are as follows: amitriptyline, H_1 (1); bupropion, DAT (526); citalopram, SET (1.6); desipramine, NET (0.83); desvenlafaxine, SET (115); duloxetine, SET (1); flibanserin, 5-HT_{1A} (1); fluoxetine, SET (1.1); fluvoxamine, SET (2.3); imipramine, SET (1.41); levomilnacipran, SET (11.2); milnacipran, SET (9); mirtazapine, Hr (0.14), nafazodone, H1 (6); nortriptyline, NET or H_1 (4.35); paroxetine, SET (0.1); reboxetine, NET (7); sertraline, SET (0.3); trazodone, 5-HT_{2A} (7.7); venlafaxine, SET (102); vilazodone, SET (0.1); vortloxetine, SET (1.6). Flibanserin and milnacipran are not labeled for antidepressant activity. They were initially developed and tested for this indication but clinical trials were not supportive. In the case of milnacipran, its active enationer, levomilnacipran, was successfully developed for an antidepressant indication[2,14]

not being discussed here. That is the reason why most of the psychiatric drugs approved from 2009 to 2016 (Table 1.3) had the same well-established mechanisms of action [9].

With that said, there have been six novel mechanisms of action drugs developed and approved over the last 25 years (Table 1.4) [7]. These drugs may point the way to the future because of common features in their development. First, they were directed at a single behavior of symptom rather than a syndrome or cluster of behaviors and symptoms which may have different mechanisms mediating them. Second, the circuitry underlying the disturbance was relatively simple and well established. Third, the outcome variable was relatively dichotomous (e.g., smoke, don't smoke) rather than a reduction in a rating scale based on a compilation of the various disparate symptoms of a syndromic diagnosis such as MDD (e.g., the Hamilton Depression Rating Scale or the Montgomery-Asberg Depression Rating Scale). As knowledge of the biology underlying MDD continues to improve, it will guide the development of mechanistically new antidepressants.

The other plus is that high-throughput screening can make new medications highly selective for their desired target. That is illustrated by the development done with tasimelteon and suvorexant which were screened against 200 targets which were not desired targets of the drug Table 1.5 [10]. The molecules, tasimelteon and suvorexant, were taken forward both because they affected their desired target at nanomolar concentrations and did not affect any of these other non-desired targets even at micromolar concentrations (i.e., 1000 times greater than the concentration needed to bind to their desired target).

Table 1.3 Psychiatric and selected CNS drugs approved 2009–2016

Year	Total drugs approved	CNS drugs	Psychiatric and selected CNS drugs	Specific psychiatric or CNS drugs generic/brand name (manufacturer)[a,b,c]	Labeled indication	Mechanism of action[d]	Substrate for drug transporter	Substrate for drug metabolizing enzymes
2009	26	4	3	Iloperidone/Fanapt (Vanda)[b]	Schizophrenia	5-HT2A > alpha-1 > D_2 (1:2:17)	Not P-gp; otherwise	CYP3A4 > CYP 2D6 ¼ carbonyl reductase
							NA	
				Asenapine/Saphris (Allergan)[b]	Schizophrenia and manic or mixed episodes of bipolar I disorder	5-HT2C > 5-HT2A > alpha-1 > D_2 (1:3:4:7)	NA	CYP1A2
				Milnacipran/Savella (Allergan)[c]	Fibromyalgia	SET > NET (1:8)	NA	Mainly excreted unchanged with little to no drug metabolism
2010	21	1	1	Lurasidone/Latuda (Sunovion)[b]	Schizophrenia and depressed phase of bipolar I disorder	5-HT2A ¼ 5-HT7 > D_2 (1:1:2)	NA	CYP3A4
2011	28	5	1	Vilazodone/Viibryd (Allergan)[b]	Major depressive disorder	SET > 5-HT1A (1:21)	NA	CYP3A4 > > 2C19 ¼ 2D6
2012	39	3	1	Locaserin.Belviq [Arena Pharm (US distributor: Eisai)][c]	Weight management/ obesity	5-HT 2C > 5-HT2A > 5-HT2B receptors (1:7:11)	NA	Multiple CYP and non-CYP pathways
2013	27	4	2	Vortioxetine/ Brintellix (Takeda)[b]	Major depressive disorder	SET > multiple 5-HT receptors ($1 \leq 10$)	NA	CYP2D6 > > CYP3A4/5, CYP2C19, CYP2C9, CYP2A6, CYP2C8, and CYP2B6

(continued)

Table 1.3 (continued)

Year	Total drugs approved	CNS drugs	Psychiatric and selected CNS drugs	Specific psychiatric or CNS drugs generic/brand name (manufacturer)[a,b,c]	Labeled indication	Mechanism of action[d]	Substrate for drug transporter	Substrate for drug metabolizing enzymes
				Levomilnacipran/ Fetzima (Allergan)[b]	Major depressive disorder	SET > NET (1:8)	NA	Mainly excreted unchanged
2014	42	4	3	Suvorexant/ Belsomra (Merck)[c]	Onset and maintenance of sleep	Orexin 1 and 2 receptors	NA	CYP3A4
				Tasimelteon/Hetlioz (Vanda)[c]	Non-24-h sleep- wake disorder	MT-1 and −2 receptors	NA	CYP1A2, 3A4, and phenolic glucuronidation
				Bupropion plus naltrexone/Contrave (Orexigen)[a]	Weight management in obesity	Bupropion: DAT > SET > NET (1:17:95); for naltrexone, see note below[e]	NA	CYP2B6 (bupropion); non-CYP enzyme (naltrexone)
2015	45	4	4	Proaripiprazole/ Aristada (Alkermes)[a]	Schizophrenia	D_2 > 5-HT1A > 5-HT2A (1:5:10)	NA	Same as aripiprazole (i.e., CYP3A4 and 2D6)
				Cariprazine/Vraylar (Allergan)[b]	Schizophrenia and manic or mixed episodes of bipolar I disorder	D3 > D2 ¼ 5-HT2B (1:6–8)	Not a substrate for multiple transporters	CYP3A4 > > 2D6
				Flibanserin/Addy (Sprout)[c]	Hypoactive sexual desire in premenopausal females	5-HT1A	NA	CYP3A4 > 2C19

				Brexpiprazole/ Rexulti (Otsuka)[b]	Schizophrenia and as an adjunct to antidepressants in major depressive disorder	5-HT1A > D_2 > 5-HT2A > alpha; 2C (1:3:4:5)	NA	CYP2D6 and 3A4
2016	26	2	1	Pimavanserin/ Nuplazid (Acadia)[b]	Psychosis in Parkinson disease	5-HT2A > 5-HT2C (1:5)	Not a substrate for multiple transporters	CYP3A4 and 3A5 > 2J2, 2D6, and various other CYP and FMO enzymes
Total	254	27	16					

Reproduced with permission from Preskorn [10]. © Preskorn 2017

CNS indicates central nervous system, *CYP* cytochrome P-450 enzyme, *D* dopamine, *DAT* dopamine transporter, *DOR* δ-opioid receptor, *FMO* flavin- containing monooxygenase, *gp* guinea pig, *H* histamine, *KOR* κ-opioid receptor, *M* muscarinic, *MOR* μ-opioid receptor, *MT* melatonin, *NA* not available, *NE* norepinephrine, *NET* norepinephrine transporter, *NME* new molecular entity, *P* porcine, *SET* serotonin transporter, *5-HT* serotonin

Of these 16 drugs:

[a]One was a combination of two existing drugs (bupropion+naltrexone); one was a prodrug of an existing molecule (proaripiprazole)

[b]Nine were NMEs approved for a psychiatric indication

[c]Five were NMEs approved for nonpsychiatric indications but which nevertheless targeted brain dysfunction (i.e., sleep, weight, or pain) with mechanisms of action that worked via the brain

[d]Numbers in parentheses in the mechanism of action column represent the relative binding affinity of the drug for these respective targets, with 1 being the highest binding affinity for the drug and the larger numbers being how much the concentration of the drug has to increase to affect the next target (with the general rule in pharmacology being that a drug is selective for a target if its next binding requires more than a tenfold increase in concentration). For more details, see Preskorn et al.[4]

Most of the drugs listed in the table are full antagonists at their respective receptors. The few exceptions are aripiprazole, brexpiprazole, and cariprazine, which are partial agonists at the D_2 receptor, and pimavanserin, which is an inverse agonist and antagonist at the 5-HT2A and 5-HT2C receptors (five times higher affinity for the 5-HT2A receptor than the 5-HT2C receptor)

[e]Naltrexone and its active metabolite 6β-naltrexol are antagonists at the MOR, to a lesser extent at the KOR, and to a far lesser and possibly insignificant extent, at the DOR. The K_i affinity values of naltrexone at the MOR, KOR, and DOR have been reported as 0.0825, 0.509, and 8.02 nM, respectively, demonstrating a MOR/KOR binding ratio of 6.17 and a MOR/DOR binding ratio of 97.2[5,6]

Table 1.4 Six central nervous system drugs with novel mechanisms of action developed in the past 25 years

Generic name	Brand name	Originator	Approval date	Latest PI revision	Indication(s)	Mechanism	Generic available
Ondansetron	Zofran	Glaxo	1/4/1991	9/18/2014	Chemotherapy-induced nausea and vomiting (CINV)	Serotonin 5-HT_3 receptor antagonism	7/2/2010
Aprepitant	Emend	Merck	3/27/2003	8/12/2014	CINV	Neurokinin (substance P)-1 receptor antagonism	7/24/2012
Ramelteon	Rozerem	Takeda	7/22/2005	3/1/2012	Insomnia[a]	Melatonin (MT_1, MT_2) receptor agonism	7/26/2013
Varenicline	Chantix	Pfizer	5/10/2006	10/15/2014	Smoking cessation	Acetylcholine nicotinic receptor alpha-4 beta-2 partial agonism	No
Lorcaserin	Belviq	Arena[b]	6/27/2012	6/27/2012	Weight loss	5-HT_{2c} agonism	No
Suvorexant	Belsomra	Merck	8/13/2014	N/A	Insomnia[c]	Dual orexin 1 and 2 receptor antagonism	No

Reproduced with permission from Preskorn [7]. © Preskorn 2014
PI package insert
[a]Difficulty with sleep onset
[b]Marketed by Eisai
[c]Difficulties with sleep onset and/or sleep maintenance

Table 1.5 Receptors for which tasimelteon (10 μm) did not inhibit or stimulate binding by >50%[a]

Adenosine A_1	Dopamine D_1	Melanocortin MC1	Rolipram
Adenosine A_{2A}	Dopamine D_{2L}	Melanocortin MC_3	Ryanodine RyR3
Adenosine A_3	Dopamine D_{2S}	Melanocortin MC_4	Serotonin 5-HT_1
Adrenergic α_{1A}	Dopamine D_3	Melanocortin MC_5	Serotonin 5-HT_{1A}
Adrenergic α_{1B}	Dopamine $D_{4.2}$	Motilin	Serotonin 5-HT_{1B}
Adrenergic α_{1D}	Dopamine D_5	Muscarinic M_1	Serotonin 5-HT_2
Adrenergic α_2	Endothelin ET_A	Muscarinic M_2	Serotonin 5-HT_{2A}
Adrenergic α_{2A}	Endothelin ET_B	Muscarinic M_3	Serotonin 5-HT_{2B}
Adrenergic α_{2C}	Epidermal growth factor	Muscarinic M_4	Serotonin 5-HT_{2C}
Adrenergic β_1	Erythropoietin EPOR	Muscarinic M_5	Serotonin 5-HT_3
Adrenergic β_2	Estrogen Erα	N-formyl peptide receptor FPR1	Serotonin 5-HT_4
Adrenergic β_3	Estrogen Erβ	N-formyl peptide receptor-like FPRL1	Serotonin 5-HT_{5A}
Adrenomedullin AM_1	G protein-coupled receptor GPR103	Neurokinin NK_1	Serotonin 5-HT_6
Adrenomedullin AM_2	G protein-coupled receptor GPR8	Neuromedin U NMU_1	Sigma σ1
Aldosterone	$GABA_B$	Neuromedin U NMU_2	Sigma σ2
Anaphylatoxin C5a	$GABA_{B1A}$	Neuropeptide Y, Y_1	Sodium channel, site 2
Androgen	$GABA_{B1B}$	Neuropeptide Y, Y_2	Somatostatin sst1
Angiotensin AT_1	Gabapentin	Neurotensin NT_1	Somatostatin sst2
Angiotensin AT_2	Galanin GAL1	Nicotinic acetylcholine	Somatostatin sst3
Apelin (APJ)	Galanin GAL2	Nicotinic acetylcholine α1	Somatostatin sst4
Atrial natriuretic factor	Glucocorticoid	Nicotinic acetylcholine α7	Somatostatin sst5
Bombesin BB1	Glutamate, AMPA	Opiate δ (OP1, DOP)	Tachykinin NK_1
Bombesin BB2	Glutamate, Kainate	Opiate κ (OP2, KOP)	Tachykinin NK_2
Bombesin BB3	Glutamate, NMDA	Opiate μ (OP3, MOP)	Tachykinin NK_3
Bradykinin B_1	Glycine, strychnine- sensitive	Orphanin ORL_1	Thromboxane A_2
Bradykinin B_2	Growth hormone secretagogue	Phorbol ester	Thyroid hormone
Calcitonin	Histamine H_1, central	Platelet activating factor	Thyrotropin releasing hormone
Calcitonin gene- related peptide $CGRP_1$	Histamine H_2	Platelet-derived growth factor	Transforming growth factor-β
Calcium channel L-type	Histamine H_3	Potassium channel [K_A]	Transporter, adenosine
Calcium channel N-type	Histamine H_4	Potassium channel [KATP]	Transporter, choline

(continued)

Table 1.5 (continued)

Cannabinoid CB_1	Hypocretin (orexin) receptor 1	Potassium channel [SK_{CA}]	Transporter, dopamine
Cannabinoid CB_2	Hypocretin (orexin) receptor 2	Potassium channel HERG	Transporter, GABA
Chemokine CCR1	Imidazoline I_2, central	Progesterone	Transporter, monoamine
Chemokine CCR2B	Inositol trisphosphate IP_3	Progesterone PR-B	Transporter, norepinephrine
Chemokine CCR4	Insulin	Prostanoid CRTH2	Transporter, serotonin
Chemokine CCR5	Interleukin IL-1	Prostanoid DP	Tumor necrosis factor
Chemokine CX3CR1	Interleukin IL-2	Prostanoid EP_2	Urotensin II
Chemokine CXCR2 (IL-$8R_B$)	Interleukin IL-6	Prostanoid EP_4	Vanilloid
Cholecystokinin CCK_1 (CCK_A)	Leptin	Prostanoid, thromboxane A_2	Vascular endothelial growth factor
Cholecystokinin CCK_2 (CCK_B)	Leukotriene (LTB_4)	Purinergic P_{2X}	Vasoactive intestinal peptide
Colchicine	Leukotriene, cysteinyl $CysLT_1$	Purinergic P_{2Y}	Vasoactive intestinal peptide 1
Corticotropin releasing factor CRF_1	Leukotriene, cysteinyl $CysLT_2$	Retinoid X receptor RXRα	Vasopressin V_{1A}
			Vasopressin V_{1B}
			Vasopressin V_2
			Vitamin D_3

[a]Reprinted from Lavedan et al. [2] under a Creative Commons license

Standard radioligand binding and enzyme inhibition assays were performed on receptors, binding sites, or enzyme systems obtained from various sources, including human, rat, mouse, guinea pig, rabbit, hamster, and bovine tissues (see Lavedan et al. [2], Supplemental Information), using the profiling screen and discovery screen panels (Panlabs) which consisted of 56 radioligand binding assays and 7 enzyme assays, respectively, and the SpectrumScreen panel (MDS Pharma Services) that included 170 pharmacological relevant targets (see Lavedan et al. [2], Supplemental Information). In addition, the $GABA_A$ benzodiazepine and $GABA_B$ binding sites were also tested independently (Panlabs biochemical pharmacology assays). Tasimelteon was used at a concentration of 10 μm except for two enzyme assays (protein kinases C: PKCα and PKCβ) where it was used at 100 μm and for the melatonin receptors in the SpectrumScreen panel where four concentrations (10 nm, 0.1 μm, 1 μm, and 10 μm) were tested. A response was considered significant if there was ≤50% inhibition or stimulation for the assays

The affinity of tasimelteon (10 μm) for the human hypocretin (orexin) receptor 1 expressed in transfected CHO cells and for the human hypocretin (orexin) receptor 2 expressed in transfected HEK-293 cells was determined in radioligand binding assays (Eurofins Cerep SA, Celle l'Evescault, France)

1.5 The Immediate Future Which is Upbeat

Between that development and the near future, ketamine and related drugs are the first legitimate hope for a new approach to treating patients with the form of MDD which is not responsive to biogenic amine antidepressants. While the antidepressant activity of ketamine and related drugs was initially discovered by chance as was the case with TCAs and MAOIs, it appears nevertheless to be robustly and rapidly effective in approximately 60% of patients whose depressive disorder is not responsive to biogenic amine antidepressants.

This new era will not simply hold the promise for treating those patients but also provide biological insights into these different forms of the major depression: (a) those responsive to biogenic amine antidepressants, (b) those not responsive to biogenic amine antidepressants but to glutaminergic antidepressants such as esketamine, and (c) those not responsive to either of these forms of treatment. The ability to divide patients with the syndrome of major depression into these three categories has the potential to permit understanding the biological reasons for why they fall into those three groups. The knowledge gained from that and from the mechanisms underlying the response to esketamine will in turn lead to new developments just as was true the development of SSRIs and SNRIs from the knowledge gained from studies of TCAS and MAOIs.

The figures and tables in this chapter come from the articles below. Each of these articles has its own reference list which the interested reader can access either through PubMed or on the Lippincott Williams & Wilkins website.

References

1. Kraepelin E. In: Barclay RM, Robertson GM, editors. Dementia praecox and paraphrenia. Huntington NY: Krieger; 1915.
2. Lavedan C, Forsberg M, Gentile AJ. Tasimelteon: a selective and unique receptor binding profile. Neuropharmacology. 2015;91:142–7.
3. Preskorn SH. The future and psychopharmacology: potentials and needs. Psychiatr Ann. 1990;20(11):625–33.
4. Preskorn SH. CNS drug development. Part I: the early period of CNS drugs. J Psychiatr Pract. 2010a;16(5):334–9.
5. Preskorn SH. CNS drug development: part II: advances from the 1960s to the 1990s. J Psychiatr Pract. 2010b;16(6):413–5.
6. Preskorn SH. CNS drug development: part III: future directions. J Psychiatr Pract. 2011;17(1):49–52.
7. Preskorn SH. CNS drug development: lessons from the development of ondansetron, aprepitant, ramelteon, varenicline, lorcaserin, and suvorexant. Part I. J Psychiatr Pract. 2014;20(6):460–5.
8. Preskorn SH. CNS drug development: lessons learned part 2. Symptoms, not syndromes as targets consistent with the NIMH research domain approach. J Psychiatr Pract. 2015;21(1):60–6.

9. Preskorn S. Psychiatric and central nervous system drugs developed over the last decade: what are the implications for the field? J Psychiatr Pract. 2017a;23(5):352–60.
10. Preskorn S. CNS drug development, lessons learned, part 4: the role of brain circuitry and genes – tasimelteon as an example. J Psychiatr Pract. 2017b;23(6):425–30.
11. Preskorn S. Drug-drug interactions with an emphasis on psychiatric medications. West Islip: Professional Communications, Inc; 2018.
12. Preskorn SH, Baker B. The overlap of DSM-IV syndromes: potential implications for the practice of polypsychopharmacology, psychiatric drug development and the human genome project. J Psychiatr Pract. 2002;8(3):170–7.

Chapter 2
The History of Drug Development in Psychiatry: A Lesson in Serendipity

Abhishek Wadhwa

Abstract The goal of this book is to provide a guide on modern day drug development in psychiatry. However, in order to understand current practices in drug development, it is important to first understand the history of psychiatry including early attempts at drug discovery and develoment. The early history of psychiatry is mired with the use of inhumane experimental treatments and the institutionalization of patients in asylums. Some of the earliest drugs used in these asylums were meant to sedate patients rather than treat underlying mental disorders. The earliest identified drugs treating mental disorders were born out of serendipitous discoveries which later led to their clinical effects being demonstrated through clinical trials and case studies. This is evident from the history of chlorpromazine, monoamine oxidase inhibitors, tricyclic antidepressants, lithium, and others. We discuss in detail about each of these psychotropic drugs, the events leading up to their discovery, and their role in formulating the biological basis of mental disorders including schizophrenia, depression, and bipolar disorder. Psychiatry, it seems has worked its way backwards from first identifying treatments before understanding the biological basis of mental disorders, in a sharp contrast to the other fields of medicine. With our growing understanding of the etiopathogenesis of mental disorders, drug development in psychiatry is evolving to develop treatments that target the underlying physiology of mental disorders.

Keywords History · Psychopharmacology · Serendipity · Mental disorders · Catecholamine · Dopamine · Antipsychotic · Antidepressants · Drug development

2.1 Introduction

Psychiatry has evolved over time in terms of how mental disorders are conceptualized and how biological treatments are used and developed. The focus of this book is on the development of biological treatments for mental disorders. In order to

A. Wadhwa (✉)
University of Alabama at Birmingham, Department of Psychiatry and Behavioral Neurobiology, Birmingham, AL, USA
e-mail: abhishekwadhwa@uabmc.edu

M. Macaluso et al. (eds.), *Drug Development in Psychiatry*, Advances in Neurobiology 30,
https://doi.org/10.1007/978-3-031-21054-9_2

understand drug development in psychiatry, one must begin with an understanding of the historical underpinnings of the field. The term psychiatry ("Psychiatrie") was first introduced by Johanne Christian Reil, professor of therapy at the University of Halle, in Germany in 1808. Reil, in his paper, argued that the cause of human disease is an essential interaction among the three domains of mental, chemical, and physical. The term "psychiatry" refers to a medical discipline rather than a philosophical or theological one [1]. Benjamin Rush is often considered the father of American psychiatry [2]. He is reported to have considered psychiatry as part of the field of medicine and devoted a large part of his teachings to the same. Building upon these concepts, Chiarugi, in Florence, for the first time suggested treating mental disorders using an approach which is respectful of patients and advocated for humane treatment. Jean-Baptiste Pussin and Philippe Pinel used this approach to institute "moral treatment" (1801), which was a psychological treatment and a sharp contrast to the violent treatments often used in asylums of the nineteenth century. Pinel further refined this approach into a "medical moral treatment." This approach included reward-based activities, physical exercise, and offering nutritious food to patients while limiting the use of physical restraints. Pinel and Pussin reported a high success rate with "mortal treatment" and inspired American psychiatry to follow the same approach [3]. The concept of "medical moral treatment" inspired Dorothea Dix who became a leading figure of national and international movements to promote the safe and humane treatment of people with mental disorders. Dix played a vital role in establishing and expanding state funded facilities for the treatment of mental disorders. By 1860, 28 out of 33 states in the USA were reported to have at least one public psychiatric hospital [4, 5].

The second and third decades of the twentieth century saw major changes in the understanding of mental disorders by the general public and medical community. It was during this time that the somatic origins of mental disorders began being systematically evaluated [5]. This was also the time where prefrontal lobotomy was developed as a treatment. Prefrontal lobotomy was first introduced in the USA in 1937 and was widely used until the early 1950s before the release of chlorpromazine. Lobotomy was heavily criticized due to being invasive, inhumane, and permanently changing the personality of patients [6, 7]. Electroconvulsive therapy (ECT) was developed in 1938 by Cerletti and was extensively used in the US in its unmodified form. The use of ECT declined in the 1960s due to a number of factors including the introduction of antidepressant drugs as well as the negative and often inaccurate depiction of ECT in the media [8]. After World War II, psychoanalysis emerged with Freud's theories gaining mainstream popularity [5].

Historically, there is documented use of psychotropics since long before the introduction of psychiatry into the practice of medicine. Ancient Greek and Indian civilizations documented the use of psychoactive substances to experience euphoria. The concept of using a drug for understanding mental disorders is reported to have been conceived by the French psychiatrist, Moreau De Tours (1845) [9]. Emil Kraeplin (1892) used the term "pharmacopsychology" to indicate the effects of drugs on psychological functioning. Kraeplin is reported to be among the first to promote the idea of treatment response to determine the clinical effect of a drug.

Sigmund Freud is also regarded as an early figure in psychopharmacology as indicated in his famous letter to Maria Bonaparte, where he predicted that the way to understanding psychosis and would be guided by organic chemistry or access to it through endocrinology. These developments took place around the same time that a paper was published to describe the antipsychotic actions of Rauwolfia which had been utilized in Indian folk medicine for a very long time. This paper was published in an Indian medical journal and was largely overlooked by Western medicine [9]. The nineteenth century also saw the use of sedatives and hypnotics including drugs such as narcotics, chloral hydrate, and bromides, primarily to sedate and calm patients but not to treat specific mental disorders [10].

Moving into the twentieth century, the serendipitous discovery of several psychopharmacologic agents lead to the development of major classes of drugs to treat mental disorders. The focus of this chapter is on historical aspects of drug development, with a focus on serendipitous discovery. Other chapters in this book will focus on the use of structure activity relationships and animal models to further develop drugs for treating mental disorders. The book will also focus on modern approaches to drug development including reverse engineering, the role of neuroimaging, and the use of biomarkers including genetic and epigenetic markers in drug development research.

2.2 Chlorpromazine

The serendipitous discovery of chlorpromazine was an early development for the budding field of psychopharmacology. Phenothiazines were developed in the late nineteenth century for use in the dye and textile industries. At the time, phenothiazines were recognized for their antiseptic and active parasitic properties and were further explored for their antihistaminic properties in the early twentieth century. In the 1930s and 1940s, there was interest in producing synthetic histamines for use in medical research. The pharmaceutical division of a French company, Rhône-Poulenc, in collaboration with research groups at the Pasteur Institute developed novel antihistamines based loosely on diphenhydramine. Paul Charpentier, a chemist at the company, modified and tested the antihistaminic properties of several phenothiazine compounds. One of his earlier successes was promethazine in 1947. Henri-Marie Laborit, a French surgeon, was experimenting with antihistamines for use in preventing surgical shock in the 1940s by creating a form of artificial hibernation by reducing hyperthermia. When Laborit administered promethazine to his patients he observed a unique psychological phenomena that he termed "euphoric quietude." The cardinal features of "euphoric quietude" included (1) weak and reversible narcosis, (2) no clouding of consciousness, and (3) emotional indifference [10–12].

Simone Corvoiser, at Rhône-Poulenc, analyzed the sedative properties of the antihistaminic agents synthesized by Paul Charpentier and his team, identifying promethazine as a promising option. Charpentier continued to work on further alterations of the core nucleus of the molecule hoping to generate greater antihistaminic effects. Among the different derivatives, a compound created by introducing a chlorine atom into one of the rings of promethazine, R.P.4560, was sent for further testing in clinical settings. Laborit tested this new drug and noted that patients were more relaxed and calm before and after surgery when taking R.P.4560. He further predicted the possible application of these agents in treating psychiatric disorders [10–13].

A French psychiatrist, Pierre Deniker, is credited for introducing chlorpromazine into psychiatry. He is regarded as one of the pioneers of pharmacological treatments in psychiatry and an advocate for seeking out the biological basis of mental disorders. He conducted clinical trials at Sainte-Anne Hospital, in Paris, where he was the head of an inpatient psychiatric unit, along with Jean Delay. He reported a calming effect of chlorpromazine on agitated patients with psychosis and recognized the therapeutic potential of this new drug. Delay suggested chlorpromazine's classification as a neuroleptic along with similar drugs producing effects of psychomotor retardation. The first controlled trial of chlorpromazine was conducted at the University of Birmingham (UK) by Joel and Charmain Elkes which was also notably the first recorded randomized, placebo controlled, trial in the history of psychiatry [10, 11, 14, 15].

Chlorpromazine was marketed under the trade name Thorazine by an American pharmaceutical company SKF who bought the North American rights to the drug and received US Food and Drug Administration (FDA) approval in May 1954. Despite its prominent use worldwide, the mechanism of action of chlorpromazine was largely unknown at the time. There was also very little evidence and consensus about the recommended dosage of the drug. In the early 1960s, it was thought that discrepancies in the earlier studies of chlorpromazine could have been due to the its hypothermic effects. Later studies on chlorpromazine controlled for the hypothermic effects of the drug and reported that it blocked dopamine receptors. This blockade induces a feedback mechanism increasing the metabolism of catecholamines. This was further substantiated by various studies demonstrating increased dopamine synthesis and metabolism induced by chlorpromazine. It was in the mid-1970s that the mechanism of action of chlorpromazine was clarified after it was established that chlorpromazine blocks dopamine receptors. These findings supported the dopamine hypothesis of schizophrenia [5, 10, 11, 16].

The dopamine hypothesis of schizophrenia emerged from the work of Carlsson and colleagues. They postulated that reserpine nonspecifically blocks the storage of all monoamines leading to their subsequent depletion by metabolic breakdown. He further established the involvement of catecholamines in the action of chlorpromazine and haloperidol. They demonstrated, using fluorometry, that antipsychotics including haloperidol and chlorpromazine cause an increase in noradrenaline and dopamine turnover in the brains of experimental rats leading to the conclusion that they might be acting by blocking dopamine receptors. In subsequent years, researchers established that antipsychotics increase the concentration of dopamine metabolites in the brain directly proportional to their potency. The dopamine hypothesis of schizophrenia has had a major influence on neurobiological research, specifically research into the mechanism of action of psychotropics [16].

2.3 Monoamine Oxidase Inhibitors

The discovery of iproniazid and MAO inhibitors is another lesson in serendipitous discovery. Hydrazines originated in the 1870s during experiments carried out by the father of organic chemistry Emil Fischer. Hans Meyer and Josef Malley later synthesized isonicotinyl-hydrazine as part of their doctoral thesis in the 1910s. Development of this drug was largely dormant over the next 40 years until the 1950s. In the early 1950s, the antitubercular properties of hydrazines were discovered partly by chance. It is reported that large stocks of hydrazine were left over after World War II in the form of rocket fuel. These stocks were then distributed to chemical and pharmaceutical industries at a low cost which might have influenced the revival of interest in these compounds [10, 12, 15, 17].

The discovery of iproniazid's antitubercular properties was mostly due to the efforts of two scientific teams working independently. These teams were led by Herbert Hyman Fox and Harry L Yale, respectively, for Hoffman La Roche laboratories and the Squibb Institute for Medical Research. In 1952, Irving J. Selikoff and Edward Robitzek published the results of their clinical study on the effectiveness of iproniazid and isoniazid in the treatment of tuberculosis. The studies were conducted at Seaview Hospital on Staten Island in New York. In addition to reporting the effectiveness of these drugs in tuberculosis, they found that iproniazid had effects on mood. The psychostimulant effects noted ranged from mood elevation to increase in social activity in patients who were being treated with iproniazid. To quote one of the observers "patients were dancing in the halls with holes in their lungs." The positive effects on mood were initially thought to be secondary to improvement in respiratory function. These positive effects on mood were also observed in patients treated with iproniazid for other chronic illnesses such as rheumatoid arthritis or cancer [10, 17].

Mary Bernheim, a researcher at the University of Cambridge, describe an enzyme that she termed tyramine oxidase in 1928 which could bring about oxidative dissemination of biogenic amines. This enzyme was named monoamine oxidase in 1937 by a group led by Herman Blaschko. They demonstrated that this enzyme was capable of metabolizing adrenaline through oxidation. Quatel and Pugh identified this enzyme in the brain, although its significance was not clear at the time. In 1952, it was at Northwestern University in Chicago where Ernest Zeller observed for the first time that iproniazid inhibited MAO. Subsequent work on the enzyme showed that MAO converted serotonin to hydroxy indoleacetic acid. This was, for a long period of time, thought to be the only path for metabolism of serotonin. It was in 1957 at the National Institute of health (NIH), where Sydney Udenfriend and colleagues demonstrated a rapid increase in levels of serotonin in experimental animals after the administration of iproniazid. These findings led to further research on brain function and possible mechanism of action of antidepressants. It is in this context that the experimental studies of Charles Scott at Warner Lambert Research Laboratories were pivotal in the characterization of these drugs as antidepressants. He hypothesized that the tranquilizing effects of reserpine, an alkaloid of Rauwolfia Serpentina, were due to the release of serotonin. Inspired by Zeller's work, he administered iproniazid to animals to limit the enzymatic destruction of serotonin to potentiate the tranquilizing effect of reserpine. However, pretreatment with iproniazid was observed to have a stimulating effect in animals. He termed this effect "marsilization" in reference to the trade name of iproniazid [15].

The clinical effects of iproniazid in non-tuberculosis depressed patients was first assessed by Nathan S. Kline and colleagues when they recruited 24 subjects (17 of which were diagnosed with schizophrenia and 7 with depression). This study was modeled after Scott's experiments to demonstrate the stimulant effects of the drug in humans. The study was conducted at Rockland State Hospital in Orangeburg, New York. The patients were administered iproniazid 50 mg three times a day. The investigators reported substantial improvement in about 70% of the patients characterized by improved mood, decreased anhedonia, and improvement in social skills. These findings were reported at the annual meeting of the American Psychiatric Association in New York. Kline proposed the term "psychic energizer" while defining this drug's action (Loomer et al. 1957). In spite of these positive results, Kline's group ran into considerable difficulties conducting further studies when they lost the support of Hoffman Laroche as the company was doubtful about the marketability of the drug. Klein was able to eventually convince Hoffman Laroche to continue funding further development of iproniazid. Despite being marketed as an antitubercular agent, iproniazid continued to be prescribed to hundreds of thousands of patients affected by depression with good clinical response. The success of this drug is attributed in part to the good response observed in tuberculosis patients and partly due to lack of therapeutic alternatives to treat depression [12, 15, 17].

Iproniazid soon fell out of favor due to safety concerns including hypertensive crisis and liver toxicity. Other MAO inhibitors were developed with greater potency to inhibit MAO and relatively better side effect profiles. Often referred to as the classic MAOI's, these constitute phenelzine, tranylcypromine, and isocarboxazid. Tranylcypromine differs from other MAOI's in that it is structurally unrelated to hydrazines. Tranylcypromine is a cyclopropylamine, which was synthesized in 1948 as an analog of amphetamines. Its MAOI action was not discovered until 1959. Tranylcypromine was initially thought to have a more acceptable safety profile than other MAOI's [10, 15].

The receptor isoforms of MAO were identified in 1968 by GP Johnston in rat brain. Further studies revealed that MAO-A, which is located preferentially in the intestinal lining, is responsible for deaminating adrenaline, noradrenaline, and serotonin. On the other hand, MAO-B isoforms are responsible for the metabolism of benzylamine and b-phenylethylamine. The classic MAOIs were shown to inhibit both isoforms. MAO-A was identified as the isoform responsible for the antidepressant action of MAOI's. Moclobemide demonstrated reversible and selective inhibition of MAO-B. Selegiline is a newer selective inhibitor of MAO-B, which is mostly prescribed for the treatment of Parkinson's disease and major depression. Transdermal selegiline has been shown to be effective in major depressive disorder. Currently, the three classic MAOIs and selegiline remain approved by FDA in the USA [15, 18].

2.4 Tricyclic Antidepressants

The history of tricyclic antidepressants could be dated back to 1883 with the development of the first phenothiazine. The first TCA was synthesized by Professor Heinrich August Bernthsen in Germany while experimenting with chemical dyes, particularly methylene blue. Bernthsen created a phenothiazine that was used by J. Thiele and O. Holzinger to synthesize iminodibenzyl. Although, at the time it was not found useful by the textile industry, it was not until 1948 that in the pursuit of a novel sedative hypnotic, iminodibenzyl would be used as the basis for synthesizing 42 derivatives. These substances were then tested on laboratory animals and volunteers; one of these substances known as G-22150 was sent to a Swiss psychiatrist, Roland Kuhn, who dismissed its use as a hypnotic agent, given its unreliable results. He noted a "peculiar calming effect" of the drug. He further experimented to test the antipsychotic effects of the substance but ran into issues due to intolerance. In 1956, Kuhn received a compound termed as G-22355, a compound with lateral chains similar to chlorpromazine. Initial studies to test this compound in patients with schizophrenia did not improve psychotic symptoms. However, it was reported that some patients with schizophrenia developed hypomania while on the drug. Kuhn, in a letter to a pharmaceutical company in February 1956 pointed out possible antidepressant effects of the drug after observing clinical improvement in three of his patients diagnosed with depressive psychosis. Hence, the discovery of the antidepressant effects of TCAs was purely serendipitous [10, 14, 15, 19].

Kuhn further administered this drug to 37 depressed patients and presented the results at the World Congress of Psychiatry in Zurich in 1957. He reported that patients appeared more animated with improved communication and positive effects on mood. This drug was later named imipramine and his results were published in the journal "Schweizerische Medizinische Wochenschrift." In addition to clinical effects, Kuhn described imipramine's delayed onset of action and side effects including its anticholinergic effects. He observed that while a few patients experienced improvement in depressive symptoms 2–3 days after starting treatment, most experienced an improvement only after 1–4 weeks. He also described the suicide of a depressed patient with schizophrenia which he attributed to self-endangerment secondary to a decrease in inhibitions. This study was peculiar as Kuhn refrained from using structured rating scales and instead relied on direct observation of patients and nursing staff reports. This could be termed as a rudimentary form of a "trial" composed of case reports of patients who received the treatment. There was a lack of a control group and no standardized rating scales or statistical analysis were used. Despite all of its flaws, the study had a major impact on how psychiatrists treat depression and the subsequent development of antidepressants [12, 14, 15, 19].

This discovery however was met with skepticism in the medical community due to widespread psychodynamic/psychoanalytic theories of depression. It was the view of leading psychiatrists and psychologists of the time that drugs might help mask a few effects of depression but would not be as helpful in the long-term treatment of the illness. Despite this resistance to pharmacological treatment of depression, imipramine was introduced under the brand name Tofranil at the beginning of 1958. In September 1958, two studies were published in the *American Journal of Psychiatry* on the use of imipramine in patients with major depressive disorder. One of the studies was a reproduction of a lecture given by Roland Kuhn at Galesburg State Hospital. Kuhn's reports were confirmed later in various studies. More than 60 studies were published worldwide by 1959 which reported the therapeutic effects and adverse reactions with imipramine. Imipramine was introduced in North America by Heinz E. Lehmann who had attended Kuhn's lecture in Zurich and used the new drug to treat a group of Canadian patients. It was Lehmann's study that demonstrated the positive clinical effects on a sample of 48 depressed patients that led to the introduction and marketing of imipramine in the USA. It was in 1959 that the first placebo-controlled clinical trial of imipramine was contacted by Ball and Kiloh who showed the efficacy of imipramine in what was then termed as endogenous depression and psychotic depression. Imipramine was approved by the FDA in 1959 for its use in major depressive disorder [10, 12, 14, 15, 17, 20].

Despite the success of imipramine, a second TCA would not come to the market until 1961. Amitriptyline was initially developed as an antipsychotic which was created after modifying the central ring of a thioxanthene. The manufacturer of amitriptyline, Merck, reached out to psychiatrists in 1958 to test for possible antipsychotic properties of the drug. Merck approached Frank J. Hyde, Jr. who was considered one of the leading biologist psychiatrists of the time from his studies on chlorpromazine. He used amitriptyline to treat 130 patients at the Baltimore Square hospital and reported on its antidepressant effects. Amitriptyline was subsequently approved by the FDA for its use in depression in 1961. The 1960s further saw the development of multiple TCAs, namely nortriptyline and desipramine, both of which were approved by the FDA in 1964. Other TCAs developed were protriptyline, doxepin, and clomipramine. The decade also witnessed the introduction of "tetra"-cyclic antidepressants, which were developed after modifying the dibenzazepine structure of imipramine. These drugs were termed as second-generation antidepressants and included maprotiline and mianserin [15, 21].

The discovery of imipramine also helped to define depression as a new disease entity. These developments were also occurring in parallel to the introduction of MAOIs. These developments helped bolster the idea of depression as a treatable illness. Similar to other psychopharmacological tools, the mechanism of action of TCAs was unclear at that time despite their popularity and clinical use. The development of these antidepressants was not only clinically relevant but was also a crucial factor for research into the etiology and pathogenesis of mood disorders. The introduction of spectrophotofluorimetry made it possible to detect changes in the level of monoamines and their metabolites in the brain and was critical in neuropsychopharmacology research [16]. It was with this technique that Bernard Brodie's group at the National Institutes of Health (NIH) demonstrated the appearance of depressive symptoms with the depletion of brain levels of serotonin in experimental animals after they were administered reserpine [14, 17]. Reserpine, in the early 1950s was also being used for the treatment of schizophrenia, mostly in Germany. It was reported to precipitate depression and soon lost favor. Brodie's team further demonstrated that pre-treatment with iproniazid was noted to nullify the effect of reserpine and there was no depletion of serotonin noted in these animals. Furthermore, if iproniazid was administered after reserpine there was no modification of its sedative effects. These observations gave rise to the hypothesis that alterations of mood could be defined by the amount of biogenic amines in the brain namely, serotonin and norepinephrine [10, 12, 15].

Although imipramine was shown to antagonize the effects of reserpine, the experiments did not yield useful findings on imipramine's mechanism of action. Brodie's team at NIH reported that imipramine inhibits the absorption of noradrenaline. Julius Axelrod led another NIH team which described that after TCA administration there was a reduction in uptake of noradrenaline at selective nerve endings. The later development of desipramine played a crucial role in understanding the modulation of norepinephrine as part of TCA's mechanism of action [10, 15].

The work of Brodie and Axelrod would influence the biological understanding of affective disorders. The "catecholamine hypothesis" of affective disorders was described in two articles in 1965, one by J.J. Schildkraut in the *American Journal of Psychiatry* [22] and the other by William Bunney and John Davis in the *Archives of General Psychiatry* [23]. They hypothesized that depression might be associated with an absolute or relative deficiency of catecholamines. Their hypothesis was reliant on the effects of antidepressants as observed clinically and in vitro [10, 15].

The other prevalent hypothesis was the serotonergic hypothesis of depression. Betty Twarog, a researcher at Harvard, identified serotonin as a neurotransmitter [10, 15]. Alec J. Coppen demonstrated the improved therapeutic effects of MAOIs after the administration of tryptophan, a precursor of serotonin, in experimental animals [15, 16]. A Dutch psychiatrist Herman M. Van Praag reported the relationship between serotonergic dysfunction and the appearance of depression. Furthermore, a postmortem study of deaths by suicide by Shaw in 1967 demonstrated decreased brain concentrations of serotonin [17]. It was in 1968 that a Swedish group led by Arvid Carlsson demonstrated inhibition of reuptake of serotonin by TCAs. This would inspire Izyaslav P. Laplin and Gregory F. Oxenkrug to theorize the serotonergic basis for depression. The hypothesis stated that there was a deficit in serotonin level in certain brain areas which was responsible for the emergence of depression [15, 24].

The scientific and medical community towards the end of 1960s were in acceptance of the idea that the therapeutic effects of antidepressants, namely TCA and MAOIs, are mediated by increasing concentrations of serotonin and catecholamines. This would pave the way for the modern era of rational drug development and the introduction of medications that would target a specific set of neurotransmitters in the brain.

2.5 "Me Too" Drugs

It is worth mentioning the phenomenon of "me too" drugs which entails the development of a drug that is chemically related to an existing drug with similar clinical outcomes and often a similar or identical mechanism of action. These new drugs rarely add in terms of therapeutic advantage but can potentially differ in terms of pharmacokinetics and side effect profiles. This phenomenon in part could be attributed to market competitiveness of pharmaceutical companies and can sometimes lead to useful treatments. The "me too" phenomenon is evident in the discovery of imipramine which was initially designed as an antipsychotic in parallel to chlorpromazine but was later found to have antidepressant effects. The development of various tricyclic antidepressants during the 1960s and 1970s could also be attributed to the "me too" phenomenon in a race by pharmaceutical companies to gain a competitive edge in the market. This "me too" phenomenon also extended to the SSRIs and other modern treatments. The "me too" phenomenon could be beneficial in cutting down costs or, in the case of SSRIs, offering alternative treatments [21].

2.6 Lithium

Lithium is one of the simplest drugs in psychiatry. It is probably the earliest used psychotropic, although its definitive use treating psychiatric illnesses is attributed to serendipitous discovery. Lithium is a naturally occurring salt, hence its name comes from Greek word "lithos" meaning stone. The lightest metal in nature, lithium is distributed widely [25].

The earliest reports of using alkaline waters for the treatment of mania dates as far back as the fifth century. It was the work of two Swedish chemists, Johann August Arfwedson and Jons Jakob Berzelius, who isolated lithium from petalite and named the metal lithium. Lithium was marketed as an effective remedy for treating nervous disorders in the form of lithium water. Despite reports of lithium's ill effects in patients with heart disease, it continued to be part of popular culture and was even once an ingredient in prominent sodas [12, 25].

Lithium was initially used as a medication for the treatment of gout through the works of Alfred Baring Garrod who was an internist working in London. Lithium was also widely used in the treatment of renal stones. Using lithium to treat psychiatric conditions was noted in 1870 by Cilas Weir Mitchell who was a neurologist in Philadelphia and used lithium bromide to treat "general nervousness" in his patients. One of the first instances of using lithium to treat mania was reported in 1971 by William Hammond at Bellevue Hospital in New York. Since the drug used was lithium bromide, he was unable to ascertain which metal was responsible for the antimanic effects. Also, worth mentioning is the work of Carl G Lange who in 1886 introduced lithium in Denmark and promoted its use for the prevention of melancholic depression. These sporadic uses failed to gain traction partly due to the toxicity associated with lithium, which was used in larger doses at the time [12, 25, 26].

The introduction of lithium in the formal practice of psychiatry is attributed to the work of John Cade, an Australian psychiatrist, who observed mood symptoms in patients with thyroid disease. He recognized manic symptoms in patients with hyperthyroidism and depressive symptoms in patients with thyroid hypofunction. Given his observations, he hypothesized that mental illnesses could be caused by toxins generated due to hormonal imbalance that were being excreted in urine. He tested his hypothesis by injecting urine from patient with psychiatric illnesses and healthy controls into the peritoneal cavity of guinea pigs at different doses. He noted through his experiments that the urine from patients with mental illnesses was toxic at lower doses compared to healthy controls further cementing his belief that the urine of these patients contained toxic substances. He postulated urea as the toxic substance initially but soon through his experiments found the urine from manic patients continued to be more toxic even when controlled for urea and creatinine. He then concluded that there might be another substance in the urine of these patients that was potentiating the toxic effects of urea and focused on uric acid. To prove his hypothesis, he decided to inject laboratory animals with a solution of urea and varying concentrations of uric acid. Since these substances were poorly soluble in water, he decided to use lithium urate which is more soluble. He was surprised to find that lithium urate decreased urea toxicity. The convulsive effects observed earlier in guinea pigs injected with urine taken from manic patients were not observed after administering lithium urate. He then decided to experiment exclusively with lithium carbonate and observed the effect of reversible lethargy in the animals. These results encouraged him to study the effects of lithium in manic patients. Given concerns with the toxicity of lithium, he self-administered lithium carbonate to assess its safety. He later decided to administer lithium carbonate to patients suffering from schizophrenia and chronic mood disorders. The results of the study showed that lithium reduced excitement and agitation in patients with schizophrenia without altering the core symptoms of the illness. Lithium was not shown to be effective in chronic depression in the studies Cade conducted at the time. He also demonstrated the efficacy of lithium in the treatment of mania with the reappearance of manic symptoms after treatment discontinuation. These results were published in the *Medical Journal of Australia* in 1949 under the title "Lithium salts in the treatment of psychotic excitement" [27]. This discovery failed to gain much interest largely due to the cardiotoxic effects observed with lithium chloride. There was an absence of commercial interest from pharmaceutical industries given that lithium is widely available in nature. The early clinical trials with lithium were poorly designed and reported negative results further reinforced the notion (at the time) of its toxicity and limited utility in medicine [12, 20, 25, 26].

Although his work failed to garner much enthusiasm at the time, Cade's original article inspired researchers to further test the use of lithium to treat mania. Most notable of which was the work of a Danish psychiatrist, Mogens Schou, who designed the first randomized controlled trial to assess the efficacy of lithium versus placebo published in 1954 [28]. He randomly assigned patients to active drug versus placebo with the flip of a coin. His results were published with reported symptomatic improvement in 40% of patients. The introduction of lithium in American psychiatry is attributed to Samuel Gershon who worked with John Cade before immigrating to the USA. He worked at the University of Michigan, and his work with lithium was pivotal in generating interest at Rowell laboratories for the commercial development of lithium. He published an article in 1960 encouraging use of lithium in Mania. Gershon eventually moved to New York where he became a well-known researcher studying lithium. Gershon worked on multiple trials of lithium which were designed to determine the clinical efficacy of lithium [29–31]. Ronald Fieve of the New York State Psychiatric Institute reported treating 19 patients with lithium who previously failed phenothiazine therapy. He reported that 11 of the 25 manic episodes experienced by these patients showed good response to lithium and the individual manic episodes were shorter compared to the duration of past episodes in these patients. The standard treatment of mania at the time was chlorpromazine and multiple studies confirmed the superiority of lithium when compared to chlorpromazine. The discovery of the Coleman photometer in 1958 also aided in the widespread use of lithium as it was now possible to measure the blood levels of lithium [12, 25, 26].

Several large clinical trials conducted at the Department of Veterans Affairs and the National Institutes of Mental Health (NIMH) in the later part of the 1960s demonstrated the efficacy and safety of lithium before it was formally approved by the FDA. Despite its proven effectiveness and widespread clinical use, obtaining FDA approval for lithium in bipolar disorder would prove to be an uphill battle. Despite lithium being widely recognized by government bodies around the world as a safe and effective treatment, it was not until 1970 that the US FDA formally approved lithium. At the time, the USA became the 50th country to officially approve lithium use in patients. In 1975 lithium was also approved by the FDA for the prophylactic treatment of mania [25, 26].

2.7 Valproate and Carbamazepine

Valproic acid was synthesized by Beverly S. Burton in 1881. It was used as an organic solvent given its good solubility in fat. It was a popular organic solvent used as a diluent for other drugs.

George Carraz and colleagues working at Laboratoire Berthier in Grenoble, France in 1963 made an unexpected discovery while studying the anticonvulsant properties of Khellin compounds in experimental animals. As was the common practice of that time, valproic acid was used as a diluent to solubilize khellins. Study results demonstrated an anticonvulsant effect but failed to establish a dose response relationship. He further used the pentylenetetrazol convulsion model and incidentally realized that the diluent used in the solution possessed anticonvulsant activity instead of the khellin derivatives. They realized through this work that all of the solutions contain valproic acid and thus serendipitously discovered the anticonvulsant activity of valproate. After establishing the anticonvulsant properties, George Carraz and his group synthesized valpromide. Carraz further demonstrated that strychnine induced epileptic convulsions were prevented by valpromide but not by valproate [12, 25].

Following successful animal studies, Carraz collaborated with Sergio Borselli and Pierre Lambert in 1965 to conduct clinical trials in patients with epilepsy to test the efficacy and safety of valpromide and valproate. These trials were conducted at the psychiatric hospital of Brittany in France. The trials demonstrated that both valpromide and valproate have anticonvulsant effects and in addition some psychotropic effects were noted in patients. The psychotropic effects ranged from improvement in depression to mild euphoria. Lambert et al. in 1966 was the first to report valproate's mood stabilizing effects in patients with bipolar disorder. Bowden et al. conducted a large multicenter, randomized, double blind, placebo controlled, and parent group trial which was supported by the Abbott pharmaceutical company. The recruited 179 acutely manic hospitalized patients who were randomly assigned to divalproex, lithium, or placebo. The investigators concluded that divalproex and lithium were significantly more effective than placebo in reducing the symptoms of acute mania. This study paved the way for FDA approval of divalproex sodium for its antimanic effects in 1995 [12, 25, 26].

Carbamazepine is a tricyclic anticonvulsant compound which was developed in the lab of J. R. Geigy in Basel, Switzerland. Its antiepileptic properties were demonstrated in 1963 by W. Theobad and H. A. Kunz. It was widely used in Japan in the 1960s as lithium was not available in the country at that time. One of the first series of trials of carbamazepine were conducted by Takezaki and Hanaoka for the treatment of bipolar patients. They published their results in 1971. Psychiatrists in Western countries were skeptical about the results because of a misconception of lower (cited as inadequate in Western literature) standards being used for the trials conducted in Japan (although the trial used similar research protocol to the trial that was used to demonstrate lithium's efficacy). It received FDA approval in 1974 as an antiepileptic agent and would be approved in 2004 for treating mania [25].

2.8 Meprobamate and Mephenesin

These sedative-hypnotics were developed by Frank Berger and William Bradley during their efforts to synthesize an antibacterial agent to target gram-negative bacteria that were resistant to penicillin. They believed that lengthening of a carbon chain of an existing disinfectant called phenoxetol could result in a drug with expected antibacterial effect. After the synthesis of this compound, they decided to assess its safety in experimental animals. To their surprise, they found that the mice injected with this drug developed a reversible flaccid paralysis of the voluntary skeletal muscles. They tested the compound, called mephenesin, in graduated doses to learn that it produces muscle relaxation and paralysis of all voluntary muscles while preserving consciousness. It was found to not affect the autonomic nervous system and resulted in spontaneous recovery. They used the term "tranquilization" to describe the effects of this drug in a report published in 1946 [10, 12].

Mephenesin is used clinically for producing muscle relaxation during light anesthesia. During use as an alternative to tubocurarine, the antianxiety effects of mephenesin were subsequently recognized. There were significant drawbacks limiting its use as an antianxiety agent: namely short duration of action, need for large doses due to low potency, and its major effect on the spinal cord [12].

Given these limitations there was a search for compounds with similar clinical properties while overcoming these shortcomings. Bernard Ludwig in 1951 introduced meprobamate, a compound similar to mephenesin. Berger in 1984 reported its selective action on the thalamus, hippocampus, and limbic system which were thought to be associated with anti-anxiety properties. Meprobamate was a widely used psychotropic which was eventually supplanted by benzodiazepines in the 1980s [10, 12].

2.9 LSD

The psychedelic effects of LSD were accidentally discovered due to a lab accident. While trying to develop uterine stimulants and hemostatics, Albert Hoffman synthesized derivatives of ergot alkaloids in 1930s. This led to the synthesis of ergometrine which was then modified to form methergine, a compound with strong oxytocic and hemostatic properties. In subsequent efforts to develop an analeptic, Hoffman synthesized LSD-25. Analeptic refers to the phenomena of stimulation of respiration and reversal of CNS depression. The structure of LSD-25 was loosely based on nikethamide, a respiratory stimulant [12, 32].

It was while working in his lab that Hoffman accidentally absorbed LSD and felt its psychotropic effects. This experience led him to plan a self-experiment with this compound. After self-administration he reported:

> *Last Friday, April 16, 1943, I was forced to stop my work in the laboratory in the middle of the afternoon and to go home, as I was seized by a peculiar restlessness associated with a sensation of mild dizziness. On arriving home, I lay down and sank into a kind of drunkenness which was not unpleasant and which was characterized by extreme activity of imagination. As I lay in a dazed condition with my eyes closed (I experienced daylight as disagreeably bright) there surged upon me an uninterrupted stream of fantastic images of extraordinary plasticity and vividness and accompanied by an intense, kaleidoscope-like play of colors. This condition gradually passed off after about two hours* [32].

Sandoz quickly worked to develop the product which was then marketed under the trade name Delysid. This had two principal indications: analytical psychotherapies to induce states of relaxation and an experimental study of psychotic states [12].

2.10 Conclusion and Future Directions

Psychiatry has come a long way since its inception and subsequent designation as an integral yet specialized area of medicine. The practice of psychiatry has grown leaps and bounds and away from primitive and inhumane experimental treatments like low-sodium diets, forced vomiting, scalp bleeding, and blistering with the aim to reduce pressure in the brain by expelling fluids [5]. The father of American psychiatry, Benjamin Rush, was part of a few of these experiments [5, 33]. Due to a lack of available treatment options for the patients, individuals with mental disorders were frequently institutionalized. Most of the patients suffering from mental disorders were confined to asylums with little to no effort to help them be a functional part of the society [2, 4, 5].

The advent of psychopharmacology provided a much-needed boost not only in terms of the clinical treatment of psychiatric illnesses but also a pathway to understanding the pathophysiology of psychiatric illnesses. Much of our understanding of the biological basis of psychiatric illnesses is derived from the mechanism of action of these drugs that were discovered accidentally. Two of the major hypotheses of psychiatric illnesses, namely the catecholamine hypothesis of affective disorders and the dopamine hypothesis of schizophrenia, were both based on the effects of drugs used for the treatment of psychiatric illnesses. The field of psychiatry unlike other areas of medicine seemed to grow in reverse from treatment to understanding of etiopathogenesis. Our understanding of psychiatric illnesses has mostly been through studying the effects of drugs used for the treatment of these illnesses. For example, the development of selective serotonin reuptake inhibitors (SSRIs) was based on our understanding of the involvement of serotonin in major depressive disorder.

As described in this chapter, most of the earliest psychotropics were discovered serendipitously. The earliest trials in psychiatry were poorly designed and mostly consisted of case series involving a small number of patients. Most of the studies lacked a control group and sometimes would only report positive outcomes. The four major drug discoveries including chlorpromazine, lithium, tricyclic antidepressants, and MAO inhibitors were all introduced in the practice of psychiatry through case series and lacked any control group. Eventually, the placebo-controlled trials of chlorpromazine, lithium, and imipramine did show favorable results. However, the antipsychotic properties of reserpine showed in earlier case series were not replicated in controlled trials. Also lacking in earlier studies of psychotropics was the use of standardized rating scales. The inclusion criteria were also broad owing to the narrative nature of DSM-I and II. There were also ethical concerns with the initial psychotropic trials as human studies were conducted before obtaining sufficient safety data and sometimes without appropriate informed consent. The introduction of randomized control trials was a shift towards an emphasis on results rather than the influence and expertise of individual clinicians. This meant that the results obtained from these trials could be replicated under standardized conditions. By the early 1970s, the FDA would judge a drug's efficacy by the results of RCTs [20].

The past few years have seen an advancement in our understanding of neurobiology due to the unprecedented growth and expansion of our understanding of neuroscience. It is now possible to develop drugs to target specific receptors in very specific areas of the brain. Modern drug development in psychiatry is discussed in the subsequent chapters of this book and seeks to achieve precision drug development through meticulously designed clinical trials focused on mechanism of action and targeting the underlying cause of the illness.

References

1. Binder DK, Schaller K, Clusmann H. The seminal contributions of Johann-Christian Reil to anatomy, physiology, and psychiatry. Neurosurgery. 2007;61(5):1091–6. https://doi.org/10.1227/01.neu.0000303205.15489.23.
2. Hirshbein L. The American Psychiatric Association and the history of psychiatry. Hist Psychiatry. 2011;22(3):302–14. https://doi.org/10.1177/0957154X10389744.
3. Pelletier J-F, Davidson L. A l'origine meme de la psychiatrie comme nouvelle specialite medicale: le partenariat Pinel-Pussin. Sante Ment Que. 2015;40(1):19–34.
4. Dix D. "I Tell What I Have Seen" – The Reports of Asylum Reformer Dorothea Dix. Am J Public Health. 2006;96(4):622–4. https://doi.org/10.2105/AJPH.96.4.622.
5. Conrad JA. A black and white history of psychiatry in the United States. J Med Humanit. 2022;43(2):247–66. https://doi.org/10.1007/s10912-020-09650-6.
6. Terrier LM, Lévêque M, Amelot A. Brain lobotomy: a historical and moral dilemma with no alternative? World Neurosurg. 2019;132:211–8. https://doi.org/10.1016/j.wneu.2019.08.254.
7. Freeman W. Lobotomy in limbo? Am J Psychiatry. 1972;128(10):1315–6. https://doi.org/10.1176/ajp.128.10.1315.
8. Clarke P. Hip hip hooray, ECT turns 80! Australas Psychiatry. 2019;27(1):53–5. https://doi.org/10.1177/1039856218815753.
9. Carlsson A. Reflections on the history of psychopharmacology. Psychopharmacol Ser. 1988;5:3–11. https://doi.org/10.1007/978-3-642-73280-5_1.
10. Braslow JT, Marder SR. History of psychopharmacology. Annu Rev Clin Psychol. 2019;15:25–50.
11. Boyd-Kimball D, Gonczy K, Lewis B, Mason T, Siliko N, Wolfe J. Classics in chemical neuroscience: chlorpromazine. ACS Chem Neurosci. 2019;10(1):79–88. https://doi.org/10.1021/acschemneuro.8b00258.
12. AlanA B, MikeF H, López-Muñoz F. Toward standardized usage of the word serendipity in the historiography of psychopharmacology. J Hist Neurosci. 2010;19(3):253–70. https://doi.org/10.1080/09647040903188205.
13. Tourney G. History of biological psychiatry in America. Am J Psychiatry. 1969;126(1):29–42. https://doi.org/10.1176/ajp.126.1.29.
14. Fangmann P, Assion HJ, Juckel G, González CÁ, López-Muñoz F. Half a century of antidepressant drugs: on the clinical introduction of monoamine oxidase inhibitors, tricyclics, and tetracyclics. Part II: tricyclics and tetracyclics. J Clin Psychopharmacol. 2008;28(1):1–4. https://doi.org/10.1097/jcp.0b013e3181627b60.
15. Lopez-Munoz F, Alamo C. Monoaminergic neurotransmission: the history of the discovery of antidepressants from 1950s until today. Curr Pharm Des. 2009;15(14):1563–86. https://doi.org/10.2174/138161209788168001.
16. Foley PB. Dopamine in psychiatry: a historical perspective. J Neural Transm. 2019;126(4):473–9. https://doi.org/10.1007/s00702-019-01987-0.

17. Hillhouse TM, Porter JH. A brief history of the development of antidepressant drugs: from monoamines to glutamate. Exp Clin Psychopharmacol. 2015;23(1):1–21. https://doi.org/10.1037/a0038550.
18. Chockalingam R, Gott BM, Conway CR. Tricyclic antidepressants and monoamine oxidase inhibitors: are they too old for a new look? Handb Exp Pharmacol. 2019;250:37–48. https://doi.org/10.1007/164_2018_133.
19. Brown WA, Rosdolsky M. The clinical discovery of imipramine. Am J Psychiatry. 2015;172(5):426–9. https://doi.org/10.1176/appi.ajp.2015.14101336.
20. Leon AC. Evolution of psychopharmacology trial design and analysis: six decades in the making. J Clin Psychiatry. 2011;72(03):331–40. https://doi.org/10.4088/JCP.10r06669.
21. Aronson JK, Green AR. Me-too pharmaceutical products: history, definitions, examples, and relevance to drug shortages and essential medicines lists. Br J Clin Pharmacol. 2020;86(11):2114–22. https://doi.org/10.1111/bcp.14327.
22. Schildkraut JJ. The catecholamine hypothesis of affective disorders: a review of supporting evidence. Am J Psychiatry. 1965;122(5):509–22. https://doi.org/10.1176/ajp.122.5.509.
23. Bunney WE, Davis JM. Norepinephrine in depressive reactions. A review Arch Gen Psychiatry. 1965;13(6):483–94. https://doi.org/10.1001/archpsyc.1965.01730060001001.
24. Feighner JP. Mechanism of action of antidepressant medications. J Clin Psychiatry. 1999;60(Suppl 4):4–11; discussion 12–3.
25. López-Muñoz F, Shen WW, D'Ocon P, Romero A, Álamo C. A history of the pharmacological treatment of bipolar disorder. Int J Mol Sci. 2018;19(7):2143. https://doi.org/10.3390/ijms19072143.
26. Ruffalo ML. A brief history of lithium treatment in psychiatry. Prim Care Companion CNS Disord. 2017;19(5):27325. https://doi.org/10.4088/PCC.17br02140.
27. Cade JF. Lithium salts in the treatment of psychotic excitement. 1949. Bull World Health Organ. 2000;78(4):518–20.
28. Schou M, Juel-Nielsen N, Stromgren E, Voldby H. The treatment of manic psychoses by the administration of lithium salts. J Neurol Neurosurg Psychiatry. 1954;17(4):250–60. https://doi.org/10.1136/jnnp.17.4.250.
29. Johnson G, Gershon S, Hekimian LJ. Controlled evaluation of lithium and chlorpromazine in the treatment of manic states: an interim report. Compr Psychiatry. 1968;9(6):563–73. https://doi.org/10.1016/s0010-440x(68)80053-7.
30. Johnson G, Gershon S, Burdock EI, Floyd A, Hekimian L. Comparative effects of lithium and chlorpromazine in the treatment of acute manic states. Br J Psychiatry J Ment Sci. 1971;119(550):267–76. https://doi.org/10.1192/bjp.119.550.267.
31. Shopsin B, Gershon S, Thompson H, Collins P. Psychoactive drugs in mania. A controlled comparison of lithium carbonate, chlorpromazine, and haloperidol. Arch Gen Psychiatry. 1975;32(1):34–42. https://doi.org/10.1001/archpsyc.1975.01760190036004.
32. Hoffman. The Discovery of LSD. Accessed April 4, 2022. https://catbull.com/alamut/Bibliothek/HOFMANN%20Albert/discovery/lsd.htm.
33. Farr CB. Benjamin rush and American psychiatry. 1944. Am J Psychiatry. 1994;151(6 Suppl):64–73. https://doi.org/10.1176/ajp.151.6.64.

Chapter 3
The Evolving Role of Animal Models in the Discovery and Development of Novel Treatments for Psychiatric Disorders

Laura B. Teal, Shalonda M. Ingram, Michael Bubser, Elliott McClure, and Carrie K. Jones

Abstract Historically, animal models have been routinely used in the characterization of novel chemical entities (NCEs) for various psychiatric disorders. Animal models have been essential in the in vivo validation of novel drug targets, establishment of lead compound pharmacokinetic to pharmacodynamic relationships, optimization of lead compounds through preclinical candidate selection, and development of translational measures of target occupancy and functional target engagement. Yet, with decades of multiple NCE failures in Phase II and III efficacy trials for different psychiatric disorders, the utility and value of animal models in the drug discovery process have come under intense scrutiny along with the widespread withdrawal of the pharmaceutical industry from psychiatric drug discovery. More recently, the development and utilization of animal models for the discovery of psychiatric NCEs has undergone a dynamic evolution with the application of the Research Domain Criteria (RDoC) framework for better design of preclinical to clinical translational studies combined with innovative genetic, neural circuitry-based, and automated testing technologies. In this chapter, the authors will discuss this evolving role of animal models for improving the different stages of the discovery and development in the identification of next generation treatments for psychiatric disorders.

Keywords Drug discovery · Animal models · RDoC · Psychiatry

Laura B. Teal, Shalonda M. Ingram and Carrie K. Jones contributed equally to this work.

L. B. Teal · S. M. Ingram · M. Bubser · C. K. Jones (✉)
Department of Pharmacology, Vanderbilt University, Nashville, TN, USA

Warren Center for Neuroscience Drug Discovery, Vanderbilt University, Nashville, TN, USA
e-mail: carrie.jones@vanderbilt.edu

E. McClure
College of Pharmacy and Health Sciences, Lipscomb University, Nashville, TN, USA

M. Macaluso et al. (eds.), *Drug Development in Psychiatry*, Advances in Neurobiology 30,
https://doi.org/10.1007/978-3-031-21054-9_3

3.1 Introduction

The current prevalence of psychiatric disorders worldwide has increased over the last two decades [1]. Within the United States, over 20% of all adults experienced mental illness in 2020 [2]. While currently approved psychiatric drugs provide documented therapeutic benefits, successful treatment outcomes are often hampered by partial response rates and/or treatment resistance, dose-related adverse side effects, little to no efficacy for many symptom clusters, and decreased patient compliance with available dosing regimens [3–8]. Collectively, these findings underscore the tremendous unmet need to develop novel treatments for psychiatric disorders.

Unfortunately, the discovery and development of novel chemical entities (NCEs) for psychiatric disorders has remained challenging with significantly lower rates of NCE progression from Phase II and Phase III efficacy trials to market launch relative to other therapeutic areas [9–11]. These industry-wide failures in the clinical development of psychiatric NCEs coupled with the limited understanding of the underlying pathophysiology for these disorders have resulted in the widespread withdrawal of pharmaceutical industry's research and development (R&D) resources [9–11]. While all stages of psychiatric drug discovery and development have received extensive examination over the last two decades, as will be discussed in subsequent chapters, the limitations of animal models for the identification and characterization of psychiatric NCEs have come under particularly intense scrutiny. Criticisms have included the lack of sufficient in vivo target validation and strong disease model construct validity, the limited understanding of the pharmacokinetic to pharmacodynamic (PK/PD) relationships for the preclinical candidate (PCC) molecules, and the failure to establish optimized translational measures of target occupancy and/or functional target engagement critical for early proof-of-concept (POC) clinical studies [12–14]. To address these major concerns, researchers within both the academic and pharmaceutical industry communities are rapidly developing new strategies for the use of animals in the identification and validation of novel CNS drug targets and characterization of NCEs for psychiatric disorders.

In this chapter, the authors will review the historical and current use of animals for modeling aspects of different psychiatric disorders. Next, we will discuss how these animal models are used to advance the progression of novel psychiatric NCEs through the stages of discovery. Finally, we will examine the ongoing application of the research domain criteria (RDoC) framework to improve preclinical to clinical translational studies coupled with novel genetic, neural circuitry-based, and automated testing technologies for novel psychiatric animal model development.

3.2 Animal Models for Psychiatric Drug Discovery

3.2.1 Historical and Current Use of Animal Models for Psychiatric Disorders

While multiple animal species are currently used in the different stages of drug discovery and development, including *Caenorhabditis elegans* [15, 16], *Drosophila melanogaster* [17], zebrafish [18, 19], dogs [20], and nonhuman primates [21], mice and rats continue to represent the predominant species for psychiatric drug discovery based on sufficient homology to humans and experimental feasibility [22]. Importantly, all research involving vertebrate animals requires institutional approval by an Institutional Animal Care and Use Committee (IACUC) at the organization where the research is conducted. The principles of the "three R's" (3R's) – replacement, reduction, and refinement – provide the critical guidelines for the ethical use of animals in academic and industry scientific research [20, 23]. These principles emphasize replacing animals when possible with in vitro assays, including immortalized or primary cell cultures; reducing the number of animals to be used with power calculation for determination of the appropriate sample size for each study; and refinement of testing procedures to minimize potential pain and stress in the test subjects.

Although the ideal animal model for a given psychiatric disorder should replicate the complete phenotype of the human condition with its underlying etiology and/or pathophysiology, the development of such models has proven to be unrealistic due to the complex heterogeneous nature of the different psychiatric disorders and the presence of substantial comorbidities. Moreover, development of animal models for psychiatric disorders has been limited by the lack of understanding of the etiology and imbalances in the neural substrates underlying the pathophysiology of each disorder. As such, researchers have focused on animal models that represent one or more defined aspects of a particular disorder in a single preclinical species. This approach has allowed the examination of different alterations in the selected target mechanism(s) and/or associated biology, including changes in cellular signaling, neural circuity, and behavior under normal and pathological conditions. In some cases, animal models for psychiatric disorders have also been based on identification of different endophenotypes or quantifiable endpoints that are linked with psychiatric symptoms with shared underlying genetic influences [24]. As we will discuss below, the strength and suitability of a given animal model for a particular stage of the drug discovery process depend on its type and degree of validity.

3.2.2 Assessing the Validity of Animal Models

The process of assessing the reliability of a particular animal model to accurately reflect a specific aspect(s) of a psychiatric disorder is based on three classic aspects of animal models, which were originally proposed by Willner in 1984 and are commonly known as construct, face, and predictive validity [25]. Construct (or etiologic) validity establishes a connection between the etiology and/or pathophysiology of the psychiatric disorder and the animal model. An animal model with high construct validity has strong mechanistic similarities to a particular psychiatric disorder etiology as demonstrated by homologous genetics and/or common disruptions in molecular and physiological signaling cascades. For example, a genetic mouse model based on a key disrupted gene, such as the chromosome 22q11.2 deletion in schizophrenia, may be considered to have high construct validity [12, 26]. However, it is also important to understand how closely the selected genetic variant being manipulated correlates with the specific psychiatric disorder. Since psychiatric disorders arise from complex interactions between multiple genetic and environmental factors, a single genetic change can often be associated with multiple psychiatric disorders. In the case of the chromosome 22q11.2 deletion, only 30% of individuals with this deletion meet the diagnostic criteria for schizophrenia or schizoaffective disorder [26, 27]. Thus, caution must be taken before declaring a single genetic deletion in mice a strong construct model of schizophrenia.

Face validity refers to shared similarities between an observed feature or phenotype of the animal model and an anatomical, neurochemical, behavioral, or neuropathological aspect of the psychiatric disorders [25]. Often models with high face validity for psychiatric disorders share a common behavioral feature with the clinical presentation of a given DSM-5 disorder. For example, patients with schizophrenia exhibit deficits in sensorimotor gating, or an inability to properly filter out irrelevant sensory stimuli. This clinical symptom can be measured by studying changes in pre-pulse inhibition (PPI) of the acoustic startle reflex, wherein an auditory tone that provokes a startle response is measured with or without a preceding auditory tone, known as a pre-pulse, which does not provoke a startle response on its own. In healthy individuals, the presentation of a pre-pulse will markedly reduce the response to a subsequent startle stimulus. However, in individuals with schizophrenia, this pre-pulse-mediated inhibition of the acoustic startle response is attenuated [28, 29]. Comparable changes in PPI of the acoustic startle response can be assessed in both human patients with schizophrenia and in rodents; thus, a model which induces PPI deficits in rats, such as amphetamine-induced disruption of PPI, has high face validity [28, 29].

Predictive validity describes the correlation between the degree of efficacy of a particular pharmacologic mechanism in an animal model and the observed efficacy for one or more symptoms of a particular psychiatric disorder [25]. One example of an animal model with high predictive validity is the rodent forced swim test used to evaluate the antidepressant-like activity of novel treatments for major depressive disorder (MDD) [30, 31]. Mice or rats are placed in an inescapable transparent testing cylinder that is filled with water and their escape-related mobility behaviors are measured with the amount of time spent immobile, not actively swimming, considered a measure of learned helplessness [30, 31]. While this assay has low construct validity and only moderate face validity, it has high predictive validity since clinically efficacious antidepressant drugs reduce immobility duration [30, 31].

As discussed below in the descriptions of representative animal models for the different psychiatric conditions, each model meets the different validity criteria to varying degrees making it important to use a variety of complementary animal models when evaluating the in vivo activity of a psychiatric NCE, to provide a broader efficacy profile that together approximates many aspects of the specific psychiatric disorder.

3.2.3 Types of Animal Models of Psychiatric Disorders

In general, models for psychiatric disorders are categorized based on how the model is established, including genetic models, lesion models, pharmacological models, or behavioral models. When appropriate, these models can also be combined to form a multimodal model which may reflect more than one aspect of a given psychiatric disorder.

Genetic models aim to understand the complex interaction between an individual's genome and their environment [32, 33]. Genetic animal models can be used to validate a pharmacological target for drug development, to understand risk factors for developing a disorder, or to test potential treatments in a disease model. The design of animal models to study the connections between genetics, neurobiological mechanisms, and psychiatric illnesses also directly aligns with the RDoC framework [34, 35]. The traditional means of studying biological implications in psychiatric illnesses has been the constitutive or conditional modification of genes associated with psychiatric diseases; see the following review for more information about specific transgenic animal models for psychiatric disorders [36].

Constitutive gene expression studies in mice focus on the loss of function, gain of function, and the constitutive expression of chromosomal rearrangements indicated in psychiatric diseases [37]. Constitutive gene expression manipulations in psychiatric animal studies have been key in understanding biochemical changes that occur as a result of genetic variants and the lifelong effects, such as constitutive 5-HTT knockouts in understanding depression [38]. Constitutive gene modifications in animal are achieved by injecting the DNA of the gene of interest into the embryos of the animal model [39]. The resulting gene modifications are expressed throughout the entire animal and inherited in the offspring. Challenges to constitutive gene expression include the manipulation of genes resulting in lethal knockouts, adverse consequences in gene modifications in off-target organ systems, limited number of gene modifications in a single animal, and animal model validation. In general, constitutive expression is permanent and lacks control beyond initial expression of the genes.

Conditional genetic expression studies in mice allow the modification of the gene based on a specific tissue or cell type and introduce inducible expression to control the onset of the genetic modifications in the animal model. Conditional models address the limitation of constitutive models in flexibility to control genetic modifications [37, 40–42]. The inducible characteristic of conditional gene expression systems is controlled by molecular switches. The tetracycline (Tet) system was one of the first established in transgenic animal models and successfully applied to neurogenomic studies in Huntington disease [43–46]. More recently, the Cre-LoxP system introduced a novel approach to tissue and site-specific modification of genes for better spatial and temporal expression of genes of interest in animal models [40, 42, 47, 48]. The advantage of the Cre-LoxP system is the ability to generate deletions, translocations, and inversions and alter gene copy number in any cell or tissue type [48]. Both approaches allow researchers to recapitulate genetic contributions of the illnesses as seen in humans with significant construct validity [12]. Furthermore, the use of these approaches has allowed researchers to identify endophenotypes that translate to human studies and the impact of genetic mutations on neurocircuitry abnormalities in models where polymorphisms are prevalent [35, 49]. However, conditional and constitutive genetic animal models are limited to single gene modifications in an animal and embryonic engineering. These limitations largely affect study designs to address multiple genetic variants that contribute to psychiatric illness comorbidities.

Another common method of generating an animal model is a pharmacological challenge. In a pharmacological model, a drug with a known mechanism is administered to an animal, and the resulting behavioral or physiological changes are examined. For example, the psychostimulant amphetamine is often used as a pharmacological challenge to mimic underlying hyperactivity of the mesolimbic dopaminergic circuit correlated with similar circuitry changes and the positive symptoms in schizophrenia. A single dose of amphetamine robustly induces hyperactivity, and repeated dosing can exaggerate this hyperactivity accompanied by increased dopamine efflux from the nucleus accumbens and dorsal striatum [50]. In addition to hyperactivity, amphetamine also produces deficits in PPI of the acoustic startle reflex comparable to those observed in patients with schizophrenia (described in more detail under section "Schizophrenia Spectrum and Other Psychotic Disorders") [51]. In both cases, amphetamine is used as a pharmacologic agent which induces behavioral changes that model circuit changes and/or symptoms of schizophrenia in rodents.

A third method of animal model generation is the lesion model. Lesion models are used when ablation of a specific brain region can mimic symptoms of a CNS disorder. For example, in Parkinson's disease, dopamine neurons within the substantia nigra deteriorate. In rodents, this can be modeled using 6-hydroxydopamine (6-OHDA), which selectively causes cell death in dopaminergic neurons. 6-OHDA lesions of the substantia nigra are therefore used to model a variety of symptoms of Parkinson's disease [52]. This model, however, differs from the progression of Parkinson's disease because the effects of the lesion develop rapidly over weeks versus mimicking the slow neuronal death in the human disorder over years. Despite this, the lesion model recreates the major neurological change seen in late-stage Parkinson's disease, making it a good model for testing potential drug treatments targeting that patient population.

Behavioral models assess a specific behavior in a certain context, often combined with another model or drug treatment. For example, a common behavioral model used in the assessment of antidepressant-like activity is the FST (defined in Sect. 3.2.2) [31]. FST is often used to assess differences in antidepressant-like activity associated with neurochemical lesion, genetic, and/or pharmacologic challenges. In addition, as will be discussed under the specific psychiatric disorder sections listed below, several stress-related behavioral models, such as chronic unpredictable mild stress and chronic social stress, are used to model different aspects of complex psychiatric disorders of depression, anxiety, and PTSD including anxiogenic-like and/or depressive-like activity.

In the following sections, we will discuss some of the classical models for multiple psychiatric disorders. Each section is accompanied by a table with the following information: (1) preclinical rodent models, (2) biological rationale/disorder etiology, (3) chronic neurobiological changes for chronic models or acute effects of pharmacologic treatment for the acute challenge models, (4) model validity, (5) utilization in stages of drug discovery, (6) FDA-approved drugs with demonstrated preclinical efficacy.

Major Depressive Disorder

Numerous genetic animal models for MDD have been designed using transgenic mice expressing different genetic variants of the following targets, including serotonin, norepinephrine, N-methyl-D-aspartate (NDMA) receptor, corticotropin-releasing hormone receptor-1, and BDNF, all of which have been hypothesized to contribute to the etiology of MDD [53]. More recently these genetic model approaches are being combined with optogenetic technologies to directly test the contributions of key disrupted brain circuits relevant to MDD [54, 55], as will be described in more detail in Sect. 3.4.2. There are two important acute models for MDD with high predictive validity that are used routinely from early-through late-stage lead optimization, specifically the forced swim [56], and tail suspension tests [57–59]. Both assays model a state of learned helplessness and are highly sensitive to the anti-depressant effects of all major approved classes of antidepressants [60] (see Table 3.1). In addition, there are a number of interesting chronic stress models with increased construct validity for modeling induction of chronic stress and anhedonic-like behavioral states, including the unpredictable chronic mild stress [61, 62] and chronic social stress models [63, 64] (see Table 3.1). In the case of the unpredictable chronic mild stress model, mice or rats are presented daily with a series of unpredictable mild stressors in a schedule that prevents habituation, including changes in housing materials, light cycles, exposure to novel odor cues, food or water restrictions, none of which induce harm to the test animal [65]. In the case of the chronic social stress model [64] (also known as the resident/intruder social defeat test) resident mice undergo daily exposure to a larger, more aggressive intruder mouse in their home cage to establish dominance (see Table 3.1). Over the course of a 3–4 day period, rodents undergoing either the unpredictable chronic mild stress model or the chronic social stress model display significant behavioral and neurochemical changes associated with the development of anhedonic-like behaviors. These include increased immobility in the FST [66–68] and decreased sucrose preference [66, 68, 69], social interaction [70], locomotor activity [68, 71], food intake [71], and intracranial self-stimulation (ICSS) [69, 72] (see Table 3.1) along with significant alterations in c-Fos expression in key brain regions that modulate stress effects including the prefrontal cortex (PFC), hippocampus, and amygdala [73], disruptions in the normal hypothalamic-pituitary-adrenal axis functions [62, 66, 74], changes in mesolimbic reward circuitry signaling [75], and decreased dendritic spine density in the PFC and hippocampus [76] (see Table 3.1). ICSS is an interesting model with high construct validity and moderate face validity for assessing the anhedonic-like behaviors and responses of the brain reward circuitry [77]. ICSS involves implanting an electrode within the ventral tegmental area (VTA) or lateral hypothalamus, which the animal can activate by lever pressing to obtain direct brain stimulation at these rewarding areas. Animals with an anhedonia-like phenotype will exhibit increased ICSS thresholds, manifesting in decreased ICSS responding, thereby assessing the anhedonic-like behaviors and responses directly in the reward circuitry of the brain [77]. Overall the majority of depressive-like, anxiogenic and anhedonic-like activity in animals exposed to either unpredictable chronic mild stress and chronic social stress can be reversed with chronic exposure to the majority of clinically approved antidepressants [38, 78–91]. For more information on animal models for MDD, see the following reviews [53, 65, 92]. The Food and Drug Administration (FDA) approved therapies for MDD include selective serotonin reuptake inhibitors (SSRIs) such as citalopram and fluoxetine, serotonin/norepinephrine reuptake inhibitors (SNRIs) such as venlafaxine and duloxetine, tricyclic antidepressants such as imipramine and desipramine, the atypical antidepressant reboxetine, the monoamine oxidase inhibitor (MAOI) selegiline, the γ-aminobutyric acid subtype A (GABA-A) receptor positive modulator brenaloxone, and the NMDA receptor antagonist esketamine. Deep brain stimulation has been studied in antidepressant-resistant rodents following chronic social stress and chronic mild stress and also demonstrates normalization of depressant behaviors [80–83]. While SSRIs remain a first line of treatment for MDD, this class of medication is also has associated dose-limiting side effects that include increased suicide risk (a side effect common to all antidepressants), serotonin syndrome, and sexual dysfunction [93]. Unfortunately, only 30% of patients receiving treatment for MDD symptoms experience full remission, and 50% have no response to treatment [94]. More recently, the FDA approval of esketamine marks an exciting approval of a novel treatment for treatment-resistant depression [95]. However, its use is strictly controlled because of potential abuse liability and sedating effects [96].

Generalized Anxiety Disorder

Behavioral models for GAD in rodents consist of freezing or hypoactivity, vigilance, and defensive behaviors, and decreased food intake [97]. Chronic social stress, chronic mild stress, and single prolonged stress models are all used to induce

Table 3.1 Animal models for major depressive disorders

Models to induce depression-like symptoms					
Preclinical rodent model	**Biological rationale/ disorder etiology**	**Chronic neurobiological changes**	**Model validity**	**Utilization in drug discovery**	**FDA-approved drugs with demonstrated preclinical efficacy**
Unpredictable chronic mild stress	Repeated unpredictable exposure to mild stressors that are not harmful, to induce anhedonic-like behaviors [61, 65, 78, 276, 277]	Cortisone ↑ [277], HDAC-5 in PFC ↓ [84], dopamine cell firing in VTA ↓↑ [75], synaptic changes in lateral amygdala [85], amygdala, hypothalamus, and brainstem c-Fos changes [73]	Face: moderate Construct: moderate Predictive: moderate	Lead optimization Preclinical candidate selection	*SSRI:* citalopram, fluoxetine, paroxetine *TCA:* imipramine, desipramine *NE/DA reuptake inhibitor:* reboxetine *NMDA antagonist:* esketamine
Chronic social stress	Repeated exposure to social stressors that are not harmful, to induce anhedonic-like behaviors [63, 65, 78, 276]	PFC-amygdala circuit [64], Na+/K+ ATPase activity in amygdala ↓ [278], NAc-VTA circuit activity ↑ [279]	Face: moderate Construct: moderate Predictive: moderate	Lead optimization Preclinical candidate selection	*SSRI:* citalopram, fluoxetine, paroxetine *TCA:* imipramine, desipramine *NE/DA reuptake inhibitor:* reboxetine
Social isolation	Isolation as a stressor to induce anxiety or depression-like behaviors [65, 86, 280, 281]	NAc-VTA circuit activity ↑ [279], synaptic spine density in PFC ↓ [76], ACTH ↓ [74], Hippocampal BDNF and 5-HT metabolism ↓ [282]	Face: low Construct: moderate Predictive: low	Disease etiology Lead optimization Preclinical candidate selection	*SSRI:* citalopram, fluoxetine, paroxetine *TCA:* imipramine desipramine, *NE/DA reuptake inhibitor:* reboxetine
Learned helplessness	Inescapable stressor such as foot shock to induce behavioral despair and helplessness [65, 87, 276]	Sensitized hippocampal norepinephrine release [283], BDNF and HDAC-5 expression in hippocampus ↓ [284]	Face: low Construct: moderate Predictive: high	Disease etiology Lead optimization	*TCA:* desipramine *SSRI:* fluoxetine, sertraline

(continued)

Table 3.1 (continued)

Pharmacodynamic models to assess antidepressant-like activity of test compounds					
Preclinical rodent model	**Biological readout**	**Antidepressant-induced neural circuitry changes**	**Model validity**	**Utilization in drug discovery**	**FDA-approved drugs with demonstrated preclinical efficacy**
Forced swim test	Latency to and/or total immobilization time as a measure of learned helplessness or behavioral despair [56, 65, 78, 88–90]	c-Fos regulation in the amygdala [91], Acute ↑ activity in DA SNc, chronic ↑ activity in DA VTA [285] Spine density in PFC ↓, GluR1 in PFC ↓ of male rats, synapsin-1 in PFC ↓ [76]	Face: low Construct: low Predictive: high	Early hit-to-lead Lead optimization	*SSRI:* citalopram, fluoxetine, and paroxetine *SNRI:* venlafaxine and duloxetine *TCA:* imipramine and desipramine *NE/DA reuptake inhibitor:* reboxetine *NMDA receptor antagonist:* esketamine *$GABA_A$ PAM:* allopregnanolone[a]
Tail suspension	Latency to and/or total immobilization time as a measure of learned helplessness or behavioral despair [57, 65, 78, 276]	Amygdala, hypothalamus, and brainstem c-Fos changes [58]	Face: low Construct: low Predictive: high	Early hit-to-lead Lead optimization	*SSRI:* citalopram, fluoxetine, paroxetine *TCA:* imipramine, desipramine, *NE/DA reuptake inhibitor:* reboxetine
Sucrose preference	Consumption of sucrose-water versus plain water as a measure of pleasure seeking behaviors [61, 65, 86, 280]	↑ PFC and hippocampus mTOR, ERK and p70S6K [61] HDAC-5 in PFC ↓, synaptic changes in lateral amygdala [85]	Face: low Construct: low Predictive: low	Lead optimization	*SSRI:* citalopram, fluoxetine, and paroxetine *TCA:* imipramine and desipramine *NE/DA reuptake inhibitor:* reboxetine *NMDA receptor antagonist:* esketamine

(continued)

Table 3.1 (continued)

Intracranial self-simulation	Lever pressing to stimulate VTA dopamine neurons, which is rewarding [77] decreased ICSS is anhedonia-like [60, 69, 96, 286, 287]	NE in BNST ↑ [72, 288]	Face: moderate Construct: moderate Predictive: high	Lead optimization Preclinical candidate selection	*SSRI:* escitalopram (acute only), fluoxetine *NE/DA reuptake inhibitor:* bupropion *BZDs:* diazepam

Abbreviations: *ACTH* adrenocorticotropic hormone, *ATPase* adenosine triphosphatase, *BDNF* brain-derived neurotrophic factor, *BDZ* benzodiazepine, *BNST* bed nucleus of the stria terminalis, *DA* dopamine, *ERK* extracellular signal-regulated kinase, *HDAC-5* histone deacetylase 5, *mTOR* mammalian target of rapamycin, *NAc* nucleus accumbens, *NE* norepinephrine, *NE/DA* norepinephrine/dopamine, *NMDA* N-methyl-D-aspartate, *PFC* prefrontal cortex, *SNc* substantia nigra pars compacta, *SNRI* serotonin/norepinephrine inhibitor, *SSRI* selective serotonin reuptake inhibitor, *TCA* tricyclic antidepressant, *VTA* ventral tegmental area, *5-HT* serotonin

[a]Allopregnanolone, a steroid hormone, was examined in preclinical assays. The FDA approved drug is brenaloxone, an IV formulation of allopregnanolone

anxiogenic-like behavior in rodents [98]. Behavioral tests used to assess anxiogenic-like behaviors acutely or after exposure to stressors, include, but are not limited to, elevated plus maze, light/dark box test, locomotor and exploratory activity, marble burying, and punished responding.

The elevated plus maze (EPM) consists of two dark (enclosed) arms and two light (open) arms arranged as a plus-shape that is raised above the floor of the testing room [99]. Time spent in the open domains is interpreted as increased anxiolytic-like activity because these areas are more exposed to natural predation and heights. The light/dark box test uses an arena with a dark chamber and a light chamber. Longer times spent in the lighter chamber indicates increased anxiolytic-like activity, while more time spent in the darker chamber indicates increased anxiety-like behavior, similar to the EPM [100]. Marble burying in rodent models has been established as an anxiogenic-like behavior [101]. Burying is a natural response by rodents when experiencing stress that is also observed when presented with a marble in their bedding [102]. Punished responding, also known as the Geller-Seifter operant test for detection of anxiolytic-like drug effects, shows particular efficacy with benzodiazepines such as diazepam [103]. Rodents are initially conditioned to press a lever for either a food pellet or liquid reward. During the second phase of the testing, a discrete foot shock is delivered after a fixed number of lever presses for rewards which suppresses further lever pressing behavior to avoid additional shocks. However, lever pressing for reward can be restored with benzodiazepine treatments [104].

The neurobiological changes in GAD preclinical models largely involve changes in plasma levels of corticosterone and glucocorticoid receptor expression in the PFC. There are also reports of sex differences in corticosterone levels and how these differences lead to differences in anxiety-like behavior [105, 106]. Additional changes in c-Fos expression in the amygdala and mPFC neurocircuitry activity are associated with inducing anxiety. Pharmacodynamic rodent models to assess the anxiety-like behaviors have been used to demonstrate the efficacy of benzodiazepines, SSRIs and SNRIs in reducing anxiety-like behaviors. Table 3.2 describes preclinical rodent models used to model and test pharmacological drugs to treat generalized anxiety disorder.

FDA-approved treatments for GAD comprise SSRIs, SNRIs, and benzodiazepines. A study evaluating a panel of SSRIs in FST, TST, light/dark box, foot shock, and social isolation demonstrated the ability of this class of drugs to reduce anxiety-related behaviors [107]. First-line treatments for GAD are SSRIs, the side effects of which are listed in section "Major Depressive Disorder". SSRIs have response rates in treating GAD of around 50–70%, which is compared to placebo response rates around 40–50% [108]. Benzodiazepines, which can be used for acute anxiety treatment, can come with physical and psychological dependence, as well as sedative effects. Drug development for anxiety is focused on treatments for non- or partial-responders as well as better overall tolerability and side-effect profiles [109].

Post-Traumatic Stress Disorder

Animal models for post-traumatic stress disorder (PTSD) are largely based on exposure to various types of stressors, single or in combination, to elicit more complex neurochemical and behaviroal changes associated with trauma. For example, in PTSD models, physical stressors such as restraint, electric shock, and single prolonged stress are used to examine the response to stress-related cues and re-exposures to stress. Single prolonged stress exposes the animal to a sequence of stressors, sometimes followed 7–14 days later by re-exposure to a single stressor [110]. The stressors in sequential order are restraint, forced swim, and diethyl ether anesthesia until the animal loses consciousness [111]. The effects of single prolonged stress can be examined using conditioned fear and learning retention paradigms. Social defeat and predator stress are also models used for PTSD. Resident intruder chronic social stress models demonstrate the social avoidance paradigm of PTSD [112]. Predator stress exposes the animals to their species-specific predators, referred to as predator-based psychosocial stress (PPS). PPS studies expose the rodents to a cat or ferret in an unescapable environment. Seven days after exposure, animals are tested in elevated plus maze or light/dark box tests, startle response test, radial arm maze and novel object recognition to assess behavioral avoidance and anxiety-like behaviors in animals [113, 114]. The single-exposure model applies to individuals that acquire PTSD after a single traumatic event. PPS models that translate to repeated exposures use chronic exposure to the predator [113]. Predator scent stress models expose the rodents to inescapable exposure of predator urine, collar, soiled cat litter, and other objects with pheromone-based signatures of the predator.

The exposure to inescapable trauma induces many changes in the PFC and hippocampus. There are also increases in activity observed upon SPS exposure in the amygdala, complemented with increases in GABA and glutamate signaling in the striatum. Stress and shock are used to induce PTSD-like symptoms in rodents, and

Table 3.2 Animal models for generalized anxiety disorder

Models to induce anxiety-like symptoms					
Preclinical rodent model	**Biological rationale/ disorder etiology**	**Chronic neurobiological changes**	**Model validity**	**Utilization in drug discovery**	**FDA-approved drugs with demonstrated preclinical efficacy**
Unpredictable chronic mild stress	Repeated unpredictable exposure to mild stressors that are not harmful, to induce anxiety-like behaviors [67, 289–291]	Corticosterone ↑ [62], glucocorticoid receptor in mPFC ↓ [292, 293], induces corticosterone ↑ and anxiety-like behaviors [62]	Face: moderate Construct: moderate Predictive: moderate	Lead optimization Preclinical candidate selection	*SSRI:* sertraline, escitalopram *SNRI:* venlafaxine *BZDs:* diazepam
Social isolation	Isolation as a stressor to induce anxiety or depression-like behaviors [291, 294–296]	DAT activity, DA release, and D_2R activity↑ [297]	Face: low Construct: moderate Predictive: low	Disorder etiology Lead optimization Preclinical candidate selection	*SSRI:* sertraline *SNRI:* duloxetine
Operant conditioning: punished responding	Contextual conditioning resulting in decreased latency to lever press or lick after a shock is delivered as a measure of anxiety-like behavior [104, 107]	Amygdala and mPFC activity↑ [62], and amygdala and mPFC expression in c-fos expression↑ [62]	Face: low Construct: moderate Predictive: high	Disorder etiology Lead optimization Preclinical candidate selection	*SSRI:* paroxetine
Chronic restraint/ immobilization	Repeated exposure to conditions of restraint or immobilization to induce anxiety-like behavior [292]	Corticosterone in PFC ↑ [62], glucocorticoid receptor expression in PFC ↓ [292], Female↑ plasma cortisone and Male ↑ pyramidal neuron activation [62, 298]	Face: moderate Construct: moderate Predictive: moderate	Lead optimization Preclinical candidate selection	*SSRI:* fluoxetine

(continued)

Table 3.2 (continued)

Pharmacodynamic models to assess anxiolytic-like actions of test compounds					
Preclinical rodent model	**Biological readout**	**Anxiolytic-induced neural circuitry changes**	**Model validity**	**Utilization in drug discovery**	**FDA-approved drugs with demonstrated preclinical efficacy**
Elevated plus maze	Total time spent in opened arms on an elevated surface as a measure of anxiety-like behavior [62, 107, 299]	Iba-1 pPI3K/PI3K and pAkt/Akt expression, IL-6, IL-1β, NF-κB and TNF-α in PFC and in HIP ↓ [289]	Face: moderate Construct: low Predictive: moderate	Early hit-to-lead Lead optimization	*SSRI:* Escitalopram *SNRI:* Venlafaxine
Open field test	Total time spent in open field as a measure of anxiety-like behavior [62, 105, 300, 301]	Changes in 5-HT and 5-HIAA expression female in PFC and hippocampus ↑, DA an 5-HT increase in HPC ↑ [302]	Face: moderate Construct: low Predictive: low	Early hit-to-lead Lead optimization	*SNRI:* Venlafaxine
Light/dark box	Total time spent in dark or light arena as a measure of anxiety-like behavior [62, 100, 107, 301]	↓ pERK1/2 and pCREB in hippocampus and PFC [303]	Face: low Construct: low Predictive: moderate	Early hit-to-lead Lead optimization	*SNRI:* Venlafaxine
Marble burying	Placement of a marble to observe burying as a measure of anxiety [101, 289, 290]	Iba-1 pPI3K/PI3K and pAkt/Akt expression, IL-6, IL-1β, NF-κB and TNF-α in PFC and in HIP ↓ [289]	Face: low Construct: low Predictive: moderate	Early hit-to-lead Lead optimization	*SSRI:* Sertraline, Escitalopram

Abbreviations: *DAT* dopamine transporter, *DA* dopamine, *D_2R* D_2 dopamine receptor, *HIP* hippocampus, *Iba-1* ionized calcium binding adaptor molecule 1, *IL* interleukin, *mPFC* medial prefrontal cortex, *NF-κB* nuclear factor kappa B, *pCREB* phosphorylated cAMP response element-binding protein, *PFC* prefrontal cortex, *PI3K* phosphoinositide 3-kinase, *SNRI* serotonin/norepinephrine inhibitor, *SSRI* selective serotonin reuptake inhibitor, *TNF-α* tumor necrosis factor alpha

the functional readouts include hiding in assays such as elevated plus maze, open field, and light dark box, as well as EEG and novel object recognition to evaluate PTSD-induced disturbances on sleep and cognition, respectively. Table 3.3 describes preclinical rodent models used to model and test pharmacological drugs to treat post-traumatic stress disorder.

Table 3.3 Animal models for post-traumatic stress disorders (PTSD)

Models to induce PTSD-like symptoms					
Preclinical rodent model	**Biological rationale/ disorder etiology**	**Acute and/or chronic neurobiological changes**	**Model validity**	**Utilization in drug discovery**	**FDA-approved drugs with demonstrated preclinical efficacy**
Single prolonged stress	Exposure to a sequence of stressors to measure conditioned fear and learning retention [115, 304–306]	*Acute*: PFC activity↓, amygdala activity↑, and striatum activity↑ [307, 308] EEG low gamma and delta power↑, and high gamma power ↓ during light phase [308]	Face: moderate Construct: moderate Predictive: low	Early hit-to-lead	*SSRI:* paroxetine, seratraline
Chronic social stress	Repeated exposure to social defeats, to induce anxiety and social avoidance behaviors [309–311]	*Chronic:* Mesolimbic dopamine circuit expression of BDNF↑ [312]	Face: moderate Construct: moderate Predictive: low	Early hit-to-lead	*SSRI:* paroxetine
Electric shock	Tests freezing behavior as a measure of cue related response [310, 311, 313, 314]	*Acute*: c-fos in PFC↑ [115], GSK3 B/AKT in hippocampus and Amygdala↑ [315], CA1 & CA3 volume ↓ in hippocampus & PFC, and GABA↑ and glutamate↑ in hippocampus [316]	Face: moderate Construct: low Predictive: low	Early hit-to-lead	*SSRI:* paroxetine
Predator-based psychosocial stress	Safe, inescapable exposure to species specific predator to induce anxiety-like and PTSD-like symptoms [304, 305, 317, 318]	*Chronic:* DA and 5-HT concentration in hippocampus & PFC↓ [319]	Face: moderate Construct: moderate Predictive: low	Early hit-to-lead Lead optimization	*SSRI:* paroxetine, sertraline

(continued)

Table 3.3 (continued)

Predator scent stress	Safe, inescapable exposure to scent of species-specific predator to induce anxiety-like and PTSD-like symptoms [115, 311, 318, 320]	*Chronic:* DA and 5-HT signaling in, hippocampus & PFC dendrite length [319]	Face: moderate Construct: moderate Predictive: low	Early hit-to-leadLead optimization	*SSRI:* paroxetine, sertraline
Pharmacodynamic models to assess anti-PTSD-like actions of test compounds					
Preclinical rodent model	**Biological readout**	**Anxiolytic-induced neural circuitry changes**	**Model validity**	**Utilization in drug discovery**	**FDA-approved drugs with demonstrated preclinical efficacy**
Light/dark box	Total time spent in dark or light arena as a measure of anxiety-like behavior [314, 321]	↓ pERK1/2 and pCREB in hippocampus and PFC [303]	Face: low Construct: low Predictive: moderate	Early hit-to-lead	*SSRI:* paroxetine
Novel object recognition	Time exploring a novel object as a measure of cognition and memory [113, 305, 322]	Perirhinal cortex, parahippocampal cortex, and entorhinal cortex activation mediates the task [323]	Face: moderate Construct: moderate Predictive: low	Early hit-to-lead Lead optimization	*SSRI:* paroxetine
EEG	Measures the electrical activity of the brain [305, 324]	Right frontal theta rhythm [325], local field potentials ↓ [326]	Face: moderate Construct: moderate Predictive: low	Early hit-to-lead Lead optimization	*SSRI:* paroxetine

(continued)

Table 3.3 (continued)

Elevated plus maze	Total time spent in opened arms on an elevated surface as a measure of anxiety-like behavior [107, 299, 306, 318]	c-fos in cingulate cortex, medial amygdala, hippocampus ↑ [327]	Face: moderate Construct: low Predictive: moderate	Early hit-to-lead	*SSRI:* paroxetine, sertraline
Open field test	Total time spent in open field as a measure of anxiety-like behavior [300, 301, 306, 320]	c-fos in PFC NAc and striatum ↑ [328]	Face: moderate Construct: low Predictive: low	Early hit-to-lead	*SSRI:* paroxetine, sertraline

Abbreviations: *5-HT* serotonin, *BDNF* brain-derived neurotrophic factor, *DA* dopamine, *GABA* gamma aminobutyric acid, *GSK3B* glycogen synthase kinase 3B, *HDAC-5* histone deacetylase 5, *NAc* nucleus accumbens, *NE/DA* norepinephrine/dopamine, *NMDA* N-methyl-D-aspartate, *PFC* prefrontal cortex, *SNRI* serotonin/norepinephrine inhibitor, *SSRI* selective serotonin reuptake inhibitor, *TCA* tricyclic antidepressant, *VTA* ventral tegmental area

FDA-approved therapies for PTSD include SSRIs (sertraline and paroxetine). In mice exposed to foot shock, paroxetine was found to prevent symptom reactivation after extinction training. This was measured in a cue-dependent conditioned freezing assay and by behavioral avoidance tests such as elevated plus maze [115]. The side effects of SSRIs are listed in section "Major Depressive Disorder". Access to treatment, particularly in veterans who comprise a major portion of PTSD patients, is a major barrier, as less than one-third of veterans needing mental health services are receiving evidence-based care [116]. The response rate to pharmacological treatment in PTSD is around 60%, with only 20–30% achieving full remission [116]. In addition, SSRIs treat the negative emotional states in PTSD but do not treat the hyperarousal. Current drug discovery is focused on drugs with novel mechanisms which show greater efficacy across PTSD symptoms [117].

Schizophrenia Spectrum and Other Psychotic Disorders

Animal models for schizophrenia include drug-induced, genetic, and developmental models. Developmental models comprise environmental causes and drug exposure during pregnancy or neonatal stages. Neonatal lesions, specifically, induce abnormalities that present with schizophrenia—in particular, social avoidance, and elevated aggression [118]. A prenatal developmental model exposes pregnant rats to methylazoxymethanol (MAM) between gestational days 15–17 to induce psychotomimetic symptoms in offspring [119, 120]. The model was confirmed by behavioral models such as social avoidance, amphetamine-induced hyperlocomotion, and a decrease in PPI [121]. Drug-induced models for schizophrenia commonly use MK-801, an NMDA receptor antagonist, and amphetamine, a dopamine releaser. MK-801, ketamine, phencyclidine, and other NMDA receptor antagonists induce psychotomimetic activity in rodents [122, 123]. Amphetamine increases dopamine activity and therefore induces psychotomimetic-like behaviors as well [124]. Genetic models of schizophrenia consist of numerous gene candidates that contribute to dopamine signaling, neuronal and synaptic plasticity, synaptogenesis, and glutamatergic function. Dysfunction of the gene disrupted-in-schizophrenia 1 (DISC1) and a deletion in 22q11.2 chromosome position are a few genetic models used to study schizophrenia. DISC1 mutations in male mouse models resulted in hyperactivity in open field tests and increased social avoidance while female mice exhibited decreased spatial memory and increased aggression [125]. The 22q11.2 chromosome deletion in mice is designed to translate to the naturally occurring human 22q11.2 chromosome deletion in the clinic [126]. This animal model recapitulates PPI deficits, spatial memory deficits, and fear conditioning modifications observed in individuals with 22q11.2 chromosome deletions [127]. Induced pluripotent stem cells (iPSCs) isolated from humans and differentiated into neurons have allowed researchers to identify relevant genetic mutations and determine differences in neurobiology contributing to the schizophrenia of individual patients [128]. Pharmacological studies using iPSCs have also been useful in understanding the mechanism of drugs such as loxapine and correlating specific drug treatments with gene modifications [129]. IPSC studies are directly translational as the same gene modifications can be mimicked in rodent animal models and studied to test behavioral, neurobiological, and physiological effects. Other behavioral models for schizophrenia are social interaction and working memory [130]. Social interaction in schizophrenia is tested in an open arena containing another unfamiliar animal. Aggression can be measured using social stress paradigms. Working memory is tested using a Y maze. More details on behavioral tests used in schizophrenia can be found in the following review [130]. A detailed overview of these animal models of schizophrenia is presented in Table 3.4.

FDA-approved antipsychotics such as clozapine, haloperidol, olanzapine, risperidone, amisulpride, and aripiprazole are efficacious in reversing ketamine-induced deficits in PPI [131]. Clozapine also restored ketamine-induced deficits in working memory and attenuated ketamine-induced hyperlocomotion [132]. First generation antipsychotics cause extrapyramidal side effects including rigidity, tremors and uncontrolled muscle movements, while second-generation antipsychotics have decreased extrapyramidal side effects but still have sedating and endocrine side effects in 86% of patients [133]. Treatment resistance to at least two medications occurs in 34% of patients [134]. In addition, treatments for schizophrenia treat only the positive symptoms, and have little efficacy on negative or cognitive symptoms, leading to recent drug discovery efforts in these domains.

Table 3.4 Animal models for schizophrenia

Models to induce schizophrenia-like symptoms					
Preclinical rodent model	**Biological rationale/disorder etiology**	**Chronic neurobiological changes**	**Model validity**	**Utilization in drug discovery**	**FDA-approved drugs with demonstrated preclinical efficacy**
Chronic pharmacologic models					
Chronic phencyclidine	NMDA receptor hypofunction	Hypofrontality, parvalbumin interneurons ↓ [329]	Face: low Construct: high Predictive: high	Lead optimization	*Atypical APDs:* clozapine, risperidone
Neurodevelopmental models					
Neonatal ventral hippocampal lesions	Hippocampal pathology in schizophrenia with adolescent onset of symptoms; hippocampal disconnection model, cause locomotion ↑ [330, 331]	D_2 dopamine receptor ↓ [332, 333], cortical disinhibition [334]	Face: low Construct: moderate Predictive: moderate	Lead optimization	*Typical APDs:* haloperidol, *Atypical APDs:* clozapine, olanzapine, and risperidone
MAM during fetal development	Neurodevelopmental hypothesis of schizophrenia. Alteration of parvalbumin neurons in schizophrenia patients [335], causes spontaneous locomotor activity ↑ [121], PPI ↓ [121]	PFC neuron excitability ↑ [336], cortical thickness ↓ [121], ventricular size ↑ [121], loss of parvalbumin interneurons in PFC and ventral subiculum [337]	Face: moderate Construct: moderate Predictive: moderate	Lead optimization	*Typical APDs*: haloperidol *Atypical APDs*: clozapine

Environmental models					
Isolation rearing	Early life environment impact on development of schizophrenia, causes PPI ↓, hyperlocomotion, auditory gating ↓ [338]	NAc dopamine ↑, PFC dopamine ↓ [339–341]	Face: moderate Construct: moderate Predictive: moderate	Lead optimization	*Atypical APDs:* clozapine, ziprasidone
Genetic models					
22q11.2 deletion	Association between 22q11.2 microdeletion and schizophrenia [342, 343], causes PPI ↓	Hippocampal volume ↓ [344, 345], COMT ↓ [346], DOPAC in PFC and striatum ↑ [347, 348]	Face: moderate Construct: high Predictive: moderate	Lead optimization	*Typical APDs:* haloperidol, *Atypical APDs:* clozapine
DISC1 mutation	Association with schizophrenia populations [349], impaired cortical development [350], causes PPI ↓	PFC dopamine innervation ↓ [351]	Face: moderate Construct: high Predictive: moderate	Lead optimization	*Atypical APDs:* clozapine
Pharmacological models					
Amphetamine-induced changes	Amphetamine hyperlocomotion and PPI deficits are psychotomimetic [352, 353]	NAc dopamine ↑↑	Face: moderate Construct: moderate Predictive: moderate	Early hit-to-lead Lead optimization	*Typical APDs*: chlorpromazine, haloperidol, perphenazine *Atypical APDs*: aripiprazole, sertindol, olanzapine, clozapine, remoxipride, risperidone, ziprasidone

(continued)

Table 3.4 (continued)

NMDA receptor antagonist-induced changes	NMDA receptor antagonists cause hyperlocomotion and PPI deficits which are psychotomimetic [354, 355]	NAc dopamine ↑	Face: moderate Construct: moderate Predictive: moderate	Early hit-to-lead Lead optimization	*Typical APDs*: eticlopride, chlorpromazine haloperidol *Atypical APDs*: clozapine, olanzapine, remoxipride, risperidone, sertindol, ziprasidone
Acute pharmacodynamic models to assess antipsychotic-like actions of test compounds					
Preclinical rodent model	**Biological readout**	**APD-induced acute neurobiological changes**	**Model validity**	**Utilization in drug discovery**	**FDA-approved drugs with demonstrated preclinical efficacy**
Hyperlocomotion	Locomotor activity [352–355]	NAc dopamine ↓	Face: low Construct: moderate Predictive: moderate	Early hit-to-lead Lead optimization	*Typical APDs*: haloperidol, eticlopride *Atypical APDs*: aripiprazole, clozapine, sertindol, olanzapine, risperidone
Pre-pulse inhibition	Inhibition of startle response [352]	NAc dopamine ↓	Face: high Construct: moderate Predictive: moderate	Early hit-to-lead Lead optimization	*Typical APDs*: chlorpromazine, haloperidol, perphenazine *Atypical APDs*: clozapine, olanzapine, remoxipride, risperidone, sertindol, ziprasidone

Abbreviations: *APD* antipsychotic drug, *DISC1* disrupted in schizophrenia 1, *MAM* methylazoxymethanol acetate, *NAc* nucleus accumbens, *NMDA* N-methyl-D-aspartate, *PFC* prefrontal cortex, *PPI* pre-pulse inhibition

Substance-Related and Addictive Disorders

There are many types of addictive disorders. Substance use disorders (SUDs) comprise, among others, alcohol, psychostimulants, opioids, benzodiazepines, and nicotine. Each area employs specific animal models to test the different stages and characteristics of different addictions, though some models are shared across types of addictive disorders. Addiction animal models can be conducted using noncontingent (involuntarily administered) treatment with substances of abuse or contingent (self-administered) drug administration, with each method examining different aspects and dynamics of addiction.

Behavioral sensitization is a noncontingent model based on the theory that repeated noncontingent administration of a drug induces drug cravings [135]. Behavioral sensitization occurs in two phases. The induction phase changes neurobiological mechanisms, while the expression phase displays the behaviors involved as a result of the neurobiological changes [135–137]. After a period of extinction of administration of an opioid, illicit drug, nicotine, or ethanol, a drug challenge test is facilitated to test the expression of behavioral sensitization. Most addiction studies apply the behavioral sensitization theory to their approach to understanding the neurobiology of the illness as well as the behavioral manifestations of the changes in the neurobiology during the induction phase.

Conditioned place preference (CPP) is a popular model to study reward processes. Noncontingent administration studies using CPP pair drug administration with a distinct location where this reward occurs, such as dosing a mouse with opioid and isolating them in a particular compartment of a box. Repeated administration of a drug of abuse induces a preference for the drug-associated compartment over the non-reward associated compartment [138]. The time spent in each compartment is compared to determine preference [138]. Extinction following CPP is used to test relapse and reinstatement in drug studies.

Contingent dosing can be assessed using self-administration (SA) to assess drug-seeking, addiction potential, reinstatement, and expression. SA is an operant response task that requires an action from the animal to receive drug or alcohol treatment. Animals are trained to self-administer the reinforcing drug to analyze the various stages of addiction. SA models are further categorized as short-access and long-access models which differentially study drug-use behaviors associated with recreational drug use versus long-term effects, respectively [139, 140]. Operant schedules can be applied to SA models using fixed ratio (FR) and progressive ratio (PR) schedules. FR schedules deliver the drug, alcohol, or sucrose after a fixed number of responses have been completed. The PR schedule increases the required number of interactions with each reinforcement earned in order to examine motivational aspects of self-administration and reward strength of a reinforcing drug. These schedules can be paired with cues and periods of extinction to study reinforcement and relapse in each schedule.

Preclinical animal models and pharmacodynamics models for SUDs comprise models to observe reward and motivation. Changes in the mesolimbic circuitry are established as the underlying neurobiology that contributes to drug-seeking behavior. Table 3.5 describes the models and relevant FDA-approved drugs used to treat substance use disorders.

SA models and CPP models are used to study the effects of drug treatments on expression, reinstatement, and relapse of addiction. Currently approved treatments for SUDs are limited and include methadone, buprenorphine, and naltrexone for opioid use disorder. Naltrexone in mouse models inhibits methamphetamine-induced hyperlocomotion, expression of CPP, and methamphetamine-cued reinstatement in CCP [141]. These treatments are approved for acute overdose and maintenance of opioid use disorder, including withdrawal symptoms and cravings. However, these treatments face issues with controlled dispensing, addiction potential, high relapse rates, and tolerance. Outside of opioid use disorder, other substances don't have many options for pharmacological treatment. In addition, access to treatment remains a major barrier, with a particularly large unmet need in homeless populations [142]. Drug discovery for treatments for SUD is focused on decreasing relapse rates, lowering addiction potential, and improving efficacy in other substance use disorders besides opioids.

Table 3.5 Animal models for substance use disorders

Preclinical rodent model	Biological readout	SUD treatment-induced acute neurobiological changes	Model validity	Utilization in drug discovery	FDA-approved drugs with demonstrated preclinical efficacy
Intravenous self-administration Fixed ratio Progressive ratio	Operant response task that requires an action to receive drug or alcohol treatment to assess addiction [356–359]	Agonists stimulate increased dopamine release to compensate for hypofunction, antagonists decrease dopamine release by drugs of abuse [360, 361]	Face: high Construct: moderate Predictive: high	Disorder etiology Early hit-to-lead Lead optimization Abuse potential	*Mu-opioid receptor agonists:* buprenorphine and methadone *Mu-opioid receptor antagonists:* naltrexone
Conditioned place preference	The time spent in the reward-associated compartment used to assess acquisition, relapse, and reinstatement [362, 363]	Agonists cause CPP because they increase dopamine, agonists cause CPA because they decrease dopamine [364, 365]	Face: moderate Construct: moderate Predictive: moderate	Disorder etiology Abuse potential	*Mu-opioid receptor agonists:* buprenorphine and methadone *Mu-opioid receptor antagonists:* naltrexone

Abbreviations: *5-HT* serotonin, *CPA* conditioned place aversion, *CPP* conditioned place preference, *VTA* ventral tegmental area

Attention Deficit Hyperactivity Disorder

Examples of animal models for ADHD are genetic models, the spontaneously hypertensive rat (SHR) model, and prenatal ethanol exposure models. Impulsivity, inattentiveness, and hyperactivity are the three hallmark behaviors of ADHD. The dopamine transporter knockout (DAT-KO) mouse model is based off the various DAT mutations that have been observed in humans with ADHD. DAT-KO models exhibit hyperactivity, increased spontaneous dopamine cell firing, and deficits in spatial memory [143–145]. Tachykinin-1 (NK1) KO mice displayed hyperactivity, impulsiveness, and inattentiveness, which can also be induced in wild-type mice using NK1 antagonists [146, 147]. Hyperactivity in NK1 models can be inhibited via psychostimulants [147]. SHR is a strain of inbred Wistar-Kyoto rats that display hyperlocomotion, impulsivity, and other key symptoms of ADHD [148–150]. Delay discounting tests are used to test impulsivity in ADHD rodent models resulting in the choice of the immediate reinforcers regardless of the larger reward obtained by delayed reinforcement [151]. The five-choice serial reaction time (5-CSRT) task is used to test attentiveness; in this assay, animals need to respond to the location of a light cue after variable delay periods to earn a food reward [152].

First-line ADHD treatments include psychostimulants such as methylphenidate and amphetamine, which treat impulsivity, hyperactivity, and inattention. Second-line treatment includes atomoxetine, an SNRI. Atomoxetine and methylphenidate were tested using 5-CSRT in Long-Evans rats and were shown to improve attentiveness and impulsivity [153]. Methylphenidate and l-amphetamine tested in adolescent SHR rats reduced hyperlocomotion and impulsivity using the 5-CSRT test [154, 155]. A third of patients don't respond to psychostimulants and are thus treated primarily with atomoxetine, which has a smaller effect size [156]. Psychostimulants can be addictive and thus face strict controls. Drug discovery for ADHD is focused on improved efficacy and decreased abuse liability.

ADHD preclinical genetic models produce changes in dopamine tone and increases in hyperactivity and impulsivity. DAT mutations, for example, alter DAT expression and activity. Pharmacodynamic models to test the efficacy of drugs to reduce hyperactivity and impulsiveness with high validity are 5-CSRT and open field test. FDA-approved drugs have been validated in these models to demonstrate a decrease in dopamine tone, hyperactivity, and impulsivity. Table 3.6 describes preclinical rodent models used to model and test pharmacological drugs to treat ADHD.

Table 3.6 Animal models for attention deficit hyperactivity disorder (ADHD)

Models to induce ADHD-like symptoms					
Preclinical rodent model	**Biological rationale/ disorder etiology**	**Chronic neurobiological changes**	**Model validity**	**Utilization in drug discovery**	**FDA-approved drugs with demonstrated preclinical efficacy**
SHR	Genetic model to induce ADHD-like behaviors [366, 367]	Hyperactivity and dopamine signaling↑ in striatum and NAc [368]	Face: high Construct: moderate Predictive: low	Disorder etiology Early hit-to-lead Lead optimization Abuse potential	*CNS Stimulant:* amphetamine *Alpha2A-adrenergic receptor agonists:* Guanfacine
DAT mutations	Genetic model to induce ADHD-like behaviors [369]	Dopamine tone, impulsivity and hyperactivity↑ [369]	Face: moderate Construct: moderate Predictive: low	Disorder etiology Early hit-to-lead Lead optimization	*CNS Stimulant:* amphetamine
Pharmacodynamic models to assess anti-ADHD-like actions of test compounds					
Preclinical rodent model	**Biological readout**	**Stimulant-induced acute neurobiological changes**	**Model validity**	**Utilization in drug discovery**	**FDA-approved drugs with demonstrated preclinical efficacy**
Five-choice serial reaction time (5-CSRT)	Nose-poke upon visual stimulus to earn a food reward as a measure to test attentiveness and impulsivity [152, 153]	↑mPFC activation for action control [370]	Face: moderate Construct: moderate Predictive: moderate	Disorder etiology Lead optimization	*CNS Stimulant:* methylphenidate *Norepinephrine Reuptake Inhibitor:* atomoxetine
Open field test	Total time spent in open field as a measure of hyperactivity [366, 371]	↑c-fos in most brain regions, ↑ dopamine in the dorsal and ventral striatum [328]	Face: low Construct: moderate Predictive: moderate	Lead optimization	*CNS Stimulant:* methylphenidate *Norepinephrine Reuptake Inhibitor:* atomoxetine

Abbreviations: *CNS* central nervous system, *NAc* nucleus accumbens

3.2.4 *Other Clinical Considerations to Modeling Psychiatric Disorders in Animals: Sex, Age, Ethnicity*

Sex, age, and ethnicity are examples of factors that impact the validity of an animal model. Sex differences in animal models include differences in metabolism, estrogen-regulated neuronal and physiological processes, symptomatic behaviors, and response to treatment. Estrogen receptor alpha (ERα) and estrogen receptor beta (ERβ) have been shown to have differences in expression depending on the brain region. ERα and ERβ have been found in the prefrontal cortex, temporal cortex, sensory motor areas, hippocampus, and the cortico-striato-thalamo-cortico (CSTC) circuit [157–163].

Age differences have various impacts on the behavior of the animals as the neurocircuitry changes, which also likely induces changes in response to treatments [164, 165]. The neurobiology and circuitry in the aged brain are different from the non-pathological adult brain, suggesting that there would be changes in the pharmacodynamics and efficacy of some psychiatric treatments [164]. Changes in neurocircuitry in the aging brain affect several brain areas and involve the dopaminergic, glutamatergic, and cholinergic transmitter systems [166–168]. Although there are more factors to consider, those mentioned above encompass some key factors to be considered with respect to animal model variability and reproducibility. Confounding results and reproducibility are highly likely in the case of poor modeling. Thus, proper protocols with specific outlines of the factors that are key to each model should be provided. Below are examples of how sex, age, and ethnicity influence symptomatology, treatment response, and pharmacokinetics in human and animal models.

Sex and Age Differences in Symptomatology of Psychiatric Illnesses

Sex- and age-related differences in prevalence and symptoms exist across all psychiatric illnesses. MDD has a higher incidence in women than in men, and mood symptoms and MDD episodes are increased during periods of low estrogen hormone levels [169]. Atypical symptoms are also increased in women with MDD compared to men [170]. Men experience more anhedonic symptoms in MDD, but women experience more anxiety [171, 172], and the presence of other depressive symptoms such as insomnia, appetite, and aggression is affected by estrogen [173]. In females, age-related changes in estrogen levels affect MDD symptoms. For example, MDD episodes and mood symptoms increase during and after menopause in women [169].

Sex-related differences are also seen in animal models of depression, such as the FST. FST in female rodents revealed an increase in immobility duration, while differences in latency to immobility are inconsistent across studies [174, 175]. In a learned helplessness model of escape following inescapable shock, female rats had faster escape times, contrary to what was seen in FST immobility [176, 177]. There is a more robust decrease in sucrose intake in male rats compared to female rats following chronic mild stress exposure [178]. These rodent phenotypes are largely consistent with the human data showing that men experience greater anhedonia and women experience greater anxiety [171, 172].

Sex- and age-related differences are demonstrated in schizophrenia as males have an earlier onset and an overall higher incidence of schizophrenia [179]. In contrast, women have less severe negative symptoms but higher levels of affective symptoms [179].

The major sex-related differences reported in post-traumatic stress disorder (PTSD) are driven by the difference in circulating hormones in men and women. Estrogen cycles in women are directly linked to changes in circulating cortisol, affecting female sensitivity to events that cause PTSD and the development of anxiety as a result of a traumatic event [180, 181]. Women are twice as likely to develop PTSD compared to men [181]. In rodent models of PTSD, female rats recapitulated the overall increased vulnerability to developing the illness. Compared to male mice, females have a decrease in fear conditioning but an increased startle response (assays described in more detail in section "Post-Traumatic Stress Disorder") [181–183].

Sex and Age Differences in Response to Psychiatric Drugs

Sex differences in the efficacy of treatment for MDD patient populations were demonstrated by showing that the selective serotonin reuptake inhibitor (SSRI) sertraline was more efficacious in women than the tricyclic antidepressant (TCA) imipramine, whereas in men, imipramine resulted in greater treatment response [184]. Similarly, female MDD patients experiencing atypical symptoms responded more favorably to SSRIs or monoamine oxidase inhibitors (MAOIs) than TCAs, while the opposite responses were observed in male MDD patients [185].

In female patients with schizophrenia, superior treatment responses to antipsychotics were observed relative to male patients, particularly with clozapine [186]. Interestingly, in the preclinical model of latent inhibition, a model of learned inattention which is impaired in schizophrenic patients, treatment with haloperidol and clozapine restored impaired latent inhibition, especially during proestrus-estrous cycles of female rats.

Sex, Age, and Ethnicity Differences in Pharmacokinetics of Psychiatric Drugs

It is well known that there are marked differences in the way distinct patient populations may respond to the same psychiatric medication. Since a comprehensive coverage of this topic is beyond the scope of this chapter, we will focus here on the way the clinical response to treatment can be affected by sex-, age-, and ethnicity-dependent changes in drug pharmacokinetics. A drug's absorption, distribution, metabolism, and excretion (ADME) properties determine if a drug (1) reaches its site of action, in the case of a psychiatric drug, the brain, (2) the brain levels that a drug reaches, and (3) a drug's duration of action. Since many drugs are metabolized by cytochrome p450 (CYP) metabolic enzymes, differences in the expression of CYP isoforms across patient populations can differentially impact drug efficacy. Examples are CYP3A, the main CYP isoform which is more highly expressed in liver microsomes from females than males [187], and

CYP1A2, an enzyme metabolizing many antipsychotic drugs, both of which have been suggested to have higher expression in females than in males and also to be influenced by age [188, 189]. Finally, CYP2D6 and CYP2B6 are enzymes that metabolize drugs used to treat psychiatric illnesses [190, 191]. Imipramine, a CYP2D6 substrate, has been shown to be metabolized faster in females than in males [191]. Furthermore, CYP2B6 polymorphisms which alter metabolic rate are influenced by sex and ethnicity [192]. Body weight, organ size, and percent body fat influence metabolism, such that patients with lower body weight responded better to SSRIs than heavier patients. Finally, females have a lower expression of the transporter protein p-glycoprotein, reducing the efflux of antidepressants in the gut and increasing the uptake of antidepressants across the blood-brain barrier [193, 194].

Ethnic differences are also seen in the distribution, activity, and expression of metabolic enzymes, such as CYP2D6 enzyme expression [195]. These differences result in differential metabolism of drugs by influencing the half-life, drug-drug interaction, and plasma and drug concentrations, thereby changing the efficacy of these drugs. To model ethnic differences, models of genetic variations in CYP enzymes are needed to assess pharmacokinetic changes to further understand differences in responses to treatment for psychiatric disorders.

Use of Female Animals for Drug Discovery in Psychiatric Disorders

Collectively, it is critical that both male *and* female animals are used not only to model the mechanisms underlying psychiatric illnesses, but the response to and efficacy of psychiatric medications. Sex differences in rodents change over the life span of the animals as estrogen levels change. Changes in the estrous cycle in female rodents, for example, have been shown to modulate behaviors related to the symptomatology of MDD [169, 196, 197], generalized anxiety disorder (GAD) [169, 196, 197], PTSD [169, 196, 197], substance use disorders (SUDs) [197], attention deficit hyperactivity disorder (ADHD) [196], and others [196, 197]. Although estrogen cycles and menopause are not directly translational in all rodent species (e.g., female mice do not undergo menopause, only estrous pause with aging), models using intact and ovariectomized female rats and mice have provided important data on the differences in onset of symptoms, mood fluctuations, and metabolism. Furthermore, the use of female models has recapitulated many important differences in the drug responses for psychiatric illnesses observed between male and female patients [197]. Hence, the use of female rodents in animal models for drug discovery is critical for detecting potential differences in treatment response in female patient populations.

3.3 Utility of Animal Models Throughout the Stages of Modern Drug Discovery Stages

3.3.1 Target Identification and Validation

As discussed in the previous sections, animal models have been and continue to be essential for understanding basic biology and developing effective treatments for psychiatric disorders. The use of animal models particularly in drug discovery requires moderate to high validity models to ensure that the effects can be recapitulated in humans effectively and safely. As shown in Fig. 3.1 and discussed below, animal models of various psychiatric disorders are used throughout the different stages of the drug discovery process. In the earliest stage, validation of a novel target for drug discovery involves establishing a strong link between the target and the disease indication for which a drug is being developed. In animal models used in drug discovery, one of the first steps is establishing gene sequence homology for the target across species of interest. If target homology is low between humans and the preclinical species, preclinical data may not accurately translate into the clinic. Sequence homology is a key factor that determines the selection of a primary preclinical species. Animal models in target validation are critical for understanding the

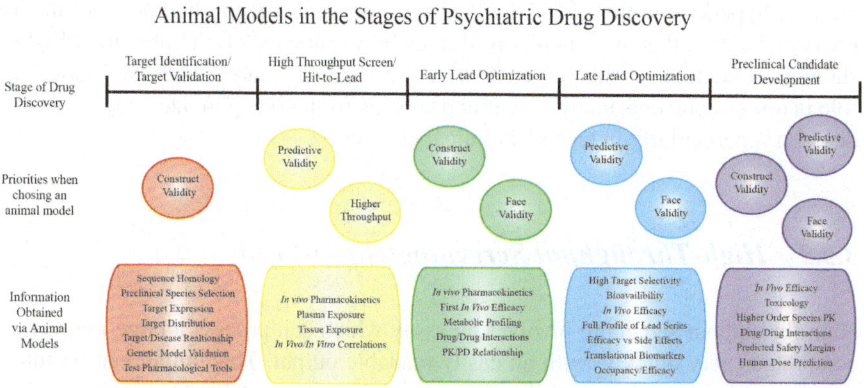

Fig. 3.1 Animal models used in the stages of psychiatric drug discovery. Critical information is gleaned from animal models throughout the different stages of drug discovery, facilitated by the choice of model and its validity. Target identification and validation prioritize models with high construct validity to establish the relationship between the target and the psychiatric disorder. During the high-throughput screen/hit-to-lead stage, higher throughput in vivo models with strong predictive validity for engagement of the target are prioritized, in order to quickly screen large numbers of compounds and establish early PK/PD relationships. Early lead optimization prioritizes construct and face validity as the in vivo efficacy in preclinical models of the psychiatric disorder are established. Late lead optimization prioritizes predictive and face validity as more in vivo efficacy is examined and translatability is the key concern. Selection of preclinical candidate involves a robust use of a variety of animal models with various validities, especially construct validity as the full preclinical package is developed. Abbreviations: PK Pharmacokinetics, PD Pharmacodynamics

target expression and distribution, throughout the CNS and the periphery, on the level of both tissue and cellular and subcellular localization. In psychiatric animal models, the expression of a target within a disease-relevant cell population in the CNS is critical for target identification. In addition, animal models are often used to establish the relationship between the target and the disease state by manipulating the target and examining disease-relevant outputs, using genetic or pharmacological manipulation, if the pharmacological tools exist. Understanding the expression, distribution, and disease-relevant function of a target in animal models before beginning the process of developing a drug can prevent a large waste of resources on an inadequately validated target.

Genetic knockdown or knockout models are often used to validate the relevance of a target to a particular disease state. In the investigation of M_5 muscarinic acetylcholine receptor antagonists for the treatment of substance use disorders (SUD), the first indication that this might be a viable target was its expression in the mesolimbic circuitry. M_5 is the only muscarinic receptor found in the ventral tegmental area (VTA), an area rich in dopamine neurons that play a role in reward processing and are a key substrate for the actions of substances of abuse [198, 199]. Generation of the M_5 knockout mouse [200] allowed for examination of the effects of decreased M_5 function on substance-use relevant behaviors and showed that these mice have decreased responsiveness to morphine-conditioned place preference [201], decreased cocaine self-administration at low doses [202], and reduced morphine-induced hyperlocomotion [203]. Collectively, these data from the knockout model provide evidence that modulation of M_5 may be a viable path for treatment of SUDs, showing validation for this target. A well-validated drug target with an established role in disease prevents many potential problems from arising in later stages of drug discovery, particularly in terms of clinical efficacy.

3.3.2 High-Throughput Screening/Hit-to-Lead

High-throughput screening involves an assay which can process a large number of compounds very quickly, with an easily readable output. This output can be functional, such as a calcium mobilization screen for a G_q-coupled G-protein coupled receptor (GPCR), or binding-based, such as radioligand displacement. Compounds that have a positive response during the high throughput screen are referred to as hits. Hit-to-lead is the stage of drug discovery that involves further analysis of these hits for validation and to assess which of multiple hits are the best prospects for continued development. These compounds with promising properties are referred to as lead compounds or lead chemical series. Most high-throughput screening and hit-to-lead is performed in vitro, but that does not mean that animal models do not have a critical role to play. The major role of animal models in these early drug development stages is in pharmacokinetic modeling. During hit-to-lead, in vivo pharmacokinetic basics are established, such as clearance, half-life, and brain penetrance. Exposure in both the plasma and the target tissue (the brain, for neuropsychiatric disorders) are established. In addition, in vitro-in vivo correlations are examined.

3.3.3 Early-, Mid-, and Late-Lead Optimization

Lead optimization involves taking the lead compounds or series from the hit-to-lead phase of drug discovery and improving the drug-like properties of the molecules. Lead optimization is often separated into early-, mid-, and late- stage lead optimization. Assays used in the early lead optimization stage are selected for high predictive validity related to the target mechanism under investigation wherever possible. For example, in the development of a novel treatment for depression, a compound may first be analyzed in the FST [56], which in addition to high predictive validity for classical antidepressants, also has relatively high throughput and can therefore be used for rapid screening of the acute effects of many compounds. However, it is worth noting that ruling out a drug based on a negative result in an assay with high predictive validity and low face/construct validity is ill-advised when working with a novel mechanism. If an assay, such as forced swim, is validated based on the mechanism of *current treatments*, a drug which may treat the symptoms of depression by a novel mechanism may show a negative result in this assay, but a positive result in other assays. Along with the first efficacy measure in vivo, animal models are also used to further understand the pharmacokinetics of early-stage compounds. This process also involves gaining an early understanding of the pharmacokinetic to pharmacodynamic relationship of key ligands. It must also be established that systemic administration of early compounds results in sufficient target tissue exposure to observe the desired effects.

During the progression from mid- to late-lead optimization, lead compounds undergo further pharmacokinetic profiling and advance into broader efficacy evaluations. By late lead optimization, compounds must have high selectivity for the target and adequate bioavailability in both the selected primary and higher-order animal species for future toxicology studies. Lead compounds are also evaluated in multiple lower-throughput models with higher etiological (construct) and face validity. For example, in the later development of a novel antidepressant drug, such models might include the unpredictable chronic mild stress model and/or chronic social stress models (described in section "Major Depressive Disorder") [204]. The relationship between stress, anhedonia, and depression is clear in humans [205], making the construct validity of this model better than an assay like forced swim. Meanwhile, the face validity in terms of decreased reward-seeking is high and can be assessed via alterations in food intake, sucrose preference, and/or intracranial self-stimulation (described in section "Major Depressive Disorder"). A combination of models with different validities allows for a full profile of the lead series in 2–4 animal models of efficacy for the related psychiatric disorder, including genetic and behavioral models, as well as efficacy versus side-effect profiling to understand the therapeutic index. Translational biomarkers are also used at this stage of drug discovery to understand the dose-related target receptor occupancy and/or changes in measures of functional target engagement, discussed in more detail in Sect. 3.3.5.

3.3.4 Preclinical Candidate Selection

Selection of the preclinical candidate (PCC) involves an extensive use of multiple animal models for the particular psychiatric disorder with various validities, especially construct validity. The effects of chronic dosing alone and in combination with psychiatric medications that potential patient populations may be taking are also evaluated for changes in PK and/or PD effects. Drug-drug interactions are also evaluated in animals to establish the therapeutic index for preclinical candidate molecules. Using higher order species pharmacokinetics, animal-to-human modeling is performed to predict the pharmacokinetic properties and relevant doses for be use for first in human studies. Predictions for margins of safety in dosing are initiated and then completed once the full toxicology package across two species has been finalized during the Investigational New Drug (IND)-enabling studies.

3.3.5 Translational Animal Models for Target Occupancy and Functional Target Engagement for Psychiatric Drug Discovery

Imaging

A critical aspect of translational drug development is an understanding of the relationship between receptor occupancy, using positron emission tomography (PET), and drug efficacy, referred to as the occupancy-to-efficacy relationship. This means understanding what percentage of receptors need to be engaged by the drug of interest in order to produce a desired behavioral effect. Understanding the occupancy-to-efficacy relationship in a preclinical species can provide critical information for the success of a clinical trial, especially when considering which doses to test in humans. If 50% target receptor inhibition is required in order to see antidepressant-like activity in the forced swim test in rats, then doses that achieve 50% target receptor occupancy in humans during clinical trials should be the initial goal. Alternatively, it may be critical to understand the relationship between changes in a measure of functional target engagement, such as functional magnetic resonance imaging (fMRI), and the efficacy observed in several preclinical species to better inform decisions of dose selection and biomarker utilization in future clinical trials with psychiatric patient populations.

fMRI

Magnetic resonance imaging (MRI) is an imaging technique that uses magnetic fields, one static and one gradient, in order to locate nuclei within the brain. Standard MRI is used to produce structural images of the brain, while functional MRI (fMRI) is used to assess brain activity. One of the major measurements of fMRI assesses the flow of oxygen throughout the brain, a signal referred to as blood-oxygen-level dependent (BOLD) contrast. Regions receiving abundant oxygen are interpreted as more active regions of the brain. Because a majority of energy consumption in the brain is due to glutamate recycling, this BOLD signal correlates with glutamate response and thus is an indirect way of examining functional target engagement at glutamate receptors [206, 207].

An example of BOLD fMRI as a measure of functional target engagement in drug discovery is the NMDAR antagonist ketamine-induced changes in BOLD signal (termed pharmacoBOLD) for measuring glutamatergic receptor functional target engagement when developing drugs for schizophrenia and other psychotic disorders. For example, the mGluR2/3 agonist TS-134 when given in humans produced a robust reduction in ketamine-induced increases in BOLD signal, specifically in the dorsal anterior cingulate cortex [208]. Pomaglumetad, another mGluR 2/3 agonist of lower potency, did not decrease ketamine-induced increases in BOLD signal at the doses that were examined in a failed clinical trial, suggesting a potential reason for this clinical failure – insufficient functional target engagement.

Ketamine-induced changes in BOLD signal can be detected in rats, as well as humans [209], providing strong translational value. Another mGluR 2/3 agonist, LY379268, robustly attenuated ketamine-induced increases in BOLD signal in the cingulate cortex [210], very similar to effects observed in humans. This direct translation of technique from preclinical animal models to humans allows for the use of pharmacoBOLD fMRI in facilitating the preclinical to clinical transition in drug discovery. Understanding the level of functional target engagement required to produce results in models of antipsychotic-like activity allows for the prediction of the levels of target engagement required to treat positive symptoms in human patients. For example, if a drug robustly attenuates 25% of pharmacoBOLD signal in rats at the doses that show efficacy in reversing a stimulant-induced locomotion assay that predicts antipsychotic-like activity, but the clinical trial dose shows no change in the pharmacoBOLD signal in humans, this may indicate that a higher dose is required to see efficacy in humans.

PET

Compared to fMRI, positron emission tomography (PET) provides a more direct measurement of receptor occupancy. Instead of using magnetic fields, PET imaging occurs by measuring radiation emitted by radiotracers, or radioactive substances that are injected into the body. PET scans then form an image based on the density of the radioactivity in different regions, which correspond to where the radioactive substance is distributed in the tissue. A wide variety of ligands can be used as PET tracers, given that they have sufficient properties to be injected into a living creature and can be labeled with a radioactive atom. Because of this diversity in PET ligands, if a PET ligand exists for the receptor under investigation, this can be used as a direct measure of binding of a drug to the receptor of interest.

One set of examples of PET and radioligand tools for a receptor of interest is [^{11}C]raclopride and [^{3}H]raclopride [211]. Raclopride is a competitive antagonist selective for D_2 dopamine receptors, and it can be radiolabeled with either tritium or carbon-11. By examining the difference in a PET scan with raclopride alone and raclopride in combination with other drugs which bind to D_2, raclopride displacement can be measured. If a drug of interest, such as a typical antipsychotic like chlorpromazine, binds to D_2 receptors in the same binding site as raclopride, the amount of raclopride binding detected will decrease proportionately to chlorpromazine binding to the D_2 receptors, allowing measurement of the amount of D_2 receptors occupied by chlorpromazine. Similar methods can be used for PET studies conducted in both humans and animal models such as nonhuman primates or rats, allowing for high translational value.

Both [^{11}C]raclopride and [^{3}H]raclopride have been used in humans and rats to examine the levels of D_2 receptor occupancy of raclopride that contribute to its antipsychotic efficacy versus the induction of extrapyramidal side effects (EPS). In rats, suppression of the conditioned avoidance response (CAR) was used to predict antipsychotic-like activity, particularly for activity through D_2 receptors [212]. In a comparison study, efficacy with raclopride in suppression of CAR occurred at a D_2 occupancy level of 70–75%, where EPS were observed at an occupancy of 80%, demonstrating the narrow window of receptor occupancy for therapeutic effects without side effects [211]. This is similar to results seen in human schizophrenia patients with typical antipsychotic drug haloperidol, where psychotic symptom alleviation became likely at a minimum of 65% D_2 occupancy and EPS at 78% receptor occupancy, as measured by [^{11}C]raclopride PET [213]. Interestingly, clozapine, a common atypical antipsychotic, produced a D_2 receptor occupancy of 38–63%, demonstrating why atypical antipsychotics show much lower rates of EPS [214].

Understanding the relationship between clinical efficacy, toxicity, and receptor occupancy using PET in preclinical animal species allows for the design of clinical trials in which it can be verified that the selected dose range is correct. If the dose range is promoting the level of receptor occupancy in a PET study which correlated to efficacy in preclinical species, then efficacy in humans would be expected. This can also aid in the prediction of dosages at which side effects would be expected. Overall, understanding receptor occupancy in preclinical species creates much more confidence in the preclinical to clinical transition.

Quantitative EEG and Event-Related Potential Measurements

Quantitative electroencephalography (qEEG) measures the electrical activity (oscillations) in the brain. Synchronous waves observed in EEG are a summation of alpha, beta, delta, theta, and gamma waves at any given time. Alpha waves (8–13 Hz) are associated with awake relaxed states with low levels of stimulation. Beta waves

(13–30 Hz) are also associated with wake states, but more focused and stimulated wake states [215]. Theta waves (4–7 Hz) are associated with rapid eye movement (REM) sleep states and light sleep. Delta waves (1–3 Hz) are highly associated with deep, restful sleep. Gamma waves (>30 Hz) are highly associated with wake, deep focus, and problem-solving. EEG is a functional means to measure changes in brain activity to assess sleep, diagnose epilepsy and in pharmacological drug studies. The comparisons of changes in resting power bands to activity related power bands is done through modulation of event-related synchronization (ERS) or desynchronization (ERD) [215]. The two types of EEG studies are spontaneous or resting state EEG, which measure the general activity of the brain, and event-related potential (ERP) studies in response to external stimuli [216]. Changes in EEG signature, sleep-wake states, and circadian rhythms are associated with many psychological diseases. Mammalian qEEG studies are highly translational to humans and provide an effective biomarker for identifying changes in sleep-wake states and associated neurocircuits that are involved [217].

Pharmacological EEG studies provide powerful insights into changes in band power frequencies that indicate alterations in arousal, learning and memory, and consciousness. For example, EEG studies in wild-type Wistar rats compared the sedative, zolpidem, to 5-HT antagonists, RO4368554 (RO), and MDL 100907 (MDL). Zolpidem-induced shorter latency to sleep onset compared to RO and MDL. All three compounds increased delta power resulting in increases in NREM. However, MDL effects were not as pronounced as RO and zolpidem. These results suggest that 5-HT antagonists may be implicated for insomnia drug development [218]. Understanding the changes that occur in sleep-wake states in psychological conditions may also allow the future development of treatments that mitigate the progression of the different psychiatric illnesses, including schizophrenia and the behavioral disturbances in Alzheimer's Disease.

3.3.6 Example of *In Vivo* Target Validation of the Selective M_4 PAM Mechanism for the Treatment of Schizophrenia

Muscarinic acetylcholine receptors (mAChRs) are comprised of five subtypes (M_1-M_5) that are differentially distributed throughout the central nervous system [219]. Since the orthosteric acetylcholine binding site is highly conserved across M_1–M_5, it has been difficult to develop subtype-selective pharmacological tools for assessing the biological role of individual mAChR subtypes. The discovery of the M_1/M_4-preferring orthosteric agonist xanomeline and its demonstrated antipsychotic efficacy in a pilot clinical trial with schizophrenia patients were breakthroughs that renewed interest in probing muscarinic mechanisms for the treatment of schizophrenia [220]. However, it was still unclear whether M_1 activation, M_4 activation, or both were required for conferring antipsychotic efficacy. To achieve the necessary mAChR subtype selectivity, compounds that targeted less conserved regions of the

M_1 or M_4 muscarinic receptor, also known as allosteric sites, were identified. In this first example, we will detail the rationale for the drug discovery of M_4 positive allosteric modulators (PAMs), i.e., compounds that potentiate the effects of acetylcholine at M_4, but have no intrinsic activity alone, for the treatment of the positive symptoms of schizophrenia. Then we will discuss the *in vivo* assays used to determine efficacy, target validation, and behavior to circuitry mapping for functional target engagement. In situ hybridization studies demonstrated high levels of M_4 mRNA expression in brain regions involved in the pathophysiology of schizophrenia such as the nucleus accumbens and striatum but also moderate expression in prefrontal cortical, hippocampal, and thalamic areas [219]. M_4 knockout mice exhibit a hyperdopaminergic phenotype with increased basal motor activity and an increased locomotor response to D_1, but not D_2 dopamine receptor agonists. Likewise, M_4 knockout mice displayed an absence of the inhibitory effect of the typical antipsychotic drug haloperidol, a D_2 dopamine receptor antagonist suggesting that potentiation of M_4 may have antipsychotic activity [221].

High-throughput cell-based screening using Ca^{2+}-mobilization and thallium flux assays followed by chemical library expansion of early hits resulted in the discovery of VU0152100, an M_4 PAM with moderate sub-micromolar potency, high selectivity versus M_1, M_2, M_3, and M_5, favorable brain bioavailability and in vivo efficacy in reversing amphetamine-induced hyperlocomotion [222]. Subsequently, we established dose-related pharmacodynamic efficacy of VU0152100 in reversing amphetamine-induced hyperlocomotion (Fig. 3.2) and demonstrated target specificity in this model by the absence of behavioral efficacy in M_4 knockout mice [223]. We then examined the relationship between behavioral efficacy and changes in the dopaminergic circuit activity underlying the behavioral readout. Using *in vivo* microdialysis and simultaneous recording of behavioral activity, we demonstrated that a dose of VU0152100 that reduced amphetamine-induced hyperlocomotion also reduced elevated extracellular dopamine level in the nucleus accumbens. As the final step, we used pharmacologic magnetic resonance imaging (phMRI) as a translational biomarker of target engagement to evaluate the effects of VU0152100 on amphetamine-induced increases in brain activity across several brain regions. This study demonstrated a brain region-dependent reversal of amphetamine-induced increases in the cerebral blood volume in the nucleus accumbens, dorsal striatum, and other forebrain regions. Taken together, these studies demonstrated that M_4 PAMs, as exemplified by VU0152100, warranted further efforts to develop preclinical candidates for subsequent testing of clinical efficacy in schizophrenia patient populations. Ongoing studies at the Warren Center for Neuroscience Drug Discovery in collaboration with Neumora are currently focused on the development of a novel M_4 PAM for the treatment of schizophrenia.

Fig. 3.2 (continued) Collectively, these data are consistent with an antipsychotic-like profile of VU0152100 and provide preclinical validation for M_4 PAMs as a potential mechanism in treating schizophrenia. Abbreviations: AHL amphetamine-induced hyperlocomotion, NA nucleus accumbens, WT wild type, KO knockout, PAM positive allosteric modulator. (This figure is modified from the original publication in Byun et al., Antipsychotic Drug-Like Effects of the Selective M4 Muscarinic Acetylcholine Receptor Positive Allosteric Modulator VU0152100, Neuropsychopharmacology, published 2014 by Springer Nature [223])

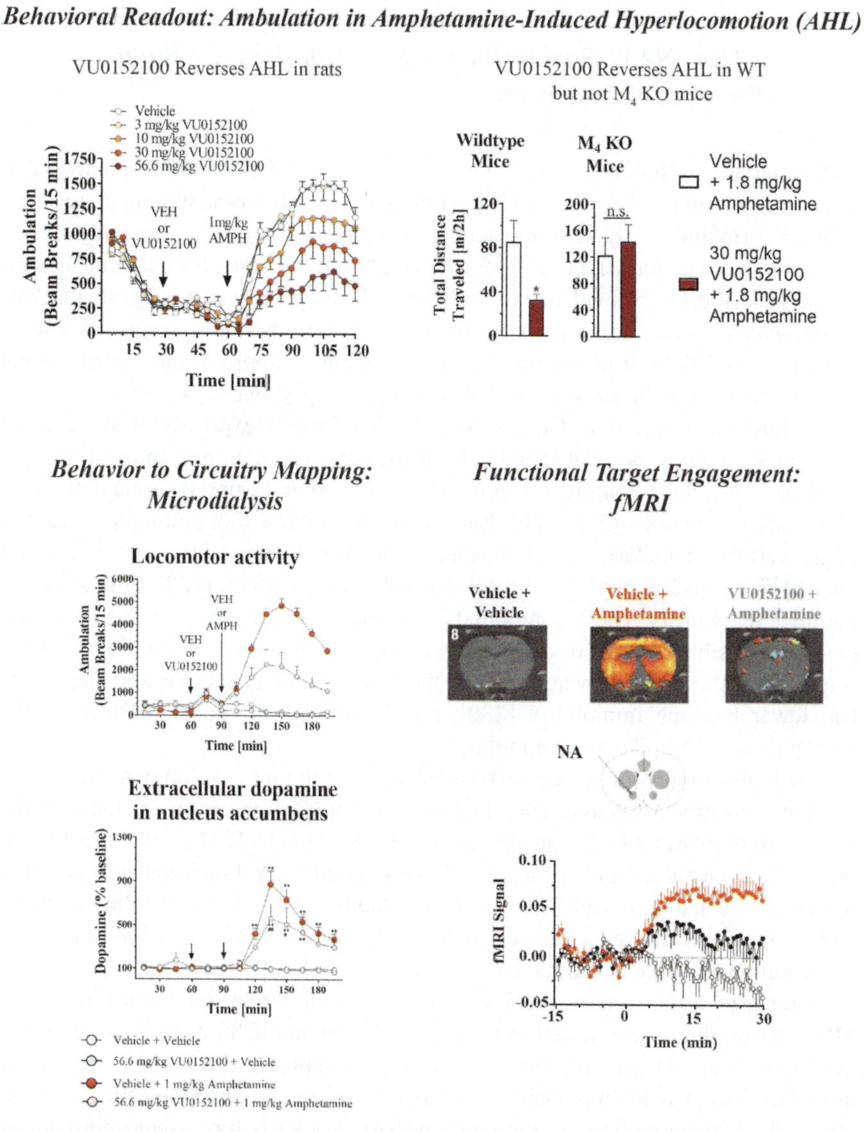

Fig. 3.2 M_4-positive allosteric modulator (PAM) VU0152100 as an example of the use of animal models in different stages of drug discovery for development of a novel antipsychotic drug. VU0152100 reversed amphetamine-induced increases in locomotion (AHL), a preclinical model predictive of antipsychotic-like activity. This reversal of AHL was absent in the M_4 knockout mice, providing validation that M_4 is mediating the observed effects. This behavior can be mapped to circuit level changes in dopamine release, as evaluated by *in vivo* microdialysis. The doses that produced a reversal of AHL also attenuated amphetamine-induced increases in extracellular dopamine in the NA. Pharmacologic magnetic resonance imaging (phMRI) was used to understand the effects of selective potentiation of M_4 on region-specific brain activation of NA in combination with amphetamine. Amphetamine induced a robust increase in cerebral blood volume in brain regions, including the nucleus accumbens, and this activation was reversed by VU0152100.

3.3.7 *Example of In Vivo Characterization of the Selective mGlu$_5$ NAM Basimglurant Through the Late Stage Preclinical Discovery*

Basimglurant (RG7090, RO4917523; 2-chloro-4-[1-(4-fluoro-phenyl)-2,5-dimethyl-1*H*-imidazol-4-ylethynyl]-pyridine) is a metabotropic glutamate receptor subtype 5 (mGluR$_5$) negative allosteric modulator (NAM), which recently underwent clinical trials for MDD. Although basimglurant failed in phase II clinical trials due to lack of efficacy, evidence in multiple secondary endpoints suggests that this compound, and more importantly the mGluR$_5$ mechanism, holds promise as a novel treatment for MDD. It also represents an excellent example of the use of animal models throughout the stage of drug discovery for a psychiatric disorder.

mGluR$_5$ is an excitatory Gq-coupled GPCR primarily expressed in the cortical and limbic regions associated with the processing of cognition, motivation, and emotion [224]. Subcellularly, it is primarily localized in the postsynaptic density on glutamatergic neurons [225, 226]. Early pharmacological tool compounds used in target validation include MPEP [2-methyl-6-(phenylethynyl)pyridine] [227] and later MTEP (3-[(2-methyl-1,3-thiazol-4-yl)ethynyl]-pyridine) [228], both selective mGluR$_5$ antagonists. These compounds reduced immobility time in the FST, suggesting possible antidepressant-like activity in wildtype mice [229, 230]. Importantly, these effects were absent in the mGluR$_5$ knockout mice, and these mice had lower baseline immobility [230], consistent with a role of mGluR$_5$ in the antidepressant-like effects seen in this assay.

The high-throughput screen performed in basimglurant development was a calcium mobilization assay assessing the function of mGluR$_5$ receptors, using mGluR$_5$ from both humans and rats transfected into HEK293 cells [231]. Pharmacokinetic profiles were assessed in Wistar rats, showing good central nervous system penetrance, a 7.5 h half-life, and 50% oral bioavailability [232]. Later profiling in cynomolgus monkeys showed a longer half-life at 20 h and a similar 50% bioavailability in fed animals, with a fasted bioavailability of 100% [232].

In regards to the *in vivo* efficacy profile of basimglurant in animal models for MDD, basimglurant was tested in both forced swim and ICSS models as described in section "Major Depressive Disorder". Using the unpredictable chronic mild stress model to disrupt ICSS motivational behaviors and model anhedonic-like activity, repeated treatment with basimglurant (3 mg/kg) once-daily for 3 weeks normalized the observed anhedonia index of the chronically stressed rats, bringing it back to a pre-stress level [232]. Acute oral basimglurant also significantly reduced immobility in the forced swim test at both 10 mg/kg and 30 mg/kg consistent with antidepressant-like activity and comparable to the effects observed with MTEP and MPEP [232]. These assays show that basimglurant has efficacy in animal models for different symptoms of MDD.

A thorough assessment of functional target engagement and receptor occupancy was also included in the preclinical candidate profiling of basimglurant. Functional MRI was used to examine the dose-related effects of basimglurant on rat brain activity patterns. Basimglurant produced robust dose-related changes in rat brain activity patterns within the dose range that also produced antidepressant-like and anti-anhendonic-like activity. In particular, basimglurant increased activity in the dorsal striatum, while it decreased activation of the medial prefrontal cortex, dorsal hippocampus, thalamus, hypothalamus, septum, accumbens, ventral pallidum, and entorhinal piriform cortex [232]. Importantly, when these regional changes in brain activity were compared with the activity of other approved antidepressants (duloxetine, reboxetine, imipramine, and bupropion), the patterns of activity as measured by fMRI were comparable, suggesting that the effects of basimglurant may be similar to other antidepressants. These data also provided a clear dose-related measurement of functional $mGluR_5$ target engagement using fMRI for translational studies in clinical populations.

To understand receptor occupancy and brain exposure, [^{3}H]ABP688 [233] (another $mGluR_5$ NAM) and [^{3}H]basimglurant were used for *in vivo* receptor occupancy studies using Sprague-Dawley rats and C57/Bl6J mice. [^{3}H]ABP688 binding was completely displaced by a 3 mg/kg oral dose of basimglurant [232], with high levels of occupancy observed in the cortex, hippocampus, striatum, amygdala, and nucleus accumbens [232]. These studies also provided a clear measurement of dose-related $mGluR_5$ target occupancy that could be adapted using a selective $mGluR_5$ NAM PET ligand like [^{18}F]FPEB for translational PET studies in clinical populations.

Insomnia is both a risk factor for and a symptom of depression, making EEG another good biomarker for the effects of antidepressants [234, 235]. In light of this, the effects of basimglurant (0.03–0.3 mg/kg) dosed daily at 2 h into the active cycle for 5 days were examined by recording EEG for 22 h after the last dose. Basimglurant significantly and dose-dependently decreased the REM/non-REM ratio and reduced time spent in REM at the two highest doses within the dose range used in the preclinical models [232]. During non-REM sleep, basimglurant also increased delta power at all doses [232]. Overall, the profile of basimglurant was consistent with wake-promoting effects. These findings also provided a dose-related measurement of functional $mGluR_5$ target engagement using EEG for future translational studies in clinical populations.

Unfortunately, despite the extensive preclinical characterization of basimglurant across behavioral, biomarker, and pharmacokinetic domains, when basimglurant was evaluated for potential efficacy in a Phase II clinical trial in patients with treatment-resistant MDD, it failed to significantly differ from placebo on the primary outcome measure of the clinician-scored Montgomery-Asberg Depression Rating Scale (MADRS) [236, 237]. There are multiple factors that may have contributed to this failed clinical trial, including the reported high placebo response and the choice of the primary outcome measure and patient population. However, one critical factor was the lack of biomarker studies to determine the dose-related levels of $mGluR_5$ target occupancy and/or functional target engagement achieved in either healthy volunteer subjects or MDD patients, despite the detailed preclinical package that provided a comprehensive understanding of the occupancy-to-efficacy relationships and measures of functional $mGluR_5$ target engagement across the different animal models. Additionally, basimglurant activity was evaluated primarily when given alone in the preclinical models, but not when administered in combination with various antidepressant drugs. Yet the clinical evaluation of basimglurant efficacy was conducted as an adjunct therapy to ongoing antidepressant treatments in the patient populations tested [237]. Taken together, these findings suggest that $mGluR_5$ NAMs remain a promising mechanism for the treatment of MMD that awaits further clinical validation. Moreover, the discovery of basimglurant provides an excellent example of the use of animals in the various stages of the psychiatric drug discovery process as well as a demonstration of the importance of using that preclinical data to better inform the clinical trial design (see [238, 239] for more on this topic).

3.4 Future Innovations for Animal Models in Psychiatric Drug Discovery

As discussed previously, while there are no perfect animal models for different psychiatric disorders, this does not mean that the current models do not have a place in drug discovery. Instead, it means that researchers in drug discovery are constantly striving to improve currently used animal models, to more closely model different aspects of psychiatric disorders and to ensure that the correct model is being used for the appropriate stage in the discovery process. Using a variety of complementary animal models which account for multidimensional analyses of a psychiatric disorder is a strategy that can improve the applicability of data collected from animal models. For example, when developing a drug for schizophrenia, this means assessing the effects of a compound in models of cognitive, positive, and negative symptoms. In addition, this may mean assessing a potential drug in a variety of genetic models and stratifying patient populations for clinical trials according to those genetic models, such as limiting initial trials and subsequent approval of a drug for patients with a particular genetic risk factor for schizophrenia. Animal models can be used to discover the full scope of contexts where a prospective drug may or may not be useful. With this information, clinical trials can be designed with the primary endpoints most relevant to the specific effects of a drug, rather than including all patients under a given diagnosis. This increased specificity may lead to a decreased overall prospective patient population for a drug but simultaneously more successful clinical trials.

3.4.1 RDoC Framework for Clinical to Preclinical Translational Studies for New Animal Model Development

In 2009, the National Institute of Mental Health (NIMH) launched the Research Domain Criteria Initiative (RDoC) to bridge gaps between translational and clinical research and to establish pathophysiological mechanisms underlying mental illnesses. The goal of RDoC is to establish a classification system for mental illnesses beyond clinical consensus alone [240]. The RDoC initiative classifies mental illnesses based on the following assumptions: (1) brain circuitry is disrupted in brain disorders, (2) neurocircuitry dysfunctions can be identified using clinical technologies such as fMRI and EEG, and (3) genetics and clinical neuroscience will reveal biosignatures that can be used for diagnosis and treatment [241].

The RDoC framework is a two-dimensional matrix to understand genetics, circuitry and neurobiology, and frame future studies [242]. The matrix compares broad constructs of behavior to units of analysis to identify areas that need to be assessed. This allows for delineation of the genetic, circuitry and neurobiological functions underlying disease-associated behaviors and cognitive functions in psychiatric illnesses. The domains are positive valence systems, negative valence systems, systems for social process, cognitive systems, sensorimotor systems, and arousal/regulatory systems [240]. The measurable units of analysis are genes, molecules, circuits, physiology, paradigm, and self-reports [240]. The intersection of each construct identifies areas of focus to develop studies which will delineate the genetic, circuitry, and neurobiological functions underlying associated behaviors and brain functions in the mental illness.

The RDoC framework to assign neurocircuitry and neurobiological mechanisms to psychiatric illnesses directly invites the use of preclinical animal models not only for novel findings, but to translate behaviors and physiological data in human studies. For example, gene variations observed in schizophrenia patients such as copy number variants have been studied in transgenic mouse models to assess the neurocircuitry dysfunctions and behavior associated with copy number variants, reviewed in the following [243]. Animal models for psychiatric illnesses aligned with the RDoC framework provide valid models that can be used in drug discovery.

3.4.2 Novel Technologies Enabling Development of Animal Models

Novel Genetic Approaches

The clustered regularly interspersed short palindromic repeat/CRISPR-associated nuclease 9 (CRISPR/Cas9) technology provides exciting novel genetic engineering strategies for the development of better animal models of psychiatric diseases. The CRISPR/Cas9 technology targets genes in a sequence specific manner using guide RNA to identify the specific target gene in any location of genomic DNA of any tissue. Once the target is identified, DNA is cut and edited to the guide RNA using nonhomologous end joining (NHEJ) or homology-directed repair (HDR) [244, 245]. CRISPR/Cas9 technology, unlike Cre-recombinase technology previously mentioned, does not require embryonic engineering. CRISPR/Cas9 also can target multiple genes at once and can be delivered directly into brain tissue, increasing its applications across studies in neurodegenerative diseases [244, 245]. A further development of this technology is the Cre-driven expression of Cas9 in mouse models which overcomes the challenge of expressing Cas9 in the brain [246]. One exciting application of this novel genetic engineering technology has been the ability to develop genetically modify nonhuman primates with genetic disruptions associated with clinical psychiatric populations for future psychiatric drug development. For example, utilization of CRISPR/Cas9 allowed disruptions of exons 6 and 12 of *SHANK3* gene in cynomolgus monkeys [247]. Similar disruptions in the *SHANK3* gene, which encodes a scaffolding protein localized in the postsynaptic density of excitatory synapses, have been shown to contribute to the pathogenesis of autism spectrum disorder (ASD) patient cases in Canada [248]. Longitudinal studies over 26 months with one male cynomolgus monkey with a 2 base pair deletion in exon 12 of *SHANK3* ($SHANK3^{M3}$) demonstrated that this genetic monkey model displayed core behavioral phenotypes of ASD, including deficits in social interaction and stereotypical locomotor activity that were improved by treatment with the SSRI fluoxetine [249].

Novel Techniques to Study Neurocircuitry Abnormalities in Psychiatric Disorders

One of the challenges to establishing the connection between psychiatric illnesses and neurocircuitry has been a lack of technologies to regulate and detect signaling in vivo. Optogenetic approaches use light-responsive proteins to probe and induce changes in circuitry. Designer Receptors Exclusively Activated by Designer Drugs (DREADDs) are a chemogenetic technology in neuroscience that takes advantage of GPCR signaling mechanisms to selectively activate specific neural circuits. GCaMP is a fluorescent calcium sensor that takes advantage of intracellular calcium release in neurotransmission. These technologies have enhanced neurocircuitry studies in alignment with the RDoC framework to identify neurocircuitry and neurobiology underlying psychiatric illnesses.

Optogenetics and dLight Signaling Strategies

Optogenetics is a novel technique that uses light to control biological activities. Opsins are light-sensitive ion channels that induce movement of ions into the intracellular environment [250–252]. Other optogenetic proteins consist of cryptochromes, light-oxygen-voltage receptors, blue-light sensor proteins, and UV-sensitive receptors. These optogenetic proteins probe second messenger interactions, protein-protein interactions, protein activation, and protein inactivation upon absorbing light [253]. Expression of these proteins using gene expression technologies allows for the targeted expression of these proteins in animal models to study the neuronal circuits via light activation. Animal studies using optogenetic technologies have provided the ability to study circuitry, neuronal signaling, and second messenger mechanisms to understand their effects on associated behaviors in freely moving mice with high temporal and spatial resolution [254–257]. Optogenetic neurotransmitter sensors have been engineered to emit a fluorescence signal upon activation or binding of the ligand [258, 259]. The dLight1.1 dopamine sensor is an optogenetic dopamine sensor engineered from the D_1 receptor fused with a circularly permutated green fluorescent protein which fluoresces when the conformation of the receptor changes after binding dopamine [259]. The advantage of this technology is the ability to detect both tonic and phasic dopamine signaling in response to various stimuli. The application of this technology then allows researchers to directly study dopamine signaling in response to biological, pharmacological, and environmental conditions under normal conditions and in psychiatric disease models. D-light technology also allows for the detection of changes in dopamine signaling evoked by optogenetic manipulations [259]. This novel application was used to differentiate neuronal firing events in the neurons of the VTA from signaling events on postsynaptic neurons in the nucleus accumbens (NAc), which are important for reward perception [260]. These studies revealed that VTA neuronal firing does not cause dopamine release in the NAc in response to reward perception, a critical mechanism involved in motivation and drug-seeking behaviors in psychiatric populations.

DREADDs

Designer receptors exclusively activated by designer drug (DREADD)-based chemogenetic tools are designed to identify underlying circuitry associated with psychiatric disorders [261, 262]. This technology, based on GPCR signaling mechanisms, relies on the conditional genetic expression of the DREADD receptor and the administration of the designer drug to activate the receptor mechanisms. The human M_3 muscarinic DREADD receptor coupled to G_q (hM_3D_q), activated specifically by bioavailable clozapine-N-oxide (CNO), is widely used to activate neuronal firing [263]. The human M_4 muscarinic DREADD receptor coupled to G_i (hM_4D_i) is used to attenuate neuronal firing also upon activation with CNO [263, 264]. *DREADD studies in transgenic mouse models have demonstrated inducible-conditional expression and subsequent behavioral and electrophysiological effects upon administration of CNO, the associated designer drugs* [264]. Advantages to this approach in comparison to optogenetic approaches are that the activation of DREADDs is noninvasive and does not rely on the availability of a light source or other technology. Limitations to the technology involve complexities with CNO kinetics in animal models and appropriate coupling of G-proteins within the various signaling pathways. One application of this technology for understanding the neural circuitry underlying symptoms of anxiety has demonstrated that chronic activation of the hM_3D_q DREADD using CNO in the excitatory forebrain neurons of mice during a critical postnatal period enhances anxiogenic-like behaviors in later development and long-term alterations in metabolic rate of glucose oxidation in the hippocampal and cortical brain regions [265].

GCaMP

In vivo calcium imaging represents an exciting direct readout of neuronal activity in animal model for psychiatric disorders. GCaMPs are members of a larger family of genetically encoded calcium indicators (GECIs) [266]. GCaMPs contain a circularly permutated green fluorescent protein, calmodulin (CaM), and muscle myosin light chain kinase (M13) [267, 268]. Calcium binding to GCaMPs changes fluorescence intensity. Similar to optogenetics, fiber optic recordings are required to obtain data for *in vivo* GCaMP calcium recordings [269–270]. GCaMP signaling technology can also serve as a reverse translational technology to measure neuronal activity compared to fMRI studies in human studies. For example, GCaMP6 conditionally expressed in excitatory neurons has been used to study signaling in healthy animals and in animals with experimentally induced strokes [271]. Changes in GCaMP in excitatory neurons in mice after stroke induction were shown to be consistent with cortical alterations detected by fMRI analysis in human stroke patients [271]. This study demonstrates the promising translational capacity of in vivo calcium imaging using GCaMP technology.

Novel High-Throughput Behavioral Screening Technologies

Behavioral phenotyping is an essential element of drug discovery for psychiatric diseases. As previously discussed, discrete behavioral assays can be used to test the acute or chronic effects of different psychiatric drugs and potential underlying mechanisms related to the associated behaviors. However, such rodent models have limited throughput. More recently, behavioral phenomics has emerged as data driven approaches to integrate automated testing systems with bioinformatics [272, 273]. The PsychoGenics smartcube is one such in vivo high-throughput testing platform that collects over half a million data points during each automated testing session, then uses bioinformatic analysis to translate these data into over 2000 different behavioral features in order to create behavioral drug signatures. It also allows for both acute and chronic monitoring of psychiatric drug effects on different social, circadian, and cognitive behaviors in mice. Successful application of this in vivo high-throughput technology to conduct a behavior-driven screen with chemically diverse compound libraries provided through a collaboration with Sunovion Pharmaceuticals, resulted in the identification of SEP-363856, a novel trace amine receptor 1 (TAAR1) agonist with serotonin 5-HT1A activity, with demonstrated statistically and clinically meaningful improvements in the Positive and Negative Symptom Scale (PANSS) of schizophrenia [274, 275].

3.5 Summary and Future of Animal Models in Psychiatric Drug Discovery

In conclusion, ongoing advances in genetic, neural circuitry-based, and automated behavioral technologies combined with the application of the RDoC platform are leading to exciting innovations in the modeling of the underlying circuitry and biologic abnormalities associated with various psychiatric disorders. These innovations represent great potential for expanding the utility and reliability of animal models for the discovery and development of NCEs for psychiatric disorders.

References

1. Patel V, Saxena S, Lund C, Thornicroft G, Baingana F, Bolton P, et al. The lancet commission on global mental health and sustainable development. Lancet. 2018;392:1553–98.
2. Center for Behavioral Health Statistics and Quality SA and MHS Administration. Substance Abuse and Mental Health Services Administration. Key substance use and mental health indicators in the United States: results from the 2020 National Survey on Drug Use and Health. Rockville, MD: Center for Behavioral Health Statistics and Quality SA and MHS Administration; 2020.
3. Lieberman JA, Stroup TS, McEvoy JP, Swartz MS, Rosenheck RA, Perkins DO, et al. Effectiveness of antipsychotic drugs in patients with chronic schizophrenia. N Engl J Med. 2005;353:1209–23.
4. Fava M, Rush AJ, Trivedi MH, Nierenberg AA, Thase ME, Sackeim HA, et al. Background and rationale for the sequenced treatment alternatives to relieve depression (STAR*D) study. Psychiatr Clin North Am. 2003;26:457–94.
5. Rush AJ, Trivedi MH, Wisniewski SR, Nierenberg AA, Stewart JW, Warden D, et al. Acute and longer-term outcomes in depressed outpatients requiring one or several treatment steps: a STAR*D report. Am J Psychiatry. 2006;163:1905–17.
6. Watts BV, Schnurr PP, Mayo L, Young-Xu Y, Weeks WB, Friedman MJ. Meta-analysis of the efficacy of treatments for posttraumatic stress disorder. J Clin Psychiatry. 2013;74:11710.
7. Garakani A, Murrough JW, Freire RC, Thom RP, Larkin K, Buono FD, et al. Pharmacotherapy of anxiety disorders: current and emerging treatment options. Front Psychiatry. 2020;11:595584.
8. Douaihy AB, Kelly TM, Sullivan C. Medications for substance use disorders. Soc Work Public Health. 2013;28:264–78.
9. Lindsley CW. New statistics on the cost of new drug development and the trouble with CNS drugs. ACS Chem Neurosci. 2014;5:1142.
10. Gribkoff VK, Kaczmarek LK. The need for new approaches in CNS drug discovery: why drugs have failed, and what can be done to improve outcomes. Neuropharmacology. 2017;120:11–9.
11. Hay M, Thomas DW, Craighead JL, Economides C, Rosenthal J. Clinical development success rates for investigational drugs. Nat Biotechnol. 2014;32:40–51.
12. Nestler EJ, Hyman SE. Animal models of neuropsychiatric disorders. Nat Neurosci. 2010;13:1161–9.
13. Tricklebank MD, Robbins TW, Simmons C, Wong EHF. Time to re-engage psychiatric drug discovery by strengthening confidence in preclinical psychopharmacology. Psychopharmacology. 2021;238:1417–36.
14. Monteggia LM, Heimer H, Nestler EJ. Meeting report: can we make animal models of human mental illness? Biol Psychiatry. 2018;84:542–5.
15. Carretero M, Solis GM, Petrascheck M. C. elegans as model for drug discovery. Curr Top Med Chem. 2017;17:2067–76.
16. Artal-Sanz M, de Jong L, Tavernarakis N. Caenorhabditis elegans: a versatile platform for drug discovery. Biotechnol J. 2006;1:1405–18.
17. Fernández-Hernández I, Scheenaard E, Pollarolo G, Gonzalez C. The translational relevance of drosophila in drug discovery. EMBO Rep. 2016;17:471–2.
18. Patton EE, Zon LI, Langenau DM. Zebrafish disease models in drug discovery: from preclinical modelling to clinical trials. Nat Rev Drug Discov. 2021;20(8):611–28.

19. Bruni G, Lakhani P, Kokel D. Discovering novel neuroactive drugs through high-throughput behavior-based chemical screening in the zebrafish. Front Pharmacol. 2014;5:153.
20. Prior H, Haworth R, Labram B, Roberts R, Wolfreys A, Sewell F. Justification for species selection for pharmaceutical toxicity studies. Toxicol Res. 2021;9:758–70.
21. Kaiser T, Feng G. Modeling psychiatric disorders for developing effective treatments. Nat Med. 2015;21:979–88.
22. Baker M, Hong SI, Kang S, Choi DS. Rodent models for psychiatric disorders: problems and promises. Lab Anim Res. 2020;36:1–10.
23. Russel WMS, Burch RL. The principles of humane experimental technique. London: Meuthen; 1959.
24. Gottesman II, Gould TD. The endophenotype concept in psychiatry: etymology and strategic intentions. Am J Psychiatr. 2003;160:636–45.
25. Willner P. The validity of animal models of depression. Psychopharmacology. 1984;83:1–16.
26. Green T, Gothelf D, Glaser B, Debbane M, Frisch A, Kotler M, et al. Psychiatric disorders and intellectual functioning throughout development in velocardiofacial (22q11.2 deletion) syndrome. J Am Acad Child Adolesc Psychiatry. 2009;48:1060–8.
27. Karayiorgou M, Simon TJ, Gogos JA. 22q11.2 microdeletions: linking DNA structural variation to brain dysfunction and schizophrenia. Nat Rev Neurosci. 2010;11(6):402–16.
28. Swerdlow NR, Braff DL, Taaid N, Geyer MA. Assessing the validity of an animal model of deficient sensorimotor gating in schizophrenic patients. Arch Gen Psychiatry. 1994;51:139.
29. Geyer MA, Krebs-Thomson K, Braff DL, Swerdlow NR. Pharmacological studies of prepulse inhibition models of sensorimotor gating deficits in schizophrenia: a decade in review. Psychopharmacology (Berl). 2001;156(2-3):117–54.
30. Yankelevitch-Yahav R, Franko M, Huly A, Doron R. The forced swim test as a model of depressive-like behavior. J Vis Exp. 2015;(97):52587.
31. Sunal R, Gümüşel B, Kayaalp SO. Effect of changes in swimming area on results of "behavioral despair test". Pharmacol Biochem Behav. 1994;49:891–6.
32. Geschwind DH, Flint J. Genetics and genomics of psychiatric disease. Science. 1979;2015(349):1489–94.
33. Rees E, Owen MJ. Translating insights from neuropsychiatric genetics and genomics for precision psychiatry. Genome Med. 2020;12(1):1–16.
34. Anderzhanova E, Kirmeier T, Wotjak CT. Animal models in psychiatric research: the RDoC system as a new framework for endophenotype-oriented translational neuroscience. Neurobiol Stress. 2017;7:47–56.
35. Walters JTR, Owen MJ. Endophenotypes in psychiatric genetics. Mol Psychiatry. 2007;12:886–90.
36. Houdebine LM. Transgenic animal models in biomedical research. Methods Mol Biol. 2007;360:163–202.
37. Skarnes WC, Rosen B, West AP, Koutsourakis M, Bushell W, Iyer V, et al. A conditional knockout resource for the genome-wide study of mouse gene function. Nature. 2011;474:337–42.
38. Ansorge MS, Zhou M, Lira A, Hen R, Gingrich JA. Early-life blockade of the 5-HT transporter alters emotional behavior in adult mice. Science. 1979;2004(306):879–81.
39. Gordon JW, Scangos GA, Plotkin DJ, Barbosa JA, Ruddle FH. Genetic transformation of mouse embryos by microinjection of purified DNA. Proc Natl Acad Sci. 1980;77:7380–4.
40. Nagy A. Cre recombinase: the universal reagent for genome tailoring. Genesis. 2000;26:99–109.
41. Kim H, Kim M, Im SK, Fang S. Mouse Cre-LoxP system: general principles to determine tissue-specific roles of target genes. Lab Anim Res. 2018;34:147–59.
42. Ramírez-Solis R, Liu P, Bradley A. Chromosome engineering in mice. Nature. 1995;378:720–4.
43. Baron U, Bujard H. Tet repressor-based system for regulated gene expression in eukaryotic cells: principles and advances. In: Methods in enzymology. Elsevier; 2000. p. 401–21.

44. Corti O, Sabaté O, Horellou P, Colin P, Dumas S, Buchet D, et al. A single adenovirus vector mediates doxycycline-controlled expression of tyrosine hydroxylase in brain grafts of human neural progenitors. Nat Biotechnol. 1999;17:349–54.
45. Corti O, Sánchez-Capelo A, Colin P, Hanoun N, Hamon M, Mallet J. Long-term doxycycline-controlled expression of human tyrosine hydroxylase after direct adenovirus-mediated gene transfer to a rat model of Parkinson's disease. Proc Natl Acad Sci. 1999;96:12120–5.
46. Mansuy IM, Bujard H. Tetracycline-regulated gene expression in the brain. Curr Opin Neurobiol. 2000;10:593–6.
47. Smith AJH, De Sousa MA, Kwabi-Addo B, Heppell-Parton A, Impey H, Rabbitts P. A site–directed chromosomal translocation induced in embryonic stem cells by Cre-loxP recombination. Nat Genet. 1995;9:376–85.
48. Orban PC, Chui D, Marth JD. Tissue- and site-specific DNA recombination in transgenic mice. Proc Natl Acad Sci. 1992;89:6861–5.
49. Chen ZY, Jing D, Bath KG, Ieraci A, Khan T, Siao CJ, et al. Genetic variant BDNF (Val66Met) polymorphism alters anxiety-related behavior. Science. 2006;314:140–3.
50. Robinson TE, Becker JB. Enduring changes in brain and behavior produced by chronic amphetamine administration: a review and evaluation of animal models of amphetamine psychosis. Brain Res Rev. 1986;11:157–98.
51. Tenn CC, Kapur S, Fletcher PJ. Sensitization to amphetamine, but not phencyclidine, disrupts prepulse inhibition and latent inhibition. Psychopharmacology. 2005;180:366–76.
52. Deumens R, Blokland A, Prickaerts J. Modeling Parkinson's disease in rats: an evaluation of 6-OHDA lesions of the nigrostriatal pathway. Exp Neurol. 2002;175:303–17.
53. Czéh B, Fuchs E, Wiborg O, Simon M. Animal models of major depression and their clinical implications. Prog Neuro-Psychopharmacol Biol Psychiatry. 2016;64:293–310.
54. Deisseroth K. Circuit dynamics of adaptive and maladaptive behaviour. Nature. 2014;505:309–17.
55. Steinberg EE, Christoffel DJ, Deisseroth K, Malenka RC. Illuminating circuitry relevant to psychiatric disorders with optogenetics. Curr Opin Neurobiol. 2015;30:9–16.
56. Porsolt RD, Le Pichon M, Jalfre M. Depression: a new animal model sensitive to antidepressant treatments. Nature. 1977;266(5604):730–2.
57. Steru L, Chermat R, Thierry B, Simon P. The tail suspension test: a new method for screening antidepressants in mice. Psychopharmacology. 1985;85:367–70.
58. Hiraoka K, Motomura K, Yanagida S, Ohashi A, Ishisaka-Furuno N, Kanba S. Pattern of c-Fos expression induced by tail suspension test in the mouse brain. Heliyon. 2017;3:e00316.
59. Can A, Dao DT, Terrillion CE, Piantadosi SC, Bhat S, Gould TD. The tail suspension test. J Vis Exp. 2012;(59):e3769.
60. Rosenberg MB, Carroll FI, Negus SS. Effects of monoamine reuptake inhibitors in assays of acute pain-stimulated and pain-depressed behavior in rats. J Pain. 2013;14:246–59.
61. Rafało-Ulińska A, Pałucha-Poniewiera A. The effectiveness of (R)-ketamine and its mechanism of action differ from those of (S)-ketamine in a chronic unpredictable mild stress model of depression in C57BL/6J mice. Behav Brain Res. 2022;418:113633.
62. Koprdová R, Bögi E, Belovičová K, Sedláčková N, Okuliarová M, Ujházy E, et al. Chronic unpredictable mild stress paradigm in male Wistar rats: effect on anxiety- and depressive-like behavior. Neuro Endocrinol Lett. 2016;37:103–10.
63. Kudryavtseva NN, Bakshtanovskaya IV, Koryakina LA. Social model of depression in mice of C57BL/6J strain. Pharmacol Biochem Behav. 1991;38:315–20.
64. Kumar S, Hultman R, Hughes D, Michel N, Katz BM, Dzirasa K. Prefrontal cortex reactivity underlies trait vulnerability to chronic social defeat stress. Nat Commun. 2014;5:1–9.
65. Planchez B, Surget A, Belzung C. Animal models of major depression: drawbacks and challenges. J Neural Transm. 2019;126:1383–408.
66. Becker C, Zeau B, Rivat C, Blugeot A, Hamon M, Benoliel JJ. Repeated social defeat-induced depression-like behavioral and biological alterations in rats: involvement of cholecystokinin. Mol Psychiatry. 2008;13:1079–92.

67. Mineur YS, Belzung C, Crusio WE. Effects of unpredictable chronic mild stress on anxiety and depression-like behavior in mice. Behav Brain Res. 2006;175:43–50.
68. Rygula R, Abumaria N, Flügge G, Fuchs E, Rüther E, Havemann-Reinecke U. Anhedonia and motivational deficits in rats: impact of chronic social stress. Behav Brain Res. 2005;162:127–34.
69. Kurre Nielsen C, Arnt J, Sánchez C. Intracranial self-stimulation and sucrose intake differ as hedonic measures following chronic mild stress: interstrain and interindividual differences. Behav Brain Res. 2000;107:21–33.
70. Barik J, Marti F, Morel C, Fernandez SP, Lanteri C, Godeheu G, et al. Chronic stress triggers social aversion via glucocorticoid receptor in dopaminoceptive neurons. Science. 2013;339(6117):332–5.
71. Meerlo P, Overkamp GJF, Daan S, van den Hoofdakker RH, Koolhaas JM. Changes in behaviour and body weight following a single or double social defeat in rats. Stress. 1996;1:21–32.
72. Moreau JL, Jenck F, Martin JR, Mortas P, Haefely WE. Antidepressant treatment prevents chronic unpredictable mild stress-induced anhedonia as assessed by ventral tegmentum self-stimulation behavior in rats. Eur Neuropsychopharmacol. 1992;2:43–9.
73. Il P, J, Zhao T, Huang GB, Sui ZY, Li CR, Han EH, et al. Effects of aripiprazole and haloperidol on Fos-like immunoreactivity in the prefrontal cortex and amygdala. Clin Psychopharmacol Neurosci. 2011;9:36–43.
74. Serra M, Pisu MG, Floris I, Biggio G. Social isolation-induced changes in the hypothalamic–pituitary–adrenal axis in the rat. Stress. 2005;8:259–64.
75. Tye KM, Mirzabekov JJ, Warden MR, Ferenczi EA, Tsai HC, Finkelstein J, et al. Dopamine neurons modulate neural encoding and expression of depression-related behaviour. Nature. 2013;493:537–41.
76. Sarkar A, Kabbaj M. Sex differences in effects of ketamine on behavior, spine density, and synaptic proteins in socially isolated rats. Biol Psychiatry. 2016;80:448–56.
77. Markou A, Koob GF. Construct validity of a self-stimulation threshold paradigm: effects of reward and performance manipulations. Physiol Behav. 1992;51:111–9.
78. Rygula R, Abumaria N, Flügge G, Hiemke C, Fuchs E, Rüther E, et al. Citalopram counteracts depressive-like symptoms evoked by chronic social stress in rats. Behav Pharmacol. 2006;17:19–29.
79. Cordner ZA, Marshall-Thomas I, Boersma GJ, Lee RS, Potash JB, Tamashiro KLK. Fluoxetine and environmental enrichment similarly reverse chronic social stress-related depression- and anxiety-like behavior, but have differential effects on amygdala gene expression. Neurobiol Stress. 2021;15:100392.
80. Dournes C, Beeské S, Belzung C, Griebel G. Deep brain stimulation in treatment-resistant depression in mice: comparison with the CRF1 antagonist, SSR125543. Prog Neuro-Psychopharmacol Biol Psychiatry. 2013;40:213–20.
81. Hamani C, Giacobbe P, Diwan M, Balbino ES, Tong J, Bridgman A, et al. Monoamine oxidase inhibitors potentiate the effects of deep brain stimulation. Am J Psychiatr. 2012;169:1320–1.
82. Hamani C, Machado DC, Hipólide DC, Dubiela FP, Suchecki D, Macedo CE, et al. Deep brain stimulation reverses anhedonic-like behavior in a chronic model of depression: role of serotonin and brain derived neurotrophic factor. Biol Psychiatry. 2012;71:30–5.
83. Hamani C, Diwan M, Macedo CE, Brandão ML, Shumake J, Gonzalez-Lima F, et al. Antidepressant-like effects of medial prefrontal cortex deep brain stimulation in rats. Biol Psychiatry. 2010;67:117–24.
84. Erburu M, Cajaleon L, Guruceaga E, Venzala E, Muñoz-Cobo I, Beltrán E, et al. Chronic mild stress and imipramine treatment elicit opposite changes in behavior and in gene expression in the mouse prefrontal cortex. Pharmacol Biochem Behav. 2015;135:227–36.
85. Zhang L, Luo J, Zhang M, Yao W, Ma X, Yu SY. Effects of curcumin on chronic, unpredictable, mild, stress-induced depressive-like behaviour and structural plasticity in the lateral amygdala of rats. Int J Neuropsychopharmacol. 2014;17:793–806.
86. Garzón J, Fuentes JA, Del Rio J. Antidepressants selectively antagonize the hyperactivity induced in rats by long-term isolation. Eur J Pharmacol. 1979;59:293–6.

87. Zazpe A, Artaiz I, Labeaga L, Lucero ML, Orjales A. Reversal of learned helplessness by selective serotonin reuptake inhibitors in rats is not dependent on 5-HT availability. Neuropharmacology. 2007;52:975–84.
88. Pereira VS, Joca SRL, Harvey BH, Elfving B, Wegener G. Esketamine and rapastinel, but not imipramine, have antidepressant-like effect in a treatment-resistant animal model of depression. Acta Neuropsychiatr. 2019;31:258–65.
89. Ciulla L, Menezes HS, Bueno BBM, Schuh A, Alves RJV, Abegg MP. Antidepressant behavioral effects of duloxetine and fluoxetine in the rat forced swimming test. Acta Cir Bras. 2007;22(5):351–4.
90. Rodríguez-Landa JF, Contreras CM, García-Ríos RI. Allopregnanolone microinjected into the lateral septum or dorsal hippocampus reduces immobility in the forced swim test: participation of the GABAA receptor. Behav Pharmacol. 2009;20:614–22.
91. Choi SH, Chung S, Cho JH, Cho YH, Im JWK, Kim JM, et al. Changes in c-Fos expression in the forced swimming test: common and distinct modulation in rat brain by desipramine and citalopram. Korean J Physiol Pharmacol. 2013;17:321–9.
92. Ma L, Xu Y, Wang G, Li R. What do we know about sex differences in depression: a review of animal models and potential mechanisms. Prog Neuro-Psychopharmacol Biol Psychiatry. 2019;89:48–56.
93. Wang SM, Han C, Bahk WM, Lee SJ, Patkar AA, Masand PS, et al. Addressing the side effects of contemporary antidepressants: a comprehensive review. Chonnam Med J. 2018;54:101–12.
94. Gaynes BN, Lux L, Gartlehner G, Asher G, Forman-Hoffman V, Green J, et al. Defining treatment-resistant depression. Depress Anxiety. 2020;37:134–45.
95. Salahudeen MS, Wright CM, Peterson GM. Esketamine: new hope for the treatment of treatment-resistant depression? A narrative review. Ther Adv Drug Saf [Internet]. 2020 [cited 2022 July 23];11:2042098620937899. Available from: https://pubmed.ncbi.nlm.nih.gov/32782779/.
96. Korte SM, Prins J, Krajnc AM, Hendriksen H, Oosting RS, Westphal KG, et al. The many different faces of major depression: it is time for personalized medicine. Eur J Pharmacol. 2015;753:88–104.
97. Walker DL, Toufexis DJ, Davis M. Role of the bed nucleus of the stria terminalis versus the amygdala in fear, stress, and anxiety. Eur J Pharmacol. 2003;463:199–216.
98. Lezak KR, Missig G, Carlezon WA Jr. Behavioral methods to study anxiety in rodents. Dialogues Clin Neurosci. 2017;19:181–91.
99. Handley SL, Mithani S. Effects of alpha-adrenoceptor agonists and antagonists in a maze-exploration model of 'fear'-motivated behaviour. Naunyn Schmiedeberg's Arch Pharmacol. 1984;327:1–5.
100. Crawley J, Goodwin FK. Preliminary report of a simple animal behavior model for the anxiolytic effects of benzodiazepines. Pharmacol Biochem Behav. 1980;13:167–70.
101. Thomas A, Burant A, Bui N, Graham D, Yuva-Paylor LA, Paylor R. Marble burying reflects a repetitive and perseverative behavior more than novelty-induced anxiety. Psychopharmacology. 2009;204:361–73.
102. Himanshu D, Sarkar D, Nutan A. A review of behavioral tests to evaluate different types of anxiety and anti-anxiety effects. Clin Psychopharmacol Neurosci. 2020;18:341–51. Available from: https://pubmed.ncbi.nlm.nih.gov/32702213/.
103. Kleven MS, Koek W. Effects of benzodiazepine agonists on punished responding in pigeons and their relationship with clinical doses in humans. Psychopharmacology. 1999;141(2):206–12.
104. Oberrauch S, Sigrist H, Sautter E, Gerster S, Bach DR, Pryce CR. Establishing operant conflict tests for the translational study of anxiety in mice. Psychopharmacology. 2019;236:2527–41.
105. Ramirez K, Sheridan JF. Antidepressant imipramine diminishes stress-induced inflammation in the periphery and central nervous system and related anxiety-and depressive-like behaviors. Brain Behav Immun. 2016;57:293–303.

106. Moreira CM, Masson S, Carvalho MC, Brandão ML. Exploratory behaviour of rats in the elevated plus-maze is differentially sensitive to inactivation of the basolateral and central amygdaloid nuclei. Brain Res Bull. 2007;71:466–74.
107. Sanchez C, Meier E. Behavioral profiles of SSRIs in animal models of depression, anxiety and aggression. Are they all alike? Psychopharmacology. 1997;129:197–201.
108. Strawn JR, Geracioti L, Rajdev N, Clemenza K, Levine A. Pharmacotherapy for generalized anxiety disorder in adults and pediatric patients: an evidence-based treatment review. Expert Opin Pharmacother. 2018;19:1057.
109. Sartori SB, Singewald N. Novel pharmacological targets in drug development for the treatment of anxiety and anxiety-related disorders. Pharmacol Ther. 2019;204:107402.
110. Colucci P, Marchetta E, Mancini GF, Alva P, Chiarotti F, Hasan MT, et al. Predicting susceptibility and resilience in an animal model of post-traumatic stress disorder (PTSD). Transl Psychiatry. 2020;10:243.
111. Knox D, George SA, Fitzpatrick CJ, Rabinak CA, Maren S, Liberzon I. Single prolonged stress disrupts retention of extinguished fear in rats. Learn Mem. 2012;19:43–9.
112. Jacobson-Pick S, Audet MC, McQuaid RJ, Kalvapalle R, Anisman H. Social agonistic distress in male and female mice: changes of behavior and brain monoamine functioning in relation to acute and chronic challenges. PLoS One. 2013;8:e60133.
113. Zoladz PR, Conrad CD, Fleshner M, Diamond DM. Acute episodes of predator exposure in conjunction with chronic social instability as an animal model of post-traumatic stress disorder. Stress. 2008;11:259–81.
114. Cohen H, Zohar J, Matar M. The relevance of differential response to trauma in an animal model of posttraumatic stress disorder. Biol Psychiatry. 2003;53:463–73.
115. Bentefour Y, Rakibi Y, Bennis M, Ba-M'hamed S, Garcia R. Paroxetine treatment, following behavioral suppression of PTSD-like symptoms in mice, prevents relapse by activating the infralimbic cortex. Eur Neuropsychopharmacol. 2016;26:195–207.
116. Reisman M. PTSD treatment for veterans: what's working, what's new, and what's next. P T. 2016;41:623.
117. Richter-Levin G, Stork O, Schmidt MV. Animal models of PTSD: a challenge to be met. Mol Psychiatry. 2019;24:1135–56.
118. Becker A, Grecksch G, Bernstein HG, Höllt V, Bogerts B. Social behaviour in rats lesioned with ibotenic acid in the hippocampus: quantitative and qualitative analysis. Psychopharmacology. 1999;144:333–8.
119. Talamini LM, Ellenbroek B, Koch T, Korf J. Impaired sensory gating and attention in rats with developmental abnormalities of the mesocortex: implications for schizophrenia. Ann N Y Acad Sci. 2000;911:486–94.
120. Lodge DJ, Grace AA. Gestational methylazoxymethanol acetate administration: a developmental disruption model of schizophrenia. Behav Brain Res. 2009;204:306–12.
121. Moore H, Jentsch JD, Ghajarnia M, Geyer MA, Grace AA. A neurobehavioral systems analysis of adult rats exposed to methylazoxymethanol acetate on E17: implications for the neuropathology of schizophrenia. Biol Psychiatry. 2006;60:253–64.
122. Jentsch JD, Taylor JR, Roth RH. Subchronic phencyclidine administration increases mesolimbic dopaminergic system responsivity and augments stress- and psychostimulant-induced hyperlocomotion. Neuropsychopharmacology. 1998;19:105–13.
123. Kocsis B, Brown RE, McCarley RW, Hajos M. Impact of ketamine on neuronal network dynamics: translational modeling of schizophrenia-relevant deficits. CNS Neurosci Ther. 2013;19:437–47.
124. Featherstone RE, Rizos Z, Kapur S, Fletcher PJ. A sensitizing regimen of amphetamine that disrupts attentional set-shifting does not disrupt working or long-term memory. Behav Brain Res. 2008;189:170–9.
125. Pletnikov MV, Ayhan Y, Nikolskaia O, Xu Y, Ovanesov MV, Huang H, et al. Inducible expression of mutant human DISC1 in mice is associated with brain and behavioral abnormalities reminiscent of schizophrenia. Mol Psychiatry. 2008;13:173–86.

126. Bassett AS, Chow EWC, Husted J, Weksberg R, Caluseriu O, Webb GD, et al. Clinical features of 78 adults with 22q11 deletion syndrome. Am J Med Genet A. 2005;138A:307–13.
127. Saito R, Miyoshi C, Koebis M, Kushima I, Nakao K, Mori D, et al. Two novel mouse models mimicking minor deletions in 22q11.2 deletion syndrome revealed the contribution of each deleted region to psychiatric disorders. Mol Brain. 2021;14:68.
128. Nakazawa T, Kikuchi M, Ishikawa M, Yamamori H, Nagayasu K, Matsumoto T, et al. Differential gene expression profiles in neurons generated from lymphoblastoid B-cell line-derived iPS cells from monozygotic twin cases with treatment-resistant schizophrenia and discordant responses to clozapine. Schizophr Res. 2017;181:75–82.
129. Brennand KJ, Simone A, Jou J, Gelboin-Burkhart C, Tran N, Sangar S, et al. Modelling schizophrenia using human induced pluripotent stem cells. Nature. 2011;473:221–5.
130. Amann LC, Gandal MJ, Halene TB, Ehrlichman RS, White SL, McCarren HS, et al. Mouse behavioral endophenotypes for schizophrenia. Brain Res Bull. 2010;83:147–61.
131. van den Buuse M, Gogos A. Differential effects of antipsychotic drugs on serotonin-1A receptor-mediated disruption of prepulse inhibition. J Pharmacol Exp Ther. 2007;320:1224.
132. Sampaio LRL, Cysne Filho FMS, de Almeida JC, dos Santos Diniz D, de Sousa CNS, Sampaio LRL, et al. Advantages of the alpha-lipoic acid association with chlorpromazine in a model of schizophrenia induced by ketamine in rats: behavioral and oxidative stress evidences. Neuroscience. 2018;373:72–81.
133. Tandon R, Lenderking WR, Weiss C, Shalhoub H, Barbosa CD, Chen J, et al. The impact on functioning of second-generation antipsychotic medication side effects for patients with schizophrenia: a worldwide, cross-sectional, web-based survey. Ann General Psychiatry. 2020;19:1–11.
134. Potkin SG, Kane JM, Correll CU, Lindenmayer JP, Agid O, Marder SR, et al. The neurobiology of treatment-resistant schizophrenia: paths to antipsychotic resistance and a roadmap for future research. NPJ Schizophr. 2020;6:1–10.
135. Robinson TE, Berridge KC. The neural basis of drug craving: an incentive-sensitization theory of addiction. Brain Res Rev. 1993;18:247–91.
136. Steketee JD, Kalivas PW. Drug wanting: behavioral sensitization and relapse to drug-seeking behavior. Sibley DR, editor. Pharmacol Rev. 2011;63:348.
137. Janetsian SS, Linsenbardt DN, Lapish CC. Memory impairment and alterations in prefrontal cortex gamma band activity following methamphetamine sensitization. Psychopharmacology. 2015;232:2083–95.
138. Mucha RF, van der Kooy D, O'Shaughnessy M, Bucenieks P. Drug reinforcement studied by the use of place conditioning in rat. Brain Res. 1982;243:91–105.
139. Ahmed SH, Koob GF. Transition from moderate to excessive drug intake: change in hedonic set point. Science. 1979;1998(282):298–300.
140. Mandt BH, Copenhagen LI, Zahniser NR, Allen RM. Escalation of cocaine consumption in short and long access self-administration procedures. Drug Alcohol Depend. 2015;149:166–72.
141. Wang Z-Y, Guo L-K, Han X, Song R, Dong G-M, Ma C-M, et al. Naltrexone attenuates methamphetamine-induced behavioral sensitization and conditioned place preference in mice. Behav Brain Res. 2021;399:112971.
142. Kaplan LM, Vella L, Cabral E, Tieu L, Ponath C, Guzman D, et al. Unmet mental health and substance use treatment needs among older homeless adults: results from the HOPE HOME study. J Community Psychol. 2019;47:1893–908.
143. Gainetdinov RR, Caron MG. An animal model of attention deficit hyperactivity disorder. Mol Med Today. 2000;6:43–4.
144. Trinh JV, Nehrenberg DL, Jacobsen JPR, Caron MG, Wetsel WC. Differential psychostimulant-induced activation of neural circuits in dopamine transporter knockout and wild type mice. Neuroscience. 2003;118:297–310.

145. Zhou M, Rebholz H, Brocia C, Warner-Schmidt JL, Fienberg AA, Nairn AC, et al. Forebrain overexpression of CK1delta leads to down-regulation of dopamine receptors and altered locomotor activity reminiscent of ADHD. Proc Natl Acad Sci U S A. 2010;107:4401–6.
146. Yan TC, Hunt SP, Stanford SC. Behavioural and neurochemical abnormalities in mice lacking functional tachykinin-1 (NK1) receptors: a model of attention deficit hyperactivity disorder. Neuropharmacology. 2009;57:627–35.
147. Yan TC, Dudley JA, Weir RK, Grabowska EM, Peña-Oliver Y, Ripley TL, et al. Performance deficits of NK1 receptor knockout mice in the 5-choice serial reaction-time task: effects of d-amphetamine, stress and time of day. PLoS One. 2011;6:e17586–6.
148. Simchon Y, Weizman A, Rehavi M. The effect of chronic methylphenidate administration on presynaptic dopaminergic parameters in a rat model for ADHD. Eur Neuropsychopharmacol. 2010;20:714–20.
149. Sagvolden T. Behavioral validation of the spontaneously hypertensive rat (SHR) as an animal model of attention-deficit/hyperactivity disorder (AD/HD). Neurosci Biobehav Rev. 2000;24:31–9.
150. Bayless DW, Perez MC, Daniel JM. Comparison of the validity of the use of the spontaneously hypertensive rat as a model of attention deficit hyperactivity disorder in males and females. Behav Brain Res. 2015;286:85–92.
151. Evenden JL. Varieties of impulsivity. Psychopharmacology. 1999;146:348–61.
152. Robbins T. The 5-choice serial reaction time task: behavioural pharmacology and functional neurochemistry. Psychopharmacology. 2002;163:362–80.
153. Navarra R, Graf R, Huang Y, Logue S, Comery T, Hughes Z, et al. Effects of atomoxetine and methylphenidate on attention and impulsivity in the 5-choice serial reaction time test. Prog Neuro-Psychopharmacol Biol Psychiatry. 2008;32:34–41.
154. Umehara M, Ago Y, Kawanai T, Fujita K, Hiramatsu N, Takuma K, et al. Methylphenidate and venlafaxine attenuate locomotion in spontaneously hypertensive rats, an animal model of attention–deficit/hyperactivity disorder, through α2-adrenoceptor activation. Behav Pharmacol. 2013;24:328–31.
155. Sagvolden T. Impulsiveness, overactivity, and poorer sustained attention improve by chronic treatment with low doses of l-amphetamine in an animal model of attention-deficit/hyperactivity disorder (ADHD). Behav Brain Funct. 2011;7:1–10.
156. Caye A, Swanson JM, Coghill D, Rohde LA. Treatment strategies for ADHD: an evidence-based guide to select optimal treatment. Mol Psychiatry. 2018;24(3):390–408.
157. Prossnitz ER, Arterburn JB. International union of basic and clinical pharmacology. XCVII. G protein-coupled estrogen receptor and its pharmacologic modulators. Pharmacol Rev. 2015;67:505–40.
158. Montague D, Weickert CS, Tomaskovic-Crook E, Rothmond DA, Kleinman JE, Rubinow DR. Oestrogen receptor alpha localisation in the prefrontal cortex of three mammalian species. J Neuroendocrinol. 2008;20(7):893–903.
159. Kritzer MF. Regional, laminar, and cellular distribution of immunoreactivity for ERα and ERβ in the cerebral cortex of hormonally intact, adult male and female rats. Cereb Cortex. 2002;12:116–28.
160. Shughrue PJ, Lane MV, Merchenthaler I. Comparative distribution of estrogen receptor-α and -β mRNA in the rat central nervous system. J Comp Neurol. 1997;388:507–25.
161. González M, Cabrera-Socorro A, Pérez-García CG, Fraser JD, López FJ, Alonso R, et al. Distribution patterns of estrogen receptor α and β in the human cortex and hippocampus during development and adulthood. J Comp Neurol. 2007;503:790–802.
162. Ostlund H, Keller E, Hurd YL. Estrogen receptor gene expression in relation to neuropsychiatric disorders. Ann N Y Acad Sci. 2003;1007:54–63.
163. Simerly RB, Swanson LW, Chang C, Muramatsu M. Distribution of androgen and estrogen receptor mRNA-containing cells in the rat brain: an in situ hybridization study. J Comp Neurol. 1990;294:76–95.

164. Gasiorowska A, Wydrych M, Drapich P, Zadrozny M, Steczkowska M, Niewiadomski W, et al. The biology and pathobiology of glutamatergic, cholinergic, and dopaminergic signaling in the aging brain. Front Aging Neurosci. 2021;13:654931.
165. Forester BP, Parikh SV, Weisenbach S, Ajilore O, Vahia I, Rothschild AJ, et al. Combinatorial pharmacogenomic testing improves outcomes for older adults with depression. Am J Geriatr Psychiatry. 2020;28:933–45.
166. Henke H, Lang W. Cholinergic enzymes in neocortex, hippocampus and basal forebrain of non-neurological and senile dementia of Alzheimer-type patients. Brain Res. 1983;267:281–91.
167. Kumar A, Foster TC. Alteration in NMDA receptor mediated glutamatergic neurotransmission in the hippocampus during senescence. Neurochem Res. 2019;44:38–48.
168. Luo Z, Ahlers-Dannen KE, Spicer MM, Yang J, Alberico S, Stevens HE, et al. Age-dependent nigral dopaminergic neurodegeneration and α-synuclein accumulation in RGS6-deficient mice. JCI Insight. 2019;5(13):e126769.
169. Backstrom T, Sanders D, Leask R, Davidson D, Warner P, Bancroft J. Mood, sexuality, hormones, and the menstrual cycle. II. Hormone levels and their relationship to the premenstrual syndrome. Psychosom Med. 1983;45(6):503–7.
170. Benazzi F. Prevalence and clinical features of atypical depression in depressed outpatients: a 467-case study. Psychiatry Res. 1999;86:259–65.
171. Marcus SM, Young EA, Kerber KB, Kornstein S, Farabaugh AH, Mitchell J, et al. Gender differences in depression: findings from the STAR*D study. J Affect Disord. 2005;87:141–50.
172. Marcus SM, Kerber KB, Rush AJ, Wisniewski SR, Nierenberg A, Balasubramani GK, et al. Sex differences in depression symptoms in treatment-seeking adults: confirmatory analyses from the sequenced treatment alternatives to relieve depression study. Compr Psychiatry. 2008;49:238–46.
173. Kornstein SG, Young EA, Harvey AT, Wisniewski SR, Barkin JL, Thase ME, et al. The influence of menopause status and postmenopausal use of hormone therapy on presentation of major depression in women. Menopause. 2010;17(4):828–39.
174. Koss WA, Einat H, Schloesser RJ, Manji HK, Rubinow DR. Estrogen effects on the forced swim test differ in two outbred rat strains. Physiol Behav. 2012;106:81–6.
175. Marvan ML, Chavez-Chavez L, Santana S. Clomipramine modifies fluctuations of forced swimming immobility in different phases of the rat estrous cycle. Arch Med Res. 1996;27:83–6.
176. Padilla E, Barrett D, Shumake J, Gonzalez-Lima F. Strain, sex, and open-field behavior: factors underlying the genetic susceptibility to helplessness. Behav Brain Res. 2009;201:257–64.
177. Shors TJ, Mathew J, Sisti HM, Edgecomb C, Beckoff S, Dalla C. Neurogenesis and helplessness are mediated by controllability in males but not in females. Biol Psychiatry. 2007;62:487–95.
178. Konkle ATM, Baker SL, Kentner AC, Barbagallo LSM, Merali Z, Bielajew C. Evaluation of the effects of chronic mild stressors on hedonic and physiological responses: sex and strain compared. Brain Res. 2003;992:227–38.
179. Jablensky A, McGrath J, Herrman H, Castle D, Gureje O, Evans M, et al. Psychotic disorders in urban areas: an overview of the study on low prevalence disorders. Aust N Z J Psychiatry. 2000;34:221–36.
180. Breslau N, Chilcoat HD, Kessler RC, Peterson EL, Lucia VC. Vulnerability to assaultive violence: further specification of the sex difference in post-traumatic stress disorder. Psychol Med. 1999;29:813–21.
181. Olff M, Langeland W, Draijer N, Gersons BPR. Gender differences in posttraumatic stress disorder. Psychol Bull. 2007;133:183–204.
182. De Jongh R, Geyer MA, Olivier B, Groenink L. The effects of sex and neonatal maternal separation on fear-potentiated and light-enhanced startle. Behav Brain Res. 2005;161:190–6.
183. Maren S, De Oca B, Fanselow MS. Sex differences in hippocampal long-term potentiation (LTP) and Pavlovian fear conditioning in rats: positive correlation between LTP and contextual learning. Brain Res. 1994;661:25–34.

184. Kornstein SG, Schatzberg AF, Thase ME, Yonkers KA, McCullough JP, Keitner GI, et al. Gender differences in treatment response to sertraline versus imipramine in chronic depression. Am J Psychiatr. 2000;157:1445–52.
185. Henkel V, Mergl R, Allgaier AK, Kohnen R, Möller HJ, Hegerl U. Treatment of depression with atypical features: a meta-analytic approach. Psychiatry Res. 2006;141:89–101.
186. Usall J, Suarez D, Haro JM. Gender differences in response to antipsychotic treatment in outpatients with schizophrenia. Psychiatry Res. 2007;153:225–31.
187. Schmidt R, Baumann F, Hanschmann H, Geissler F, Preiss R. Gender difference in ifosfamide metabolism by human liver microsomes. Eur J Drug Metab Pharmacokinet. 2001;26:193–200.
188. Cuzzolin L, Schinella M, Tellini U, Pezzoli L, Lippi U, Benoni G. The effect of sex and cardiac failure on the pharmacokinetics of a slow-release theophylline formulation in the elderly. Pharmacol Res. 1990;22:137–8.
189. Prior TI, Baker GB. Interactions between the cytochrome P450 system and the second-generation antipsychotics. J Psychiatry Neurosci. 2003;28:99–112.
190. Timmer CJ, Sitsen JM, Delbressine LP. Clinical pharmacokinetics of mirtazapine. Clin Pharmacokinet. 2000;38:461–74.
191. Abernethy DR, Greenblatt DJ, Shader RI. Imipramine and desipramine disposition in the elderly. J Pharmacol Exp Ther. 1985;232:183–8.
192. Ilic K, Hawke RL, Thirumaran RK, Schuetz EG, Hull JH, Kashuba ADM, et al. Influences on CYP2B6 activity using bupropion. Drug Metab Dispos. 2013;41:575.
193. Benet LZ, Izumi T, Zhang Y, Silverman JA, Wacher VJ. Intestinal MDR transport proteins and P-450 enzymes as barriers to oral drug delivery. J Control Release. 1999;62:25–31.
194. O'Brien FE, O'Connor RM, Clarke G, Dinan TG, Griffin BT, Cryan JF. P-glycoprotein inhibition increases the brain distribution and antidepressant-like activity of escitalopram in rodents. Neuropsychopharmacology. 2013;38(11):2209–19.
195. Bradford LD. CYP2D6 allele frequency in European Caucasians, Asians, Africans and their descendants. Pharmacogenomics. 2002;3:229–43.
196. Hwang WJ, Lee TY, Kim NS, Kwon JS. The role of estrogen receptors and their Signaling across psychiatric disorders. Int J Mol Sci. 2020;22:373.
197. Kokras N, Dalla C. Sex differences in animal models of psychiatric disorders. Br J Pharmacol. 2014;171:4595–619.
198. Weiner DM, Levey AI, Brann MR. Expression of muscarinic acetylcholine and dopamine receptor mRNAs in rat basal ganglia. Proc Natl Acad Sci. 1990;87:7050–4.
199. Vilaró MT, Palacios JM, Mengod G. Localization of m5 muscarinic receptor mRNA in rat brain examined by in situ hybridization histochemistry. Neurosci Lett. 1990;114:154–9.
200. Yamada M, Lamping KG, Duttaroy A, Zhang W, Cui Y, Bymaster FP, et al. Cholinergic dilation of cerebral blood vessels is abolished in M5 muscarinic acetylcholine receptor knockout mice. Proc Natl Acad Sci. 2001;98:14096–101.
201. Basile AS, Fedorova I, Zapata A, Liu X, Shippenberg T, Duttaroy A, et al. Deletion of the M5 muscarinic acetylcholine receptor attenuates morphine reinforcement and withdrawal but not morphine analgesia. Proc Natl Acad Sci. 2002;99:11452–7.
202. Fink-Jensen A, Fedorova I, Wörtwein G, Woldbye DPD, Rasmussen T, Thomsen M, et al. Role for M5 muscarinic acetylcholine receptors in cocaine addiction. J Neurosci Res. 2003;74:91–6.
203. Steidl S, Yeomans JS. M5 muscarinic receptor knockout mice show reduced morphine-induced locomotion but increased locomotion after cholinergic antagonism in the ventral tegmental area. J Pharmacol Exp Ther. 2009;328:263–75.
204. Katz RJ. Animal model of depression: pharmacological sensitivity of a hedonic deficit. Pharmacol Biochem Behav. 1982;16:965–8.
205. Kessler RC. The effects of stressful life events on depression. Annu Rev Psychol. 1997;48:191–214.

206. Javitt DC, Carter CS, Krystal JH, Kantrowitz JT, Girgis RR, Kegeles LS, et al. Utility of imaging-based biomarkers for glutamate-targeted drug development in psychotic disorders: a randomized clinical trial. JAMA Psychiat. 2018;75:11–9.
207. De Simoni S, Schwarz AJ, O'Daly OG, Marquand AF, Brittain C, Gonzales C, et al. Test–retest reliability of the BOLD pharmacological MRI response to ketamine in healthy volunteers. NeuroImage. 2013;64:75–90.
208. Kantrowitz JT, Grinband J, Goff DC, Lahti AC, Marder SR, Kegeles LS, et al. Proof of mechanism and target engagement of glutamatergic drugs for the treatment of schizophrenia: RCTs of pomaglumetad and TS-134 on ketamine-induced psychotic symptoms and pharmacoBOLD in healthy volunteers. Neuropsychopharmacology. 2020;45(11):1842–50.
209. Littlewood CL, Jones N, O'Neill MJ, Mitchell SN, Tricklebank M, Williams SCR. Mapping the central effects of ketamine in the rat using pharmacological MRI. Psychopharmacology. 2006;186:64–81.
210. Chin CL, Upadhyay J, Marek GJ, Baker SJ, Zhang M, Mezler M, et al. Awake rat pharmacological magnetic resonance imaging as a translational pharmacodynamic biomarker: metabotropic glutamate 2/3 agonist modulation of ketamine-induced blood oxygenation level dependence signals. J Pharmacol Exp Ther. 2011;336:709–15.
211. Wadenberg MLG, Kapur S, Soliman A, Jones C, Vaccarino F. Dopamine D2 receptor occupancy predicts catalepsy and the suppression of conditioned avoidance response behaviour in rats. Psychopharmacology. 2000;150:422–9.
212. Wadenberg MLG. Conditioned avoidance response in the development of new antipsychotics. Curr Pharm Des. 2010;16:358–70.
213. Kapur S, Zipursky R, Jones C, Remington G, Houle S. Relationship between dopamine D2 occupancy, clinical response, and side effects: a double-blind PET study of first-episode schizophrenia. Am J Psychiatr. 2000;157:514–20.
214. Farde L, Nordström AL, Wiesel FA, Pauli S, Halldin C, Sedvall G. Positron emission tomographic analysis of central D1 and D2 dopamine receptor occupancy in patients treated with classical neuroleptics and clozapine: relation to extrapyramidal side effects. Arch Gen Psychiatry. 1992;49:538–44.
215. Pfurtscheller G, Lopes da Silva FH. Event-related EEG/MEG synchronization and desynchronization: basic principles. Clin Neurophysiol. 1999;110:1842–57.
216. Medithe JWC, Nelakuditi UR. Study of normal and abnormal EEG. In: 2016 3rd international conference on advanced computing and communication systems (ICACCS). IEEE; 2016. p. 1–4.
217. Sullivan D, Mizuseki K, Sorgi A, Buzsáki G. Comparison of sleep spindles and theta oscillations in the hippocampus. J Neurosci. 2014;34:662.
218. Morairty SR, Hedley L, Flores J, Martin R, Kilduff TS. Selective 5HT2A and 5HT6 receptor antagonists promote sleep in rats. Sleep. 2008;31:34–44.
219. Bubser M, Byun NE, Wood MR, Jones CK. Muscarinic receptor pharmacology and circuitry for the modulation of cognition. Handb Exp Pharmacol. 2011;208:121–66.
220. Shekhar A, Potter WZ, Lightfoot J, Lienemann J, Dubé S, Mallinckrodt C, et al. Selective muscarinic receptor agonist xanomeline as a novel treatment approach for schizophrenia. Am J Psychiatr. 2008;165:1033–9.
221. Gomeza J, Zhang L, Kostenis E, Felder C, Bymaster F, Brodkin J, et al. Enhancement of D1 dopamine receptor-mediated locomotor stimulation in M(4) muscarinic acetylcholine receptor knockout mice. Proc Natl Acad Sci U S A. 1999;96:10483–8.
222. Brady AE, Jones CK, Bridges TM, Kennedy JP, Thompson AD, Heiman JU, et al. Centrally active allosteric potentiators of the M4 muscarinic acetylcholine receptor reverse amphetamine-induced hyperlocomotor activity in rats. J Pharmacol Exp Ther. 2008;327:941–53.
223. Byun NE, Grannan M, Bubser M, Barry RL, Thompson A, Rosanelli J, et al. Antipsychotic drug-like effects of the selective M4 muscarinic acetylcholine receptor positive allosteric modulator VU0152100. Neuropsychopharmacology. 2014;39:1578–93.

224. Shigemoto R, Mizuno N. Chapter III Metabotropic glutamate receptors — immunocytochemical and in situ hybridization analyses. In: Handbook of chemical neuroanatomy, vol. 18. Elsevier; 2000. p. 63–98.
225. Luján R, Roberts JDB, Shigemoto R, Ohishi H, Somogyi P. Differential plasma membrane distribution of metabotropic glutamate receptors mGluR1α, mGluR2 and mGluR5, relative to neurotransmitter release sites. J Chem Neuroanat. 1997;13:219–41.
226. Kuwajima M, Hall RA, Aiba A, Smith Y. Subcellular and subsynaptic localization of group I metabotropic glutamate receptors in the monkey subthalamic nucleus. J Comp Neurol. 2004;474:589–602.
227. Gasparini F, Lingenhöhl K, Stoehr N, Flor PJ, Heinrich M, Vranesic I, et al. 2-Methyl-6-(phenylethynyl)-pyridine (MPEP), a potent, selective and systemically active mGlu5 receptor antagonist. Neuropharmacology. 1999;38:1493–503.
228. Cosford NDP, Tehrani L, Roppe J, Schweiger E, Smith ND, Anderson J, et al. 3-[(2-Methyl-1,3-thiazol-4-yl)ethynyl]-pyridine: a potent and highly selective metabotropic glutamate subtype 5 receptor antagonist with anxiolytic activity. J Med Chem. 2003;46:204–6.
229. Liu CY, Jiang XX, Zhu YH, Wei DN. Metabotropic glutamate receptor 5 antagonist 2-methyl-6-(phenylethynyl)pyridine produces antidepressant effects in rats: role of brain-derived neurotrophic factor. Neuroscience. 2012;223:219–24.
230. Li X, Need AB, Baez M, Witkin JM. Metabotropic glutamate 5 receptor antagonism is associated with antidepressant-like effects in mice. J Pharmacol Exp Ther. 2006;319:254–9.
231. Lindemann L, Jaeschke G, Michalon A, Vieira E, Honer M, Spooren W, et al. CTEP: a novel, potent, long-acting, and orally bioavailable metabotropic glutamate receptor 5 inhibitor. J Pharmacol Exp Ther. 2011;339:474–86.
232. Lindemann L, Porter RH, Scharf SH, Kuennecke B, Bruns A, Von Kienlin M, et al. Pharmacology of basimglurant (RO4917523, RG7090), a unique metabotropic glutamate receptor 5 negative allosteric modulator in clinical development for depression. J Pharmacol Exp Ther. 2015;353:213–33.
233. Ametamey SM, Kessler LJ, Honer M, Wyss MT, Buck A, Hintermann S, et al. Radiosynthesis and preclinical evaluation of 11 C-ABP688 as a probe for imaging the metabotropic glutamate receptor subtype 5. J Nucl Med. 2006;47:698–705.
234. Steiger A, Pawlowski M. Depression and sleep. Int J Mol Sci. 2019;20(3):607.
235. Mayers AG, Baldwin DS. Antidepressants and their effect on sleep. Hum Psychopharmacol. 2005;20:533–59.
236. Montgomery SA, Asberg M. A new depression scale designed to be sensitive to change. Br J Psychiatry. 1979;134:382–9.
237. Quiroz JA, Tamburri P, Deptula D, Banken L, Beyer U, Rabbia M, et al. Efficacy and safety of basimglurant as adjunctive therapy for major depression: a randomized clinical trial. JAMA Psychiat. 2016;73:675–84.
238. Belzung C. Innovative drugs to treat depression: did animal models fail to be predictive or did clinical trials fail to detect effects. Neuropsychopharmacology. 2014;39:1041–51.
239. McGonigle P, Ruggeri B. Animal models of human disease: challenges in enabling translation. Biochem Pharmacol. 2014;87:162–71.
240. Cuthbert BN, Insel TR. Toward the future of psychiatric diagnosis: the seven pillars of RDoC. BMC Med. 2013;11:126.
241. Insel T, Cuthbert B, Garvey M, Heinssen R, Pine DS, Quinn K, et al. Research domain criteria (RDoC): toward a new classification framework for research on mental disorders. Am J Psychiatr. 2010;167:748–51.
242. Cuthbert BN. Research domain criteria: toward future psychiatric nosologies. Dialogues Clin Neurosci. 2015;17:89–97.
243. Hyman SE. Use of mouse models to investigate the contributions of CNVs associated with schizophrenia and autism to disease mechanisms. Curr Opin Genet Dev. 2021;68:99–105.

244. Mali P, Esvelt KM, Church GM. Cas9 as a versatile tool for engineering biology. Nat Methods. 2013;10:957–63.
245. Jinek M, Chylinski K, Fonfara I, Hauer M, Doudna JA, Charpentier E. A programmable dual-RNA–guided DNA endonuclease in adaptive bacterial immunity. Science. 2012;337:816–21.
246. Platt RJ, Chen S, Zhou Y, Yim MJ, Swiech L, Kempton HR, et al. CRISPR-Cas9 knockin mice for genome editing and cancer modeling. Cell. 2014;159:440–55.
247. Zhao H, Tu Z, Xu H, Yan S, Yan H, Zheng Y, et al. Altered neurogenesis and disrupted expression of synaptic proteins in prefrontal cortex of SHANK3-deficient non-human primate. Cell Res. 2017;27:1293–7.
248. Moessner R, Marshall CR, Sutcliffe JS, Skaug J, Pinto D, Vincent J, et al. Contribution of SHANK3 mutations to autism spectrum disorder. Am J Hum Genet. 2007;81:1289–97.
249. Tu Z, Zhao H, Li B, Yan S, Wang L, Tang Y, et al. CRISPR/Cas9-mediated disruption of SHANK3 in monkey leads to drug-treatable autism-like symptoms. Hum Mol Genet. 2019;28:561–71.
250. Boyden ES, Zhang F, Bamberg E, Nagel G, Deisseroth K. Millisecond-timescale, genetically targeted optical control of neural activity. Nat Neurosci. 2005;8:1263–8.
251. Hong SI, Kang S, Chen JF, Choi DS. Indirect medium spiny neurons in the dorsomedial striatum regulate ethanol-containing conditioned reward seeking. J Neurosci. 2019;39:7206.
252. Chow BY, Han X, Dobry AS, Qian X, Chuong AS, Li M, et al. High-performance genetically targetable optical neural silencing by light-driven proton pumps. Nature. 2010;463:98–102.
253. Rost BR, Schneider-Warme F, Schmitz D, Hegemann P. Optogenetic tools for subcellular applications in neuroscience. Neuron. 2017;96:572–603.
254. Tye KM, Prakash R, Kim SY, Fenno LE, Grosenick L, Zarabi H, et al. Amygdala circuitry mediating reversible and bidirectional control of anxiety. Nature. 2011;471:358–62.
255. Carter ME, Yizhar O, Chikahisa S, Nguyen H, Adamantidis A, Nishino S, et al. Tuning arousal with optogenetic modulation of locus coeruleus neurons. Nat Neurosci. 2010;13:1526–33.
256. Kravitz AV, Tye LD, Kreitzer AC. Distinct roles for direct and indirect pathway striatal neurons in reinforcement. Nat Neurosci. 2012;15:816–8.
257. Wentz CT, Oettl LL, Kelsch W. Optogenetics in psychiatric animal models. Cell Tissue Res. 2013;354(1):61–8.
258. Leopold AV, Shcherbakova DM, Verkhusha VV. Fluorescent biosensors for neurotransmission and neuromodulation: engineering and applications. Front Cell Neurosci. 2019;13:474.
259. Patriarchi T, Cho JR, Merten K, Howe MW, Marley A, Xiong WH, et al. Ultrafast neuronal imaging of dopamine dynamics with designed genetically encoded sensors. Science [Internet]. 2018 [cited 2022 June 19];360(6396):eaat4422. Available from: https://www.ncbi.nlm.nih.gov/pmc/articles/PMC6287765/.
260. Mohebi A, Pettibone JR, Hamid AA, Wong JMT, Vinson LT, Patriarchi T, et al. Dissociable dopamine dynamics for learning and motivation. Nature. 2019;570:65–70.
261. Sternson SM, Roth BL. Chemogenetic tools to interrogate brain functions. Annu Rev Neurosci. 2014;37:387–407.
262. Urban DJ, Roth BL. DREADDs (designer receptors exclusively activated by designer drugs): chemogenetic tools with therapeutic utility. Annu Rev Pharmacol Toxicol. 2015;55:399–417.
263. Armbruster BN, Li X, Pausch MH, Herlitze S, Roth BL. Evolving the lock to fit the key to create a family of G protein-coupled receptors potently activated by an inert ligand. Proc Natl Acad Sci U S A. 2007;104(12):5163–8.
264. Alexander GM, Rogan SC, Abbas AI, Armbruster BN, Pei Y, Allen JA, et al. Remote control of neuronal activity in transgenic mice expressing evolved G protein-coupled receptors. Neuron. 2009;63:27–39.
265. Pati S, Saba K, Salvi SS, Tiwari P, Chaudhari PR, Verma V, et al. Chronic postnatal chemogenetic activation of forebrain excitatory neurons evokes persistent changes in mood behavior. Cheer JF, Wassum KM, Bolton J, Eisch AJ, editors. elife. 2020;9:e56171–1.
266. Mank M, Griesbeck O. Genetically encoded calcium indicators. Chem Rev. 2008;108:1550–64.

267. Nakai J, Ohkura M, Imoto K. A high signal-to-noise Ca2+ probe composed of a single green fluorescent protein. Nat Biotechnol. 2001;19:137–41.
268. Ohkura M, Matsuzaki M, Kasai H, Imoto K, Nakai J. Genetically encoded bright Ca2+ probe applicable for dynamic Ca2+ imaging of dendritic spines. Anal Chem. 2005;77:5861–9.
269. Sato M, Kawano M, Ohkura M, Gengyo-Ando K, Nakai J, Hayashi Y. Generation and imaging of transgenic mice that express G-CaMP7 under a tetracycline response element. PLoS One. 2015;10:e0125354.
270. Chen X, Sobczak F, Chen Y, Jiang Y, Qian C, Lu Z, et al. Mapping optogenetically-driven single-vessel fMRI with concurrent neuronal calcium recordings in the rat hippocampus. Nat Commun. 2019;10:1–12.
271. Cramer JV, Gesierich B, Roth S, Dichgans M, Düring M, Liesz A. In vivo widefield calcium imaging of the mouse cortex for analysis of network connectivity in health and brain disease. NeuroImage. 2019;199:570–84.
272. Alexandrov V, Brunner D, Hanania T, Leahy E. High-throughput analysis of behavior for drug discovery. Eur J Pharmacol. 2015;750:82–9.
273. Brunner D, Nestler E, Leahy E. In need of high-throughput behavioral systems. Drug Discov Today. 2002;7(18 Suppl):S107–12.
274. Dedic N, Jones PG, Hopkins SC, Lew R, Shao L, Campbell JE, et al. SEP-363856, a novel psychotropic agent with a unique, non-D 2 receptor mechanism of action. J Pharmacol Exp Ther. 2019;371:1–14.
275. Koblan KS, Kent J, Hopkins SC, Krystal JH, Cheng H, Goldman R, et al. A non-D2-receptor-binding drug for the treatment of schizophrenia. N Engl J Med. 2020;382:1497–506.
276. Seligman MEP. Learned helplessness. Annu Rev Med. 1972;23:407–12.
277. Katz RJ, Roth KA, Carroll BJ. Acute and chronic stress effects on open field activity in the rat: implications for a model of depression. Neurosci Biobehav Rev. 1981;5:247–51.
278. Crema L, Schlabitz M, Tagliari B, Cunha A, Simão F, Krolow R, et al. Na+, K+ ATPase activity is reduced in amygdala of rats with chronic stress-induced anxiety-like behavior. Neurochem Res. 2010;35:1787–95.
279. Qi G, Zhang P, Li T, Li M, Zhang Q, He F, et al. NAc-VTA circuit underlies emotional stress-induced anxiety-like behavior in the three-chamber vicarious social defeat stress mouse model. Nat Commun. 2022;13:1–19.
280. Pinna G, Agis-Balboa RC, Zhubi A, Matsumoto K, Grayson DR, Costa E, et al. Imidazenil and diazepam increase locomotor activity in mice exposed to protracted social isolation. Proc Natl Acad Sci U S A. 2006;103:4275–80.
281. Mumtaz F, Khan MI, Zubair M, Dehpour AR. Neurobiology and consequences of social isolation stress in animal model- a comprehensive review. Biomed Pharmacother. 2018;105:1205–22.
282. Rilke O, Will K, Jähkel M, Oehler J. Behavioral and neurochemical effects of anpirtoline and citalopram in isolated and group housed mice. Prog Neuro-Psychopharmacol Biol Psychiatry. 2001;25:1125–44.
283. Petty F, Chae Y, lae, Kramer G, Jordan S, Wilson LA. Learned helplessness sensitizes hippocampal norepinephrine to mild restress. Biol Psychiatry. 1994;35:903–8.
284. Su CL, Su CW, Hsiao YH, Gean PW. Epigenetic regulation of BDNF in the learned helplessness-induced animal model of depression. J Psychiatr Res. 2016;76:101–10.
285. Sekine Y, Suzuki K, Ramachandran PV, Blackburn TP, Ashby CR. Acute and repeated administration of fluoxetine, citalopram, and paroxetine significantly alters the activity of midbrain dopamine neurons in rats: an in vivo electrophysiological study. Synapse. 2007;61:72–7.
286. Schwienteck KL, Li G, Poe MM, Cook JM, Banks ML, Negus S, S. Abuse-related effects of subtype-selective GABA A receptor positive allosteric modulators in an assay of intracranial self-stimulation in rats. Psychopharmacology. 2017;234:2091–101.
287. Vogel G, Neill D, Hagler M, Kors D, Hartley P. Decreased intracranial self-stimulation in a new animal model of endogenous depression. Neurosci Biobehav Rev. 1990;14:65–8.

288. Park J, Bucher ES, Fontillas K, Owesson-White C, Ariansen JL, Carelli RM, et al. Opposing catecholamine changes in the bed nucleus of the stria terminalis during intracranial self-stimulation and its extinction. Biol Psychiatry. 2013;74:69–76.
289. Ye Y, Yao S, Wang R, Fang Z, Zhong K, Nie L, et al. PI3K/Akt/NF-κB signaling pathway regulates behaviors in adolescent female rats following with neonatal maternal deprivation and chronic mild stress. Behav Brain Res. 2019;362:199–207.
290. Bourke CH, Stowe ZN, Neigh GN, Olson DE, Owens MJ. Prenatal exposure to escitalopram and/or stress in rats produces limited effects on endocrine, behavioral, or gene expression measures in adult male rats. Neurotoxicol Teratol. 2013;39:100–9.
291. Ramanathan M, Ashok Kumar SN, Suresh B. Evaluation of cognitive function of fluoxetine, sertraline and tianeptine in isolation and chronic unpredictable mild stress-induced depressive Wistar rats - PubMed. Indian J Exp Biol. 2003;41:1269–72.
292. Chiba S, Numakawa T, Ninomiya M, Richards MC, Wakabayashi C, Kunugi H. Chronic restraint stress causes anxiety- and depression-like behaviors, downregulates glucocorticoid receptor expression, and attenuates glutamate release induced by brain-derived neurotrophic factor in the prefrontal cortex. Prog Neuropsychopharmacol Biol Psychiatry. 2012;39:112–9.
293. McKlveen JM, Myers B, Flak JN, Bundzikova J, Solomon MB, Seroogy KB, et al. Role of prefrontal cortex glucocorticoid receptors in stress and emotion. Biol Psychiatry. 2013;74:672–9.
294. Skelly MJ, Chappell AE, Carter E, Weiner JL. Adolescent social isolation increases anxiety-like behavior and ethanol intake and impairs fear extinction in adulthood: possible role of disrupted noradrenergic signaling. Neuropharmacology. 2015;97:149.
295. Barnes TD, Rieger MA, Dougherty JD, Holy TE. Group and individual variability in mouse pup isolation calls recorded on the same day show stability. Front Behav Neurosci. 2017;11:243.
296. Huang Q, Zhou Y, Liu LY. Effect of post-weaning isolation on anxiety- and depressive-like behaviors of C57BL/6J mice. Exp Brain Res. 2017;235:2893–9.
297. Yorgason JT, España RA, Konstantopoulos JK, Weiner JL, Jones SR. Enduring increases in anxiety-like behavior and rapid nucleus accumbens dopamine signaling in socially isolated rats. Eur J Neurosci. 2013;37:1022–31.
298. Galea LAM, McEwen BS, Tanapat P, Deak T, Spencer RL, Dhabhar FS. Sex differences in dendritic atrophy of CA3 pyramidal neurons in response to chronic restraint stress. Neuroscience. 1997;81:689–97.
299. Pellow S, Chopin P, File SE, Briley M. Validation of open:closed arm entries in an elevated plus-maze as a measure of anxiety in the rat. J Neurosci Methods. 1985;14:149–67.
300. Prut L, Belzung C. The open field as a paradigm to measure the effects of drugs on anxiety-like behaviors: a review. Eur J Pharmacol. 2003;463:3–33.
301. Hall C, Ballachey EL. A study of the rat's behavior in a field. A contribution to method in comparative psychology. Univ Calif Publ Psychol. 1932;6:1–12.
302. Xu Y, Ma L, Jiang W, Li Y, Wang G, Li R. Study of sex differences in duloxetine efficacy for depression in transgenic mouse models. Front Cell Neurosci. 2017;11:344.
303. Flores-Ramirez FJ, Themann A, Sierra-Fonseca JA, Garcia-Carachure I, Castillo SA, Rodriguez M, et al. Adolescent fluoxetine treatment mediates a persistent anxiety-like outcome in female C57BL/6 mice that is ameliorated by fluoxetine re-exposure in adulthood. Sci Rep. 2021;11:7758.
304. Wu YP, Gao HY, Ouyang SH, Kurihara H, He RR, Li YF. Predator stress-induced depression is associated with inhibition of hippocampal neurogenesis in adult male mice. Neural Regen Res. 2019;14:298–305.
305. Wang H, Zuo D, He B, Qiao F, Zhao M, Wu Y. Conditioned fear stress combined with single-prolonged stress: a new PTSD mouse model. Neurosci Res. 2012;73:142–52.
306. Xu JN, Chen LF, Su J, Liu ZL, Chen J, Lin QF, et al. The anxiolytic-like effects of estazolam on a PTSD animal model. Psychiatry Res. 2018;269:529–35.

307. Piggott VM, Bosse KE, Lisieski MJ, Strader JA, Stanley JA, Conti AC, et al. Single-prolonged stress impairs prefrontal cortex control of amygdala and striatum in rats. Front Behav Neurosci. 2019;13:18.
308. Nedelcovych MT, Gould RW, Zhan X, Bubser M, Gong X, Grannan M, et al. A rodent model of traumatic stress induces lasting sleep and quantitative electroencephalographic disturbances. ACS Chem Neurosci. 2015;6:485–93.
309. Kvetňanský R, Mikulaj L. Adrenal and urinary catecholamines in rats during adaptation to repeated immobilization stress. Endocrinology. 1970;87:738–43.
310. van der Kolk B, Greenberg M, Boyd H, Krystal J. Inescapable shock, neurotransmitters, and addiction to trauma: toward a psychobiology of post traumatic stress. Biol Psychiatry. 1985;20:314–25.
311. Verbitsky A, Dopfel D, Zhang N. Rodent models of post-traumatic stress disorder: behavioral assessment. Transl Psychiatry. 2020;10:132.
312. Berton O, McClung CA, DiLeone RJ, Krishnan V, Renthal W, Russo SJ, et al. Essential role of BDNF in the mesolimbic dopamine pathway in social defeat stress. Science (1979). 2006;311:864–8.
313. Gellhorn E. Interruption of behavior, inescapable shock, and experimental neurosis: a neurophysiologic analysis. Cond Reflex. 1967;2:285–93.
314. Bentefour Y, Bennis M, Garcia R, M'Hamed SB. Effects of paroxetine on PTSD-like symptoms in mice. Psychopharmacology. 2015;232:2303–12.
315. Dahlhoff M, Siegmund A, Golub Y, Wolf E, Holsboer F, Wotjak CT. AKT/GSK-3beta/beta-catenin signalling within hippocampus and amygdala reflects genetically determined differences in posttraumatic stress disorder like symptoms. Neuroscience. 2010;169:1216–26.
316. Gao J, Wang H, Liu Y, Li YY, Chen C, Liu LM, et al. Glutamate and GABA imbalance promotes neuronal apoptosis in hippocampus after stress. Med Sci Monit. 2014;20:499–512.
317. Adamec RE, Shallow T. Lasting effects on rodent anxiety of a single exposure to a cat. Physiol Behav. 1993;54:101–9.
318. Brad Wilson C, McLaughlin LD, Ebenezer PJ, Nair AR, Dange R, Harre JG, et al. Differential effects of sertraline in a predator exposure animal model of post-traumatic stress disorder. Front Behav Neurosci. 2014;8:256.
319. Wilson CB, Ebenezer PJ, McLaughlin LD, Francis J. Predator exposure/psychosocial stress animal model of post-traumatic stress disorder modulates neurotransmitters in the rat hippocampus and prefrontal cortex. PLoS One. 2014;9:e89104.
320. Goswami S, Rodríguez-Sierra O, Cascardi M, Paré D. Animal models of post-traumatic stress disorder: face validity. Front Neurosci. 2013;7:89.
321. Bourin M, Hascoët M. The mouse light/dark box test. Eur J Pharmacol. 2003;463:55–65.
322. Eagle AL, Fitzpatrick CJ, Perrine SA. Single prolonged stress impairs social and object novelty recognition in rats. Behav Brain Res. 2013;256:591–7.
323. Cohen SJ, Stackman RW. Assessing rodent hippocampal involvement in the novel object recognition task. A review. Behav Brain Res. 2015;285:105–17.
324. Philbert J, Beeské S, Belzung C, Griebel G. The CRF_1 receptor antagonist SSR125543 prevents stress-induced long-lasting sleep disturbances in a mouse model of PTSD: comparison with paroxetine and d-cycloserine. Behav Brain Res. 2015;279:41–6.
325. Shadli SM, Ando LC, McIntosh J, Lodhia V, Russell BR, Kirk IJ, et al. Right frontal anxiolytic-sensitive EEG 'theta' rhythm in the stop-signal task is a theory-based anxiety disorder biomarker. Sci Rep. 2021;11:1–12.
326. McKillop LE, Fisher SP, Milinski L, Krone LB, Vyazovskiy VV. Diazepam effects on local cortical neural activity during sleep in mice. Biochem Pharmacol. 2021;191:114515.
327. Troakes C, Ingram CD. Anxiety behaviour of the male rat on the elevated plus maze: associated regional increase in c-fos mRNA expression and modulation by early maternal separation. Stress. 2009;12:362–9.

328. Badiani A, Oates MM, Day HEW, Watson SJ, Akil H, Robinson TE. Amphetamine-induced behavior, dopamine release, and c-fos mRNA expression: modulation by environmental novelty. J Neurosci. 1998;18:10579–93.
329. Cochran SM, Kennedy M, McKerchar CE, Steward LJ, Pratt JA, Morris BJ. Induction of metabolic hypofunction and neurochemical deficits after chronic intermittent exposure to phencyclidine: differential modulation by antipsychotic drugs. Neuropsychopharmacology. 2003;28:265–75.
330. Lipska BK, Weinberger DR. Delayed effects of neonatal hippocampal damage on haloperidol-induced catalepsy and apomorphine-induced stereotypic behaviors in the rat. Brain Res Dev Brain Res. 1993;75:213–22.
331. Lipska BK, Weinberger DR. Subchronic treatment with haloperidol and clozapine in rats with neonatal excitotoxic hippocampal damage. Neuropsychopharmacology. 1994;10:199–205.
332. Lipska BK, Lerman DN, Khaing ZZ, Weickert CS, Weinberger DR. Gene expression in dopamine and GABA systems in an animal model of schizophrenia: effects of antipsychotic drugs. Eur J Neurosci. 2003;18:391–402.
333. Le Pen G, Moreau JL. Disruption of prepulse inhibition of startle reflex in a neurodevelopmental model of schizophrenia: reversal by clozapine, olanzapine and risperidone but not by haloperidol. Neuropsychopharmacology. 2002;27:1–11.
334. O'Donnell P. Cortical disinhibition in the neonatal ventral hippocampal lesion model of schizophrenia: new vistas on possible therapeutic approaches. Pharmacol Ther. 2012;133:19–25.
335. Fiore M, Di Fausto V, Aloe L. Clozapine or haloperidol in rats prenatally exposed to methylazoxymethanol, a compound inducing entorhinal-hippocampal deficits, alter brain and blood neurotrophins' concentrations - PubMed. Ann Ist Super Sanita. 2008;44:167–77.
336. Lavin A, Moore HM, Grace AA. Prenatal disruption of neocortical development alters prefrontal cortical neuron responses to dopamine in adult rats. Neuropsychopharmacology. 2005;30:1426–35.
337. Lodge DJ, Behrens MM, Grace AA. A loss of parvalbumin-containing interneurons is associated with diminished oscillatory activity in an animal model of schizophrenia. J Neurosci. 2009;29:2344–54.
338. Stevens KE, Johnson RG, Rose GM. Rats reared in social isolation show schizophrenia-like changes in auditory gating. Pharmacol Biochem Behav. 1997;58:1031–6.
339. Lapiz MD, Fulford A, Muchimapura S, Mason R, Parker T, Marsden CA. Influence of post-weaning social isolation in the rat on brain development, conditioned behaviour and neurotransmission. Rossiiskii fiziologicheskii zhurnal imeni IM Sechenova/Rossiĭskaia akademiia nauk. 2001;87:730–51.
340. Fone KCF, Porkess MV. Behavioural and neurochemical effects of post-weaning social isolation in rodents-relevance to developmental neuropsychiatric disorders. Neurosci Biobehav Rev. 2008;32:1087–102.
341. Heidbreder CA, Foxton R, Cilia J, Hughes ZA, Shah AJ, Atkins A, et al. Increased responsiveness of dopamine to atypical, but not typical antipsychotics in the medial prefrontal cortex of rats reared in isolation. Psychopharmacology. 2001;156:338–51.
342. Karayiorgou M, Morris MA, Morrow B, Shprintzen RJ, Goldberg R, Borrow J, et al. Schizophrenia susceptibility associated with interstitial deletions of chromosome 22q11. Proc Natl Acad Sci U S A. 1995;92:7612–6.
343. Xu B, Roos JL, Levy S, Van Rensburg EJ, Gogos JA, Karayiorgou M. Strong association of de novo copy number mutations with sporadic schizophrenia. Nat Genet. 2008;40:880–5.
344. Campbell LE, Daly E, Toal F, Stevens A, Azuma R, Catani M, et al. Brain and behaviour in children with 22q11.2 deletion syndrome: a volumetric and voxel-based morphometry MRI study. Brain. 2006;129:1218–28.
345. Simon TJ, Bish JP, Bearden CE, Ding L, Ferrante S, Nguyen V, et al. A multilevel analysis of cognitive dysfunction and psychopathology associated with chromosome 22q11.2 deletion syndrome in children. Dev Psychopathol. 2005;17:753–84.

346. Gothelf D, Law AJ, Frisch A, Chen J, Zarchi O, Michaelovsky E, et al. Biological effects of COMT haplotypes and psychosis risk in 22q11.2 deletion syndrome. Biol Psychiatry. 2014;75:406–13.
347. Didriksen M, Fejgin K, Nilsson SRO, Birknow MR, Grayton HM, Larsen PH, et al. Persistent gating deficit and increased sensitivity to NMDA receptor antagonism after puberty in a new mouse model of the human 22q11.2 microdeletion syndrome: a study in male mice. J Psychiatry Neurosci. 2017;42:48–58.
348. Tripathi A, Spedding M, Schenker E, Didriksen M, Cressant A, Jay TM. Cognition- and circuit-based dysfunction in a mouse model of 22q11.2 microdeletion syndrome: effects of stress. Transl Psychiatry. 2020;10:1–15.
349. Hodgkinson CA, Goldman D, Jaeger J, Persaud S, Kane JM, Lipsky RH, et al. Disrupted in schizophrenia 1 (DISC1): association with schizophrenia, schizoaffective disorder, and bipolar disorder. Am J Hum Genet. 2004;75:862–72.
350. Kamiya A, Kubo KI, Tomoda T, Takaki M, Youn R, Ozeki Y, et al. A schizophrenia-associated mutation of DISC1 perturbs cerebral cortex development. Nat Cell Biol. 2005;7:1067–78.
351. Niwa M, Kamiya A, Murai R, Kubo K, Gruber AJ, Tomita K, et al. Knockdown of DISC1 by in utero gene transfer disturbs postnatal dopaminergic maturation in the frontal cortex and leads to adult behavioral deficits. Neuron. 2010;65:480–9.
352. Andersen MP, Pouzet B. Effects of acute versus chronic treatment with typical or atypical antipsychotics on d-amphetamine-induced sensorimotor gating deficits in rats. Psychopharmacology. 2001;156:291–304.
353. Mavrikaki M, Nomikos GG, Panagis G. Efficacy of the atypical antipsychotic aripiprazole in d-amphetamine-based preclinical models of mania. Int J Neuropsychopharmacol. 2010;13:541–8.
354. Kitaichi K, Yamada K, Hasegawa T, Nabeshima T, Furukawa H. Effects of risperidone on phencyclidine-induced behaviors: comparison with haloperidol and ritanserin. Jpn J Pharmacol. 1994;66:181–9.
355. Hoffman DC. Typical and atypical neuroleptics antagonize MK-801-induced locomotion and stereotypy in rats. J Neural Transm Gen Sect. 1992;89:1–10.
356. Minnaard AM, Ramakers GMJ, Vanderschuren LJMJ, Lesscher HMB. Baclofen and naltrexone, but not N-acetylcysteine, affect voluntary alcohol drinking in rats regardless of individual levels of alcohol intake. Behav Pharmacol. 2021;32:251–7.
357. Mello NK, Negus SS. Preclinical evaluation of pharmacotherapies for treatment of cocaine and opioid abuse using drug self-administration procedures. Neuropsychopharmacology. 1996;14:375.
358. Weeks JR. Experimental morphine addiction: method for automatic intravenous injections in unrestrained rats. Science. 1979;1962(138):143–4.
359. Leri F, Tremblay A, Sorge RE, Stewart J. Methadone maintenance reduces heroin- and cocaine-induced relapse without affecting stress-induced relapse in a rodent model of polydrug use. Neuropsychopharmacology. 2004;29:1312–20.
360. Mello NK, Mendelson JH, Bree MP, Lukas SE. Buprenorphine and naltrexone effects on cocaine self-administration by rhesus monkeys. J Pharmacol Exp Ther. 1990;254:926–39.
361. Mello NK, Lukas SE, Mendelson JH, Drieze J. Naltrexone–buprenorphine interactions: effects on cocaine self-administration. Neuropsychopharmacology. 1993;9:211–24.
362. Jordan CJ, Cao J, Newman AH, Xi ZX. Progress in agonist therapy for substance use disorders: lessons learned from methadone and buprenorphine. Neuropharmacology. 2019;158:107609.
363. Beach HD. Morphine addiction in rats. Can J Psychol. 1957;11:104–12.
364. Bespalov AY, Tokarz ME, Bowen SE, Balster RL, Bear. Effects of test conditions on the outcome of place conditioning with morphine and naltrexone in mice. Psychopharmacology. 1999;141:118–22.
365. Houj M, Bisaga A, Popik P. Conditioned rewarding effects of morphine and methadone in mice pre-exposed to cocaine. Pharmacol Rep. 2013;65:1176–84.

366. Myers MM, Musty RE, Hendley ED. Attenuation of hyperactivity in the spontaneously hypertensive rat by amphetamine. Behav Neural Biol. 1982;34:42–54.
367. Sagvolden T. The alpha-2A adrenoceptor agonist guanfacine improves sustained attention and reduces overactivity and impulsiveness in an animal model of attention-deficit/hyperactivity disorder (ADHD). Behav Brain Funct. 2006;2:41.
368. Miller EM, Pomerleau F, Huettl P, Russell VA, Gerhardt GA, Glaser PEA. The spontaneously hypertensive and Wistar Kyoto rat models of ADHD exhibit sub-regional differences in dopamine release and uptake in the striatum and nucleus accumbens. Neuropharmacology. 2012;63:1327–34.
369. Zhuang X, Oosting RS, Jones SR, Gainetdinov RR, Miller GW, Caron MG, et al. Hyperactivity and impaired response habituation in hyperdopaminergic mice. Proc Natl Acad Sci U S A. 2001;98:1982–7.
370. Hardung S, Jäckel Z, Diester I. Prefrontal contributions to action control in rodents. Int Rev Neurobiol. 2021;158:373–93.
371. Moon SJ, Kim CJ, Lee YJ, Hong M, Han J, Bahn GH. Effect of atomoxetine on hyperactivity in an animal model of attention-deficit/hyperactivity disorder (ADHD). PLoS One. 2014;9:e108918.

Chapter 4
Discovery and Development of Monoamine Transporter Ligands

Shaili Aggarwal and Ole Valente Mortensen

Abstract Monoamine transporters (MATs) are targets of a wide range of compounds that have been developed as therapeutic treatments for various neuropsychiatric and neurodegenerative disorders such as depression, ADHD, neuropathic pain, anxiety disorders, stimulant use disorders, epilepsy, and Parkinson's disease. The MAT family is comprised of three main members – the dopamine transporter (DAT), the norepinephrine transporter (NET), and the serotonin transporter (SERT). These transporters are through reuptake responsible for the clearance of their respective monoamine substrates from the extracellular space. The determination of X-ray crystal structures of MATs and their homologues bound with various substrates and ligands has resulted in a surge of structure-function-based studies of MATs to understand the molecular basis of transport function and the mechanism of various ligands that ultimately result in their behavioral effects. This review focusses on recent examples of ligand-based structure-activity relationship studies trying to overcome some of the challenges associated with previously developed MAT inhibitors. These studies have led to the discovery of unique and novel structurally diverse MAT ligands including allosteric modulators. These novel molecular scaffolds serve as leads for designing more effective therapeutic interventions by modulating the activities of MATs and ultimately their associated neurotransmission and behavioral effects.

Keywords Dopamine · Serotonin · Norepinephrine · Antidepressants · Psychostimulants · Allosteric modulators

---

S. Aggarwal (✉) · O. V. Mortensen (✉)
Department of Pharmacology and Physiology, Drexel University College of Medicine, Philadelphia, PA, USA
e-mail: sa3295@drexel.edu; ovm23@drexel.edu

M. Macaluso et al. (eds.), *Drug Development in Psychiatry*, Advances in Neurobiology 30,
https://doi.org/10.1007/978-3-031-21054-9_4

4.1 Introduction and Overview of Monoamine Transporters

The monoamine neurotransmitter transporters (MATs) are part of the solute carrier 6 (SLC6) family of transporters [1, 2] and are namely the dopamine transporter (DAT, SLC6A3), the norepinephrine transporter (NET, SLC6A2), and the serotonin transporter (SERT, SLC6A4) which mediate the reuptake of monoamine neurotransmitters dopamine (DA), norepinephrine (NE), and serotonin (5-HT), respectively, from the extracellular space into the intracellular presynaptic compartment. The MATs are membrane-bound transporters expressed in presynaptic neuronal terminals of their respective pathways within the CNS and mediate a rapid reuptake of neurotransmitters from the synaptic cleft into the presynaptic neurons, where the neurotransmitters are sequestered into synaptic vesicles (via vesicular monoamine transporters or VMAT) for recycling and release or are degraded by monoamine oxidase enzymes. The transport of substrates by the transporters is favored by the energy gradient produced by the movement of Na^+ ions into the cell and therefore driven by the concentration gradient created by Na^+/K^+ ATPase. DAT and NET transport one dopamine or norepinephrine molecule along with 2 Na^+ ions and one Cl^- ion, whereas SERT cotransports one 5-HT molecule with one Na^+ and one Cl^- along with the counter-transport of one K^+. Thus, the MATs are sometimes also referred to as Na^+/ Cl^--symporters [3].

4.2 Therapeutic Relevance of MATs

Since MATs control the signal amplitude and duration of monoaminergic neurotransmission, they remain a prominent therapeutic target of interest for the treatment of several neurological disorders [4]. In addition, the MATs are also the primary targets of action of a number of psychostimulants and recreational drugs of abuse such as cocaine, methamphetamine, 3,4-methylenedioxy-methamphetamine ("ecstasy" or MDMA), cathinones (or "bath salts") which all either block or reverse the transport of neurotransmitters and as a result increase the synaptic neurotransmission leading to stimulatory effects.

Tricyclic antidepressants (e.g., clomipramine and amitriptyline) and the selective inhibitors of SERT (also known as selective serotonin-reuptake inhibitors or SSRIs) such as fluoxetine, sertraline and paroxetine are commonly prescribed for depression, anxiety, and panic disorders. Bupropion, a DAT inhibitor, is an antidepressant and smoking cessation aid. Methylphenidate, another DAT-inhibitor is marketed for attention-deficit hyperactivity disorder (ADHD). Reboxetine is used in depression, panic disorder, and ADHD. Different from targeting the inhibition of MATs for use in mood- or cognition-related disorders, synthetic compounds that act as non-selective substrates of MATs and promote monoamine efflux have also found application for clinical use. For example, mixed amphetamine salts (racemic mixture of amphetamine isomers) are prescribed in low doses to treat ADHD and are commonly used off-label to improve cognition and to help in narcolepsy by promoting wakefulness. However, such drugs along with many other cognition-enhancing drugs that interact with MATs are strictly controlled and regulated because of their abuse liability.

4.3 Structural Insights and Transport Mechanism

Breakthrough discoveries such as crystal structures of homologous bacterial (*Aquifex aeolicus*) leucine transporter (LeuT) [5–7], *Drosophila melanogaster* dopamine transporter (dDAT) [8–10] and human serotonin transporter (hSERT) [11] have been pivotal in enhancing our understanding of the structural biology of the MATs and in guiding further studies to elucidate underlying molecular mechanisms of ligand transport and interaction.

The X-ray crystal structure of the bacterial leucine transporter (LeuT) from *Aquifex aeolicus* that shares 20–25% overall homology with the MATs [6] gave insights into the tertiary arrangement of the transporters and their functioning, especially of the core region which shares ~60% homology with the MATs. The co-crystal structures of *Drosophila melanogaster* DAT (dDAT) bound to the substrate DA, D-amphetamine, methamphetamine, cocaine, and many other ligands which has 50–55% homology with the human MATs [8–10] and the human SERT in complex with paroxetine and escitalopram [11–13] further enhanced our understanding of mechanism of transport and ligand induced structural conformations of MATs.

The structure of all MATs is comprised of 12 alpha-helical transmembrane spanning domains connected with flexible intracellular and extracellular loops. The N- and C-termini lie in the intracellular region. The high affinity primary substrate binding site, also referred to as the S1 or orthosteric site, lies at the core of the translocation pathway located between TM1 and TM6. The S1 site holds a substrate molecule along with one or two Na^+ ions. The S1 pocket is comprised of a hydrophobic region, which holds the aromatic substituents of the substrates, and a hydrophilic or a polar region that contains a conserved aspartate residue (SERT: D98, DAT: D79, NET: D75) to form ionic interactions with the amine group of the substrates. The transporters translocate the substrate via a three-state "alternating access" mechanism through sequential binding and conformational changes [1]. According to the "alternating access" mechanism, the transporter adopts distinct conformations where the access to the central binding site is alternatingly sealed from the extracellular and the intracellular side facilitated by the opening and closing of the gating networks present above and below the S1 site. The substrate and the ions bind to the transporter when it adopts an outward-open conformation, which further triggers the closing of the transporter, and thereby occluding the substrate and the ions from either side of the membrane. Next, the occluded state is followed by the opening of the intracellular gate, leading to an inward-facing conformation, which releases the ions and substrate into the cytoplasm via diffusion facilitated by hydration of the substrate binding site. The recently reported SERT cryo-EM structure in complex with ibogaine provided further insights into the structural rearrangements that occur during the translocation cycle. Ibogaine is a hallucinogenic natural product with psychoactive and anti-addictive properties and a non-competitive inhibitor of transport. It is bound within the S1 site of SERT and the complexes were captured in outward-open, occluded, and inward-open conformations providing a snapshot of the neurotransmitter transport mechanism and inhibition [14].

4.4 Central Binding Site Versus Allosteric Binding Sites in MATs

LeuT crystal structures bound with leucine and tryptophan substrates within the S1 site are known in the literature. There are several known co-crystal structures of dDAT (shares 55% homology with human DAT at the amino acid level) in complex with dopamine, amphetamine, cocaine, tricyclic antidepressants, and several other inhibitors bound within the S1 site [10]. In addition, several crystal structures of LeuBAT (LeuT with key residues in the central binding site mutated to corresponding amino acids of human SERT) have been published with mazindol, TCAs, SSRIs, SNRIs, etc., all occupying the S1 site [15]. SERT crystal structures have recently been reported bound to the antidepressants sertraline, fluvoxamine, and paroxetine [12] occupying the central S1 site. Moreover, several homology models and computational studies, that have employed the known crystal structures as templates, have aided in the discovery and development of MAT ligands. Many novel MAT ligands have been identified as promising lead compounds through virtual screening studies using the computational models of MATs. Xue and colleagues [16] provide a comprehensive list of reported computational models known to date. Together, these structures have provided detailed insights into the molecular and structural details of the MAT central binding sites and have led to the advancement in the study of transporter interactions with their ligands and of structural differences among DAT, SERT, and NET [17]. Increasing structural and biochemical evidence have indicated the presence of additional ligand binding sites within the MATs other than the S1 central binding site. These sites most likely are present in the extracellular vestibular region of the transport channel above the S1 site. Several studies in the literature have endeavored to determine the exact molecular and structural features of ligand binding to this region of the transporters. One of the earliest evidences indicating the presence of ligand binding site in the extracellular region comes from the X-ray crystal structures of bacterial transporter LeuT in complex with several TCAs and SSRIs bound in the putative secondary binding site region [5–7, 18, 19]. More recently, hSERT crystal structure with two (S)-citalopram molecules simultaneously bound (one in the S1 site and the other in the extracellular vestibule region) [11] as well as the drastic increase in affinity of some bivalent ligands of SERT and DAT [20] compared to the common monovalent ligands strongly indicate the presence of distinct low-affinity ligand binding sites in the extracellular region within MATs other than the high-affinity primary binding site (i.e., S1). These results add to the earlier data on antagonist dissociation experiments of SERT that showed that there is a low-affinity allosteric site in SERT that slows the dissociation of inhibitors from a separate high-affinity site [21]. In this study, it was shown that S-citalopram or clomipramine impeded the dissociation of [^{3}H]-citalopram, although with a relatively low potency of ~5 μM while it is known to bind to the central binding site with high potency. Later, Plenge et al. used simulations to show that ligand binding in the S1 propagates conformational changes in the S2 site. Recently, the same group reported cryo-EM structure of SERT bound with the antidepressant vilazodone in the extracellular vestibular [22]. In another study, molecular dynamics simulations and comparative genomics techniques identified another allosteric pocket [23, 24] in the extracellular vestibule of SERT and a virtual screening of a database against this site identified a few low potency ligands that specifically modulate the function of SERT ligands known to bind to the high-affinity S1 site. We employed a similar method of hybrid-based pharmacophore approach and virtual screening to identify a novel allosteric binding pocket within the extracellular region of DAT and discovered a screening hit, KM822, which modulates the activity of DAT in beneficial ways [25]. Thus, in addition to the primary binding site S1 observed in the crystal structures of LeuT and SERT, there is a possibility that multiple low affinity binding sites are present in regions distinct from the direct translocation pathway that could serve as allosteric sites for modulating conformational changes in the transporter and affect its function. The relevance of these allosteric sites as functional binding sites for monoamine substrates during translocation and for modulators is a topic of great interest currently.

4.5 Medicinal Chemistry of MAT Ligands

A wide range of MAT-interacting compounds have been developed to date as pharmacological and therapeutic tools to modulate and regulate neurotransmission. Different classes of non-selective or selective drugs that can inhibit, modulate, or promote the activity of the three transporters have been extensively developed via design and synthesis efforts to evaluate their efficacy in the treatment of many CNS-related diseases such as depression and the abuse of psychostimulants like cocaine and amphetamines. The following provides an overview of structural details of the most prominent ligands of MATs and summarizes the recent medicinal chemistry and structure-activity relationship studies of some of the MAT ligands that have been deemed "unique" and novel in terms of their mode of mechanism, site of action, neurochemical properties, and/or their behavioral effects.

4.5.1 Structure-Activity Relationship Studies of DAT Ligands

Cocaine (**1**, Fig. 4.1) is a highly addictive drug of abuse which binds to DAT and inhibits the reuptake of dopamine from the synapse into presynaptic neurons. This results in an increase in extracellular dopamine levels which is the primary mechanism of cocaine's psychostimulant effects and abuse liability. Structurally, cocaine belongs to the tropane alkaloid class of drugs [26] and nonspecifically binds to DAT, NET, and SERT with similar potencies. Numerous substituted derivatives of cocaine with DAT selectivity have been evaluated as potential pharmacotherapies for psychostimulant abuse and other DAT-related disorders. In addition, SAR studies on cocaine have been of interest for many years to characterize the binding sites of cocaine and identify the potency and selectivity requirements of moieties within cocaine for its interaction with DAT. Carroll et al. reported several cocaine analogs through systematic structure activity relationship on cocaine's structure [27, 28].

The 3-aryl analogues of cocaine were among the first series that provided highly potent and selective compounds against DAT. The position and type of substituents on the 3-aryl ring further improved the affinity and selectivity of these tropanes. WIN-35428 (**1a**, Fig. 4.1) emerged from the 3-phenyltropanes series with higher affinity and selectivity for DAT over NET and SERT as compared to cocaine. Although this series was explored to find analogs that could block the psychostimulant effects of cocaine by specifically binding to DAT and be useful as anti-addiction therapeutics, most of the analogs produced cocaine-like behavioral effects in animals. Nevertheless, WIN-35428 is extensively used to date as a pharmacological

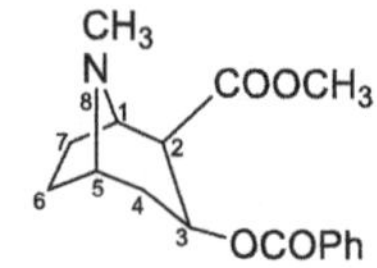

1, Cocaine

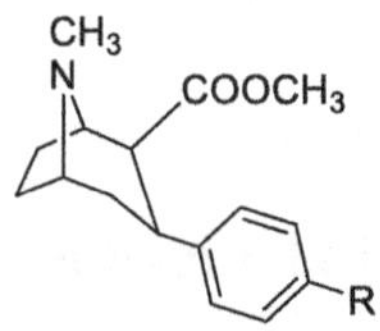

1a, WIN 35,428 R = F
1b, RTI-55 R = I

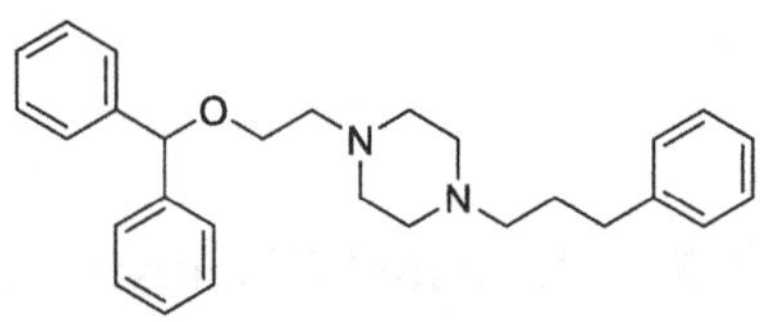

2, GBR-12935

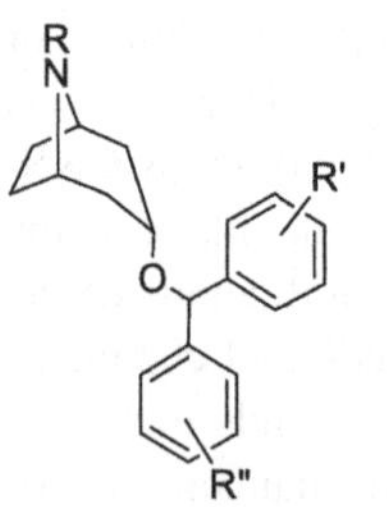

Compd	R	R', R"
3, benztropine	CH_3	H
3a	H	4'-Cl
3b	H	3'-Cl
3c	H	4',4"-diF
3d	CH_3	4',4"-diF
3e	benzyl	4',4"-diCl
3f, JHW007	n-butyl	4',4"-diF

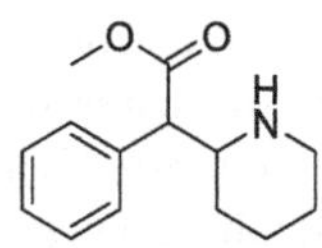

4, Methylphenidate

Structures of select benztropine analogs

Fig. 4.1 Chemical structures of DAT inhibitors

tool in DAT-binding assays and comparative studies as the removal of the phenyl ester group in cocaine's structure imparted stability in aqueous solutions.

RTI-55 (**1b**) is another compound in this series which has selectivity for DAT and SERT over NET and is commonly used as a reference tool in pharmacological assays. Based on such results, it was hypothesized that cocaine analogs that inhibit DAT to block dopamine transport will have reinforcing effects similar to that of cocaine [29]. However, GBR 12909 (**2**), a piperazine-based selective DA uptake inhibitor which was in clinical evaluation for cocaine abuse, piqued interest in further discovery of compounds with a desirable behavioral profile [30, 31].

One of the most pivotal discoveries among the tropane class of DAT inhibitors was the compound benztropine (**3**), with a diphenylether substitution, similar to GBR 12909, at the 3α position and a lack of substituent at the 2-position. Interestingly, benztropine has high affinity and selectivity for DAT, but displayed "atypical" effects where it did not share cocaine's stimulatory behavioral profile in animal models [32]. Although benztropine was a potent DAT ligand, its binding towards off-target sites confounded the interpretation of its behavioral effects in animal models of addiction. Further exploration of benztropine SAR studies to mitigate the off-target activity showed that the addition of halogen groups at 3- and 4-position of the phenyl rings and modification of the *N*-Methyl moiety could improve DAT affinity and selectivity of the benztropine analogs and minimize off-target engagement (Table 4.1) [33]. These efforts led to the discovery of JHW007 (**3f**), a potent and selective DAT inhibitor with a unique in vivo profile. JHW007, although inhibiting DAT with similar potency as cocaine, did not show cocaine-like stimulant or subjective effects. In addition, JHW007 is also shown to antagonize the behavioral effects of cocaine or methamphetamine [34–36]. Although JHW007 has never been evaluated clinically, benztropine did not produce a desirable response to acute cocaine administration in a clinical setting.

Because of the desirable behavioral effects in the animal models, atypical DAT inhibitors have continued to be actively pursued and investigated to elucidate their

Table 4.1 Binding data of tropane-based analogs at DAT, NET, and SERT

Compound	DAT (K_i, nM)	NET (K_i, nM)	SERT (K_i, nM)
1, cocaine	187 ± 18.7	3300 ± 170	172 ± 15
1a, WIN35428	11	NA	160
1b, RTI-55	1.26	NA	4.21
2, GBR12935	12.0 ± 1.9	497 ± 17.0	105 ± 11.4
3, benztropine	118 ± 18.7	1390 ± 134	>10,000
3a	36.8 ± 3.3	1010 ± 116	1320 ± 194
3b	26.2 ± 2.1	508 ± 70.0	2100 ± 285
3c	11.2 ± 1.2	NA	NA
3d	11.8 ± 1.3	610 ± 81	3260 ± 110
3e	399 ± 27.9	7660 ± 1240	3610 ± 214
3f, JHW007	24.6 ± 2.0	1730 ± 232	1330 ± 151
4, methylphenidate	60 ± 0.01	100 ± 0.01	132.43 ± 10.71

mechanism and find their potential use as psychostimulant abuse medications. Their unique effects are suggestive of a different mode of action indicating a ligand-specific control of DAT function. Structure-function studies of typical and atypical DAT ligands have hypothesized that DAT ligands prefer different DAT conformations which subsequently dictate their behavioral effects. Continued efforts in the discovery of atypical ligands have shifted focus to a new generation of atypical DAT ligands comprised of modafinil and its analogs. Modafinil (**5**, Table 4.2), which is used as a wake-promoting agent for the treatment of narcolepsy and other sleep disorders [37], also has low affinity for DAT and low abuse potential according to the available clinical data [38, 39]. To discover modafinil analogs with improved pharmacological and behavioral profiles that could potentially be developed into addiction therapeutics [40], several SAR studies have been reported. Although modafinil is available as a racemic mixture, studies have suggested that R-(−)-modafinil is more metabolically stable, longer-acting than the (S)-enantiomer [41] and slightly more potent (threefold) than the S-(+) counterpart. In addition, although the racemic, R-, and S-enantiomers all stimulated locomotor activity in mice, they were much less effective and less potent than cocaine [37]. Nevertheless, in clinical trials, modafinil has demonstrated limited effectiveness in treating cocaine abuse, although there is some evidence of benefit in nonalcohol-dependent cocaine abusers [42]. Several attempts have been made to further increase the therapeutic efficacy for stimulant use disorders by the design, synthesis, and evaluation of several modafinil analogs to improve its water solubility and DAT affinity.

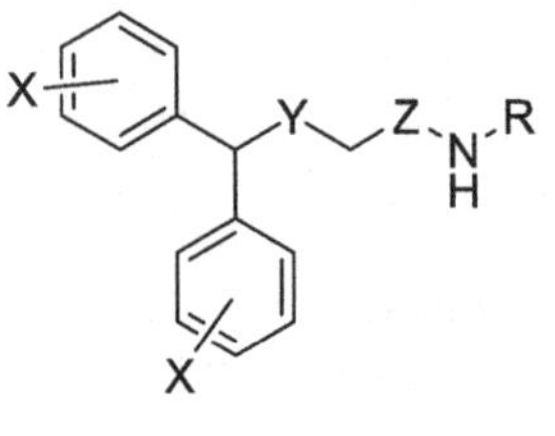

In one of the earlier SARs explorations of modafinil, it was found that reducing the sulfoxide with sulfide decreased DAT affinity and improved SERT binding if the diphenyl rings are unsubstituted (see Table 4.2) [43]. Addition of alkyl groups on terminal amide nitrogen reduced DAT affinity. Binding affinity at DAT increased with halogen substitutions at *para* position of the aryl rings in the order of Br > Cl > F ≥ H. In addition, this halogen substitution order was also followed by halogens at *meta*-positions of the diphenyl rings [43]. Compound **5e** displayed remarkable selectivity towards DAT as compared to SERT and NET with 3,3′-di-Cl substitution with thioacetamide scaffold. 4,4′-dimethyl substitution on the diphenyl rings (**5k**) decreased DAT affinity by fivefold but did not affect the DAT selectivity of the compound. For those analogues that contained unsubstituted diphenyl rings, it was generally observed that for the thioethanamine scaffold, DAT affinity increased with more bulky substituents on terminal amine nitrogen whereas in thioacetamides, DAT affinity decreased with an increase in bulk at the terminal amine

Table 4.2 Binding affinities of Modafinil and its analogs against DAT, NET, and SERT

Compound	X	Y	R	Z	DAT (K_i, nM)	NET (K_i, nM)	SERT (K_i, nM)
5, modafinil	H	S=O	H	C=O	2600 [2430–2780]	NA	NA
5a	H	S	H	C=O	12400 [10800–14300]	NA	NA
5b	H	S=O	Methyl	C=O	13100 [12600–13700]	NA	NA
5c	3,3′-di-F	S=O	H	C=O	5930 [4990–7060]	NA	NA
5d	3,3′-di-Cl	S=O	H	C=O	881 [763–1020]	NA	NA
5e	3,3′-di-Cl	S	H	C=O	275 [257–295]	45400 [39600–52000]	NA
5f	4′,4″-di-Br	S	Methyl	C=O	3010 [2770–3260]	11000 [9540–12600]	5720 [5320–6150]
5g	4′,4″-di-Cl	S	Methyl	C=O	4130 [3620–4710]	9770 [9170–10400]	10700 [7310–15700]
5h	H	S	n-butyl	C=O	23600 [20500–27100]	NA	NA
5i	4′,4″-di-Br	S	n-butyl	C=O	722 [659–792]	7580 [7210–7970]	7090 [6990–8180]
5j	4′,4″-di-F	S	n-butyl	C=O	6400 [5820–7050]	56100 [53900–58500]	25500 [23300–28000]
5k	4′,4″-di-CH_3	S=O	H	C=O	12700 [12400–13100]	NA	NA
5l	H	S	H	CH_2	142 [131–155]	980 [938–1020]	221 [191–257]
5m	H	S	Cyclopropylmethyl	CH_2	435 [406–466]	17300 [15400–19400]	10000 [9570–10400]
5n	H	S	n-butyl	CH_2	310 [275–350]	11500 [10700–12300]	5700 [5040–6440]
5o	H	S	3-phenylpropyl	CH_2	295 [268–325]	5500 [5140–5880]	927 [786–1090]

(continued)

Table 4.2 (continued)

Compound	X	Y	R	Z	DAT (K_i, nM)	NET (K_i, nM)	SERT (K_i, nM)
5p (JJC8-016)	4′,4″-di-F	S	3-phenylpropyl	CH_2	114 [97.4–132]	3850 [3830–3870]	354 [312–402]

Table 4.3 Binding affinities of piperazine-based modafinil analogs

Compound	X	Y	Z	R	DAT (K_i, nM)	NET (K_i, nM)	SERT (K_i, nM)
5q	H	S=O	C=O	3-phenylpropyl	33.0 ± 2.83	54300 ± 3210	15200 ± 1100
5r	4,4′-diF	S=O	C=O	3-phenylpropane	37.6 ± 1.86	12000 ± 1430	1320 ± 152
5s	4,4′-diF	S=O	C=O	2-OH-propyl	752 ± 87.4	NA	32800 ± 4430
5t	3,3′-diCl	S=O	C=O	2-OH-propyl	1380 ± 191	>50 uM	NA
5u	H	S	C	2-OH-propyl	49.6 ± 4.31	44500 ± 2400	26700 ± 2630
5v (JJC8-091)	4,4′-diF	S	C	2-OH-propyl	16.7 ± 1.22	17800 ± 885	1770 ± 234
5w (JJC8-088)	4,4′-diF	S	C	2-OH, 3-phenylpropyl	6.72 ± 0.977	1950 ± 227	213 ± 13.2

nitrogen [43]. Compound **5l** with the thioethanamine scaffold and unsubstituted diphenyl rings was one of the most potent compound towards DAT, however, lost significant DAT selectivity. This indicated that the replacement of amide with amine to increase water solubility was well tolerated. The affinity of compound **5o**, with *N*-propylphenylamine substituent, for DAT increased by tenfold as compared to modafinil and was found to be more selective towards DAT than NET and SERT [41]. The DAT affinity further increased 2.5-fold with 4,4′-difluoro substituted diphenyl-moiety (**5p**, JJC8-016). In addition, *N*-cyclopropylmethyl and *N*-*n*-butyl substituted analogs (compounds **5m** and **5n**) were also among the most DAT-selective compounds in this series. Further functionalization of the terminal nitrogen was attempted by incorporating the piperazine ring in the main scaffold (Table 4.3) [42].

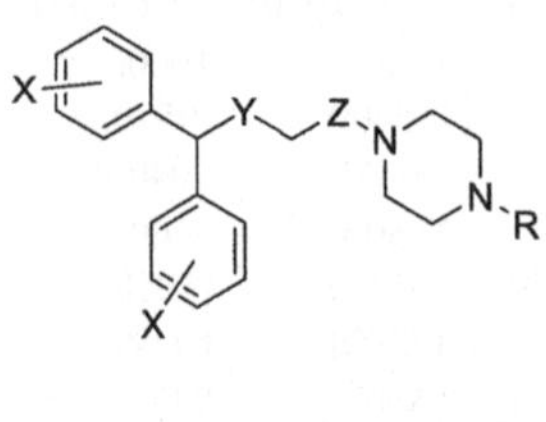

Compound **5q** with a piperazine amide substitution retaining the 3-phenylpropyl moiety of compound **5p** displayed 100-fold higher potency than modafinil and high selectivity for DAT as compared to SERT and NET, with additional 4,4′-diF groups imparting a modest change in affinity (**5r**). This series was further diversified by hydroxylating the *N*-bearing propyl chain. Compounds **5s** and **5t** with 2-OH-propyl side chain lost potency for DAT but retained high DAT selectivity and were almost inactive at SERT and NET. The removal of amide carbonyl C=O to get compounds like **5u** and **5v** was well tolerated as well [42]. Comparison of compounds 2-hydroxypropyl **5v** (JJC8-091) and 2-hydroxyphenylpropyl **5w** (JJC8-088) revealed that phenyl moiety imparted greater affinity to the compound with DAT affinity threefold higher for **5w** than **5v** with similar selectivity profile for DAT versus SERT. Compounds JJC8-016, JJC8-088, and JJC8-091 emerged as the most promising compounds with superior DAT affinity compared to modafinil and better aqueous solubility [44]. These compounds were evaluated for their effects on the neurochemistry via brain microdialysis and fast-scan cyclic voltammetry and behavior through ambulatory activity of male Swiss-Webster mice. The results of these studies are summarized in the literature [44–46]. Briefly, compound JJC8-016 reduced the psychostimulant effects in various animal models of addiction. For example, cocaine enhanced locomotion, cocaine self-administration, and cocaine-induced reinstatement of drug seeking behavior were dose-dependently reduced by JJC8-016 pretreatment [47] although there is evidence that it might possess cardiotoxic adverse effects similar to that seen with its analog, GBR12909 [48]. Compound JJC8-088 displayed cocaine-like effects and preferred open DAT conformation similar to that of cocaine. Compound JJC8-091 showed promising results in behavioral testing in a murine locomotor assay where it was effective in reducing stimulant effects in rats exposed to long-access (6 h) methamphetamine [48, 49]. Furthermore, JJC8-091 reduced cocaine self-administration and cocaine-primed reinstatement of cocaine seeking [49]. Hence, JJC8-091 emerged as a promising lead compound and structural modifications of this scaffold continue to be reported that have better profile in behavioral models of cocaine abuse and are more metabolically stable [50]. Giancola and colleagues have recently reported a new series of modafinil where they have replaced the metabolically susceptible piperazine ring of compounds like JJC8-016, JJC8-088, and JJC8-091 with more stable bioisosteric moieties aminopiperidine and piperidine amine (Table 4.4). The importance of the tertiary amine was explored through compounds with either a tertiary amine or amide, and the oxidation state of the sulfide was varied. Of note, the bis(4-F-phenyl)methyl moiety of the previous lead compounds (JJC8-091, JJC8-088, and JJC8-089) was retained in these new series [51]. Several compounds were designed and synthesized in this series with terminally attached alkyl-phenyl and *para*-substituted phenyl moieties.

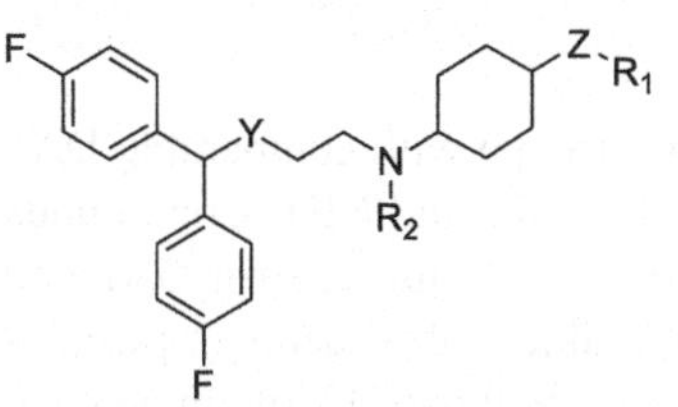

Aminopiperidine scaffold (A)

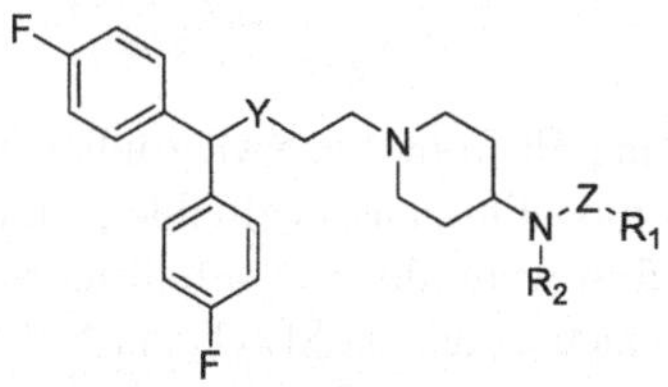

Piperidine amine scaffold (B)

Table 4.4 Binding affinities of aminopiperidine and piperidine amine-containing modafinil analogs

						DAT	SERT	σ_1
Compound	Scaffold	Y	Z	R_1	R_2	$K_i \pm$ SEM (nM)		
5aa	A	S=O	C=O	Phenyl	H	79.1 ± 20.6	7780 ± 734	585 ± 21.5
5bb	A	S=O	C=O	4-fluorophenyl	H	77.2 ± 4.54	4640 ± 381	1440 ± 131
5cc	A	S=O	CH_2	4-fluorophenyl	H	50.6 ± 11.2	373 ± 23.8	26.5 ± 3.88
5dd	A	S	C=O	2,4-difluorophenyl	H	30.0 ± 8.25	296 ± 35.3	20.6 ± 2.38
5ee	A	S	CH_2	Phenyl	H	32.9 ± 5.86	409 ± 58.2	3.81 ± 0.639
5ff	A	S=O	CH_2	1-OH-2-phenylethyl	H	91.8 ± 21.3	599 ± 54.4	351 ± 25.9
5gg	A	S	CH_2	1-OH-2-phenylethyl	H	31.4 ± 9.64	129 ± 33.4	309 ± 38.3
5hh	B	S	CH_2	Phenyl	H	108 ± 17.5	331 ± 34.6	60.9 ± 8.90
5ii	B	S	CH_2	4-fluorophenyl	H	55.9 ± 6.08	268 ± 9.33	41.4 ± 10.9
5jj	B	S	CH_2	1-OH-2-phenylethyl	H	47.7 ± 2.62	66.8 ± 2.82	88 ± 4.02
5kk	A	S	C=O	2,4-difluorophenyl	H	30.0 ± 8.25	296 ± 35.3	20.6 ± 2.38

The aminopiperidines were slightly more potent than the piperidine amine analogs; however, the trend was opposite with 2-OH-propylphenyl-substituted analogs. Reduced sulfides were slightly more potent than their sulfoxide counterparts with a moderate loss in DAT selectivity compared to SERT (**5ff** versus **5gg**). Overall, all compounds resulted in moderate binding affinity for DAT and relatively improved potency towards SERT. This series of compounds was also explored for activity against σ_1 receptors due to preclinical precedence related to the therapeutic value of dual DAT/ σ_1 inhibitors [52]. All compounds were inactive towards NET. The most promising among this series of compounds were **5bb**, **5cc**, and **5dd**, with highest stability, which were carried forward for further in vitro and in vivo evaluation. None of these analogues showed a cocaine-like profile as systemic administration produced only minimal stimulation of ambulatory activity. Kalaba et al. [53] have reported another series where the carboxamide of modafinil is replaced with a number of five- and six-membered aromatic heterocycles to improve activity and selectivity and in vitro/in vivo profiles (Table 4.5).

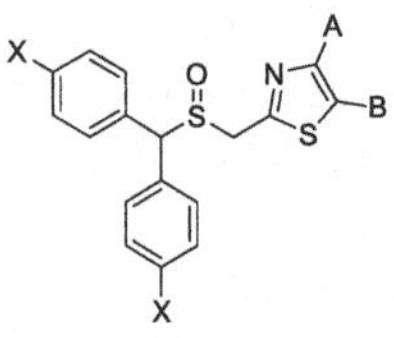

A

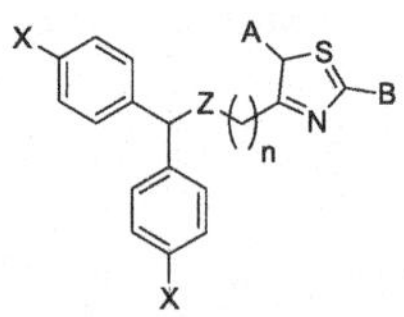

B

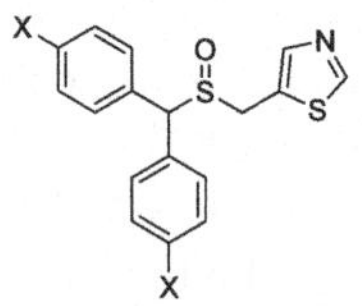

C

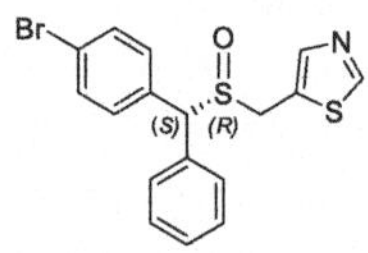

5uu

Compound **5ll** from this SAR studies was one prototype with comparable DAT potency as modafinil, and negligible activity towards SERT and NET. Compounds **5mm** and **5nn** were almost fivefold more potent than modafinil against DAT and also were inactive against SERT and NET. Larger, bulkier substituents in place of methyl or cyclopropyl of **5mm** and **5nn** resulted in two- to threefold loss in potency (**5oo** to **5pp**). However, a chloro substituent was well tolerated (**5oo**). Para-substitutions at the diphenyl rings (dimethyl in **5qq** and difluoro in **5rr**) continued

Table 4.5 Inhibition of DAT, NET, and SERT-mediated uptake of respective radioligands by thiazole-containing modafinil analogues

Compound	Scaffold	X	Z	*n*	A	B	DAT (IC_{50} ± SD, μM)	NET (IC_{50} ± SD, μM)	SERT (IC_{50} ± SD, μM)
5ll	A	H	–	–	H	H	14.73	420.1	NA
5mm	B	H	S=O	1	H	CH_3	3.25	174	NA
5nn	B	H	S=O	1	H	Cyclopropyl	4.1 ± 0.8	687.2 ± 152.8	436.1 ± 129
5oo	B	H	S=O	1	H	*n*-propyl	28.43 ± 1.93	NA	NA
5pp	B	H	S=O	1	H	*n*-butyl	13.50 ± 2.76	164 ± 56.88	NA
5qq	B	CH_3	S=O	1	H	H	7.32 ± 0.04	NA	NA
5rr	B	F	S=O	1	H	H	6.76 ± 0.67	NA	NA
5ss	C	CH_3	–	–	–	–	10.57 ± 0.80	215.50 ± 66.01	NA
5tt	C	OCH_3	–	–	–	–	71.59 ± 8.28	NA	NA
5uu	–	–	–	–	–	–	0.65 ± 0.07	73.25 ± 9.15	NA

to provide highly potent and selective compounds in this series. Other analogs with thiazole ring linked to the main scaffold via 2- and 5-position led to reduced potency. The authors have also reported results of chiral separation of several analogs with promising potency and selectivity. One of the chiral HPLC separation products (**5uu**, Table 4.5) displayed highest DAT activity (IC_{50} = 0.65 ± 0.07 μM) and 100-fold less active on NET (IC_{50} = 73 ± 9 μM) and inactive for SERT. Hence, this series of compounds also represent potential new leads for developing psychostimulant therapeutics. Overall, the SAR studies of modafinil analogs provide promising new leads to develop drugs with an atypical DAT inhibitor profile that could have utility as anti-addiction therapies. Discovery of such atypical DAT inhibitors that lack "cocaine-like" profile piqued further interest in determining their molecular interactions with DAT. It is suggested that these ligands bind and stabilize a DAT conformation which is disparate from that of cocaine-bound DAT conformation resulting in contrasting psychostimulant effects. Experimental evidence suggest that cocaine stabilizes outward-open DAT conformation whereas JHW-007, GBR-12909, modafinil, and its analogs tend to be less affected by conformational changes and most likely favor a more inward-facing occluded conformation [54]. This conformational "preference" could attribute to the lack of addiction potential of atypical inhibitors [55]. Another reason for reduced addiction liability of atypical DAT inhibitors could be their slow rate of onset of action in the brain or their off-target effects that might also contribute to their reduced behavioral reinforcing effects [55]. Hence, further studies related to specific structural basis of interaction of DAT ligands are needed to promote the discovery and development of improved medications for psychostimulant addiction.

With the growing interest in the possible therapeutic potential of allosteric modulators of DAT, a library of diphenylmethyl (benzhydryl)-containing compounds were screened for their allosteric activity against the MATs (Fig. 4.2) [55, 56]. Among these, SRI-9804 (**6a**) was reported to inhibit both uptake and efflux of [^{3}H]-dopamine, but only with partial efficacy (40–60%). Similar to allosteric activity of S-citalopram in SERT, SRI-9804 reduced the dissociation of a radioligand prebound in the orthosteric site of DAT. Two other compounds, SRI-20040 (**6b**) and SRI-20041 (**6c**), were eventually reported to partially inhibit [^{125}I]RTI-55 binding and [^{3}H]-DA uptake slow the dissociation rate of [^{125}I]RTI-55 from the DAT, and allosterically modulate d-amphetamine-induced, DAT-mediated DA release. Subsequently, a series of SRI-20041 analogs were designed, synthesized, and evaluated in order to increase the affinity for DAT. This resulted in SRI-29574 (**6d**), one of the high-affinity and selective DAT inhibitors in this series, showing similar pharmacological features to its parent compound. SRI-29574 partially inhibited DAT uptake ($IC_{50} = 2.3 \pm 0.4$ nM) while being inactive in inhibiting DAT binding. The binding site of these compounds is still unknown, but regardless of their potential clinical application, their unconventional transporter pharmacology makes them potentially interesting candidates to further explore DAT allosteric modulators.

Our group recently reported a novel allosteric inhibitor of DAT known as KM822 (**7**, Fig. 4.2) [25, 57]. KM822 was identified using in silico screening against the putative allosteric site present in the extracellular region of the DA translocation pathway. Biochemical investigation revealed it to be a noncompetitive inhibitor of DAT with relatively good selectivity. It was also shown to reduce psychostimulant-mediated stimulatory effects in various ex vivo and in vivo models of addiction and thus represent a possible candidate for exploring another novel class of drugs to treat addiction.

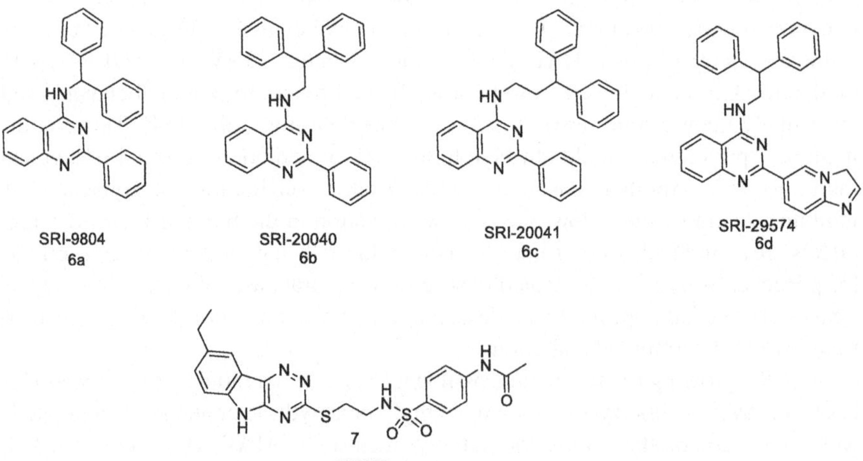

Fig. 4.2 Chemical structures of DAT allosteric ligands

4.5.2 Structure-Activity Relationship Studies of SERT Ligands

Several classes of antidepressant drugs that target SERT have been developed and have been classified as tricyclic antidepressants (TCAs), selective serotonin reuptake inhibitors (SSRIs), and serotonin-norepinephrine reuptake inhibitors (SNRIs) (Fig. 4.3). The first-generation antidepressants included the TCAs like imipramine (**8**), amitriptyline (**9**), desipramine (**10**), and clomipramine (**11**). However, due to their nonspecific effect on cholinergic, histaminergic, and α-adrenergic pathways and their low therapeutic index, their use now is limited to treatment-resistant patients who have failed to respond to newer generations of antidepressants [58]. In the search for highly efficient antidepressant drugs with an improved safety and tolerability profile, the considerable effort of the last five decades led to the development of the selective monoamine reuptake inhibitors, such as SSRIs. SSRIs are currently the most prescribed antidepressant pharmacotherapy and the most common SSRIs used till date are fluvoxamine (**12**), fluoxetine (**13**), citalopram (**16**), paroxetine (**14**), and sertraline (**15**). SSRIs have a diverse chemical structure, and, in many instances, they do not share common structural motifs. The structural explanation of how these diverse ligands bind to SERT have been summarized by Coleman and Gouaux [12] by obtaining and comparing X-ray crystal structures of SERT in complex with paroxetine, citalopram, sertraline, and fluvoxamine. The diversity of SSRI chemical structures, in turn, results in compounds with varied affinities towards SERT and substantial pharmacological differences. The most

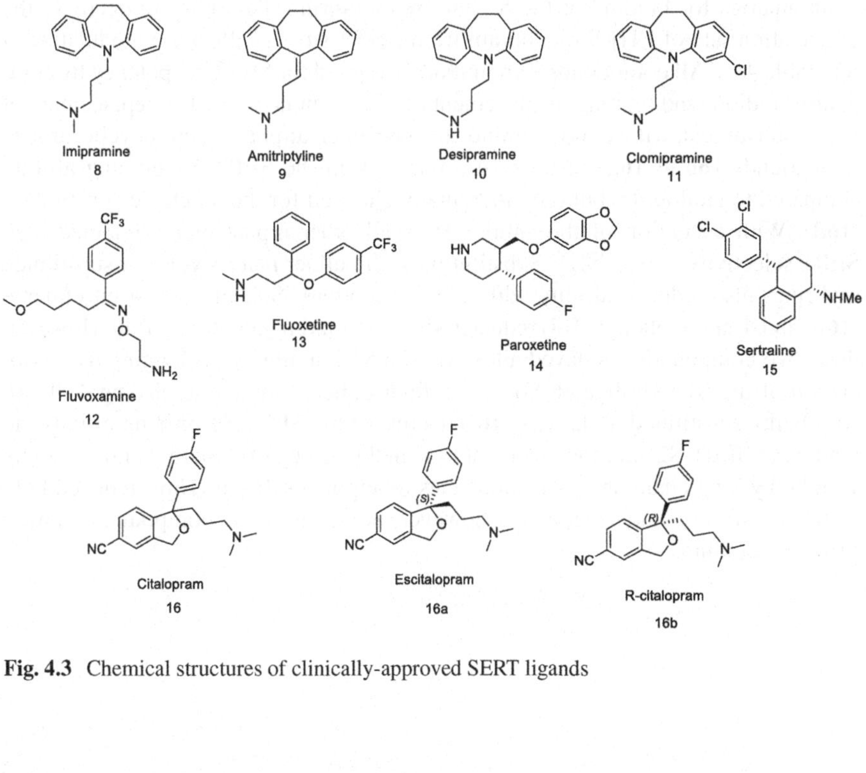

Fig. 4.3 Chemical structures of clinically-approved SERT ligands

studied SSRI to date is citalopram. It is the most SERT-selective SSRI. The S-enantiomer (**16a**) is responsible for the high selectivity and affinity for SERT and is ~30-fold more potent than R-citalopram (**16b**). In the past, several SAR studies on citalopram have been reported [59–61]. These SAR studies employing the analogues of citalopram have been pivotal in characterizing the S1 site for SERT [62, 63, 65, 76, 61]. More recently, the interest in achieving higher drug target selectivity to minimize side effects have led to an impetus in the study and development of SERT inhibitors that target the allosteric sites. The SSRIs like sertraline, paroxetine, clomipramine, and citalopram have been shown to possess allosteric activity as they can impair dissociation of a pre-bound high affinity radioligand to the transporter [74]. However, the allosteric potencies of these compounds are all in the micromolar range, while they bind to the orthosteric site with low-nanomolar affinity [74]. The co-crystal structure of S-citalopram bound in the allosteric region of SERT has made it feasible to further explore and understand the allosteric regulation mechanism of MATs [11].

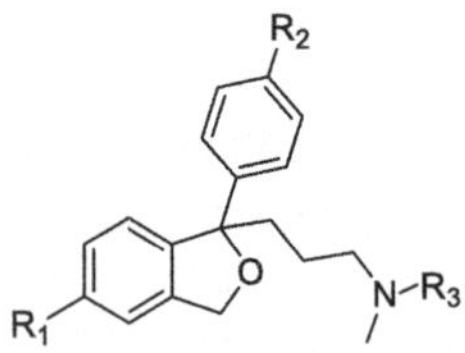

Banala and colleagues [64] were the initial ones to explore the SAR of citalopram analogs for binding at the S2 site by measuring the ability to decrease the dissociation rate of [^{3}H]-S-citalopram from the S1 site via allosteric modulation at S2 (Table 4.6). All compounds were initially assessed for SERT-S1 potency by competitive radioligand binding displacement of [^{3}H]-S-citalopram. The replacement of 5-CN substituent with amide, amino, or any other aliphatic heterocyclic groups (compounds **16c** to **16g**) resulted in a decrease in the SERT S1 binding affinity compared to citalopram but only marginally changed for the aldehyde substitution (**16d**). With exception of the amide (**16c**), all other replacements retained high SERT selectivity versus NET. Substitution with bulkier heterocycles or substituted aryl rings also reduced affinity with a few exceptions. Substitutions 4-nitrophenyl (**16h**) and 4-aminophenyl (**16i**) reduced SERT-S1 binding almost 20-fold. However, these two compounds displayed allosteric site S1 binding by prolonging the dissociation of the S1 radioligand. For the 4′-fluoro phenyl moiety, replacing 4′-fluoro with bulky substituted aryls (e.g., **16j**) decreased the SERT-S1 binding affinity, no binding at SERT-S2 site. Replacement of a methyl group of dimethyl amine moiety with bulky long chain aryl or heterocycles (compounds **16k** to **16m**) reduced both SERT-S1 affinity and selectivity; however, some of these compounds showed SERT-S2 binding.

Table 4.6 Citalopram analogues assessed by radioligand binding displacement of [^{3}H]-escitalopram (for SERT), tritiated-nisoxetine (for NET), and potency of 30 μM compound in inhibiting dissociation of [^{3}H]-escitalopram from SERT

Compound	R_1	R_2	R_3	SERT (K_i ± SEM, nM)	NET (K_i ± SEM, nM)	[^{3}H] escitalopram dissoc. $t_{1/2}$ (min)
Citalopram (16)	CN	F	CH_3	1.94 ± 0.198	5950 ± 77.4	16.1 ± 1.0
S-citalopram (16a)	CN	F	CH_3	0.89 ± 0.132	10500 ± 893	68 ± 12
16c	$CONH_2$	F	CH_3	17.7 ± 1.80	123 ± 17.7	NA
16d	CHO	F	CH_3	4.3 ± 0.096	NA	NA
16e	CH_2NH_2	F	CH_3	41.9 ± 5.73	NA	17.4 ± 0.8
16f	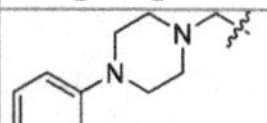	F	CH_3	3.24 ± 0.328	11100 ± 1490	16.4 ± 0.3
16g	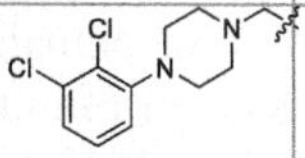	F	CH_3	22.7 ± 2.62	NA	24.8 ± 1.1
16h	4-nitrophenyl	F	CH_3	18.6 ± 1.65	8280 ± 825	40.2 ± 7.0
16i	4-aminophenyl	F	CH_3	29.2 ± 4.11	ND	52.6 ± 4.1
16j	CN	3-cyanophenyl	CH_3	51.5 ± 6.63	2800 ± 353	NA
16k	CN	F	C4H9	61.6 ± 6.19	276 ± 41.1	NA
16l	CN	F	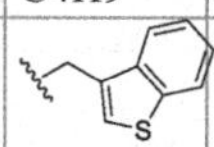	22.8 ± 2.29	ND	59.6 ± 1.0
16m	CN	F	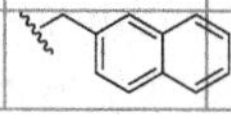	98.8 ± 13.3	272 ± 48.5	60.1 ± 1.0

NA not active

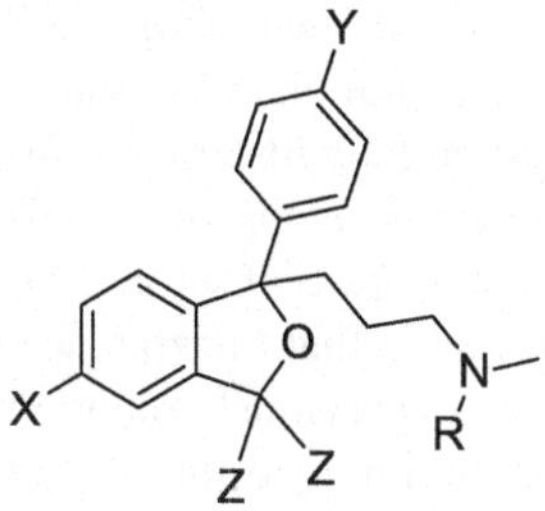

Larsen et al. [68] performed a systematic SAR study on citalopram (which is >1000-fold selective for S1 than S2) to identify S2 selective compounds by exploring all possible combinations of substituents between citalopram and talopram (Table 4.7, **16n**, a close analog of citalopram and only possess low SERT-S1 affinity) [63]. The authors measured the IC_{50} for inhibition of [^{3}H]-S-citalopram dissociation by determining the [^{3}H]-S-citalopram dissociation rates in the presence of increasing concentrations of allosteric inhibitor [21]. The authors concluded that –CN group (absent in talopram) is absolutely necessary for S2 activity. Compounds

Table 4.7 Systematic SAR exploration of citalopram and talopram analogs

Compound	X	Y	Z	R	SERT S1 binding IC_{50} (μM)	$T_{1/2}$ for [^{3}H]S-CIT dissociation (min)	SERT allosteric potency IC_{50} (μM)
Escitalopram (16a)	CN	F	H	CH_3	0.010 [0.0008; 0.013]	790 ± 160	5.8 [5.4; 6.3]
Talopram (16n)	H	H	CH_3	H	0.718	100 ± 3.8	ND
16o	CN	F	H	H	0.041 [0.032; 0.051]	380 ± 26	10.1 [10.0; 10.2]
16p	CN	F	CH_3	CH_3	6.4 [4.7; 8.8]	1090 ± 130	3.6 [3.3; 3.8]
16q	CN	H	CH_3	CH_3	10 [9.2; 12]	440 ± 30	12 [10; 13]

ND not determined

without –CN displayed a fivefold reduction in S2 potency compared to their –CN congeners. The presence of para-fluoro was also important for S2 activity but its removal did not affect the S1 activity much. The addition of phthalane dimethyl groups in ##3 as compared to citalopram markedly reduced S1 affinity with a significant increase in the allosteric potency. Changing the dimethyl amine to monosubstitution showed almost no change in S2 activity but a reduction in S1 activity. The compound with the highest allosteric potency and lowest orthosteric affinity reported was dimethyl citalopram (**16p**), with a twofold increase in the allosteric potency compared to the orthosteric affinity [68]. The tested compounds were all racemic mixtures. To date, investigation of the enantioselectivity of these compounds has not been reported.

A recent in silico screening of a library of citalopram analogs and subsequent in vitro evaluation for their allosteric potency led to the discovery of Lu AF60097 (**17**, Fig. 4.4) with an allosteric potency of 31 nM (measure of imipramine dissociation inhibition at S1), which is more than a 150-fold increase compared with citalopram [74]. Lu AF60097 also possess a S1 binding component with SERT S1 binding potency of ~265 nM and hence could be another template for future SAR studies to isolate S2 selective compounds. Another potential allosteric modulator of recent interest is vilazodone (**18**, Fig. 4.4). Vilazodone is an approved antidepressant with partial agonist activity at 5-HT_{1A} receptor. Recent molecular pharmacology and cryo-EM structural elucidation have revealed that vilazodone's primary binding site is the allosteric site S2 [22], as opposed to previously reported results of MD simulations that had suggested that vilazodone binds to the S1 site similarly to all other TCAs and SSRIs [80]. Vilazodone inhibits [^{3}H]5-HT transport with an apparent K_i of 1.1 nM and causes a decrease in the V_{MAX} of [^{3}H]5-HT without having any significant effect on K_M suggesting a noncompetitive binding with 5-HT. Vilazodone also inhibits [^{3}H]-imipramine dissociation with an allosteric potency of 14 nM and [^{3}H]S-citalopram dissociation with an allosteric potency of 250 nM. The high-affinity profile of such allosteric compounds along with the elucidation of the molecular binding structures opens the possibility of developing compounds with beneficial therapeutic profiles.

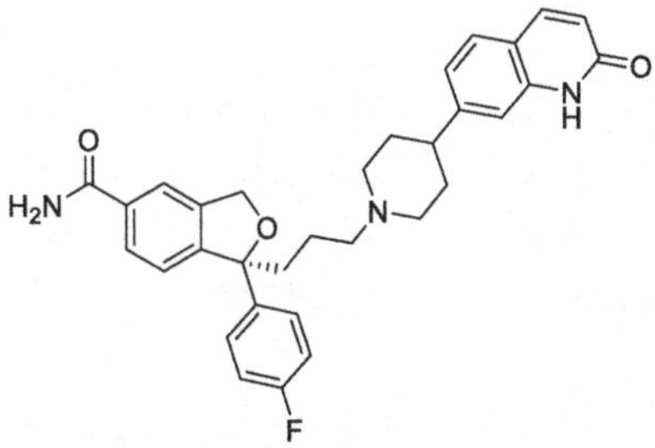

Lu AF60097

17

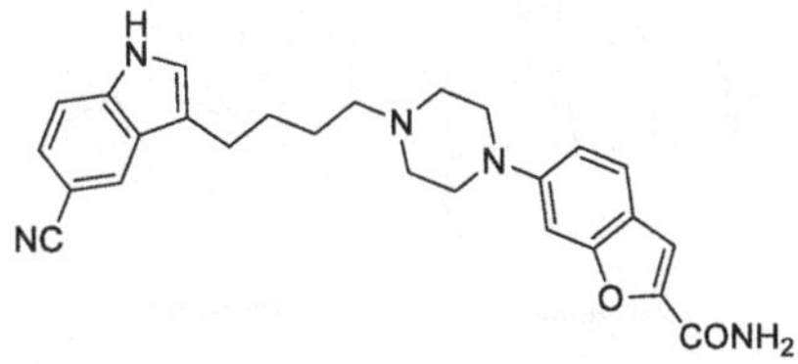

Vilazodone

18

Fig. 4.4 Structures of SERT allosteric ligands

4.5.3 Structure-Activity Relationship Studies of NET Ligands

Ligands of NET are commonly classified into: selective norepinephrine reuptake inhibitors (sNRIs), serotonin-norepinephrine reuptake inhibitors (SNRIs), norepinephrine-dopamine reuptake inhibitors, and triple reuptake inhibitors (TRIs). Selective NRIs (Fig. 4.5) such as nisoxetine (**19**), reboxetine (**20**), and atomoxetine (**21**) are useful in the treatment of depression and ADHD. Nisoxetine is a selective, potent NRI with (R)-nisoxetine much more potent than (S)-enantiomer. Reboxetine is marketed as a racemic mixture, with (+)-(S,S)-enantiomer is much more potent for NET inhibition as compared to SERT, with (−)-(R,R)-enantiomer gaining potency for SERT. Substitutions at the aryloxy ether group affected the selectivity of compounds between NET and SERT, and in general, the (S,S)-enantiomers were more potent for NET than SERT. Several SAR studies have been reported for reboxetine where the aryloxy ether group is replaced by a number of substituted arylthiolether functionalities which in general retain the potency against NET. Atomoxetine was developed and launched by Eli Lilly as a treatment for ADHD and is a potent, selective inhibitor of NET. It is an (−)-isomer of an ortho-methylphenoxy analog of nisoxetine, and is a derivative of phenoxypropylamine. Its mechanism of action in the treatment of ADHD is unclear but is thought to be related to its selective inhibition of presynaptic norepinephrine reuptake in the prefrontal cortex, resulting in increased noradrenergic transmission, important for attention, learning, memory, and adaptive response [66]. The X-ray crystal structure of dDAT with nisoxetine and reboxetine exhibit outward-open conformations with inhibitors bound in the central S1 pocket provide insights into the pharmacophores that dictate selectivity towards hNET [9]. The new generation of antidepressants has seen a paradigm shift where a mixed action on both SERT and NET is desirable to achieve additional therapeutic benefits such as improved antidepressant efficacy and faster onset of action. Approved SNRIs such as duloxetine (**22**), milnacipran (**23**), and venlafaxine (**25**) may have improved antidepressant efficacy. In another study, X-ray crystal structures of dDAT with NE as well as SNRIs duloxetine and milnacipran were

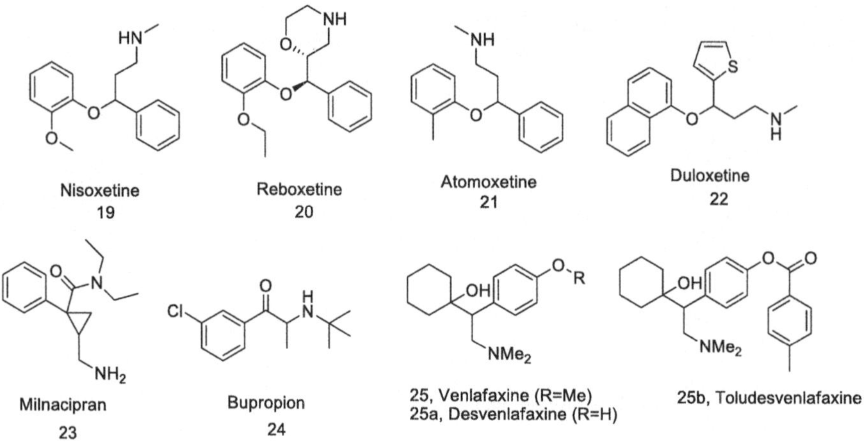

Fig. 4.5 Structures of NET ligands

identified. Bupropion (**24**) belongs to the NDRI class of dual reuptake inhibitors and has been approved to treat major depressive disorder (MDD) and for smoking cessation [70]. Despite the approval of several antidepressants, they still lack satisfactory therapeutic effect due to the low remission rates and delayed therapeutic onset, especially in the case of major depressive disorder (MDD) [67]. TRIs which inhibit all three transporters (DAT, NET, and SERT) are currently of great interest and several compounds have advanced into clinical trials (Fig. 4.6). Such polypharmacological profile of TRIs has led to their better efficacy, safety, and less tolerance in clinical evaluation for MDD, ADHD, binge eating disorder, and cocaine addiction [77].

One of the earliest TRIs developed was compound **26a** that reached clinical trials for pain treatment [79]. Extensive SAR analysis of **26a** was conducted by industrial groups to develop compounds with superior potency at all three transporters. Compounds **26b**, **26c**, and **26d** that emerged from the SAR studies on **26a** are also in the clinical development stage for ADHD and depression. Amitifadine (**26c**) is currently in a Phase 3 trial for MDD, and centanafadine (**26d**) is in a Phase 3 trial for MDD and ADHD. NS2359, another promising TRI currently in a Phase 2 clinical trial for cocaine addiction, MDD, and ADHD, was developed through structural modifications on naturally occurring TRI, cocaine. Dasotraline, **27** was discovered as part of SAR development around *cis*-sertraline. Tu et al. [78] have recently compared the binding modes of dasotraline (**27**), NS2359 (**28**), and centanafadine (**26d**) by docking them into the central binding sites of the crystal structure of SERT, and the homology models of DAT and NET, and created a pharmacophore model for TRI binding to understand their polypharmacological profile which could be useful in discovering new and improved TRIs.

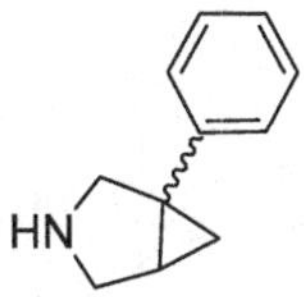

bicifadine
26a

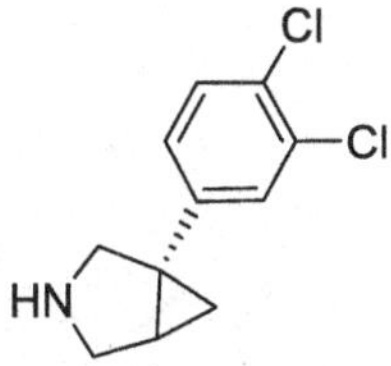

26b

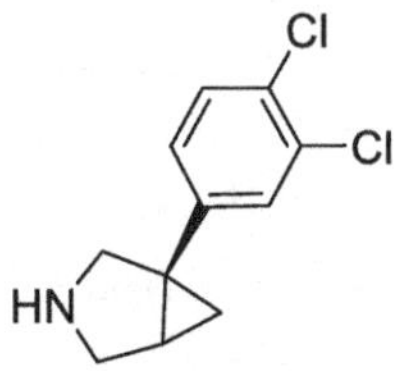

Amitifadine
26c

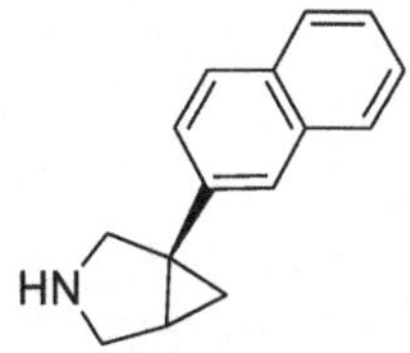

Centanafadine
26d

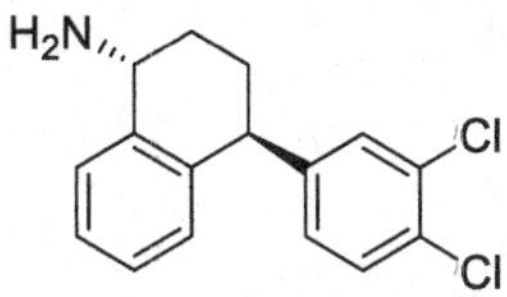

Dasotraline
27

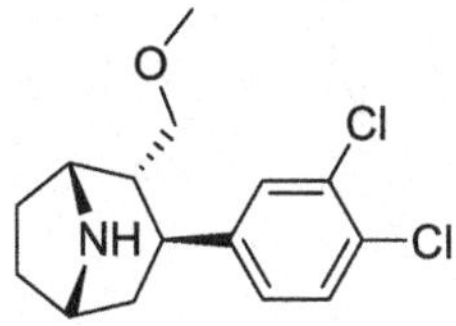

NS2359
28

Fig. 4.6 Structures of TRIs currently under clinical investigation

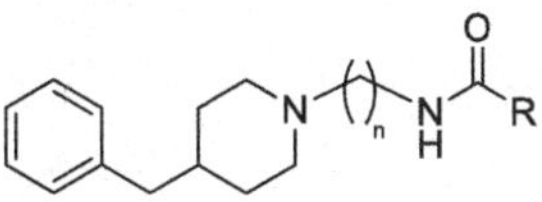

Recently, a number of 4-benzylpiperidine carboxamides were explored as TRIs (Table 4.8) [72]. The potency was compared to venlafaxine, the control drug for SERT and NET activity, and GBR12909 to compare DAT potency. In general, changing 3-carbon linker to 2-carbon drastically improved DAT inhibition with no change in NET and SERT inhibition (example, compound **29c** versus **29d**). The authors found that they could modulate the inhibition of DAT, NET, and SERT by changing the R substituents. Although most R substituents provided high potency at NET, 4-biphenyl substitution (with 3-carbon linker) provided exceptionally high inhibitory potency against NET and SERT (**29a**). Another novel compound, LPM580098 (toludesvenlafaxine, Fig. 4.5, compound **25b**), was designed and synthesized based on the structure of venlafaxine and was reported to show promising preclinical results in neuropathic pain models [69]. It is a prodrug that can quickly convert to the SNRI desvenlafaxine (**25a**) under the hydrolysis of ubiquitous esterase in vivo [81]. Another series of TRIs reported recently contain the asymmetric pyran scaffold (Table 4.9) [75].

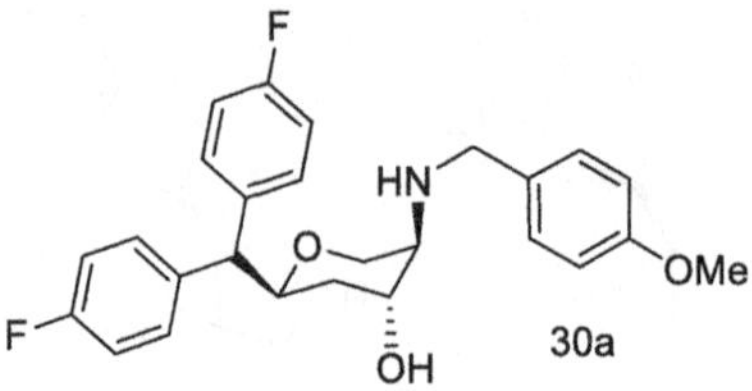
30a

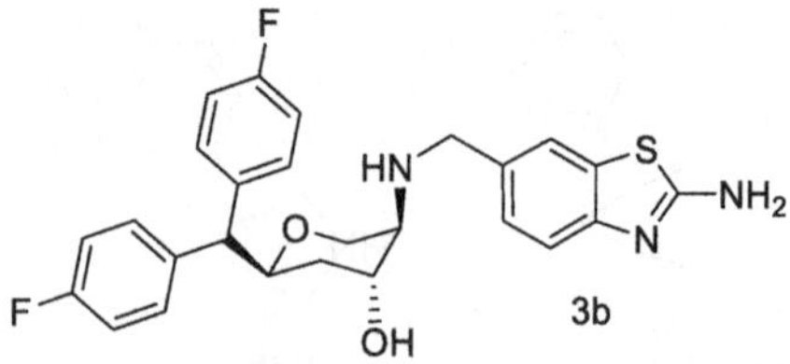
3b

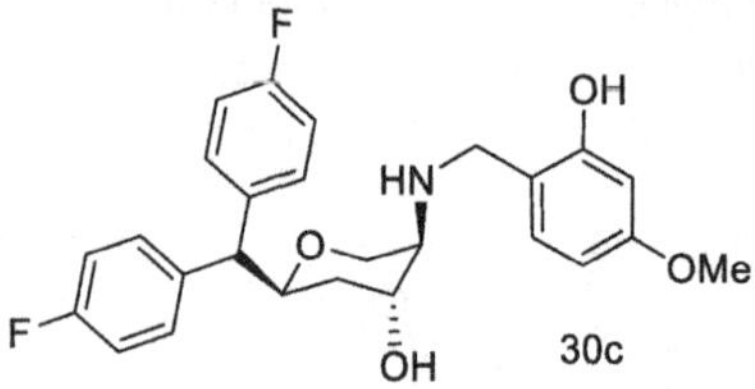
30c

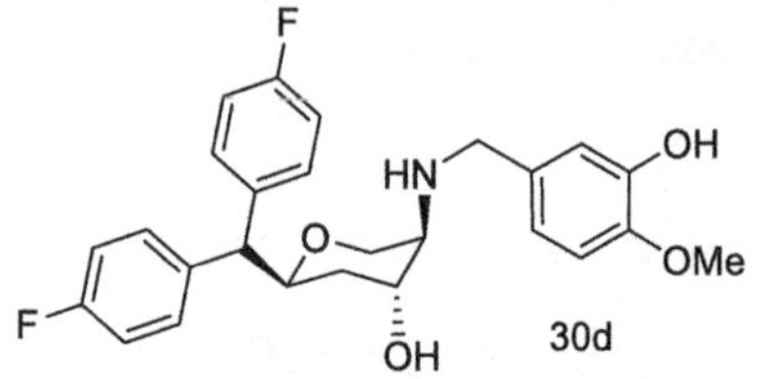
30d

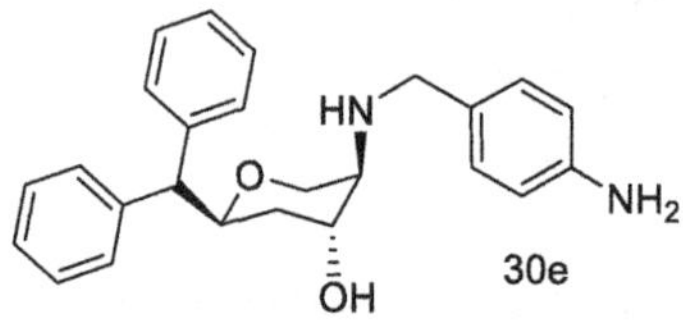
30e

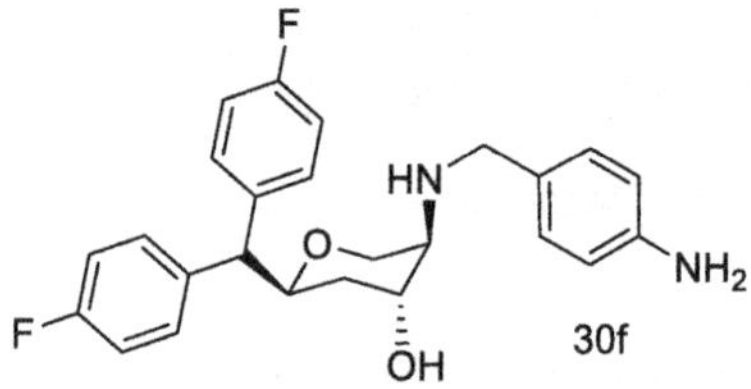
30f

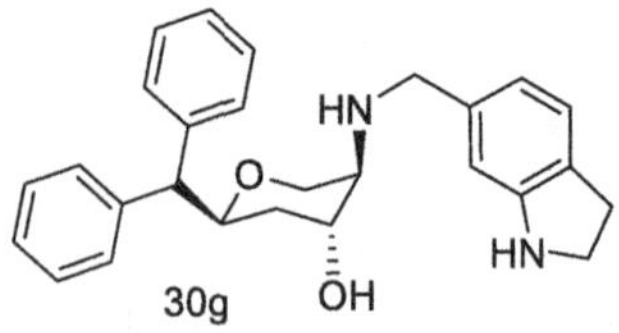
30g

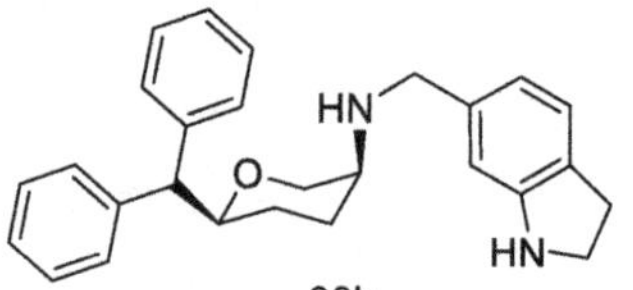
30h

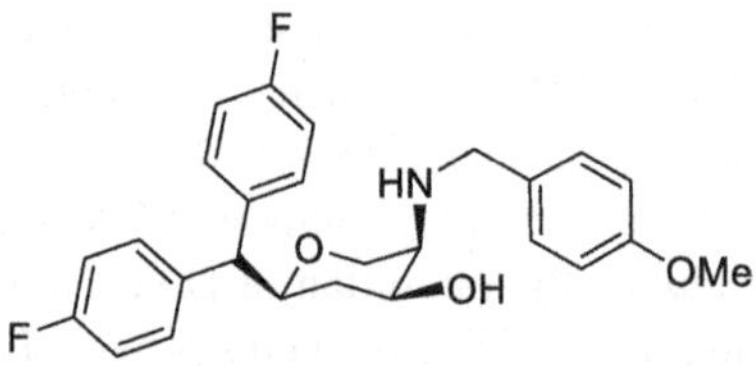
30i

Table 4.8 SAR exploration of 4-benzylpiperidine carboxamide containing TRIs

Compound	N	R	SERT	NET	DAT
			IC_{50} (in μM) are shown as 95% confidence intervals		
29a	3	4-biphenyl	0.04–0.07	0.01–2.12	>10
29b	2	4-biphenyl	0.986	0.17–1.28	>10
29c	3	2-naphthyl	0.06–0.18	0.15–0.76	>10
29d	2	2-naphthyl	0.39–1.31	0.28–0.57	0.71–3.70
Venlafaxine	–	–	0.07–0.52	2.16–3.01	–
GBR12909	–	–	–	–	0.02–0.08

Table 4.9 Systematic SAR exploration of pyran-based TRIs

Compound	DAT (K_i, nM)	NET (K_i, nM)	SERT (K_i, nM)
30a	13.3 ± 2.0	13.2 ± 3.5	46.7 ± 17.0
30b	16.2 ± 1.5	3.23 ± 0.99	16.2 ± 1.5
30c	29.8 ± 4.1	524 ± 132	259 ± 77
30d	7.94 ± 0.66	14.6 (6) ± 2.9	367 ± 52
30e	182 ± 50	27.9 ± 4.9	1540 ± 299
30f	24.5 ± 1.2	3.92 (5) ± 0.71	339 ± 39
30g	29.6 ± 6.9	2.14 ± 0.58	1.72 ± 0.48
30h	56.3 ± 1.1	20.8 ± 4.6	13.6 ± 1.6
30i	6.29 ± 1.52	6.74 ± 1.80	30.1 ± 5.3

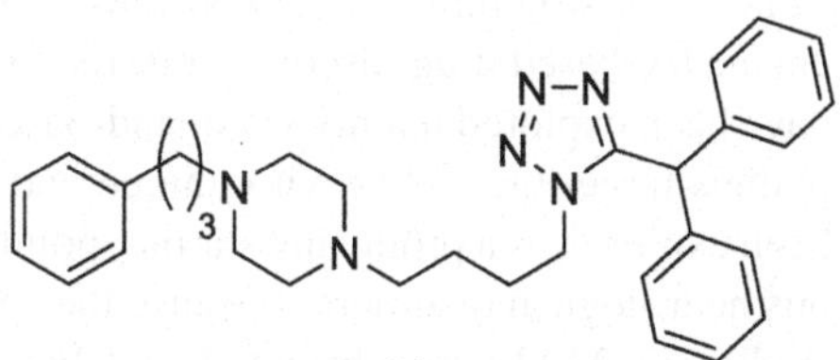

31

Fig. 4.7 Prototype of 1,5-disubstituted tetrazole series of TRIs

The authors used extensive SAR studies to map out the structural requirements for interaction of the pyran derivatives with MATs to design superior TRIs (Table 4.9). The group had previously reported compounds **30a** and **30b** as orally active TRIs with high efficacy in the rat animal model of depression. Molecular docking studies have revealed that **30b** makes strong, conserved H-bonds and hydrophobic interactions within the S1 site of all three transporters. Further modification of the aryl ring of N-benzyl moieties of **30a** and **30b** provided compounds with varying degrees of potency against the three transporters. Subtle changes in the aryl ring substitution, e.g., comparing **30c** and **30d**, and drastic changes in potency are observed within the two positional isomers, with DAT and SERT potencies only moderately affected. The *para*-fluoro substitutions on the biphenyl rings were important for TRI activity as their removal (**30e** versus **30f**) resulted in tenfold reduction in NET potency, sevenfold reduction in DAT activity, and fivefold decrease in SERT inhibition. The importance of (S)-hydroxyl was evident when it was removed from **30g** to get **30h** resulting in tenfold reduction in NET and SERT activity, but only a moderate decrease in DAT potency. Inverting the stereochemistry of hydroxyl from axial position in **30a** to equatorial position to get **30i** slightly improved the TRI activity as compared to **30a**.

Paudel et al. have designed, synthesized, and analyzed the SAR of another promising class of TRIs with the 1,5-disubstituted tetrazole scaffold [71, 73]. The prototype compound **31** (Fig. 4.7) in this series showed potent inhibitory activity against the three reuptake transporters (IC_{50}; 158.7 nM for 5-HT; 99 nM for NE; 97.5 nM for DA).

In conclusion, several SAR studies are currently in place to characterize novel TRI compounds that can be further utilized for developing better therapeutic agents against neurological disorders.

4.6 Conclusion

DAT, NET, and SERT represent important therapeutic targets for a number of CNS-related pathophysiological conditions. The discovery of crystal structures of transporter homologs bound with substrate and inhibitors and corresponding homology-based computational studies have substantially progressed our understanding of structure-function profiles of MATs and has aided in several medicinal chemistry-based drug discovery efforts. Over the years, a multitude of drug classes have been explored mainly via ligand-based drug design and discovery. These compounds target the MATs with unique and interesting behavioral profiles and have been explored as a promising starting point for developing therapeutics against various neurological disorders. Because the vastness of the literature available on SAR studies of MATs is substantial and beyond the scope of this review, we have attempted to summarize the most recent developments that have been made targeting each of the transporters with a particular focus on drug developments that have resulted in unique and superior behavioral effects.

References

1. Kristensen AS, Andersen J, Jorgensen TN, Sorensen L, Eriksen J, Loland CJ, Stromgaard K, Gether U. SLC6 neurotransmitter transporters: structure, function, and regulation. Pharmacol Rev. 2011;63(3):585–640. https://doi.org/10.1124/pr.108.000869.
2. Rudnick G, Kramer R, Blakely RD, Murphy DL, Verrey F. The SLC6 transporters: perspectives on structure, functions, regulation, and models for transporter dysfunction. Pflugers Arch. 2014;466(1):25–42. https://doi.org/10.1007/s00424-013-1410-1.
3. Aggarwal S, Mortensen OV. Overview of monoamine transporters. Curr Protoc Pharmacol. 2017;79:12.16.11–7. https://doi.org/10.1002/cpph.32.
4. Howell LL, Negus SS. Monoamine transporter inhibitors and substrates as treatments for stimulant abuse. Adv Pharmacol. 2014;69:129–76. https://doi.org/10.1016/B978-0-12-420118-7.00004-4.
5. Singh SK, Yamashita A, Gouaux E. Antidepressant binding site in a bacterial homologue of neurotransmitter transporters. Nature. 2007;448(7156):952–6. https://doi.org/10.1038/nature06038.
6. Yamashita A, Singh SK, Kawate T, Jin Y, Gouaux E. Crystal structure of a bacterial homologue of Na+/Cl--dependent neurotransmitter transporters. Nature. 2005;437(7056):215–23. https://doi.org/10.1038/nature03978.
7. Zhou Z, Zhen J, Karpowich NK, Goetz RM, Law CJ, Reith ME, Wang DN. LeuT-desipramine structure reveals how antidepressants block neurotransmitter reuptake. Science. 2007;317(5843):1390–3. https://doi.org/10.1126/science.1147614.

8. Penmatsa A, Wang KH, Gouaux E. X-ray structure of dopamine transporter elucidates antidepressant mechanism. Nature. 2013;503(7474):85–90. https://doi.org/10.1038/nature12533.
9. Penmatsa A, Wang KH, Gouaux E. X-ray structures of drosophila dopamine transporter in complex with nisoxetine and reboxetine. Nat Struct Mol Biol. 2015;22(6):506–8. https://doi.org/10.1038/nsmb.3029.
10. Wang KH, Penmatsa A, Gouaux E. Neurotransmitter and psychostimulant recognition by the dopamine transporter. Nature. 2015;521(7552):322–7. https://doi.org/10.1038/nature14431.
11. Coleman JA, Green EM, Gouaux E. X-ray structures and mechanism of the human serotonin transporter. Nature. 2016;532(7599):334–9. https://doi.org/10.1038/nature17629.
12. Coleman JA, Gouaux E. Structural basis for recognition of diverse antidepressants by the human serotonin transporter. Nat Struct Mol Biol. 2018;25(2):170–5. https://doi.org/10.1038/s41594-018-0026-8.
13. Coleman JA, Navratna V, Antermite D, Yang D, Bull JA, Gouaux E. Chemical and structural investigation of the paroxetine-human serotonin transporter complex. elife. 2020;9:e56427. https://doi.org/10.7554/eLife.56427.
14. Coleman JA, Yang D, Zhao Z, Wen PC, Yoshioka C, Tajkhorshid E, Gouaux E. Serotonin transporter-ibogaine complexes illuminate mechanisms of inhibition and transport. Nature. 2019;569(7754):141–5. https://doi.org/10.1038/s41586-019-1135-1.
15. Wang H, Goehring A, Wang KH, Penmatsa A, Ressler R, Gouaux E. Structural basis for action by diverse antidepressants on biogenic amine transporters. Nature. 2013;503(7474):141–5. https://doi.org/10.1038/nature12648.
16. Xue W, Fu T, Zheng G, Tu G, Zhang Y, Yang F, Tao L, Yao L, Zhu F. Recent advances and challenges of the drugs acting on monoamine transporters. Curr Med Chem. 2020;27(23):3830–76. https://doi.org/10.2174/0929867325666181009123218.
17. Ortore G, Orlandini E, Betti L, Giannaccini G, Mazzoni MR, Camodeca C, Nencetti S. Focus on human monoamine transporter selectivity. New human DAT and NET models, experimental validation, and SERT affinity exploration. ACS Chem Neurosci. 2020;11(20):3214–32. https://doi.org/10.1021/acschemneuro.0c00304.
18. Loland CJ. The use of LeuT as a model in elucidating binding sites for substrates and inhibitors in neurotransmitter transporters. Biochim Biophys Acta. 2015;1850(3):500–10. https://doi.org/10.1016/j.bbagen.2014.04.011.
19. Singh SK, Piscitelli CL, Yamashita A, Gouaux E. A competitive inhibitor traps LeuT in an open-to-out conformation. Science. 2008;322(5908):1655–61. https://doi.org/10.1126/science.1166777.
20. Andersen J, Ladefoged LK, Kristensen TN, Munro L, Grouleff J, Stuhr-Hansen N, Kristensen AS, Schiott B, Stromgaard K. Interrogating the molecular basis for substrate recognition in serotonin and dopamine transporters with high-affinity substrate-based bivalent ligands. ACS Chem Neurosci. 2016;7(10):1406–17. https://doi.org/10.1021/acschemneuro.6b00164.
21. Plenge P, Shi L, Beuming T, Te J, Newman AH, Weinstein H, Gether U, Loland CJ. Steric hindrance mutagenesis in the conserved extracellular vestibule impedes allosteric binding of antidepressants to the serotonin transporter. J Biol Chem. 2012;287(47):39316–26. https://doi.org/10.1074/jbc.M112.371765.
22. Plenge P, Yang D, Salomon K, Laursen L, Kalenderoglou IE, Newman AH, Gouaux E, Coleman JA, Loland CJ. The antidepressant drug vilazodone is an allosteric inhibitor of the serotonin transporter. Nat Commun. 2021;12(1):5063. https://doi.org/10.1038/s41467-021-25363-3.
23. Kortagere S, Fontana AC, Rose DR, Mortensen OV. Identification of an allosteric modulator of the serotonin transporter with novel mechanism of action. Neuropharmacology. 2013;72:282–90. https://doi.org/10.1016/j.neuropharm.2013.04.026.
24. Mortensen OV, Kortagere S. Designing modulators of monoamine transporters using virtual screening techniques. Front Pharmacol. 2015;6:223. https://doi.org/10.3389/fphar.2015.00223.
25. Aggarwal S, Liu X, Rice C, Menell P, Clark PJ, Paparoidamis N, Xiao YC, Salvino JM, Fontana ACK, Espana RA, Kortagere S, Mortensen OV. Identification of a novel allosteric

modulator of the human dopamine transporter. ACS Chem Neurosci. 2019;10(8):3718–30. https://doi.org/10.1021/acschemneuro.9b00262.
26. Kohnen-Johannsen KL, Kayser O. Tropane alkaloids: chemistry, pharmacology, biosynthesis and production. Molecules. 2019;24(4):796. https://doi.org/10.3390/molecules24040796.
27. Carroll FI, Gao Y, Abraham P, Lewin AH, Lew R, Patel A, Boja JW, Kuhar MJ. Probes for the cocaine receptor. Potentially irreversible ligands for the dopamine transporter. J Med Chem. 1992;35(10):1813–7. https://doi.org/10.1021/jm00088a017.
28. Cline EJ, Scheffel U, Boja JW, Carroll FI, Katz JL, Kuhar MJ. Behavioral effects of novel cocaine analogs: a comparison with in vivo receptor binding potency. J Pharmacol Exp Ther. 1992;260(3):1174–9.
29. Kuhar MJ, Ritz MC, Boja JW. The dopamine hypothesis of the reinforcing properties of cocaine. Trends Neurosci. 1991;14(7):299–302. https://doi.org/10.1016/0166-2236(91)90141-G.
30. Lewis D, Zhang Y, Prisinzano T, Dersch CM, Rothman RB, Jacobson AE, Rice KC. Further exploration of 1-[2-[bis-(4-fluorophenyl)methoxy]ethyl]piperazine (GBR 12909): role of N-aromatic, N-heteroaromatic, and 3-oxygenated N-phenylpropyl substituents on affinity for the dopamine and serotonin transporter. Bioorg Med Chem Lett. 2003;13(7):1385–9. https://doi.org/10.1016/s0960-894x(03)00108-2.
31. Prisinzano T, Greiner E, Johnson EM 2nd, Dersch CM, Marcus J, Partilla JS, Rothman RB, Jacobson AE, Rice KC. Piperidine analogues of 1-[2-[bis(4-fluorophenyl)methoxy]ethyl]-4-(3-phenylpropyl)piperazine (GBR 12909): high affinity ligands for the dopamine transporter. J Med Chem. 2002;45(19):4371–4. https://doi.org/10.1021/jm020264r.
32. Newman AH, Agoston GE. Novel benztropine [3a-(diphenylmethoxy)tropane] analogs as probes for the dopamine transporter. Curr Med Chem. 1998;5(4):305–19.
33. Newman AH, Kulkarni S. Probes for the dopamine transporter: new leads toward a cocaine-abuse therapeutic – A focus on analogues of benztropine and rimcazole. Med Res Rev. 2002;22(5):429–64. https://doi.org/10.1002/med.10014.
34. Dassanayake AF, Canales JJ. Replacement treatment during extinction training with the atypical dopamine uptake inhibitor, JHW-007, reduces relapse to methamphetamine seeking. Neurosci Lett. 2018;671:88–92. https://doi.org/10.1016/j.neulet.2018.02.025.
35. Ferragud A, Velazquez-Sanchez C, Canales JJ. Modulation of methamphetamine's locomotor stimulation and self-administration by JHW 007, an atypical dopamine reuptake blocker. Eur J Pharmacol. 2014;731:73–9. https://doi.org/10.1016/j.ejphar.2014.03.015.
36. Velazquez-Sanchez C, Ferragud A, Murga J, Carda M, Canales JJ. The high affinity dopamine uptake inhibitor, JHW 007, blocks cocaine-induced reward, locomotor stimulation and sensitization. Eur Neuropsychopharmacol. 2010;20(7):501–8. https://doi.org/10.1016/j.euroneuro.2010.03.005.
37. Loland CJ, Mereu M, Okunola OM, Cao J, Prisinzano TE, Mazier S, Kopajtic T, Shi L, Katz JL, Tanda G, Newman AH. R-modafinil (armodafinil): a unique dopamine uptake inhibitor and potential medication for psychostimulant abuse. Biol Psychiatry. 2012;72(5):405–13. https://doi.org/10.1016/j.biopsych.2012.03.022.
38. Myrick H, Malcolm R, Taylor B, LaRowe S. Modafinil: preclinical, clinical, and post-marketing surveillance – a review of abuse liability issues. Ann Clin Psychiatry. 2004;16(2):101–9. https://doi.org/10.1080/10401230490453743.
39. Vosburg SK, Hart CL, Haney M, Rubin E, Foltin RW. Modafinil does not serve as a reinforcer in cocaine abusers. Drug Alcohol Depend. 2010;106(2–3):233–6. https://doi.org/10.1016/j.drugalcdep.2009.09.002.
40. Hersey M, Bacon AK, Bailey LG, Coggiano MA, Newman AH, Leggio L, Tanda G. Psychostimulant use disorder, an unmet therapeutic goal: can modafinil narrow the gap? Front Neurosci. 2021;15:656475. https://doi.org/10.3389/fnins.2021.656475.
41. Cao J, Prisinzano TE, Okunola OM, Kopajtic T, Shook M, Katz JL, Newman AH. Structure-activity relationships at the monoamine transporters for a novel series of modafinil (2-[(diphenylmethyl)sulfinyl]acetamide) analogues. ACS Med Chem Lett. 2010;2(1):48–52. https://doi.org/10.1021/ml1002025.

42. Cao J, Slack RD, Bakare OM, Burzynski C, Rais R, Slusher BS, Kopajtic T, Bonifazi A, Ellenberger MP, Yano H, He Y, Bi GH, Xi ZX, Loland CJ, Newman AH. Novel and high affinity 2-[(Diphenylmethyl)sulfinyl]acetamide (modafinil) analogues as atypical dopamine transporter inhibitors. J Med Chem. 2016;59(23):10676–91. https://doi.org/10.1021/acs.jmedchem.6b01373.
43. Okunola-Bakare OM, Cao J, Kopajtic T, Katz JL, Loland CJ, Shi L, Newman AH. Elucidation of structural elements for selectivity across monoamine transporters: novel 2-[(diphenylmethyl)sulfinyl]acetamide (modafinil) analogues. J Med Chem. 2014;57(3):1000–13. https://doi.org/10.1021/jm401754x.
44. Tanda G, Hersey M, Hempel B, Xi ZX, Newman AH. Modafinil and its structural analogs as atypical dopamine uptake inhibitors and potential medications for psychostimulant use disorder. Curr Opin Pharmacol. 2021;56:13–21. https://doi.org/10.1016/j.coph.2020.07.007.
45. Keighron JD, Giancola JB, Shaffer RJ, DeMarco EM, Coggiano MA, Slack RD, Newman AH, Tanda G. Distinct effects of (R)-modafinil and its (R)- and (S)-fluoro-analogs on mesolimbic extracellular dopamine assessed by voltammetry and microdialysis in rats. Eur J Neurosci. 2019;50(3):2045–53. https://doi.org/10.1111/ejn.14256.
46. Keighron JD, Quarterman JC, Cao J, DeMarco EM, Coggiano MA, Gleaves A, Slack RD, Zanettini C, Newman AH, Tanda G. Effects of (R)-modafinil and modafinil analogues on dopamine dynamics assessed by voltammetry and microdialysis in the mouse nucleus accumbens shell. ACS Chem Neurosci. 2019;10(4):2012–21. https://doi.org/10.1021/acschemneuro.8b00340.
47. Zhang HY, Bi GH, Yang HJ, He Y, Xue G, Cao J, Tanda G, Gardner EL, Newman AH, Xi ZX. The novel modafinil analog, JJC8-016, as a potential cocaine abuse pharmacotherapeutic. Neuropsychopharmacology. 2017;42(9):1871–83. https://doi.org/10.1038/npp.2017.41.
48. Tunstall BJ, Ho CP, Cao J, Vendruscolo JCM, Schmeichel BE, Slack RD, Tanda G, Gadiano AJ, Rais R, Slusher BS, Koob GF, Newman AH, Vendruscolo LF. Atypical dopamine transporter inhibitors attenuate compulsive-like methamphetamine self-administration in rats. Neuropharmacology. 2018;131:96–103. https://doi.org/10.1016/j.neuropharm.2017.12.006.
49. Newman AH, Cao J, Keighron JD, Jordan CJ, Bi GH, Liang Y, Abramyan AM, Avelar AJ, Tschumi CW, Beckstead MJ, Shi L, Tanda G, Xi ZX. Translating the atypical dopamine uptake inhibitor hypothesis toward therapeutics for treatment of psychostimulant use disorders. Neuropsychopharmacology. 2019;44(8):1435–44. https://doi.org/10.1038/s41386-019-0366-z.
50. Slack RD, Ku TC, Cao J, Giancola JB, Bonifazi A, Loland CJ, Gadiano A, Lam J, Rais R, Slusher BS, Coggiano M, Tanda G, Newman AH. Structure-activity relationships for a series of (bis(4-fluorophenyl)methyl)sulfinyl alkyl alicyclic amines at the dopamine transporter: functionalizing the terminal nitrogen affects affinity, selectivity, and metabolic stability. J Med Chem. 2020;63(5):2343–57. https://doi.org/10.1021/acs.jmedchem.9b01188.
51. Giancola JB, Bonifazi A, Cao J, Ku T, Haraczy AJ, Lam J, Rais R, Coggiano MA, Tanda G, Newman AH. Structure-activity relationships for a series of (bis(4-fluorophenyl)methyl)sulfinylethyl-aminopiperidines and -piperidine amines at the dopamine transporter: bioisosteric replacement of the piperazine improves metabolic stability. Eur J Med Chem. 2020;208:112674. https://doi.org/10.1016/j.ejmech.2020.112674.
52. Hiranita T, Hong WC, Kopajtic T, Katz JL. Sigma receptor effects of N-substituted benztropine analogs: implications for antagonism of cocaine self-administration. J Pharmacol Exp Ther. 2017;362(1):2–13. https://doi.org/10.1124/jpet.117.241109.
53. Kalaba P, Ilic M, Aher NY, Dragacevic V, Wieder M, Zehl M, Wackerlig J, Beyl S, Sartori SB, Ebner K, Roller A, Lukic N, Beryozkina T, Gonzalez ERP, Neill P, Khan JA, Bakulev V, Leban JJ, Hering S, Pifl C, Singewald N, Lubec J, Urban E, Sitte HH, Langer T, Lubec G. Structure-activity relationships of novel thiazole-based modafinil analogues acting at monoamine transporters. J Med Chem. 2020;63(1):391–417. https://doi.org/10.1021/acs.jmedchem.9b01938.

54. Schmitt KC, Rothman RB, Reith ME. Nonclassical pharmacology of the dopamine transporter: atypical inhibitors, allosteric modulators, and partial substrates. J Pharmacol Exp Ther. 2013;346(1):2–10. https://doi.org/10.1124/jpet.111.191056.
55. Reith ME, Blough BE, Hong WC, Jones KT, Schmitt KC, Baumann MH, Partilla JS, Rothman RB, Katz JL. Behavioral, biological, and chemical perspectives on atypical agents targeting the dopamine transporter. Drug Alcohol Depend. 2015;147:1–19. https://doi.org/10.1016/j.drugalcdep.2014.12.005.
56. Rothman RB, Ananthan S, Partilla JS, Saini SK, Moukha-Chafiq O, Pathak V, Baumann MH. Studies of the biogenic amine transporters 15. Identification of novel allosteric dopamine transporter ligands with nanomolar potency. J Pharmacol Exp Ther. 2015;353(3):529–38. https://doi.org/10.1124/jpet.114.222299.
57. Refai O, Aggarwal S, Cheng MH, Gichi Z, Salvino JM, Bahar I, Blakely RD, Mortensen OV. Allosteric Modulator KM822 Attenuates Behavioral Actions of Amphetamine in Caenorhabditis elegans through interactions with the dopamine transporter DAT-1. Mol Pharmacol. 2022;101(3):123–31. https://doi.org/10.1124/molpharm.121.000400.
58. Little A. Treatment-resistant depression. Am Fam Physician. 2009;80(2):167–72.
59. Bigler AJ, Bøgesø K, et al. Quantitative structure-activity relationship in a series of selective 5-HT uptake inhibitors. Eur J Med Chem. 1977;12:289–95.
60. Eildal JNN, Andersen J, Kristensen AS, Jørgensen AM, Bang-Andersen B, Jørgensen M, Strømgaard K. From the selective serotonin transporter inhibitor citalopram to the selective norepinephrine transporter inhibitor talopram: synthesis and structure–activity relationship studies. J Med Chem. 2008;51(10):3045–8. https://doi.org/10.1021/jm701602g.
61. Zhang P, Cyriac G, Kopajtic T, Zhao Y, Javitch JA, Katz JL, Newman AH. Structure–activity relationships for a novel series of citalopram (1-(3-(dimethylamino)propyl)-1-(4-fluorophenyl)-1,3-dihydroisobenzofuran-5-carbonitrile) analogues at monoamine transporters. J Med Chem. 2010;53(16):6112–21. https://doi.org/10.1021/jm1005034.
62. Andersen J, Olsen L, Hansen KB, Taboureau O, Jørgensen FS, Jørgensen AM, Bang-Andersen B, Egebjerg J, Strømgaard K, Kristensen AS. Mutational mapping and modeling of the binding site for (S)-citalopram in the human serotonin transporter. J Biol Chem. 2010;285(3):2051–63. https://doi.org/10.1074/jbc.M109.072587.
63. Andersen J, Stuhr-Hansen N, Zachariassen L, Toubro S, Hansen SM, Eildal JN, Bond AD, Bogeso KP, Bang-Andersen B, Kristensen AS, Stromgaard K. Molecular determinants for selective recognition of antidepressants in the human serotonin and norepinephrine transporters. Proc Natl Acad Sci U S A. 2011;108(29):12137–42. https://doi.org/10.1073/pnas.1103060108.
64. Banala AK, Zhang P, Plenge P, Cyriac G, Kopajtic T, Katz JL, Loland CJ, Newman AH. Design and synthesis of 1-(3-(dimethylamino)propyl)-1-(4-fluorophenyl)-1,3-dihydroisobenzofuran-5-carboni trile (citalopram) analogues as novel probes for the serotonin transporter S1 and S2 binding sites. J Med Chem. 2013;56(23):9709–24. https://doi.org/10.1021/jm4014136.
65. Coleman JA, Green EM, Gouaux E. Thermostabilization, expression, purification, and crystallization of the human serotonin transporter bound to S-citalopram. J Vis Exp. 2016;(117):54792. https://doi.org/10.3791/54792.
66. Garnock-Jones KP, Keating GM. Atomoxetine: a review of its use in attention-deficit hyperactivity disorder in children and adolescents. Pediatr Drugs. 2009;11(3):203–26.
67. Lane RM. Antidepressant drug development: focus on triple monoamine reuptake inhibition. J Psychopharmacol. 2015;29(5):526–44. https://doi.org/10.1177/0269881114553252.
68. Larsen MA, Plenge P, Andersen J, Eildal JN, Kristensen AS, Bøgesø KP, Gether U, Strømgaard K, Bang-Andersen B, Loland CJ. Structure-activity relationship studies of citalopram derivatives: examining substituents conferring selectivity for the allosteric site in the 5-HT transporter. Br J Pharmacol. 2016;173(5):925–36. https://doi.org/10.1111/bph.13411.

69. Li N, Li C, Han R, Wang Y, Yang M, Wang H, Tian J. LPM580098, a novel triple reuptake inhibitor of serotonin, noradrenaline, and dopamine, attenuates neuropathic pain. Front Pharmacol. 2019;10:53. https://doi.org/10.3389/fphar.2019.00053.
70. Patel K, Allen S, Haque MN, Angelescu I, Baumeister D, Tracy DK. Bupropion: a systematic review and meta-analysis of effectiveness as an antidepressant. Ther Adv Psychopharmacol. 2016;6(2):99–144. https://doi.org/10.1177/2045125316629071.
71. Paudel S, Acharya S, Yoon G, Kim KM, Cheon SH. Design, synthesis and in vitro activity of 1,4-disubstituted piperazines and piperidines as triple reuptake inhibitors. Bioorg Med Chem. 2017;25(7):2266–76. https://doi.org/10.1016/j.bmc.2017.02.051.
72. Paudel S, Kim E, Zhu A, Acharya S, Min X, Cheon SH, Kim KM. Structural requirements for modulating 4-benzylpiperidine carboxamides from serotonin/norepinephrine reuptake inhibitors to triple reuptake inhibitors. Biomol Ther (Seoul). 2021;29(4):392–8. https://doi.org/10.4062/biomolther.2020.233.
73. Paudel S, Min X, Acharya S, Khadka DB, Yoon G, Kim KM, Cheon SH. Triple reuptake inhibitors: design, synthesis and structure-activity relationship of benzylpiperidine-tetrazoles. Bioorg Med Chem. 2017;25(20):5278–89. https://doi.org/10.1016/j.bmc.2017.07.046.
74. Plenge P, Abramyan AM, Sørensen G, Mørk A, Weikop P, Gether U, Bang-Andersen B, Shi L, Loland CJ. The mechanism of a high-affinity allosteric inhibitor of the serotonin transporter. Nat Commun. 2020;11(1):1491. https://doi.org/10.1038/s41467-020-15292-y.
75. Santra S, Kortagere S, Vedachalam S, Gogoi S, Antonio T, Reith MEA, Dutta AK. Novel potent dopamine–norepinephrine and triple reuptake uptake inhibitors based on asymmetric pyran template and their molecular interactions with monoamine transporters. ACS Chem Neurosci. 2021;12(8):1406–18. https://doi.org/10.1021/acschemneuro.1c00078.
76. Sorensen L, Andersen J, Thomsen M, Hansen SM, Zhao X, Sandelin A, Stromgaard K, Kristensen AS. Interaction of antidepressants with the serotonin and norepinephrine transporters: mutational studies of the S1 substrate binding pocket. J Biol Chem. 2012;287(52):43694–707. https://doi.org/10.1074/jbc.M112.342212.
77. Subbaiah MAM. Triple reuptake inhibitors as potential therapeutics for depression and other disorders: design paradigm and developmental challenges. J Med Chem. 2018;61(6):2133–65. https://doi.org/10.1021/acs.jmedchem.6b01827.
78. Tu G, Fu T, Yang F, Yang J, Zhang Z, Yao X, Xue W, Zhu F. Understanding the polypharmacological profiles of triple reuptake inhibitors by molecular simulation. ACS Chem Neurosci. 2021;12(11):2013–26. https://doi.org/10.1021/acschemneuro.1c00127.
79. Zhang M, Jovic F, Vickers T, Dyck B, Tamiya J, Grey J, Tran JA, Fleck BA, Pick R, Foster AC, Chen C. Studies on the structure–activity relationship of bicifadine analogs as monoamine transporter inhibitors. Bioorg Med Chem Lett. 2008;18(13):3682–6. https://doi.org/10.1016/j.bmcl.2008.05.077.
80. Zhang Y, Zheng G, Fu T, Hong J, Li F, Yao X, Xue W, Zhu F. The binding mode of vilazodone in the human serotonin transporter elucidated by ligand docking and molecular dynamics simulations. Phys Chem Chem Phys. 2020;22(9):5132–44. https://doi.org/10.1039/c9cp05764a.
81. Zhu H, Wang W, Sha C, Guo W, Li C, Zhao F, Wang H, Jiang W, Tian J. Pharmacological characterization of Toludesvenlafaxine as a triple reuptake inhibitor. Front Pharmacol. 2021;12:741794–4. https://doi.org/10.3389/fphar.2021.741794.

Chapter 5
Drug Development for New Psychiatric Drug Therapies

M. Lynn Crismon, Janet Walkow, and Roger W. Sommi

Abstract Drug development is an expensive, high risk, and highly regulated process. Only about 6.2% of new molecules tested for mental disorders eventually achieve Food and Drug Administration (FDA) approval. New molecular entities are produced, and extensive in vitro animal testing is performed before they are evaluated in humans. The compound is used in animals to predict clinical effects in humans, and studies addressing pharmacodynamics, pharmacokinetics, toxicology, and mutagenicity are conducted. Human research proceeds in three stages with the ultimate goal of proving that a new agent is efficacious and safe for a treatment of a specific disease in humans. If efficacy and safety are demonstrated in two Phase III studies, then the sponsor can submit a new drug application (NDA) to the FDA. The FDA oversees each step of the process to assure that good research practices are followed, data integrity is assured, and human research subjects are protected.

Keywords Drug development · Clinical trial · FDA · Biomarker · Animal models · Psychotropic · Regulation · Pharmacokinetics · Pharmacodynamics · Efficacy · Safety

Innovations in healthcare products or services have the potential to positively impact society; however, taking an early discovery or idea to the clinic is a daunting challenge. Along the way, these technologies face roadblocks unique to healthcare. The drug development pathway (Fig. 5.1) details the various steps and regulatory requirements that guide the development of new psychiatric drug therapies. A strong

M. L. Crismon (✉) · J. Walkow
The University of Texas at Austin, Austin, TX, USA
e-mail: lynn.crismon@austin.utexas.edu; jwalkow@austin.utexas.edu

R. W. Sommi
University of Missouri at Kansas City, Kansas City, MO, USA
e-mail: SommiR@umkc.edu

M. Macaluso et al. (eds.), *Drug Development in Psychiatry*, Advances in Neurobiology 30,
https://doi.org/10.1007/978-3-031-21054-9_5

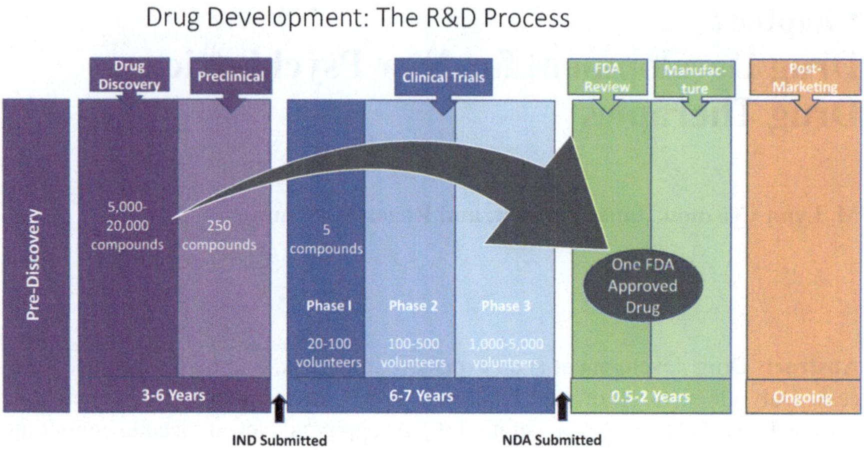

Fig. 5.1 The research and development process. Adapted from the American Association for Cancer Research [3]

understanding of the processes, regulatory requirements, timelines, and the impact they have on developing and commercializing new psychiatric drug products is essential for understanding the timing, cost, and complexity involved for each product that is commercialized. The goal of drug development is to find and provide new drugs that we can depend on to improve our health and quality of life.

Psychotropic medications are used to treat the symptoms associated with a variety of mental disorders. These therapies exert effects on the brain and nervous system, as well as other organ systems, and they undergo extensive development and testing before becoming a prescription product. The pathway for new drug therapies to be approved by the Food & Drug Administration (FDA) is highly regulated to ensure the efficacy, safety, and quality of these medicines. Understanding the steps involved, the clinical components, and the FDA's role in approving new psychiatric therapies provides important context for researchers, clinicians, and other health professionals.

The information presented in this chapter reflects the current US practices and regulatory guidelines unless otherwise noted.

5.1 The Drug Development Pathway

Drug development starts with discovery of a new molecule which, if successful, proceeds through in vitro studies, animal studies, studies in healthy individuals, and proceeding to studies in people with the disorder of interest. If found to be both efficacious and safe, then a new drug application is submitted to the FDA.

5.1.1 Pathway Overview

There are four distinct stages of the pathway for developing new drug therapies. The process for evaluating and identifying the best candidates to advance is time-consuming and expensive. It can take between 10 and 14 years and cost billions to develop a therapy with the characteristics to achieve a specific, desired result.

The earliest stage, *discovery/preclinical*, includes all of the activities before a new drug therapy advances into human clinical trials. The first part, *drug discovery*, involves researchers scanning up to tens of thousands of potential compounds before identifying lead candidates for a specific disease. This process identifies up to 250 compounds to move forward for more detailed evaluation.

The lead candidates progress to the *preclinical* phase for further characterization, experiments, and formulation development into the most suitable forms for the specific treatment, including a tablet, capsule, orally dissolving tablet, injectable, intravenous solution, topical patch, or inhaled product. As potential drug candidates go through this stage, the number narrows to a limited number (five or fewer) that will advance into clinical trials. Once the preclinical work has progressed sufficiently, an *investigational new drug (IND)* application is filed. The FDA reviews the application and decides whether there is enough information to allow a product to proceed to human clinical trials. Only those with the best potential to achieve specific clinical results will move into this stage.

Once a product is approved by the FDA to enter *clinical trials*, a *clinical development plan (CDP)* is developed and reviewed with the FDA to define how the clinical trials will be conducted. The clinical trials consist of three stages, typically starting with a small group of healthy volunteers (Phase I) to determine the safety and basic pharmacokinetics of the test product, to large studies involving hundreds to a thousand or more patients at multiple study sites (Phase III). In order to progress from one phase to the next, the FDA must review the data and decide whether to allow the studies to progress to the next phase.

The third stage that encompasses regulatory review by the FDA is the manufacturing processes. These activities take place as early as the drug discovery and preclinical phases. Throughout the development pathway, regulatory activities and efforts to scale up production of the *active therapeutic ingredient (ATI)* as well as the final formulation are occurring. The *investigational new drug (IND)* application is filed during the preclinical phase, and a *new drug application (NDA)* can be filed with the FDA for review and potential approval following completion of Phase III clinical trials. From the original thousands of compounds identified in the discovery phase, only one will typically make it all the way through the process and be approved by the FDA for commercialization.

Following the approval of a new drug product, ongoing monitoring of compliance with manufacturing and distribution processes must occur. In addition, information concerning adverse effects is collected over a long period of time. Clinical trials are limited to hundreds to perhaps over a thousand people, and it is often not obvious which adverse effects are most prominent or problematic until many more patients have been treated. It may take years to answer whether certain adverse effects, particularly those that are rare but potentially serious, are problematic and those effects are addressed in the fourth stage that is often referred to as Phase IV, *post-marketing surveillance programs*, which assess patient safety and quality of life as well as drug product effectiveness.

5.1.2 Drug Development Costs

As indicated in Fig. 5.2, developing new drug therapies is a massive investment in time and resources. In the 1980s, the cost to develop a new drug product was approximately $100 million. A 2014 report found the cost to develop a prescription drug had reached $2.6 billion [20]. Over four decades the cost has risen exponentially. Examination of the cost per development stage reveals small increases in discovery, accelerating costs in preclinical, and the most significant increases occurring in the clinical trials (Phases I, II, and III). Only launch and discovery costs appear to have decreased, likely attributed to the reduced costs of digital advertising and newer, high-throughput screening tools used in early discovery [20]. Part of this cost is driven by the fact that only about 9.6% of all compounds across all therapeutic areas are FDA approved and reach the market. This is even lower in psychiatry (6.2%) and neurology (8.4%). The Phase III clinical trials' failure rate with central nervous system (CNS) acting agents is significantly higher than with other types of agents, and it is estimated that the Phase III clinical trial failure rate is 20% higher than in other therapeutic areas [39]. This has prompted pharmaceutical companies to be both cautious and strategic in deciding the types of compounds to develop. [38, 69].

Since the 1990s, drug development costs have more than doubled, with the largest cost increase being in human clinical trials. Adapted from O'Hagan [59], Life Size VC [53], Policy and Medicine [61].

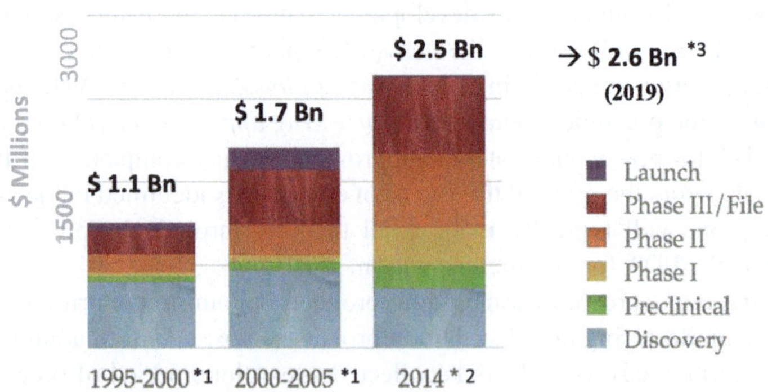

Fig. 5.2 Increase in drug development costs over time

Between 2009 and 2018, it was estimated to cost between $314 million and $2.8 billion to bring a new therapeutic entity to market, including the capitalized R&D cost per product and expenditures on failed trials [77].

5.1.3 Regulatory Overview

The approval pathways for drug therapies can be complex, time-consuming, and expensive. The regulatory tools that have been developed to certify compliance and efficacy of new medicines are designed to ensure the safety and integrity of every product before it can be approved for commercialization. Most take for granted that drug therapies are safe if they are approved for commercialization and sale – a good assumption since FDA's involvement starts early in the development process and continues after a product is commercialized.

The FDA becomes involved in the preclinical phase, when the sponsor identifies a lead molecule and plans to prepare for testing its therapeutic potential in human clinical trials. This is when regulations become a prominent part of developing the strategy for advancing a product.

The regulations that are required or authorized by statute are published in the Federal Register, the US government's official publication for notifying the public of agency actions. The procedures that health innovators follow come from US laws, executive orders, and FDA regulations that can be found in the Code of Federal Regulations (CFR). Regulations that apply to the FDA's oversight of food and drugs are found in Section 21 of the CFR [15]. These regulations document all actions required of drug sponsors under federal law.

To better understand how FDA regulations align with the R&D timeline, the development pathway is depicted in Fig. 5.3, including the key regulatory elements as they relate to the development pathway.

Drug development is a highly regulated process, with the Food and Drug Administration carefully monitoring and evaluating each step of the process. Developed from [28].

These will be discussed in greater depth in the preclinical, clinical, regulatory, and post-marketing sections.

5.1.4 Types of Drug Therapies

Drug therapies fall into several categories: new molecular entities (NME), therapeutic biologics, generics, biosimilars, over-the-counter (OTC) drugs, vaccines, blood products, and cellular and gene therapy products. New psychiatric drug products are currently prescription products in the NME category.

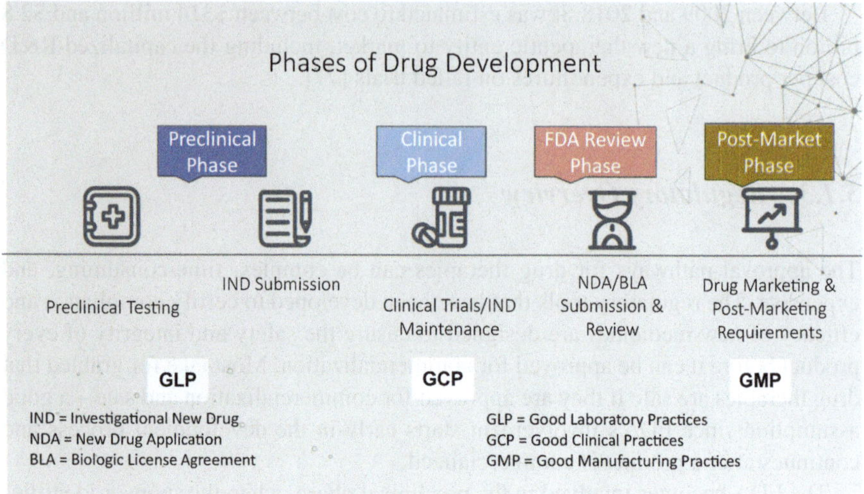

Fig. 5.3 Regulatory components of drug development

New Molecular Entities

The development and approval of new molecules and biological therapies are handled by CDER. This FDA branch also oversees nonprescription, or OTC drugs as well as the process for moving an approved prescription (Rx) product to OTC status, called an *Rx-to-OTC switch.*

Generics

Generics are copies of innovator or brand-name prescription drugs and comprise almost 90 percent of all prescriptions dispensed in the United States [7]. Generic drug developers are allowed to use data generated by brand-name companies for efficacy and safety, resulting in significantly lower development costs and lower prices for patients. The Hatch-Waxman Act of 1984 established patent exclusivity protections for drug innovators and granted generic companies access to regulatory approval by filing an abbreviated NDA (ANDA). To achieve FDA approval, the manufacturer must demonstrate that the generic product has bioavailability similar to the innovator product. To be considered bioequivalent, the generic product must have a peak plasma concentration, time to peak plasma concentration, and total amount of drug absorbed (area under the curve) that is not statistically different from the innovator's product. These studies are typically crossover studies in a small number of healthy male subjects, using the subject as his own control

5.2 Preclinical Drug Development Phase

The preclinical phase is a robust transition with the overall goal of predicting whether a compound will be beneficial in treating a particular disease. Activities focus on evaluating drug molecules from the discovery phase.

5.2.1 Characterization

A primary job of the pharmaceutical formulation scientist is determining how best to formulate and deliver a compound to its site of action in the body, while ensuring that it remains stable over time. The starting point of this process focuses on obtaining a greater understanding of the lead compound itself and determining the physical and chemical properties of the lead compound. Larger quantities of the drug molecules are made so that they can be tested to gauge their solubility in various liquids, sensitivity to light or heat, chemical stability, and interactions with other materials.

All of these important properties need to be evaluated and addressed before attempting to incorporate an active compound into a dosage form. These answers play a large role in the stability of a dosage form, how available it is at its target site of action, the most appropriate storage conditions, and how the drug product should be manufactured.

5.2.2 Developing a Formulation Prototype

Characterizing the drug molecules, coupled with knowledge of the disease or condition being treated, as well as the patient population, informs the type of formulation that should be considered, whether it is preferred as a tablet, capsule, orally disintegrating tablet, injectable, inhaled product, nasal spray, topical cream/ointment, or transdermal patch. The selection is based on the drug's chemical and physical properties with consideration of patient population needs. Another consideration is the release rate – whether a rapid release or extended release formulation is preferred to reach appropriate dosing levels in a specific clinical condition. Patient adherence may be impacted by the dosing schedule, and extended release products that require only daily or twice daily doses can be helpful for medication adherence. Similarly, long-acting injectable products may allow parenteral administration every few weeks or months.

In psychiatry, several interesting developments occur beyond oral formulations that are directed at specific clinical management challenges. Specifically, it is important to address administration of medications during an acute crisis where patients may be combative and uncooperative and treatment nonadherent. Other formulations may address either tolerability or unique oral absorption challenges.

Some examples of strategies to counter the acute crisis situation are the use of oral solutions and orally disintegrating tablets – where the clinician has some confirmation that the patient has actually swallowed the medication or that the drug will be absorbed through the oral mucosa. Immediate acting injections allow quick onset but may be less acceptable to patients. Other novel formulations include inhaled loxapine [19] which results in rapid absorption and onset of action.

Challenges with overcoming barriers with medication adherence have been addressed principally with improving the drug's tolerability and the development of long-acting injectable (LAI) formulations. In psychiatry most of the LAI development has been with antipsychotics. LAI development incorporates molecular strategies including using prodrugs (esters) and/or physical strategies of using oils, polymers, gels, and manipulation of particle size with the ultimate goal of delaying absorption to the point the drug can be administered in intervals of weeks to months. Marketed products include the use of esterified molecules delivered in sesame oil, drug molecules trapped in glycolide/lactide microspheres, crystallization and grinding processes to create various particle sizes of relatively insoluble ester salts, and the use of gels which essentially form an implant that slowly dissolves and releases drug.

Challenges to counter absorption and tolerability include the use of patches and sublingual formulations. Selegiline patches allow for the absorption of the drug with clinically relevant CNS concentrations while minimizing the inhibition of MAO-A in the gut – improving the tolerability and limiting the risk of dietary tyramine interactions [18]. Asenapine is available as a sublingually absorbed formulation as the bioavailability is drastically reduced if swallowed. More recently, an asenapine patch formulation allows for absorption of the drug without the issue of dysgeusia limiting its use. Other antipsychotic patch formulations are currently in development [1]. In addition to solving issues around oral absorption, administration in a crisis, or enhancing adherence, the use of oral solutions, orally disintegrating tablets, and patches provides a pathway to drug administration for patients unable to swallow capsules or tablets or who are restricted from taking medications orally.

The use of prodrugs is another method to potentially avoid drug related complications. A prodrug is inactive pharmacologically but is metabolized in the body to the active form. An example is lisdexamfetamine which is hydrolyzed in the blood to amphetamine. Lisdexamfetamine has potentially less abuse potential than amphetamine, even when administered intravenously [54].

5.2.3 In Vitro-in Vivo Testing

The material characterization and early formulation studies of a lead compound are performed using a variety of equipment and in vitro testing methods. These provide important information to assess whether to proceed to in vivo studies. This is a noteworthy transition point, as the in vivo studies require extensive knowledge and experience to design, conduct, and analyze the results.

Significant effort is taken in selecting formulations to include in the in vivo studies, as they require substantial time and cost commitments. The early and later stage pharmacokinetic-pharmacodynamic (PK-PD) studies will often begin with rodents and may expand to other, more specific animal models for the intended disease target. It is common for one or more formulations to be evaluated in the early in vivo studies to observe differences in effectiveness, tolerability, and adverse effects that can lead to choosing a lead formulation or deciding that it needs to be re-engineered.

At the same time, the in vivo studies are occurring, the formulations being tested in animals are often undergoing stability studies at various temperatures, light conditions, and humidity. These parallel activities can serve to expedite moving through the preclinical phase.

5.2.4 Pharmacokinetic-Pharmacodynamic (PK-PD) Analysis

The PK-PD analysis examines the drug concentration in a body compartment – most commonly venous blood – and relates it to the drug's effect. The bioanalysis provides an estimate of the molecule's safety and efficacy. Animal studies can provide a pharmacokinetic profile that shows how well the drug products are absorbed, distributed, metabolized, and excreted. Drug absorption refers to the percent of administered drug that ultimately reaches the circulation, also referred to as the drug's bioavailability. For example, intravenous medications have 100% bioavailability; however, other routes of administration and formulation are generally lower since they may be absorbed in the intestine or be subject to first pass metabolism in the liver. Bioavailability in preclinical and clinical studies is determined by plotting the blood concentration as a function of time and referred to as the area under the curve (AUC).

Biomarker selection and correlation with clinical endpoints are important for successful PK-PD modeling and provide predictive value in drug development, if they reflect the mechanism of action for intervention, whether or not they are surrogate endpoints [16]. Identification of biomarkers that can be used for predictive clinical assessment of disease progression can help measure the effect of drug interventions.

The results of the PK-PD studies provide information on how to refine drug administration, evaluating whether the dose needs to be adjusted or deciding if the formulation needs to be changed. The studies may show that the drug or its metabolites stay in the body a short time, making it necessary to take several doses a day or develop a sustained release formulation.

Psychotropic drug development has been limited in that a mental disorder diagnosis is based upon clinical phenotypes which likely represent substantial heterogeneity in etiology and pathophysiology [9]. In fact, the specific pathophysiology of mental disorders is currently unknown. Therefore, specific pharmacological mechanisms of action are used for heterogenous disorders. Serendipity has played a major role in the discovery of psychotropic medications, particularly during their first few decades of development. This is best represented by the fact that chlorpromazine was first investigated as a potential medication for "surgical shock" largely based upon its antihistaminic properties. Although ineffective for this purpose, the French surgeon Henri Laborit noted that patients experienced "no loss of consciousness, no change in the patient's mentality but a slight tendency to sleep and above all 'disinterest' for all that goes on around him" [71]. Based upon his observations, he encouraged psychiatrists to use it in patients with psychosis, and it was subsequently developed as the first modern era antipsychotic.

Animal Models

Animal models have significant limitations when used to study mental disorders and the effects of medications. Rodents do not express the same range of emotions as humans, and the neural circuits are not nearly as complex [65]. Traditionally, the animal models used to predict efficacy of a medication for a particular mental disorder were developed based upon a behavior, and if it affected that behavior, then the mechanism of action was explored [17]. For example, early animal models of antipsychotic effect were dependent on a drug's ability to produce catalepsy [58]. Thus, developed antipsychotics were almost guaranteed to produce extrapyramidal adverse effects in humans. More recently, computational chemistry has been used to predict the mechanisms that specific compounds will have. However, psychotropic drugs are largely still not developed based upon a known pathophysiology for a given disorder. Although it is hoped that genetics may ultimately allow psychotropics to be developed for homogenous disorders that is currently not reality [9].

Animal models typically involve rodents – rats or mice. The early developed animal models were based upon drugs that had been proven clinically effective. Thus, the models tended to predict efficacy for compounds that had similar mechanisms of action (e.g., action on the benzodiazepine receptor for anxiety and dopamine receptor antagonism for psychosis) [10]. Animal models are developed with the goal of having the following [10]:

- Face validity – the behavioral and physiological response in the animals is identical or at least very similar to that seen in humans.

- Predictive validity – drugs with clinical efficacy in a given disorder should produce the response in the animal model.
- Construct validity – the etiology of the behavior and the pathophysiology are similar in both the human and the animal model.

Animal models are challenging in that our understanding of the pathophysiology of mental disorders is inadequate, and human and rodent behaviors are much different. Many of the symptoms and behaviors seen in human mental disorders are not present in other animals. Much of the time, the animal model does not correspond with human emotions such as depression or anxiety, but rather examines certain kinds of locomotor behavior in different conditions. Thus, construct validity is highly suspect, and this may be one potential explanation for the high failure rate of CNS acting compounds in clinical trials. Similar challenges exist for face validity since the etiopathophysiology of most mental disorders is unclear, and humans and other species have different behaviors. Predictive validity is complicated by the fact that animal models are validated based upon existing FDA-approved psychotropics that for the most part have only modest efficacy. This is further complicated by compounds being assessed in animals acutely while pharmacotherapy in humans is typically several months to years. [39].

Select animal models are briefly discussed below.

Animal models for antipsychotics – Animal models involving the administration of either amphetamine or phencyclidine (PCP) are commonly used to screen compounds for potential antipsychotic effects. The amphetamine model is based upon the observation that amphetamine produces positive symptoms of psychosis in humans. Amphetamine produces excessive mesolimbic dopaminergic activity, resulting in spontaneous locomotor activity and stereotypy in animals. D_2 receptor antagonists block these effects [51]. Chronic amphetamine administration produces desensitization with resulting deficits in learning and attention. Both first- and second-generation antipsychotics have been shown to block desensitization.

Phencyclidine (PCP) produces both positive and negative psychotic symptoms in humans and may be a better model for the symptoms associated with schizophrenia. The attentional set-shifting test (ASST) is used to assess executive function in rodent models. The animals must learn a rule and then shift their attention to a previously irrelevant stimulus. NMDA antagonists such as ketamine produce positive and negative symptoms as well as cognitive deficits. These effects are attenuated by antipsychotics as well as nicotinic agonists. Some deficits are reversed by second generation but not first-generation antipsychotics. Animal models of cognition are more predictive of human cognition than models for symptoms such as delusions or hallucinations. Prepulse inhibition (PPI) looks at suppression of a strong stimulus by a small stimulus. Antipsychotics lessen the worsening of this response induced by NMDA receptor antagonists [51]. The Morris Water MAZE (MWM) assesses multiple cognitive functions, including learning, memory, and retention. The animal learns where a submerged platform is in a tank of water. Then the platform is moved or removed from the tank. PCP negatively affects performance on this test, and it is reversed by SGAs. Social interaction is assessed by placing unfamiliar rodents in a lighted arena, and then the amount of time they spend interacting is measured. PCP impairs their social interaction [52].

Animal models for anxiety disorders – Anxiety disorders are complex, and no one animal model is appropriate for all anxiety disorders. Anxiety animal models can be divided into two groups. Conditional response models, such as the Geller-Seifter conflict model, involve the animal's response to painful or stressful stimuli. Unconditioned response models, such as the elevated plus maze model, involve the animal's natural response to stimuli that do not involve stress or pain [10]. Animal models for potential anxiolytic effect are limited by being based upon suppression of normal behavioral response rather than on decreasing pathological anxiety [39].

Animal models for depression – The forced swim test is a highly predictive animal model used to screen for antidepressant effect. The rat is placed in a vat of water and forced to swim to stay alive. Eventually, the rat will cease to swim and will only move enough to keep its head above water. This is commonly referred to as a "state of despair," but may more accurately be a learned adaptation [39].

Over the past 50 years, new psychotropics have primarily differed clinically by having different side effect profiles than older medications. In particular, SSRIs and other newer antidepressants are much less toxic in overdose situations than the tricyclic antidepressants. Clozapine is the one major exception, and it is unclear why clozapine reduces psychotic symptoms in many patients with schizophrenia who have not responded with other antipsychotics. If we want psychotropic drug development to produce more effective agents, it is critical that we have a better understanding of the basic neurological (as well as other systems) mechanisms underlying different human mental disorders. If successful, mental disorders could be classified based upon pathophysiology rather than symptom presentation [39].

New approaches may assist in our understanding of the pathophysiology of mental disorders and the development of future medications. Behavioral assessments that target a single neural circuit in both humans and other animals increase the utility of animal models. For example, stop signal reaction time has been used in both rodents and humans to demonstrate the effects of atomoxetine on decreasing impulsivity, and this is associated with activation of homologous areas of the inferior frontal cortex [73]. Thus, it is important to utilize animal behavioral models that have a human behavioral or cognitive counterpart. It has also been suggested that objective behavioral assessments should be incorporated into Phase II and III trials because they reflect the pharmacology of the drug and not just its effect on clinical symptoms [73].

Optogenetics can be used to visualize a genetically targeted neural circuit. This allows one to turn a neural circuit on or off. Artificial intelligence may also hold promise in being able to identify new biological targets and compounds for investigation [39]. The future may lie in the use of induced pluripotent stem cells (iPS cells) from patients with the disease of interest. These can be reprogrammed into almost any cell line and allow for the study of individual neurons with the same genetics as the patient. These can be further developed into cerebral organoids that resemble the developing human brain [65]. Patient-specific iPS cells can be transplanted into rodent brains. In a schizophrenia model, human glial cells produced from iPS cells are transplanted into mice producing an animal where the majority of the glial cells are human. The chimeric mice develop decreased social interaction, anxiety, and decreased prepulse inhibition. This has potential for producing animal models to study drug action. Advances in our ability to study brain function and evaluate behavioral and cognitive function may lead to advances in our understanding of mental disorders as well as their treatment. This could potentially result in patient specific drug development. However, the cost of this would likely be prohibitive.

5.2.5 Mutagenicity

One of the nonclinical safety tests that must be performed prior to a new drug candidate's approval is assessing the drug's ability to cause changes to a cell's DNA sequence or mutagenicity [31]. These changes have been linked to a drug product's potential to cause cancer. These tests are performed using bacteria as well as mammalian cells and animals. The inactive ingredients in a formulation may also generate impurities as by-products of manufacturing or degradation during storage. This has led to the development of computer models based on chemical structure to aid in predicting the mutagenicity of both drug impurities and drug substance. These statistically-based Quantitative Structure-Activity Relationship correlations, or QSAR models, have been the subject of vigorous research by the FDA for predicting mutagenicity as well as other drug toxicities. There are now international guideliens [46] ratified by FDA and regulatory counterparts, that allow QSAR models to substitute for traditional laboratory tests for determining mutagenic of drug impurities.

5.2.6 Toxicology Considerations

Animal models are used in preclinical drug development to simulate what occurs in human biologic systems and evaluate endpoints of interest. Selecting the most appropriate animal models for conducting the formal Good Laboratory Practices (GLP) toxicology studies is critical for predicting human toxicity. While animal models are rarely 100 percent predictive, those that are physiologically similar to the endpoint of interest can guide researchers as long as there is a grasp of the similarities and variances of the model as it relates to humans.

Dose escalation studies can be used to explore the drug's toxicology. Specific negative effects on major organ systems (e.g., brain, heart, kidney, liver) can be examined at each dose. In addition, the lethal dose in 50% of the animals studied (LD_{50}) can be determined. Toxicology studies are typically conducted in rodents (mice or rats) plus one other species (e.g., dogs, minipigs, primates, rabbits). Even when testing in multiple species, human organ toxicity can occur in the absence of toxicity in other species. [60].

Establishing the safety of lead products is a culminating activity of the preclinical phase. The toxicology studies – also referred to as safety assessment – represent a pivotal component of the information that undergoes regulatory review and leads to the FDA's go/no go decision for allowing a product to move forward into human clinical trials. Using appropriate animal models and methods for generating robust, reliable data will establish a toxicology profile for new drug candidates.

5.2.7 Regulatory Pathway: Preclinical to Clinical Trials

New drug development is a highly regulated, complicated process that requires specialists and intense research and development skill sets in the medical research community. All regulations and safety indications must be observed carefully, and human and animal clinical trial subjects treated professionally and with the utmost care.

The FDA's [33] evaluates new drugs before they are approved to be marketed and sold. The regulations that CDER enforces work to ensure that drug products are safe and effective and that their benefits outweigh any known risks. The formal animal toxicity studies (preclinical) and the application to enter clinical trials are subject to FDA regulations: Good Laboratory Practices (GLP) and the Investigational new Drug Application (IND) are discussed below.

The toxicity studies intended to support applications for entering human clinical trials must be conducted under GLP. The GLP describes the FDA regulations for in vivo and in vitro experiments subject to FDA safety review. Investigational drugs being evaluated in nonclinical safety studies must comply with GLPs. The regulation embodies a set of principles that provides a framework within which laboratory studies are planned, performed, monitored, reported, and archived, covering the personnel, facilities, operations, and records that are involved in GLP studies.

GLPs are designed to provide regulatory guidance for ensuring the integrity of data from nonclinical studies. In the USA, the GLPs are administered by the FDA and are found in the Code of Federal Regulations [15]. These regulations cover the definitive preclinical studies that FDA reviews prior to making the final decision regarding approval to start testing in humans.

5.2.8 Investigational New Drug (IND) Application

Preclinical data are used as a basis to design the pivotal safety studies that will be included in the IND submission that is reviewed by CDER. As a new product undergoes preclinical evaluation, the sponsoring company collects data from a variety of prescribed testing to demonstrate that the drug is safe and effective for its intended use. Prior to being tested on people, laboratory and animal testing is conducted to determine how the drug works, whether it is successfully treating the targeted disease, and if it appears to be safe.

Sponsors are required to submit an IND application to FDA prior to initiating clinical research. The IND must include animal study data and toxicity data, manufacturing information, clinical protocols, data from any prior human research, and information about the investigator. The FDA has 30 days to review the original IND submission. Clinical trials are allowed to proceed once the IND is approved by FDA. The FDA may decide a clinical hold to delay or stop the investigation if it is believed that participants are exposed to unreasonable or significant risks, investigators are not qualified, materials for the volunteer participants are misleading, or the application does not include sufficient information about the trial's risks.

The formal regulations that pertain to INDs can be found in Section 21 of the Code of Federal Regulations [15]. There are several parts referring to INDs, including the following: drug labeling, orphan drugs, protection of human subjects, financial disclosure by clinical investigators, IRBs, and GLPs for nonclinical lab animal studies.

5.3 Clinical Development Phase

Once FDA approves the IND, a drug candidate can move forward to human clinical trials. CDER oversees the clinical trials necessary for the FDA to determine whether a new medication will be approved for use. The ultimate goal is to determine the efficacy, safety and quality of drug candidates and ensure the right dose and dosing schedule for a specified patient population. Other desirable properties of drug therapies include good bioavailability with low variability, distribution to the site(s) of action, limited metabolism, and a broad therapeutic index.

The sponsoring company may conduct the early studies with clinical research companies or academic institutions with dedicated facilities for the conduct of early phase research until they reach the later phase larger clinical trials. Pharmaceutical companies often contract with Contract Research Organizations (CROs) to conduct and oversee their clinical trials. The CROs contract with private research clinics, academic research centers, physician groups, and hospitals to conduct the studies in order to enroll the number of patients required. The CROs provide oversite and monitor the sites to assure that they are completing research documentation in an appropriate manner. In this phase, regulatory compliance and oversight is in place and the clinical trials must comply with FDA Good Clinical Practices (GCP) [34]. These regulations are designed to ensure the integrity of clinical data and protect the rights, safety, and welfare of human subjects.

There are three phases of clinical trials, typically starting with a small group of healthy volunteers (Phase I) to determine the safety and basic pharmacokinetics of the test product, to large studies involving hundreds to over one thousand patients at multiple study sites (Phase III). In order to progress from one phase to the next, the FDA must review the data and decide whether to allow the studies to progress to the next phase. This is considered "IND maintenance."

Clinical trials are designed to establish the efficacy and safety of investigational drug products and to answer other questions about the agent as needed. Protocols are developed that explain what will occur in the trial, how it will be conducted, and why each step is necessary for answering scientific questions about the product as well as safeguarding the health of participants. The trials must follow the study plan that is developed by the researcher or manufacturer and approved by the FDA. Trial participants must meet the eligibility criteria defined in the protocol in order to qualify, and the plan stipulates how many people will participate in the study and how long it will last. The research questions and objectives will dictate whether the trial will include a control group, how the drug will be administered, what dose(s) will be studied, as well as how the data will be reviewed and analyzed.

5.3.1 Phase I Clinical Trials

Traditionally, Phase I psychotropic clinical trials have focused on pharmacokinetics (PK), safety, tolerability, and definition of the maximum tolerated dose. The PK profile is impacted by the drug's physicochemical properties, formulation, and route of administration. In addition, extrinsic factors including diseases, other medications, and food can impact the PK profile, and traditionally only healthy male volunteers have been included as research subjects in psychotropic Phase I trials. The clinical pharmacology information from Phase 1 informs the design of Phase II and III trials.

More recently, individuals with the disease state of interest are being incorporated into Phase I studies. For example, it is known that individuals with schizophrenia tolerate D2/D3 antagonists better than healthy individuals. Studying subjects with the target illness in Phase I studies may better predict the dosages to be used in Phase II trials [23]. In addition, psychotropics have significant adverse effects, and medications with the potential for use in difficult-to-treat mental illnesses may have Phase I trials conducted in the population that would receive the medication. An example is clozapine. After clozapine was found to cause agranulocytosis, its bioavailability study for the NDA for use in treatment-resistant schizophrenia was shifted to be conducted in males with treatment resistant schizophrenia [13].

Phase I studies may be either open label or randomized, placebo-controlled, double-blind trials, escalating single and multiple-dose studies in a small number (12 to 100) of healthy volunteers. There may be two phases to these studies: a single ascending dose (SAD) phase and a multiple ascending dose (MAD) phase; the SAD phase is conducted first, followed by the MAD phase.

There has been increasing interest in collecting pharmacodynamic data as part of Phase 1 trials with the hope of informing the design of Phase II trials, particularly as it relates to dose. Increasingly, potential biomarkers are being introduced into early clinical trials. In order to be useful as an outcomes measure in clinical trials, the biomarker needs to confirm the diagnosis or predict response to treatment. For example, it has been suggested to perform targeted neuroimaging and collection of potential biomarkers during Phase I [56]. The effects of drugs on neural circuits can be studied using such techniques as function magnetic resonance imaging 9fMRI) and high-density electroencephalography [37]. The use of cognitive challenges during fMRI in individuals with schizophrenia may allow for the identification of medications with pro-cognitive effects.

Mismatch negativity (MMN) has been described as possessing nearly all of the features of a translational research biomarker. It is a neurophysiological response usually measured by auditory evoked potentials that involve assessing the effects of an uncommon stimulus that follows repeated normal stimuli. It is impaired in schizophrenia and involves both dysfunction in neural circuits and clinical outcome. Dysfunction in schizophrenia involves impairment in both auditory sensory perception and cognition. It can be used in both rodents and humans, and NMDA receptor antagonists such as ketamine or phencyclidine negatively affect it in both preclinical and clinical models. Both glycine and the glycine agonist D-serine reverse the negative effects of ketamine on auditory evoked potential [8].

In perhaps the best example of the use of a biomarker in antidepressant studies, the effects of deep learning on the fMRI was associated with an R^2 of 0.48 on predicting Hamilton Depression Rating Scale (HAMD) improvement with a number needed to treat (NTT) of 4.86 patients [57]. It has also been suggested that the use of fMRI during Phase I trials could actually decrease the overall cost of drug development. Another example is the use of positron emission tomography (PET) to study molecular interactions and potentially determine the dosages associated with efficacy or adverse effects.

Currently marketed antidepressants cause neurogenesis in the hippocampus. MRI measured changes in hippocampal volume have been proposed as a screen for potential antidepressant effect. If increases in hippocampal volume predict antidepressant activity, this could be used to explore the potential of compounds not acting on monoamine receptors as antidepressants. This technique is also appealing because it can be used in both humans and nonhuman species. Although this has been explored in a Phase 1b clinical trial, it is too early to confirm its predictability [25].

Thus, the use of multiple pharmacodynamic measures during Phase I studies may help identify unique mechanisms and the clinical profile as well as to better predict the dosages to use in Phase II trials, and perhaps even efficacy [23]. However, there is a need for harmonization and standardization of the methods used. If biomarkers that predict a drug's efficacy can be developed, these could be used in Phase I clinical trials in patients with the disease of interest. If no positive effect on the biomarker is found, then costly Phase II and Phase III clinical trials could be avoided. Although not yet definitive, brain-derived neurotrophic factor (BDNF) is potentially such a biomarker for antidepressant activity [73].

Increased use of Model Informed Drug Development (MIDD) is guiding clinical trial development, optimizing dosing, and providing supportive evidence for efficacy and in policy development. The use of modeling has the potential to speed up the development process and increase efficiency of conducting clinical trials to support drug approval. Extrapolation of efficacy and safety from small data sets in difficult-to-study populations (children, rare diseases, hepatic dysfunction, drug interactions etc.) and making model-based inferences in lieu of pivotal clinical data to make efficacy, dosing, and safety recommendations are two high impact areas for use of modeling in drug development. Population pharmacokinetic computer modeling (in silico) in CNS drug development is increasingly being used, especially in the area of dose formulation [43, 75]. The FDA and EMA have developed several guidance documents for modeling in the drug approval process. [24, 26, 36, 67].

The importance of early pharmacodynamic studies is clearly recognized. Assessing "pharmacodynamic target-based measures" early in clinical research may not only help dose response relationships to be used in Phase II and III trials; it potentially helps better define mechanism of action. The NIMH through its contracting process has created research teams comprised of pharma researchers, academics, and NIMH researchers to support early pharmacodynamic assessment of candidate agents [41].

FDA reviews the Phase I data and if they find sufficient evidence to support continuing clinical trials, the drug candidate is allowed to move into Phase II. Phase I studies typically last several months, and 70% of drugs move to the next clinical phase [29].

5.3.2 *Phase II Clinical Trials*

Phase II studies focus on the drug's efficacy and adverse effects. They can last from several months to 2 years in duration. Phase II trials involve testing the experimental drug on a larger number (up to several hundred) patients who have the disease of interest. Phase II trials are dose finding studies looking for the appropriate doses for evaluating safety and efficacy in larger Phase III trials. These trials generally include a study arm which uses up to the maximum tolerated dose (MTD) from the Phase I trials. Because of the small number of subjects in Phase I, the dose may need to be adjusted either down if excessive toxicity is observed or increased if the desired biological effect is not seen.

Early Phase II trials are often open label, but randomized placebo-controlled trials are required by the FDA to demonstrate efficacy. Because of the large placebo response rates seen in most studies of mental disorders, a placebo group must be included to prove efficacy. The FDA does not allow the use of standardized treatment comparison studies (i.e., noninferiority trials) without a placebo group [29].

It is critical that a priori sample sizes be calculated to assure that adequate number of subjects are enrolled to be able to show a statistical difference between groups. Conventionally, an $\alpha \leq 0.05$ is used to demonstrate a difference between groups. Although β may vary, it is typically $\beta \leq 0.2$.

It is essential to assure that patients enrolled into clinical trials have the disorder of interest, and the Structured Clinical Interview for DSM (SCID) or MINI International Neuropsychiatric Interview (MINI) are generally required to verify the patient's diagnosis. In general, patients with varying duration of illness or numbers of episodes are included in Phase II and III trials. However, if an indication for a treatment resistant disorder is being sought, then the patient subject population must reflect the intended indication. In some cases, an indication as an adjunctive agent is being sought, then the patient population would be comprised of individuals who failed to obtain adequate response with a standard treatment, and patients are randomized to receive either the investigational drug or placebo plus the standard treatment.

Psychiatric diagnoses are based on a clinical syndrome which is represented by a set of symptoms or behaviors, severity of symptoms, duration of symptoms, impairment in psychosocial functioning, and other clinical characteristics. Symptom overlap occurs from syndrome to syndrome (e.g., anxiety is often present in depression, bipolar disorder, and schizophrenia, as well as in anxiety disorders). Because of the spectrum of symptoms and the overlap, it is challenging to tie any one syndrome to a specific neural circuit. However, it is often possible to tie one specific symptom to a given neurocircuit, and with the correct biomarkers, perhaps to a specific dysfunction in that circuit. Preskorn argues that clinical drug trials should be based upon treating specific symptoms that are associated with dysfunction in a specific neural circuit [64]. Research Domain Criteria (RDoc), established by the NIMH, is an attempt to integrate scientific data with clinical phenomena. This transdiagnostic approach incorporates such data as biomarkers, genetics, neurocircuitry, imagining, and neuropsychology with five domains of human emotion and behavior [21, 64]. Human behavior is divided into five categories based upon positive or negative valence. This has been referred to as a "systems neuroscience approach" to drug development with the goal of understanding the basic neuroscience associated with human behavior [73]. Collection and compilation of these types of data into large datasets may allow for computational modeling of future new medication [21].

Approximately 33% of drug candidates in Phase II trials move into Phase III studies [29]. Use of a different approach to elucidating the clinical effects of potential drug therapies has the potential to improve the percentage of drugs moving into Phase III testing.

5.3.3 Phase III Clinical Trials

Following FDA review and approval of Phase II trial data, sponsors are allowed to continue into Phase III clinical trials. Study participants who have the disease or condition are enrolled and can involve hundreds to a few thousand volunteers. A priori power analysis calculations are performed in order to determine the number of research subjects that must be enrolled in order to evaluate efficacy. However, much larger numbers of individuals treated with the investigational drug are needed to determine safety. These studies focus on efficacy and monitoring of adverse reactions, dose-response, wider populations, efficacy at various stages of disease, and use in combination with other agents. These large and extensive trials take significant time and are expensive. Increasingly, quality of life and social functioning assessments are being incorporated into Phase III trials.

In the USA, Phase III trials are double-blind and typically placebo-controlled. This helps to eliminate bias when interpreting results. Patients enrolled in Phase III trials are typically between 18 and 64 years of age, do not have a history of nonresponse to psychotropic treatment, have no serious general medical disorders and no co-occurring mental disorders, including alcohol or substance abuse. This "clean" clinical trial patient population is different than that often seen by clinicians in practice, and thus, generalizability of clinical trial results can be limited [51].

Phase III trials are the final clinical phase of test before the drug product's details and clinical trial results are submitted to the FDA for consideration of approval. They generate important data that reflect the test product's efficacy and adverse reactions.

In general, the FDA requires two placebo-controlled, randomized clinical trials indicating that the psychotropic medication is efficacious and safe. If the medication is already FDA approved for use in adults, then only one efficacy and safety trial is required for the same indication in children or adolescents. There are exceptions to the use of the investigational drug compared with only a placebo control group. For example, in the esketamine clinical trials, patients were randomized to receive either esketamine or placebo added to a newly started antidepressant [40]. In Kane's classic study of clozapine in treatment-resistant schizophrenia, chlorpromazine plus benztropine was used as the control [50].

In addition to large randomized placebo-controlled trials to demonstrate efficacy, tolerability, and safety in treating the acute phase of a mental disorder, the FDA requires continuation studies. Since most mental disorders are either chronic (e.g., schizophrenia) or recurring (e.g., major depressive disorder [MDD]), continuation trials must be of sufficient duration for the specific disorder to demonstrate that the drug has continued efficacy and that it is safe and well tolerated. The design of such trials is variable. They may enroll drug responders, and then after the designated continuation phase, patients may be randomized to continue on the investigational agent or receive placebo. This allows comparison of relapse rates following discontinuation [29]. These studies may be performed in either Phase III or IV.

In general, the FDA has required efficacy of psychotropics to be demonstrated by two independent clinical measures, typically both a validated rating scale for the disorder [e.g., Positive and Negative Symptom Scale (PANSS) for schizophrenia and the HAMD for MDD] and a clinical global impression (CGI) scale. The CGI-Severity (CGI-S) is preferred because it is subject to less recall bias than the CGI-Improvement (CGI-I) [29]. Although multiple measures of clinical outcome may be used in a clinical trial, it is critical that the primary measures of efficacy be determined when the trial is designed, or in an a priori fashion. Primary outcomes must show statistical significance (typically $\alpha \leq 0.05$), between active compound and placebo to demonstrate efficacy of the test drug. Efficacy cannot be determined based upon statistically significant improvements in secondary measures if a statistically significant difference is not found in the primary outcome measure.

Although there is great interest in biomarkers, no universally accepted biomarkers for drug efficacy in mental disorders currently exist. For example, in Alzheimer's dementia, which is often considered both a neurological and a psychiatric disorder, B-amyloid plaques are associated with both the diagnosis and the severity of Alzheimer's dementia. Several anti-β-amyloid monoclonal antibodies have been developed that decrease β-amyloid plaques in the brain, but they have not been shown to improve clinical symptoms or slow clinical deterioration. The Food and Drug Administration's (FDA's) approval of aducanumab is an example of a drug that produced great controversy regarding the use of a biomarker to demonstrate efficacy in Alzheimer's disease. In clinical trials, aducanumab decreased B-amyloid plaques in the brains of patients with Alzheimer's, but it did not consistently improve clinical symptoms or delay cognitive decline compared with placebo [66].

It could be useful to utilize biomarkers to exclude patients from clinical trials who are unlikely to respond to treatment. For example, elevated baseline C-reactive protein (CRP) (> 1 mg/L), a marker of inflammation, has been shown to predict poor response to SSRIs in females, but not in males. It could be useful to use this as a biomarker to exclude women with depression from clinical trials with serotonergic antidepressants. It also appears that patients with elevated CRP serum concentrations respond better with noradrenergic and dopaminergic antidepressants [48, 49].

Challenges exist with the use of rating scales to evaluate efficacy in clinical trials. The total score is the sum of the severity ratings on individual items. The data are not continuous, and ideally, should not be evaluated using parametric statistics. There is not necessarily a linear relationship between change in the total score and the actual clinical status of the patient. For example, depending on the severity of the individual items, a patient with schizophrenia and a PANSS score of 48 could actually be just as or more psychotic than a patient with a PANSS score of 90 [64].

Adaptive design trials have been recommended as one way of improving the demonstration of efficacy. In adaptive designs, prospectively planned interim analyses are performed and the trial adapted without damaging the scientific integrity of the study [56, 69]. This can allow for enrichment of treatment responsive patient populations and may actually decrease the enrollment requirements and allow an earlier decision to be made about success or failure [69].

Placebo response is a huge challenge in clinical psychopharmacology. Placebo response has increased over the past 40–50 years, and in some studies 50% or more of patients have experienced clinically significant improvement with placebo. High placebo response rates can make it challenging to distinguish active medication from placebo. Several reasons have been proposed for the increase in placebo response including: the move by the industry to conduct clinical trials in private clinics rather than academic settings, varying research experience of clinical investigators, the rise of professional subjects who repeatedly enroll in clinical trials, variable intra-rater and inter-rater reliability in performing psychometric assessments, and lack of rigor of clinical investigators in adhering to study enrollment criteria [51, 56]. In addition, the amount of time that clinical investigators and their staff spend with clinical research subjects at frequent visits potentially serves as a form of supportive psychotherapy. Although it is important to collect appropriate data for efficacy and safety purposes, perhaps data collection should be streamlined in order to shorten the visits. A placebo run-in period, with placebo responders eliminated before randomization, has been frequently used in attempt to decrease placebo response rates. However, a recent large meta-analysis of antidepressant clinical trials found that the use of a placebo run-in did not alter the difference between active drug and placebo response rates [68].

Approximately 25–30% of drugs complete Phase 3 trials and move forward to be considered by the FDA for marketing approval and continuation to Phase IV studies.

5.3.4 Pediatric Considerations

Additional considerations must be given when conducting clinical trials of psychotropic medications in children or adolescents. Obviously, the agent must be used in clinical trials for a mental disorder seen in pediatric populations. A valid scientific rationale must exist for studying the agent in children and the possibility of different pharmacodynamics and pharmacokinetics must be considered.

Both physiological and emotional development are dynamic processes, and potential differences in efficacy and tolerability must be studied for each age group to be included in the FDA product labeling. Juvenile animal studies may also be needed to examine the effects of the agent on growth and development [29]. Medications may have developmentally dependent adverse effects that are different than those seen in adults, and it is important that this be systematically assessed. Currently, no standardized assessments for adverse events in children are required [14]. The three approaches to soliciting adverse effects are general inquiry, checklist of adverse effects previously reported with this drug class, and a systematic potential adverse effects checklist of a review of systems [14]. The limitations of not using a standardized approach to safety assessment challenge became readily apparent in the FDA's assessment of potential suicidality of antidepressants in children and adolescents [63]. Although there is a suicidality item on most depression rating scales, suicidality in antidepressant trials had been primarily assessed by general inquiry. This would result in the clinical investigator recording a narrative note regarding reports of suicidal ideation or attempt. When the FDA saw a potential signal of suicidality in antidepressant trials in children and adolescents, they contracted with an independent group of suicidality experts to blindly evaluate the case reports. They developed a standardized assessment of the case reports, the Columbia Classification Algorithm of Suicide Assessment (C-CASA). After blindly reviewing the case reports, the panel found that the pharmaceutical companies tended to over identify cases as suicide attempts and under identify overall suicidality. This led the FDA to require the use of the C-CASA in evaluating case reports of possible suicidality [63]. In addition, the FDA recommends the prospective use of the Columbia Suicide Severity Rating Scale (CSSRS) in clinical trials of antidepressants [62, 63].

The clinical failure rate of psychotropic medication is higher in children and adolescents than in adults. There are likely multiple reasons for this. Youth studies share the challenges of high placebo response rates and the limitations of symptom-based assessments of efficacy which are also a bane for psychotropic studies in adults. Placebo response may be larger in children because of parent or other caregiver's expectations for improvement. Children and adolescents tend to have significant state dependency with regard to symptoms such as depression and anxiety, and thus, spontaneous improvement may occur depending on the youth's environmental situation. Studies in children and adolescents commonly include a broad age range. CNS as well as other organ development is a dynamic process throughout childhood, and including broad age ranges introduces greater heterogeneity into the subject population. The sample sizes in these studies are typically too small to perform subgroup analyses [40].

5.4 Regulatory Review Process

The FDA ensures that US consumers have access to safe medicines and treatments. The FDA's CDER evaluates new drugs before they are approved to be marketed and sold [33]. The regulations that CDER enforces ensure that drug products are safe and effective and that their benefits outweigh any known risks. There are mechanisms in place, including labeling, dosing directions, patient package inserts, and other education materials, to ensure that healthcare professionals and patients have the information they need to use medicines appropriately.

CDER teams composed of clinicians, chemists, statisticians, pharmacologists, and other scientists to review the sponsor's submitted data and make the determination whether there is sufficient proof that the drug should be allowed to move into the next phase in the pathway. These are scientific, unbiased, independent reviews by CDER teams – they review data to assess whether there is sufficient evidence to either approve or reject the company's application to approve a new drug product.

The FDA commonly issues guidance documents to assist sponsors in designing studies that are appropriate for NDA submission. However, these guidance documents often lag behind advances in research in the discipline. For example, the last FDA-approved guidance document for the development of drugs for MDD was approved in 1977. A draft of a revised guidance document for MDD was released for comment in 2018, and it had yet to be approved by submission of this chapter [30].

Because of the pharmaceutical industry's reluctance to perform studies in children, the Congress passed two important pieces of legislation – the Best Pharmaceuticals for Children Act (BPCA) and the Pediatric Research Equity Act

(PREA). PREA requires the industry to perform efficacy and safety studies of certain drug categories in children if the product has the potential to be beneficial for children with the given disorder. The BPCA grants the company an additional 6 months of patent protection if the company conducts studies in children. It is noteworthy that the BPCA also provides the NIH with funds to study drugs FDA approved for use in adults if the medication would be potentially beneficial for children and adolescents [40].

5.4.1 Regulatory Pathway: Clinical Trials to Commercialization

A new drug application (NDA) can be filed with the FDA for review and approval following completion of clinical trials. Throughout the development pathway, efforts to scale up production of the active therapeutic ingredient (ATI) and the final formulation are occurring.

During every development phase, sponsors collect information and data to include in the NDA. These data will need to be sufficient and compelling so that FDA reviewers can determine whether the drug's safety, efficacy, labeling, and manufacturing align with FDA's goal of approving products that meet strict criteria and whose benefits outweigh their risks, ultimately ensuring product quality and integrity.

The NDA document presents a story of the drug's development, from the ingredients in the dosage form to the results of animal testing and stability studies, including results from clinical trials, profiles of drug absorption, distribution, and metabolism, and pharmacodynamics, toxicology, manufacturing and packaging processes, and labeling requirements.

The timeline for reviewing NDAs has been significantly decreased in the past 20 years. The median FDA review and approval times have decreased from 20.9 months and 26.9 months in 1993 to 10.1 months and 10.1 months in 2016 [27].

5.4.2 NDA Review and Approval

The first step in the NDA approval process is for the FDA review team to determine whether the application is complete. If it fails to meet this standard, the review team can refuse to file the NDA. For complete NDA's, the team has 6–10 months to decide if the application meets the standards for approval. In addition to reviewing the various sections, FDA inspectors visit clinical sites to ensure there is no evidence of fabrication, manipulation, or withholding of data. The various reviews and other documents are consolidated into an action package or the formal record of the FDA review.

For drug products shown to be safe and effective for their intended use, the FDA will work with the sponsor company to develop and refine the prescribing,or labeling information. Labeling is a prominent feature that must accurately and objectively describe the basis for approval and the best use for the drug.

It is common for an application to have issues, both major and minor, that need to be resolved prior to approving a drug for marketing. These can range from questions pertaining to existing data to the FDA requiring additional studies. It is up to the sponsor company to decide whether they want to continue development of the product.

5.4.3 Abbreviated NDAs

The Drug Price Competition and Patent Term Restoration Act of 1984, also known as Hatch-Waxman, established the approval pathway for generic drug products and provisions that involve patents and exclusivities related to new drug applications. The approval of generic drug products requires submission of an Abbreviated New Drug Application (ANDA) [32].

These are considered abbreviated since generic drug approvals generally require limited preclinical and clinical data to establish safety and efficacy since they are allowed to refer to the innovator company's data that were submitted in the original NDA. Generic applications must scientifically perform in the same way as the innovator company's drug product. This can be accomplished using in vitro and human in vivo testing to prove that there is bioequivalence – the same rate and extent of absorption, distribution, metabolism, and excretion – when compared with the innovator product.

The FDA has an additional "hybrid" or "bridge" pathway called the 505(b)(2) pathway, for new products that aren't necessarily a copy of the originator product. Examples might include new formulations, new indications, new combinations, new route of administration, minor changes to the molecule such as prodrugs, etc. The application uses all of the safety and efficacy data from the originator product and creates a bridge between their new product and the original product.

5.4.4 Advisory Committees

The FDA has established advisory committees to provide FDA with independent opinions and recommendations from outside experts on applications to market new drugs. These committees are composed of outside experts, who receive a summary of the information from the sponsor application and have access to the FDA's review of the application documents. Advisory committees recommend approval or rejection of an NDA; however, FDA makes the final decision.

5.4.5 *Expedited Review Programs*

The FDA's expedited review programs are intended to facilitate and expedite development and review of new drugs that address unmet medical needs in the treatment of a serious or life-threatening condition. The four programs are fast track designation, breakthrough therapy designation, accelerated approval, and priority review designation. The purpose of the expedited review guidance is to provide a single resource for information on the FDA's policies and procedures for these four programs as well as threshold criteria generally applicable to concluding that a drug is a candidate for these expedited development and review programs. In particular, "breakthrough therapy" designation allows for more rigorous feedback and expedited review than earlier expedited review programs [11, 32, 37].

5.4.6 *Prescription Drug Labeling Information*

FDA approves labeling information provided to patients in medication guides (MG), patient package inserts (PPI), and instructions for use (IFU). These are available for consumers and patients and include facts about adverse effects, drug interactions, proper storage, and other useful information. These various types of labeling information align with the FDA's Risk Evaluation and Mitigation Strategy (REMS) drug safety program for medicines with serious safety concerns. REMS are created with the goal of ensuring proper use, monitoring, and safe use of medications.

PPIs provide the information that is included in patient labeling, providing comprehensive and considerable detail about a drug product. These are developed by the manufacturer, approved by FDA, and required for specific products or classes of products. For other products, PPIs may be submitted to FDA on a voluntary basis by the manufacturer and approved; however, distribution is not mandated.

MGs are primarily for outpatient prescription products with potential serious public health concerns and are provided to the patient when the product is dispensed. They are required if labeling could help prevent serious adverse events, enhance patient adherence to directions or affect a patient's decision to use a product. IFUs are FDA approved labeling developed by the manufacturer that are dispensed with specified products with complicated dosing instructions. These are intended to help the patient properly use the product.

5.5 Phase IV Activities

FDA oversight continues after a drug product is approved for marketing and is referred to as post-marketing surveillance. Following NDA approval, compliance with manufacturing and distribution of new drug products is closely monitored. During Phase IV, additional clinical trials may be required for observing long-term efficacy and ongoing monitoring of adverse effects and assessment of health outcomes.

5.5.1 Phase IV Clinical Trials

Phase IV clinical trials may be required of the drug product sponsor as a condition for approving the marketing application. Pre-NDA clinical trials only monitor hundreds to a thousand or more patients while Phase IV trials monitor how a drug product performs in a broader population. These studies aim to assess the long-term risks and benefits, monitor known adverse effects, and identify any rare but serious adverse effects. The Phase IV trials can increase confidence in a product. However, in some instances, the product may be withdrawn from the market based on what is discovered when it is used in larger populations.

Pharmaceutical companies may decide to perform comparative efficacy studies as part of Phase IV. These may be noninferiority or superiority trials. These are often performed in hope of providing the company with a marketing advantage for the drug. The company may decide to pursue additional indications for the drug or obtain FDA approval for the use of the drug in other populations (e.g., children or adolescents), and this results in a return to Phase III trials.

5.5.2 Monitoring Adverse Effects

The FDA has created a post marketing safety surveillance program for all marketed drug products. The regulations require manufacturers to report adverse events received from healthcare professionals and consumers for inclusion in the FDA Adverse Event Reporting System (FAERS). This allows the FDA to identify and monitor new safety concerns, evaluate a manufacturers' compliance with FAERS regulations, and provide information in response to outside requests. Additionally, voluntary reports can come directly from healthcare professionals and consumers. This database allows interested parties to find information regarding adverse events for drug products.

5.5.3 Phase IV Health Outcomes/Quality of Life

Following the approval of a new drug product, the Phase IV trial data may also study other characteristics of the drug therapy, including the impact on a patient's quality of life or the cost effectiveness of the treatment. These questions may take years to answer and are influenced by other factors including the quality of medical care, treatment outcomes, hospital readmissions, the patient experience, and mortality.

5.6 Bioethical Issues

Protection of human subjects should always be at the forefront of designing, implementing, and conducting clinical trials. The FDA [15] and the European Medicines Agency (EMA) have both promulgated rules around human subjects' protection – most of which are very similar in nature and structure with key differences. The International Council for Harmonization of Technical Requirements for Pharmaceuticals for Human Use (ICH) has developed guidelines to help navigate these regulations commonly referred to as Good Clinical Practice [47]. Much of the regulations and guidance is derived from key foundational documents – the Nuremberg Code [2], the Belmont report [72], and the updated report from the Declaration of Helsinki [76]. The FDA also has guidance for sponsors, investigators, and institutional review boards. [35].

Each of these outlines the necessary steps for assuring the rights and welfare of human subjects – including the review by institutional review boards, informed consent procedures, the processes involved to mitigate risks to subjects, and clinical trial design expectations with respect to assuring the quality of the work, the feasibility of the study, and the reliability of the results. Study design is a critical component to assuring the ethical construct of a clinical trial. That is, making sure the trial is designed to produce results that are meaningful and reliable and have utility. ICH has developed multiple guidelines for the design and conduct of clinical trials [45].

There is also the issue of people from racial and ethnic minorities being underrepresented in research. Diversity, which includes race, ethnicity, gender, socioeconomic status, age, and stage of disease, is essential to the clinical trial industry to improve the safety and efficacy of new treatments being developed and the generalizability of results. The lack of diversity in clinical trials is an obstacle to understanding the safety and efficacy of novel therapies across population subgroups, which is important for reducing disparities.

Development of drugs for psychiatric conditions has all the same ethical challenges that developing drugs for other conditions would pose. Psychiatric drug development also presents some ethical challenges unique to the conduct of psychotropic clinical trials. Psychiatry and patients with mental illness are subject to higher degrees of stigma, and some individuals are critical of the very nature of the field, leading to a higher level of negative attention from many sources. Patients with mental illness are often perceived as being particularly vulnerable – even to the point where even the notion of conducting trials in the population is questioned. Clearly, there is a need to better understand the biological and physiological underpinnings of psychiatric illnesses and develop a better taxonomy that would facilitate study of more homogeneous subpopulations of patients.

At the forefront of ethical issues in psychiatry is the use of placebos versus active comparator study designs. The double-blind placebo controlled study design has long been considered the gold standard in creating evidence of efficacy in psychiatric clinical trials. Notably, the FDA still provides guidance that the use of placebos is one of the more certain strategies for showing efficacy of a new molecule. The

EMA uses the Declaration of Helsinki as a guide for the drug development process. The Declaration of Helsinki is clear about the use of placebos in clinical trials:

Use of Placebo

"The benefits, risks, burdens and effectiveness of a new intervention must be tested against those of the best proven intervention(s), except in the following circumstances: [76]

- Where no proven intervention exists, the use of placebo, or no intervention, is acceptable; or
- Where for compelling and scientifically sound methodological reasons the use of any intervention less effective than the best proven one, the use of placebo, or no intervention is necessary to determine the efficacy or safety of an intervention and the patients who receive any intervention less effective than the best proven one, placebo, or no intervention will not be subject to additional risks of serious or irreversible harm as a result of not receiving the best proven intervention.
- Extreme care must be taken to avoid abuse of this option."

The EMA by in large does not allow placebo-controlled research for psychotropic drug development. The divergence of guidance across the continents is challenging in psychiatry as it is difficult to satisfy both guidance documents when simultaneously developing a drug in Europe and the USA. Unfortunately, the solution is to conduct separate trials to meet the specifications, exposing greater numbers of subjects to potentially nontherapeutic and noxious agents, increasing the cost of drug development, and delaying the development of the products.

The ethical issues surrounding the use/nonuse of placebos are multifaceted. On the surface, the concern would seem to be the use of placebos amounts to withholding treatment in patients with serious disorders for the duration of the trial. For psychiatric disorders, this increases the risk for symptom exacerbation and relapse with potential consequences of suicidal ideation/attempt, death by suicide, reincarceration, and a host of social/educational/employment consequences.

Clearly placebos are unethical when the consequence of withholding standard treatment is lethal or has the potential to cause irreversible morbidity. Placebos are also considered ethical – as guided by the Declaration of Helsinki when there are no standard treatments and, in the situation where the experimental therapy is added to standard therapy. Exploring the issue further, one must ask – what are the alternative study designs to placebo-controlled studies and the ethical consequences of those alternatives?

Active control studies would seem to be the logical alternative – comparing the test compound with currently available treatments. The challenge with this approach in psychiatry is the high rate of placebo response in clinical trials (30% or greater), resulting in a high number of failed studies. The results of active control studies would be challenged on the basis of the reliability of the results and risks approving a therapy that might in actuality be no better than placebo. For example, if a comparator study showed an investigational drug to be equally effective as diazepam in generalized anxiety disorder, one would conclude that the investigational drug is effective. However, if a placebo group was included in this study, and placebo response was equivalent to the investigational drug and diazepam, then it is a failed study. If a study design produces unreliable results – then the ethics of enrolling subjects into the study comes into question. Furthermore, the active control design would need to use a superiority analysis design – the test compound would need to be better than the standard. Superiority is a high hill to climb in developing drugs for psychiatric conditions that are highly variable along many constructs. Most new agents to market are no more efficacious than existing therapies.

Noninferiority analysis, where the test compound is compared to a standard treatment, is another design to be considered. The question that arises with these analyses is whether the endpoint of noninferiority has any clinical utility – again bringing to question the ethics of enrolling subjects in such a study. Additionally, noninferiority studies often need to have significantly more subjects enrolled in the trial, potentially increasing the exposure to a test product that could have serious adverse effects or may not be efficacious.

Clearly there are good arguments on both sides of the placebo/antiplacebo debate. At this point, the FDA continues to press for "adequate and well-controlled" studies to demonstrate the effectiveness prior to approval. It is likely we will continue to see a mix of randomized placebo-controlled trials (RPCT) and other designs used to meet this standard. When placebos are utilized in a trial, Miller offers four ethical considerations for design and implementation [12, 55]:

1. Placebo-controlled trials must be scientifically sound and the results of the study have potential significant clinical utility.
2. Risks should be minimized and justified relative to the anticipated benefits to clinical care and individual subjects – subject selection should consider risks and benefits to the individual; there should be a plan for managing distress associated with temporary exacerbations and removal of the subject from the trial if severe. Monitoring/assessment should target known risks such as suicidal ideation, and duration of exposure to placebo should be as short as possible.
3. Subjects must give adequate informed consent – free from coercion, with clear delineation that it is research and not treatment, the chances of receiving placebo are understood, and individual vulnerabilities are attended to.
4. Subjects should be offered short-term individualized treatment after completion of research participation to maintain their symptom stabilization and be referred for continuation of care

Another key challenge to clinical trial work in the psychiatric population is the perception that patients are a vulnerable population and therefore must receive higher levels of protection. Vulnerability within the population of people with mental health disorders can be grouped in to two major categories – capacity-based vulnerability and power-based vulnerability. Capacity-based vulnerability refers to the inability to make an informed consent decision based on the subject's impaired capacity. Power-based vulnerability refers to the power differential between the subject and the subject's provider/investigator. There also exists a wide range of vulnerability within this population and not all subjects are considered vulnerable. For example, studies in depression found that 90% of subjects had full comprehension and that the levels of comprehension were similar to patients from the general community [4, 70]. Clearly there are subgroups within the population that are more at risk for capacity-based vulnerability that need to be identified [78].

The ability to provide voluntary consent often comes into question in trials of psychiatric conditions. Much has been written about the decisional capacity of psychiatric patients in clinical trials. The four key principles of decisional capacity with respect to providing informed consent are understanding – ability to know the meaning of information, appreciation – relating the information to oneself, reasoning – using information to weigh the options, and expressing a clear choice – making a clear decision [5, 42]. There are many different approaches to assessing capacity. The use of the MacArthur Competence Assessment Tools for Clinical Research and for Treatment (MacCAT) is an assessment tool with good empirical data to support use in populations of subjects where the subjects' ability to understand, appreciate, reason, and express a choice are questioned. The MacCAT has been utilized in the consent process to verify decisional capacity [22] L). Additionally, there are considerations, again not unique to psychiatry, surrounding the use of a legal guardian, spouse, parent, or other surrogate to make decisions for the subject who may not be capable of deciding about trial participation.

Including patients in clinical trials that are involuntarily detained is a relatively unique challenge in psychiatry. These are patients admitted to a psychiatric facility under the provision of civil involuntary detention laws – where a clinician and judge are making decisions about treatment. States and local treatment culture vary in how to handle cases where the patient qualifies for enrolling in a clinical trial but is involuntarily detained. The willingness of guardians to allow their wards to participate in clinical trials also varies. Public administrator guardians may be less likely to agree to participation – especially in higher risk studies – and often request judicial review prior to signing the consent form. Family member guardians may also be more protective of their loved one. Interestingly, family members may also be more likely to agree to allow participation under the notion that they are helping the patient find a better treatment – especially if the patient has had difficulty managing symptoms. In any case, there should be an agreement by the subject to participate (assent) if a surrogate is providing consent.

Research subjects and guardians may have higher than reasonable hopes that participation in the clinical trial will provide benefit – so called therapeutic misconception. These misconceptions where the subjects fail to appreciate the differentiation between the requirements and obligations of clinical research and treatment as usual can undermine the potential for the subject to provide a true informed consent.

The perceived benefit can come in many forms – increased attention or potential for improvement in long-standing symptoms. In the minds of patients, these may be prioritized over the weighing of potential risks, disadvantages and incumbrances associated with the study procedures. The therapeutic alliance a provider has with the patient is an important bias that has the potential to increase this misconception as patients may be more likely to follow the recommendation of their provider and the provider, out of an obligation to do what's best for the patient, may be more likely to refer to a clinical trial if it appears to offer a benefit that cannot be currently provided [6]. Clinical trials designed with randomization, and which use strong informed consent processes that do not involve the provider as the one explaining the study, that make a clear distinction between the objectives of the research, and which acknowledge that research by definition has inherent risks and may not result in the patient deriving any benefit from the study are the ones more likely to be perceived as being ethically designed and implemented. Potential therapeutic benefit to the patient is a secondary issue to consider and discuss with the subject [12].

Another challenge in designing clinical trials is creating a set of experiments that answers pressing clinical issues and the results of which are translatable to clinical practice. Few molecules in development have the potential to revolutionize the practice of psychiatry. Often, drug development clinical trials in psychiatry are directed more toward developing molecules that provide a better tolerability profile or provide some incremental improvement in drug delivery. As such, the inclusion/exclusion criteria are relatively narrow. Subjects with significant medical problems, drug interactions, multiple comorbid psychiatric conditions, suicidal ideation, substance use disorders, and significant social-economic burdens such as homelessness are often excluded from participation. In addition, pediatric and geriatric populations are often not included in clinical trials. Pediatric studies are often included in the additional studies required by the FDA in the post-marketing phase of drug development. Excluding these patients from the trials serves the purpose of creating a set of results that favor the developing drug product but produce results that do not necessarily serve the greater good as the complicated patients that comprise a significant portion of patients in clinical practice are not reflected in the results. [44, 46]

Along these same lines, the US Department of Justice estimates that up to 43% of inmates may have a history of mental illness and up to 40% of inmates are receiving treatment for mental health concerns [74]. In addition, growth in the forensic psychiatric population within the behavioral healthcare system continues to grow. Clinical trials with these populations are fraught with ethical complications, and it is important to remember the IRB regulations regarding the use of prisoners in clinical trials. Notwithstanding the complications of getting informed consent from judges, and the challenges of working with a population with highly complex illnesses, the potential for coercion in this population is high. Clearly, the same standards of treatment for incarcerated subjects should exist as for any other trial participant. Similarly, there should be no special incentives provided such as gaining access to certain privileges or leverage in decisions around parole. Subjects who are incarcerated may be motivated/coerced by perceived incentives – simply spending time outside of the cell or prison milieu may be viewed as an incentive. Clearly monetary incentives are viewed differently in prison relative to the community. A prisoner representative of the IRB must attend the meeting where the study is reviewed [12].

Financial payments for participation are often reviewed with higher levels of scrutiny by IRBs. A perception exists that patients with psychiatric diagnoses, especially those that are unemployed or on disability, may be coerced into participation by smaller amounts of remuneration than the general population. Very little empiric evidence is available to support this notion. However, concern does exist about providing cash payments to subjects that might be at risk for substance use. But again, there is no evidence to support that providing non-cash payments reduces this risk. IRBs have wide ranging policies and guidance for providing remuneration to vulnerable subjects.

5.7 Conclusions

As evidenced in this chapter, drug development is an expensive, high risk, and long process. Once molecules with potential therapeutic effect are developed, they must have extensive in vitro and animal testing before entering trials in humans. A sequence of studies is conducted in humans to determine the characteristics of the drug in the body and its efficacy and safety. It is a highly regulated process, and ethical considerations are given paramount importance.

References

1. Abruzzo A, Cerchiara T, et al. Transdermal delivery of antipsychotics: rationale and current status. CNS Drugs. 2019;33:849–65. https://doi.org/10.1007/s40263-019-00659-7.
2. Allied Control Council. Permissible medical experiments. trials of war criminals before the Nuremberg Military Tribunals under Control Council Law No. 10. Nuremberg October 1946 – April 1949. 1946–1949. Washington. U.S. Government Printing Office, pp 181–182.
3. American Association for Cancer Research. 2011. https://www.cancerprogressreport-digital.org/cancerprogressreport/2011?pg=52#pg52. Accessed 20 May 2021.
4. Appelbaum PS, Grisso T, et al. Competence of depressed patients for consent to research. Am J Psychiatry. 1999;156:1380–4. https://doi.org/10.1176/ajp.156.9.1380.
5. Appelbaum PS, Roth LH. Competency to consent to research: a psychiatric overview. Arch Gen Psychiatry. 1982;39:951–8. https://doi.org/10.1001/archpsyc.1982.04290080061009.
6. Appelbaum PS, Roth LH, et al. False hopes and best data: consent to research and the therapeutic misconception. Hast Cent Rep. 1987;17:20–4.
7. Association for Accessible Medicines. The U.S. generic & biosimilar medicines savings report. 2021. https://accessiblemeds.org/sites/default/files/2021-10/AAM-2021-US-Generic-Biosimilar-Medicines-Savings-Report-web.pdf. Accessed 10 Dec 2021.
8. Avissar M, Javitt D. Mismatch negativity: a simple and useful biomarker of N-methyl-o-aspartate receptor (NMDAR)-type glutamate dysfunction in schizophrenia. Schizophr Res. 2018;191:1–4. https://doi.org/10.1016/j.schres.2017.02.027.
9. Ban TA. Towards a clinical methodology for neuropsychopharmacological research. Neuropsychopharmacol Hung. 2007;9:81–90.
10. Bourin M. Animal models for screening anxiolytic-like drugs: a perspective. Dialogues Clin Neurosci. 2015;17:295–303. https://doi.org/10.31887/DCNS.2015.17.3/mbourin.
11. Brady LS, Potter WZ, Gordon JA. Redirecting the revolution: new developments in drug development for psychiatry. Expert Opin Drug Discovery. 2019;14:1213–9. https://doi.org/10.1080/17460441.2019.1666102.
12. Carrier F, Banayan D, et al. Ethical challenges in developing drugs for psychiatric disorders. Prog Neurobiol. 2017;152:58–69. https://doi.org/10.1016/j.pneurobio.2017.03.002.
13. Choc M, Hsuan F, et al. Single vs. multiple dose pharmacokinetics of clozapine in psychiatric patients. Pharm Res. 1990;7:347–51. https://doi.org/10.1023/a:1015859103824.

14. Coates M, Spanos M, et al. A review of methods for monitoring adverse events in pediatric psychopharmacology clinical trials. Drug Saf. 2018;41:465–71. https://doi.org/10.1007/s40264-017-0633-z.
15. Code of Federal Regulations. Code of federal regulations, title 21, U.S. FDA. 2021. https://www.ecfr.gov/current/title-21. Accessed 10 Dec 2021.
16. Colburn WA, Lee JW. Biomarkers, validation and pharmacokinetic-pharmacodynamic modelling. Clin Pharmacokinet. 2003;42:997–1022. Acccesed 10 Dec 2021. https://doi.org/10.2165/00003088-200342120-00001.
17. Conley AC, Newhouse PA. Advances in drug discovery and development in geriatric psychiatry. Curr Psychiatry Rep. 2018;20:10–9. https://doi.org/10.1007/s11920-018-0871-5.
18. Culpepper L, Kovalick LJ. A review of the literature on the selegiline transdermal system: an effective and well-tolerated monoamine oxidase inhibitor for the treatment of depression. Prim Care Companion J Clin Psychiatry. 2008;10:25–30. https://doi.org/10.4088/pcc.v10n0105.
19. de Berardis D, Fornaro M, et al. The role of inhaled loxapine in the treatment of acute agitation in patients with psychiatric disorders: a clinical review. Int J Mol Sci. 2017;18:349–64. https://doi.org/10.3390/ijms18020349.
20. DiMasi JA, Grabowski HG, Hansen RW. Innovation in the pharmaceutical industry: new estimates of R&D costs. J Health Econ. 2016;47:20–30. https://doi.org/10.1016/j.jhealeco.2016.01.012.
21. Drukarch B, Jacobs GE, Wilhelmus MMM. Solving the crisis in psychopharmacological research: cellular-membrane(s) pharmacology to the rescue? Biomed Pharmacother. 2020;130:110545. https://doi.org/10.1016/j.biopha.2020.110545.
22. Dunn LB. Assessing decisional capacity for clinical research or treatment: a review of instruments. Am J Psychiatry. 2006;163:1323–34. https://doi.org/10.1176/ajp.2006.163.8.1323.
23. English BA, Thomas K, et al. Use of translational pharmacodynamic biomarkers in early-phase clinical studies for schizophrenia. Future Med. 2014;8:29–49.
24. European Medicines Agency. Role of modelling and simulation in regulatory decision making in Europe. 2011. https://www.ema.europa.eu/en/documents/presentation/presentation-role-modelling-simulation-regulatory-decision-making-europe_en.pdf. Accessed 8 Jan 2022.
25. Fava M, Johe K, et al. A phase 1B, randomized, double blind, placebo controlled, multiple-dose escalation study of NSI-189 phosphate, a neurogenic compound, in depressed patients. Mol Psychiatry. 2016;21:1372–80. https://doi.org/10.1038/mp.2015.178.
26. Food & Drug Administration. Guidance for industry exposure-response relationships — study design, data analysis, and regulatory applications. 2003. https://www.fda.gov/media/71277/download. Accessed 10 Jan 2022.
27. Food & Drug Administration. CDER approval times for priority and standard NDAs and BLAs calendar years 1993-2016. 2016. https://wwwfdagov/ media/102796/download. Accessed 10 Dec 2021.
28. Food & Drug Administration. The drug development process. 2018a. https://www.fda.gov/patients/learn-about-drug-and-device-approvals/drug-development-process. Accessed 10 Dec 2021.
29. Food & Drug Administration. Step 3: Clinical research 2018b. https://www.fda.gov/patients/drug-development-process/step-3-clinical-research#Clinical_Research_Phase_Studies. Accessed 10 Dec 2021.
30. Food & Drug Administration. Major depressive disorder: developing drugs for treatment guidance for industry DRAFT GUIDANCE. 2018c. https://www.fda.gov/regulatory-information/search-fda-guidance-documents/major-depressive-disorder-developing-drugs-treatment. Accessed 10 Dec 2021.
31. Food & Drug Administration. Impact story: CDER assessment of drug impurity mutagenicity by quantitative structure-activity relationship modeling, regulatory science in action. 2020a. https://www.fda.gov/drugs/regulatory-science-action/impact-story-cder-assessment-drug-impurity-mutagenicity-quantitative-structure-activity-relationship#:~:text=Among%20

these%20tests%20are%20those%20that%20assess%20a,as%20well%20as%20in%20mammalian%20cells%20and%20animals. Accessed 10 Dec 2021.
32. Food & Drug Administration. Expedited programs for Serious Conditions—Drugs and Biologics. 2020b. https://www.fda.gov/regulatory-information/search-fda-guidance-documents/expedited-programs-serious-conditions-drugs-and-biologics. Accessed 17 Dec 2021.
33. Food & Drug Administration. New drugs at FDA: CDER's new molecular entities and new therapeutic biological products, development & approval process. 2021a. https://www.fda.gov/drugs/development-approval-process-drugs/new-drugs-fda-cders-new-molecular-entities-and-new-therapeutic-biological-products. Accessed 10 Dec 2021
34. Food & Drug Administration. Regulations: Good Clinical Practice and clinical trials 2021. 2021b. https://www.fda.gov/science-research/clinical-trials-and-human-subject-protection/regulations-good-clinical-practice-and-clinical-trials. Accessed 10 Dec 2021.
35. Food & Drug Administration. Information sheet guidance for sponsors, clinical investigators, and IRBs. Frequently asked questions statement of investigator(form FDA 1572, Revision 1). 2021c. https://www.fda.gov/media/148810/download.
36. Food & Drug Administration. Population pharmacokinetics guidance for industry. (2022). https://www.fda.gov/media/128793/download. Accessed 8 Jan 2022.
37. Fornaro M, De Berardis D, et al. Novel psychopharmacology for depressive disorders. Adv Exp Med Biol. 2021;1305:449–61. https://doi.org/10.1007/978-981-33-6044-0_22.
38. Gamo NJ, Birknow MR, et al. Valley of death: a proposal to build a "translational bridge" for the next generation. Neurosci Res. 2017;115:1–4. https://doi.org/10.1016/j.neures.2016.11.003.
39. Hendriksen H, Groenink L. Back to the future of psychopharmacology: a perspective on animal models in drug discovery. Eur J Pharmacol. 2015;759:30–41. https://doi.org/10.1016/j.ejphar.2015.03.020.
40. Grabb MC, Gobburu JVS. Challenges in developing drugs for pediatric CNS disorders: a focus on psychopharmacology. Prog Neurobiol. 2017;152:38–57. https://doi.org/10.1016/j.pneurobio.2016.05.003.
41. Grabb MC, Hillefors M, Potter WZ. The NIMH 'fast-fail trials' (fast) initiative: rationale, promise, and progress. Pharmaceut Med. 2020;34:233–45. https://doi.org/10.1007/s40290-020-00343-y.
42. Grisso T, Appelbaum PS. Assessing competence to consent to treatment: a guide for physicians and other health professionals. New York: Oxford University Press; 1998.
43. Gutiérrez-Casares JR, Quintero J, et al. Methods to develop an *in silico* clinical trial: computational head-to-head comparison of lisdexamfetamine and methylphenidate. Front Psych. 2021;12:741170. https://doi.org/10.3389/fpsyt.2021.741170.
44. Iltis A, Misra S, et al. Addressing risks to advance mental health research. JAMA Psychiat. 2013;70:1363–71. https://doi.org/10.1001/jamapsychiatry.2013.2105.
45. International Council for Harmonisation of Technical Requirements for Pharmaceuticals for Human Use (ICH). Efficacy guidelines. 2022 .https://www.ich.org/page/efficacy-guidelines. Accessed 9 Jan 2022.
46. International Council for Harmonisation of Technical Requirement for Pharmaceuticals for Human Use. Assessment and control of DNA reactive (Mutagenic) impurities in pharmaceuticals to limit potential carcinogenic risk, ICH Harmonized Guideline. 2017. Step 4 Version; 21 March 2017. https://database.ich.org/sites/default/files/M7_R1_Guideline.pdf . Accessed 20 Dec 2021.
47. International Council for Harmonisation of Technical Requirements for Pharmaceuticals for Human Use (ICH). ICH-E6 Good Clinical Practice (GCP). 2021. https://database.ich.org/sites/default/files/ICH_E6-R3_GCP-Principles_Draft_2021_0419.pdf. Accessed 6 Jan 2022.
48. Jha MK, Minhajuddin A, et al. Sex differences in the association of baseline c-reactive protein (CRP) and acute-phase treatment outcomes in major depressive disorder: findings from the EMBARC study. J Psychiatr Res. 2019;113:165–71. https://doi.org/10.1016/j.jpsychires.2019.03.013.

49. Jha MK, Trivedi MH. Pharmacogenomics and biomarkers of depression. Handb Exp Pharmacol. 2018;250:101–13. https://doi.org/10.1007/164_2018_171.
50. Kane J, Honigfeld G, et al. Clozapine for the treatment resistant schizophrenic: a double-blind comparison with chlorpromazine. Arch Gen Psychiatry. 1988;45:789–96. https://doi.org/10.1001/archpsyc.1988.01800330013001.
51. Keshavan MS, Lawler AN, et al. New drug developments in psychosis: challenges, opportunities and strategies. Prog Neurobiol. 2017;152:3–20. https://doi.org/10.1016/j.pneurobio.2016.07.004.
52. Koszla O, Targowska-Duda KM. In in vitro and in vivo models for the investigation of potential drugs against schizophrenia. Biomol Ther. 2020;10:24. https://doi.org/10.3390/biom10010160.
53. Life Size VC 2014. https://lifescivc.com/2014/11/a-billion-here-a-billion-there-the-cost-of-making-a-drug-revisited/. Accessed 10 Dec 2021.
54. Madaan V, Kolli V, et al. Update on optimal use of lisdexamfetamine in the treatment of ADHD. Neuropsychiatr Dis Treat. 2013;9:977–83. https://doi.org/10.2147/NDT.S34092.
55. Miller FG. Placebo-controlled trials in psychiatric research: an ethical perspective. Biol Psychiatry. 2000;47:707–16. https://doi.org/10.1016/s0006-3223(00)00833-7.
56. Millan MJ, Goodwin GM, et al. Learning from the past and looking to the future: emerging perspectives for improving the treatment of psychiatric disorders. Eur Neuropsychopharmacol. 2015;25:599–656. https://doi.org/10.1016/j.euroneuro.2015.01.016.
57. Nguyen KP, Fatt CC, et al. Patterns of pretreatment reward task brain activation predict individual antidepressant response: key results from the EMBARC randomized clinical trial. Biol Psychiatry. 2022;91:550–60. https://doi.org/10.1016/j.biopsych.2021.09.011.
58. Nucifora FC, Mihaljevic M, et al. Clozapine as a model for antipsychotic development. Neurotherapeutics. 2017;14:750–61. https://doi.org/10.1007/s13311-017-0552-9.
59. O'Hagan P, Farkas C. Bringing Pharma R&D back to health. Bain & Company. 2009. https://www.bain.com/contentassets/6c3bdd8a984647ec9e4e6a4791fd9cdd/bb_managing_randd_hc.pdf. Accessed 10 Dec 2021.
60. Peters TS. Do preclinical testing strategies help predict human. Hepatotoxic Potentials? Toxicologic Pathology. 2005;33:146–54. https://doi.org/10.1080/01926230590522121.
61. Policy and Medicine. 2014. https://www.policymed.com/2014/12/1-tough-road-cost-to-develop-one-new-drug-is-26-billion-approval-rate-for-drugs-entering-clincal-de.html. Accessed 10 Dec 2021
62. Posner K, Brown GK, et al. The Columbia-suicide severity rating scale: initial validity and internal consistency findings from three multisite studies with adolescents and adults. Am J Psychiatry. 2011;168:1266–77. https://doi.org/10.1176/appi.ajp.2011.10111704.
63. Posner K, Oquendo M, et al. Columbia classification algorithm of suicide assessment (C-CASA): classification of suicidal events in the FDA's pediatric suicidal risk analysis of antidepressants. Am J Psychiatry. 2007;164:1035–43. https://doi.org/10.1176/ajp.2007.164.7.1035.
64. Preskorn SH. CNS drug development: lessons learned part 2. Symptoms, not syndromes as targets consistent with the NIMH research domain approach. Psychiatr Pract. 2015;21:60–6. https://doi.org/10.1097/01.pra.0000456594.66363.6f.
65. Rossetti AC, Koch P, Ladewig J. Drug discovery in psychopharmacology: from 2D models to cerebral organoids. Dialogues Clin Neurosci. 2019;21:203–10.
66. Rubin R. Recently approved Alzheimer drug raises questions that might never be answered. JAMA. 2021;326:469–72. https://doi.org/10.1001/jama.2021.11558.
67. Rusten IS, Musuamba FT. Scientific and regulatory evaluation of empirical pharmacometric models: an application of the risk informed credibility assessment framework. CPT Pharmacometrics Syst Pharmacol. 2021;10:1281–96. https://doi.org/10.1002/psp4.12708.
68. Scott AJ, Sharpe L, et al. Association of single-blind placebo run-in periods with the placebo response in randomized clinical trials of antidepressants: a systematic review and meta-analysis. JAMA Psychiat. 2022;79:42–9. https://doi.org/10.1001/jamapsychiatry.2021.3204.

69. Simon NG, Brownstein MJ. Challenges in developing drugs for neurological and psychiatric disorders. Prog Neurobiol. 2017;152:1–2. https://doi.org/10.1016/j.pneurobio.2017.04.001.
70. Stiles PG, Poythress NG, et al. Improving understanding of research consent disclosures among persons with mental illness. Psychiatr Serv. 2001;52:780–5. https://doi.org/10.1176/appi.ps.52.6.780.
71. Swazey JP. Chlorpromazine in psychiatry: a study in therapeutic innovation. Cambridge: MIT Press; 1974.
72. The National Commission for the Protection of Human Subjects of Biomedical and Behavioral Research. The Belmont report. 1979. Ethical principles and guidelines for the protection of human subjects of research Department of Health, Education, and Welfare https://wwwhhs-gov/ohrp/sites/default/files/the-belmont-report-508c_FINALpdf.. Accessed 6 Jan 2022.
73. Tricklebank MD, Robbins TW, et al. Time to re-engage psychiatric drug discovery by strengthening confidence in preclinical psychopharmacology. Psychopharmacology. 2021;238:1417–36. https://doi.org/10.1007/s00213-021-05787-x.
74. Maruschak LM, Bronson J, Alper M. Indicators of mental health problems reported by prisoners. U.S. Department of Justice. 2021. https://bjs.ojp.gov/sites/g/files/xyckuh236/files/media/document/imhprpspi16st.pdf. Accessed 9 Jan 2022.
75. Wang Y, Zhu H, et al. Model-informed drug development: current US regulatory practice and future considerations. Clin Pharmacol Ther. 2019;105:899–911. https://doi.org/10.1002/cpt.1363.
76. World Medical Organization. Declaration of Helsinki. Medical research involving human subjects. 2013. https://wwwwmanet/what-we-do/medical-ethics/declaration-of-helsinki. Accessed 8 Jan 2022.
77. Wouters OJ, McKee M, Luyten J. Estimated research and development investment needed to bring a new medicine to market, 2009-2018. JAMA. 2020;323:844–53. https://doi.org/10.1001/jama.2020.1166.
78. Yanos PT. Research risk for persons with psychiatric disorders: a decisional frameworkto meet the ethical challenge. Psychiatr Serv. 2009;60:374–83. https://doi.org/10.1176/appi.ps.60.3.374.

Chapter 6
Post-Approval Research in Drug Development: Priorities and Practices

David Williamson, Jack Sheehan, and Ella Daly

Abstract A prescriber might ask if a new medication is a good option for use in the patients he or she sees in clinic, with their particular blends of demographic and comorbid clinical characteristics. Is this medicine more effective, safe, tolerable, or affordable than the options used in the past? A payer may ask if the new medication offers a more effective, cost-efficient, or convenient alternative to those treatments already being covered. These are the types of questions that are often difficult to answer on the basis of the clinical trials used to support a medication's initial approval, which are generally designed to evaluate a medication's efficacy, safety, and tolerability in narrowly defined patient populations. Consequently, in order to answer the questions most relevant to key stakeholders (i.e., regulators, patients, and clinicians), it is important to continue to examine a medication's impact and characteristics after it has received regulatory approval. Such studies vary in their purpose, scope, and methodology. In this chapter, we review the types of questions most likely to be investigated after regulatory approval, the methods generally used to investigate them, and the characteristics typically considered when prioritizing the allocation of resources.

Keywords Phase 4 trials · Post-approval research

The process by which a new drug becomes a candidate for approval by the US Food and Drug Administration (FDA) and similar regulatory authorities has been extensively characterized throughout this volume. In seeking approvalw from

D. Williamson
University of South Alabama College of Medicine, Departments of Psychiatry and Neurology, Mobile, AL, USA

J. Sheehan
Independent Medical Consultant, Doylestown, PA, USA
e-mail: jsheeha2@its.jnj.com

E. Daly
Janssen Scientific Affairs, LLC, Titusville, NJ, USA

M. Macaluso et al. (eds.), *Drug Development in Psychiatry*, Advances in Neurobiology 30,
https://doi.org/10.1007/978-3-031-21054-9_6

regulatory authorities, drug developers must weigh a host of considerations, including precise definitions of the diseases being treated, populations being studied, trial designs and statistical methods most appropriate to characterize efficacy, and the medication's safety and tolerability. This is a costly, resource-intensive process; for central nervous system medications, recent estimates suggest the total development expenditure is approximately $766 million [23]. Manufacturers seek to discover and develop a medication that addresses a key unmet medical need and is efficacious, safe, tolerable, and convenient for patients. However, for many products, the process of gaining regulatory approval is only part of the story. Much of what is known about the medications we use is based on research performed after approval by regulatory authorities. This research can include comparisons to other drugs or modalities of treatment, investigations of "real-world" samples that focus on effectiveness rather than efficacy, and examinations of drug performance in different populations or indications. These studies may be conducted independently from the pharmaceutical industry, but pharmaceutical companies are often active in these post-approval efforts. In this chapter, we will provide one perspective on the types of priorities, decisions, and practices that drive these investigations, with a particular focus on considerations that would be relevant for drug development in neurology or psychiatry.

6.1 Priorities

When seeking regulatory approval for a medication, the top priority is to provide sufficient evidence of benefit for the condition being treated in the target patient population. After approval, a variety of questions concerning the medication may be asked, but all are generally driven by a central consideration: *what do patients, prescribers, and/or payers* (e.g., *insurance companies*) *need to know about the medication, and how does it fit into the existing or anticipated marketplace*? The responses to this question may vary dramatically depending on the condition or patients being treated, the number and characteristics of existing competing products or treatment modalities, availability of treatment guidelines for the disease, and the resources of the company funding or collaborating in the research.

6.2 Regulatory Commitments

During the post-approval process, regulatory agencies may require additional evidence to better understand questions that were outstanding at the time of approval. Such studies may focus on aspects such as long-term safety, efficacy, or optimal dosing in specific populations. For example, additional studies may be planned to evaluate the efficacy and safety of a medication in a pediatric population after the initial approval was granted for adults only. Similarly, some medications are initially approved as adjunctive therapies, and following approval, regulatory authorities may pose questions about a medication's potential utility as a monotherapy agent.

For all products, once regulatory approval has been received, the manufacturer must submit periodic safety update reports (or, in the United States, periodic adverse drug experience reports, better known as PADER), where case reports with serious unlisted events are presented in a narrative or tabular format. For a limited number of products, there may be a risk evaluation and mitigation strategy (REMS) required at the time of approval. A REMS is a drug safety program that the FDA may require for products with serious safety concerns to help ensure the benefits of a medication outweigh its risks [21]. These types of programs or registries are typically conducted by the pharmaceutical company manufacturing the drug, often with the assistance of a specialized vendor, with specific metrics regularly shared with the FDA on a predetermined schedule. The FDA may in turn seek additional data or clarifications in the form of information requests, to which the manufacturer must respond to within a given timeframe.

6.3 Further Clinical Considerations

A variety of clinical questions may stimulate investigations after a medication's initial regulatory approval for a specific indication. The fundamental questions of general efficacy, safety, and tolerability of a product are addressed by pivotal trials, phase 3 clinical trials that generate the core evidence that informs the initial approval decision by regulatory authorities. However, these data never answer every clinical question, and the patients included in phase 3 studies may not be broadly representative of real-world patient populations. Additionally, placebo effects observed in trials of drugs targeted at neurological or psychiatric conditions can be substantial [9, 11, 12]. Consequently, in order to maximize the likelihood of finding valid evidence that a treatment is working in the population for whom it is intended, inclusion and exclusion criteria for phase 3 clinical trials are designed to create a relatively homogenous clinical population. However, this strategy may exclude potentially relevant strata of disease severity, comorbidities, or other aspects of clinical presentation that inform everyday clinical decision-making on a regular basis. Furthermore, behavioral or operational aspects of everyday clinical situations (e.g., nonadherence, missing scheduled appointments, unexpected delays in treatment coverage or supply) are generally minimized or eliminated by the structure of formal clinical trials. In addition, for ethical reasons, pivotal trials may exclude patients perceived to have elevated safety concerns so as not to confound the safety signal, and there may be reluctance to use an investigational agent in these populations during drug development. Finally, many trials are international in nature, both for reasons of cost and varying regulatory requirements [16].

The cumulative impact of these clinical trial features is that some populations may be relatively understudied, despite there being sufficient evidence that a medication is effective, safe, tolerable, and convenient enough in the intended clinical population to receive regulatory approval. Such patient characteristics may include gender, age (e.g., those aged ≥65 years or <18 years), race, ethnicity, hepatic or renal impairment, pregnancy, breastfeeding, relevant comorbidities, and differences in disease presentation or subpopulations (e.g., positive or negative symptoms in schizophrenia; suicidality, irritability, or anxiety in major depressive disorder; focal or generalized onset in epilepsy, presence or absence of aura in migraine). Although the restriction of sample characteristics is understandable, especially given the high failure rate of phase 3 clinical trials [2], it has implications for the translation of findings to clinical practice. For instance, Zimmerman and colleagues have observed that when applying inclusion and exclusion criteria used in 158 antidepressant efficacy trials to a sample of 1271 outpatients seeking treatment for depression at an outpatient center, anywhere from 44.4 to 99.8% of patients would have been excluded from trials. Moreover, this trend appears to be increasing; the mean percentage of patients that would have been excluded by clinical trial criteria from 1995 to 2009 was 83.8%, whereas this percentage rose to 91.4% in trials performed between 2010 and 2014 [24].

These issues overlap with considerations of efficacy versus effectiveness. Registration trials are predominantly designed to demonstrate efficacy; that is, they are designed to "investigate the benefits and harms of an intervention under highly controlled conditions" [19]. Effectiveness trials, in contrast, "examine interventions under circumstances that more closely approach real-world practice, with more heterogeneous patient populations, less standardized treatment protocols, and delivery in routine clinical settings" [19]. Neither of these approaches is inherently superior to the other; they simply answer different questions. Post-approval trials are more likely to incorporate elements of effectiveness trials because of the broader array of questions that they are often designed to answer.

Other clinical considerations that may arise include benefits or adverse events associated with a medication that may not have been anticipated at the time the clinical trials were designed. For example, previously unidentified adverse events may emerge when the medicine is used outside of the controlled trial setting, or rare adverse events may emerge when the medicine is used widely post-approval. Additional gaps in the current understanding of the medication or its use may be identified after approval, or new research or interest in the field may suggest that a drug may be an efficacious treatment for related or unrelated indications. This work may be conducted independently or potentially in collaboration with a pharmaceutical company as part of the investigator-initiated study (IIS) or collaborative study.

For instance, in the field of neurology, multiple medications initially approved for epilepsy have since received FDA approval for migraine prophylaxis years after they first became available on the market. In contrast, such medications may also manifest tolerability or safety issues (e.g., impaired cognition) that spur further investigation so clinicians and patients can be fully informed of both potential benefits and risks of the medications across different populations.

Clinicians must also consider operational requirements of medications and how to incorporate them into current treatment paradigms, particularly those that may be used in smaller practice settings. This may be particularly pertinent for novel treatments. For example, if a product is not an oral medication for which a patient can fill a prescription at a local pharmacy, issues of logistics are likely to emerge. For instance, is refrigeration required for a medication that needs to be stored on site, or is the medication a controlled substance? Does a practice have the infrastructure required to support administration of the product at the site (e.g., office space, staffing)? Are there additional requirements associated with the medication (e.g., mandatory enrollment of the patient in a REMS and required training and documentation)? Although some sites and settings have experience with interventional treatment in psychiatry, such as transcranial magnetic stimulation (TMS) and electroconvulsive therapy (ECT), most are not equipped to administer novel, non-oral treatments with requirements that depart significantly from routine psychiatric practice, and as a result, such treatments are not easily implemented. Post-approval research can examine these types of issues (e.g., exploring how access to care may vary across different types of practice settings or geographies).

Finally, there is often interest in understanding different perspectives regarding the efficacy, safety, tolerability, and convenience of a product. For many disorders in the fields of psychiatry and neurology, the manifestation of pathology (e.g., mood disturbance, hallucinations, thought disorder, pain) is inherently subjective. Complicating this dynamic is the robust finding that the relationship between the patient's perception of symptoms that can be objectively quantified (e.g., quantity of sleep, ability to learn and recall new information) and what is measured or observed by others is, at best, inconsistent [1, 3, 4, 6, 8, 13, 15, 17, 18, 22]. Therefore, these disparate sources (patients, observers, clinicians, objective measurements) often provide different types of information that may be better suited to answering different types of questions.

For instance, clinician-rated measures of depressive symptoms appear to be more sensitive to change within the context of clinical trials than patient-rated measures are [5]. However, although patient-reported difficulties with memory are often more closely tied to depressive symptoms than to objectively measured deficits in learning [10, 13], the subjective impact of these difficulties or the extent to which a patient interprets an intervention to be useful may relate differently to what is perceived rather than what is objectively measured. Similarly, the extent to which adherence to a medication regimen correlates more closely with patient-perceived improvement rather than to improvement perceived by an informed observer may be a relevant concern for some populations. Dhillon and colleagues have noted that patient-perceived cognitive impairment contributes to disability much more than objectively measured cognitive impairment and depression do [7]. Importantly, patient-clinician dialogue is critical to optimal patient management, particularly when subjective symptoms are core to the presenting diagnosis. As a consequence, there is a growing emphasis on obtaining the perspective of patients, often via self-report scales (now generally referred to as "patient-reported outcomes" or PROs) [14, 20].

6.4 Payer Considerations

In addition to the clinical considerations driving post-approval research priorities, questions from payers are relevant as well. For example, in the United States, commercial payers compete to provide employer-sponsored insurance. To effectively compete, the payer must balance a set of healthcare benefits that are both compelling for employees and competitively priced for their employer. Payers must consider the potential budgetary impact of making a medication available to their customers. This may extend beyond the simple question of cost per unit of treatment; it is important to expound upon the context in which the treatment will be provided and the costs of administering the treatment.

Fundamental considerations include the size of the treatment-eligible population, the expected uptake of the medication within this population, the cost of other treatments the new treatment may displace, and the burden of the condition being treated. This burden includes both healthcare outcomes and societal impact. For instance, is there evidence of cost offset from using the medication (e.g., decreasing the frequency and/or length of required inpatient care)? Is there evidence that the medication attenuates the need for concomitant medications currently used to treat symptoms related to the clinical condition or adverse events associated with other medications? There are questions about financial impact on a societal level as well that may be of interest to employers. For example, is there evidence that use of a product increases the likelihood of employment or decreases absenteeism or negative work outcomes?

Additionally, payers will often incorporate treatment guidelines into coverage decisions. Guidelines may have implications for how a new product is used or its recommended place in line of treatment. Payer questions may also be framed by the number and types of competing products available to treat a condition. For instance, are there similar products in the same class available for a lower price? If so, how different are these medications from one another? Can formulary management techniques like step-edit or prior authorization be used to steer patients who might benefit from a less expensive product toward the less expensive product? How much money will this formulary management strategy save? How much will managing this program cost? How will it impact members and clinicians?

All these questions are likely to be considered from several different perspectives. How certain are the estimates? Are these estimates based on similar products that are used in the same or very similar therapeutic conditions, or is this a medication establishing a new paradigm of treatment or treating a condition for which there are few (if any) alternatives? If there are other available treatment options, are the effects or cost offsets of sufficient magnitude to justify a product's cost in comparison to the cost of competing products that may provide similar benefits? Are there subsets of patients that are likely to benefit differentially from the treatment, and do costs vary systematically in those subgroups? The answers to these questions are often not simple "yes-or-no" responses in terms of providing access to medications; rather, they generate discussions about how policy designs can manage the use of a new medication without imposing undue burden on members of a healthcare plan, clinicians, or payer resources.

6.5 Decisions: What Gets Studied?

As is evident from the preceding sections, there is a wealth of potential questions to be investigated after the regulatory approval of a medication. The questions most relevant to individual products will vary based upon clinical, regulatory, and market concerns. Thus, decisions must be made about which questions will be pursued, in what order, and in what fashion.

Regulatory commitments are mandatory and will always be given priority. After the commitments made to regulatory authorities, priorities are generally driven by the extent to which clinical use or availability of the medication is being impacted by unanswered questions such as those enumerated in the Clinical and Payer sections above. These are not the only important issues influencing these decisions, however. In addition to answering the question "what do we most want to know?," companies must also answer the questions "how can we best answer this question?" and "what resources are available to obtain these answers?"

Several considerations enter the decision-making process in determining which questions to pursue. An important issue is availability of relevant data. For instance, it may appear that, based on case reports or small studies published in the literature, a particular dose, pattern of dosing, or the use with an adjunctive medication may be clinically relevant. Alternatively, clinicians or patients may hear of claims made by competitors suggesting comparative advantages or disadvantages of one product relative to others. Authoritative responses to each of these concerns will nearly always involve reference to data that have not been collected in any large-scale, systematic fashion.

Consequently, choices must be made as to whether the questions are widespread or compelling enough to justify the time, expense, and effort of collecting the relevant data. Unanswered questions become more compelling as they become more important to relevant decision-makers, be it patients, clinicians, or those considering whether medications should be included on formularies. Once a question is considered important, attention turns to whether the answer is tractable, i.e., is it possible to reliably gather enough data so that the question can be answered with a reasonable degree of confidence? In some situations (e.g., rare adverse events of a medication used to treat a rare condition), there simply may not be enough data of sufficient quality to accomplish this. In other cases, answers may be feasible but only within the context of a very large dataset or a dataset in which large numbers of patients are tracked over periods of months or years.

6.6 Practices

For the questions that rise to the level of sufficient importance, pharmaceutical companies have a few different tools available to them to inform answers (Table 6.1). The most frequently used strategy to answer clinical questions is post hoc analysis of clinical trial data. These data have the advantages of (a) being collected in a rigorous manner, (b) having already been deemed suitable to support a regulatory approval, (c) generally, but not always, being large enough to provide sufficient statistical power to examine an issue, and (d) being readily available. Limitations to this approach include the efficacy/effectiveness issues discussed previously and the unknown extent to which the results generalize to an independent set of data not used to obtain regulatory approval.

Another strategy is collaboration. This can take a variety of forms. Companies might sponsor a collaboration with vendors with access to large databases relevant to the questions being asked (e.g., a commercial insurer or the Veterans Health Administration). Alternatively, in investigator-initiated studies (IIS), an independent investigator might propose an idea consistent with a product's labeling that fits

Table 6.1 Types of studies often performed in post-approval drug development

Typical criteria	Randomized explanatory trial	Randomized pragmatic trial	Non-comparative registry	Retrospective database analysis/ economic modeling	Post hoc analysis of existing trial data
Time and effort					
Total cost	+++++	++++	+++	++	+
Cost per patient	+++	++	+	Trivial	Trivial
Time from concept to results	1–7 years	2–7 years	3–10 years	6–24 months	6 months
Scientific considerations					
Sensitivity to convenience/ tolerability differences	Low	Moderate	Moderate	High	Low (assuming data are from an explanatory trial)
Outcomes	Researcher defined	Researcher defined	Researcher defined	Must be available in database	Must be available in trial database
Consistent follow-up	Yes	Variable	Usually no	No	Yes
Causal inference	Yes	Yes	Limited	Partial	Partial
Project timing					
Conducted pre-launch	Yes	Potentially	Usually no	No	Usually no

within a known area of interest to the company, and the company then funds a research study performed by this independent investigator. Another approach is to retrospectively permit access to specific data collected during the course of a clinical trial to an independent investigator who then independently analyzes that data pertaining to a specific question. Such an approach is fostered by such platforms as the Yale University Open Data Access Project (https://yoda.yale.edu/johnson-johnson). For example, the following information is provided on the YODA platform for those interested in accessing data related to Johnson & Johnson products:

> The YODA Project performs independent scientific reviews of investigator requests for Clinical Study Reports and de-identified participant-level data from Johnson & Johnson's pharmaceutical, medical device, and consumer product sectors. Johnson & Johnson has conferred on the YODA Project the authority to make decisions about the release of their clinical trial data.

A final strategy is to design and carry out prospective trials targeted at answering the questions of interest. This is the least common approach because it is the most resource and time intensive. However, such programs might be considered should an issue emerge that consistently and significantly impacts the availability or use of a product or points to an unanticipated benefit that may lead to a new indication or an important change to existing labeling.

When considering questions most relevant to payers, other methodologies come to the forefront. Such studies typically employ three potential sources: (1) direct trial results, (2) economic modeling that uses trial data as an input, or (3) retrospective database analysis of insurance claims or electronic health record data after the product is approved. These latter data are often termed "real-world evidence" (RWE). When relying on clinical trial data, it is often useful to either translate the value of clinical outcome measures used in the trials to measures of increased relevance to payers or to directly measure outcomes that are relevant to payers (e.g., within the context of major depressive disorder, facilitating the understanding of how obtained test scores translate to more easily understood outcomes such as response, remission, and quality of life). As an illustrative example, a common goal of manufacturers is to understand how their product may impact disease-specific hospitalization. To do this, different approaches are possible with different trade-offs in timing, cost, and confidence in the evidence. For example, a clinical trial may directly measure disease-specific hospitalization as an outcome. Alternatively, a trial may measure symptom improvement, and this information can then be linked to other data to estimate how disease-specific hospitalization may change as part of a modeling exercise. Finally, a retrospective database analysis may compare a new product with another product by focusing on disease-specific hospitalization. When outcomes such as hospitalization are measured directly in a randomized trial, confidence in the evidence is high. However, this approach is not always feasible because the number of hospitalization events is too few or the trial protocol itself may affect hospitalization decisions in ways that are inconsistent with real-world practice.

Modeling estimates from trial data can preserve the advantages of randomization and, because events can be extrapolated from the trial data over a large population and longer period of time, these estimates are less subject to concerns that can occur when applying trial data directly, such as a small number of events or trial design decisions impacting hospitalization. However, the researcher must make assumptions about how changes in symptoms are related to hospitalization to construct and quantify these estimates. Using real-world data has the advantage that the outcome analyzed (e.g., disease-specific hospitalization) is not driven by a trial protocol but reflects typical practice. It also usually includes a much larger sample of patients than the pivotal trials. However, this approach is generally not available at the time of medication approval. Moreover, because the patients using the products being compared are not randomly assigned to them, the patient populations treated by two different medications may differ in risk of disease-specific hospitalization independent of treatment. For this reason, advanced mathematical techniques are used to adjust for this risk to isolate the impact of treatment on the outcome of interest. Despite these efforts, it is usually not possible for the researcher to have total certainty they have completely isolated the impact of treatment; therefore, confidence in causal conclusions from this evidence is usually lower than in evidence from randomized trials.

Note that both the importance of questions and the most appropriate and/or feasible way of addressing them may evolve over time. For instance, a company may receive a question deemed to be very important at product launch and realize that a new prospective, randomized controlled trial will need to be performed to provide the most robust answers. The study is subsequently designed and executed early in the product life cycle. In the interim, the company answers the identified question in the best way available with a post hoc analysis of existing clinical trial data. As the use of the product increases, the company may conduct retrospective database analysis of insurance claims data or electronic health record data. Finally, after the prospective trial results are available, the company subsequently uses the trial data to address the question.

6.7 Conclusions

The evidence required to obtain regulatory approval for a medication is comprehensive and costly to obtain; nonetheless, it invariably leaves some questions unanswered. Due to the limitations of what was known about the medication when the registration trials were initiated or unforeseeable changes in the healthcare environment, some unanticipated questions will emerge after approval is obtained. The need to narrowly define clinical populations in registration trials to demonstrate efficacy, safety, and tolerability for a specific therapeutic state inevitably leads to challenges in generalizability; in real-world populations, comorbidities, access challenges, variations in adherence, and variations in treatments (and their combinations) abound. Payers will need to know how medications fit within the options they make available to their members. There will therefore always be a need for research that addresses the aforementioned questions after a drug has obtained regulatory approval.

In this chapter, we have endeavored to provide some insight into how a company may determine the questions to be answered; prioritize those most important to patients, clinicians, and payers; and implement different strategies and collaborations to provide suitable answers. Our perspective has been informed by our respective roles within the company in which we work; the perspectives of authors who hold different positions or work in different companies or settings (e.g., academia) may vary. Because the healthcare industry has sometimes seen a striking change in relatively short periods of time, it is likely that our own perspectives will evolve along with the industry and healthcare environment more broadly.

References

1. Adelantado-Renau M, Beltran-Valls MR, Migueles JH, Artero EG, Legaz-Arrese A, Capdevila-Seder A, Moliner-Urdiales D. Associations between objectively measured and self-reported sleep with academic and cognitive performance in adolescents: DADOS study. J Sleep Res. 2019;28(4):e12811. https://doi.org/10.1111/jsr.12811.
2. Arrowsmith J, Miller P. Trial watch: phase II and phase III attrition rates 2011-2012. Nat Rev Drug Discov. 2013;12(8):569. https://doi.org/10.1038/nrd4090.
3. Baeza-Velasco C, Guillaume S, Olie E, Alacreu-Crespo A, Cazals A, Courtet P. Decision-making in major depressive disorder: subjective complaint, objective performance, and discrepancy between both. J Affect Disord. 2020;270:102–7. https://doi.org/10.1016/j.jad.2020.03.064.
4. Butzbach M, Fuermaier ABM, Aschenbrenner S, Weisbrod M, Tucha L, Tucha O. Metacognition in adult ADHD: subjective and objective perspectives on self-awareness of cognitive functioning. J Neural Transm (Vienna). 2021;128(7):939–55. https://doi.org/10.1007/s00702-020-02293-w.
5. Cuijpers P, Li J, Hofmann SG, Andersson G. Self-reported versus clinician-rated symptoms of depression as outcome measures in psychotherapy research on depression: a meta-analysis. Clin Psychol Rev. 2010;30(6):768–78. https://doi.org/10.1016/j.cpr.2010.06.001.
6. Davis DA, Mazmanian PE, Fordis M, Van Harrison R, Thorpe KE, Perrier L. Accuracy of physician self-assessment compared witih observed measures of competence: a systematic review. JAMA. 2006;296:1094–102.
7. Dhillon S, Videla-Nash G, Foussias G, Segal ZV, Zakzanis KK. On the nature of objective and perceived cognitive impairments in depressive symptoms and real-world functioning in young adults. Psychiatry Res. 2020;287:112932. https://doi.org/10.1016/j.psychres.2020.112932.
8. John SE, Evans SA, Hanfelt J, Loring DW, Goldstein FC. Subjective memory complaints in white and African American participants. J Geriatr Psychiatry Neurol. 2020;33(3):135–43. https://doi.org/10.1177/0891988719868305.
9. Jones, B. D. M., Razza, L. B., Weissman, C. R., Karbi, J., Vine, T., Mulsant, L. S., . . . Daskalakis, Z. J. (2021). Magnitude of the placebo response across treatment modalities used for treatment-resistant depression in adults: a systematic review and meta-analysis. JAMA Netw Open, 4(9), e2125531. doi:https://doi.org/10.1001/jamanetworkopen.2021.25531.
10. Jorm, A. F., Butterworth, P., Anstey, K. J., Christensen, H., Easteal, S., Maller, J., . . . Sachdev, P. (2004). Memory complaints in a community sample aged 60-64 years: associations with cognitive functioning, psychiatric symptoms, medical conditions, APOE genotype,

hippocampus and amygdala volumes, and white-matter hyperintensities. Psychol Med, 34(8), 1495. Retrieved from http://www.cup.org.

11. Khan A, Fahl Mar K, Schilling J, Brown WA. Magnitude and pattern of placebo response in clinical trials of antiepileptic medications: data from the Food and Drug Administration 1996-2016. Contemp Clin Trials. 2018;64:95–100. https://doi.org/10.1016/j.cct.2017.10.017.
12. Khan A, Faucett J, Brown WA. Magnitude of placebo response and response variance in antidepressant clinical trials using structured, taped and appraised rater interviews compared to traditional rating interviews. J Psychiatr Res. 2014;51:88–92. https://doi.org/10.1016/j.jpsychires.2014.01.005.
13. Konen T, Karbach J. Self-reported cognitive failures in everyday life: a closer look at their relation to personality and cognitive performance. Assessment. 2020;27(5):982–95. https://doi.org/10.1177/1073191118786800.
14. McNaughton, E. C., Curran, C., Granskie, J., Opler, M., Sarkey, S., Mucha, L., . . . McCue, M. (2019). Patient attitudes toward and goals for MDD treatment: a survey study. Patient Prefer Adherence, 13, 959–967. doi:https://doi.org/10.2147/PPA.S204198.
15. Park, S., Lee, J. H., Lee, J., Cho, Y., Park, H. G., Yoo, Y., . . . Lee, J. Y. (2019). Interactions between subjective memory complaint and objective cognitive deficit on memory performances. BMC Geriatr, 19(1), 294. doi:https://doi.org/10.1186/s12877-019-1322-9.
16. Qiao Y, Alexander GC, Moore TJ. Globalization of clinical trials: variation in estimated regional costs of pivotal trials, 2015-2016. Clin Trials. 2019;16(3):329–33. https://doi.org/10.1177/1740774519839391.
17. Reukers, D. F. M., Aaronson, J., van Loenhout, J. A. F., Meyering, B., van der Velden, K., Hautvast, J. L. A., . . . Kessels, R. P. C. (2020). Objective cognitive performance and subjective complaints in patients with chronic Q fever or Q fever fatigue syndrome. BMC Infect Dis, 20(1), 397. doi:https://doi.org/10.1186/s12879-020-05118-z.
18. Rezaie L, Fobian AD, McCall WV, Khazaie H. Paradoxical insomnia and subjective-objective sleep discrepancy: a review. Sleep Med Rev. 2018;40:196–202. https://doi.org/10.1016/j.smrv.2018.01.002.
19. Singal AG, Higgins PD, Waljee AK. A primer on effectiveness and efficacy trials. Clin Transl Gastroenterol. 2014;5:e45. https://doi.org/10.1038/ctg.2013.13.
20. U.S. Food and Drug Administration. FDA patient-focused drug development guidance series for enhancing the incorporation of the patient's voice in medical product development and regulatory decision making. 2021a. Retrieved from https://www.fda.gov/drugs/development-approval-process-drugs/fda-patient-focused-drug-development-guidance-series-enhancing-incorporation-patients-voice-medical.
21. U.S. Food and Drug Administration. Risk evaluation and mitigation strategies. 2021b. Retrieved from https://www.fda.gov/drugs/drug-safety-and-availability/risk-evaluation-and-mitigation-strategies-rems.
22. Vaskivuo, L., Hokkanen, L., Hanninen, T., Antikainen, R., Backman, L., Laatikainen, T., . . . Ngandu, T. (2019). Self and informant memory reports in FINGER: associations with two-year cognitive change. J Alzheimers Dis, 71(3), 785–795. doi:https://doi.org/10.3233/JAD-190133.
23. Wouters OJ, McKee M, Luyten J. Estimated research and development investment needed to bring a new medicine to market, 2009-2018. JAMA. 2020;323(9):844–53. https://doi.org/10.1001/jama.2020.1166.
24. Zimmerman M, Balling C, Chelminski I, Dalrymple K. Have treatment studies of depression become even less generalizable? Applying the inclusion and exclusion criteria in placebo-controlled antidepressant efficacy trials published over 20 years to a clinical sample. Psychother Psychosom. 2019;88(3):165–70. https://doi.org/10.1159/000499917.

Chapter 7
Discovery of New Transmitter Systems and Hence New Drug Targets

Tiffany Schwasinger-Schmidt and Sheldon H. Preskorn

Abstract The development of medications used to treat psychiatric conditions has largely proceeded through serendipity, where a potential drug to treat mental illness is identified by chance. This approach is based on a limited understanding of the underlying pathophysiology of mental illness and brain disorders. Identification of novel neurotransmitter systems has allowed for new molecular-based approaches for drug development that identify specific receptor targets to treat a specific symptom. An example of this approach includes the development of suvorexant, which is a dual orexin receptor antagonist FDA approved in 2014 for the treatment of insomnia. This chapter will discuss challenges in psychiatric drug development; the importance of identifying discrete neurotransmitter systems that target a specific symptom, not a syndrome; the orexin pathway and targets within this pathway that can be used to modulate sleep; and a high-throughput approach to streamlining drug development.

Keywords Psychiatric drug development · Orexin neurotransmitters · Dual orexin receptor antagonists · Suvorexant · Targeted drug development

7.1 Introduction

The drug discovery process in psychiatry has largely proceeded through serendipity, where medications used for the treatment of psychiatric conditions were discovered by chance [1, 2]. While this approach has yielded hundreds of medications used to

T. Schwasinger-Schmidt (✉)
Department of Internal Medicine, University of Kansas School of Medicine-Wichita, Wichita, KS, USA
e-mail: tschwasinger-schmidt@kumc.edu

S. H. Preskorn
Department of Psychiatry and Behavioral Science, University of Kansas School of Medicine-Wichita, Wichita, KS, USA
e-mail: spreskon@kumc.edu

M. Macaluso et al. (eds.), *Drug Development in Psychiatry*, Advances in Neurobiology 30,
https://doi.org/10.1007/978-3-031-21054-9_7

treat mental illness, starting from the early 1800s [3, 4], the process of drug discovery has evolved to newer approaches of targeted drug design that focus on the underlying pathophysiology of disease [1]. This newer approach utilizes neurotransmitter systems that are correlated with specific symptoms to help maximize clinical utility and minimize the cost of development [5]. The average cost of developing a new medication from initial idea to clinical implementation ranges from $1 to 3 billion with thousands of compounds tested to produce one potential treatment [3, 5]. Given the significant risk and cost associated with drug development utilizing chance discovery, it is clear why only 9 of the 254 drugs developed between 2009 and 2016 were approved for psychiatric indications [5]. This also sheds light on the reasoning behind why most pharmaceutical companies have reduced or eliminated research programs in central nervous system (CNS) therapies due to the low clinical success rates [1].

7.1.1 Limitations in Psychiatric Drug Development

To fully understand the challenges in psychiatric drug development, the limitations in knowledge of the underlying pathophysiology of mental illness and the classification schemes used in diagnosis must be explored. The human brain is a highly complex system comprised of over 160 billion neurons that connect to over 10,000 additional neurons resulting in nearly 1,000 trillion synaptic connections that are monitored and maintained by numerous non-neuronal cells [1]. The complexity underlying these connections and alterations that occur through neuroplasticity of the brain present a significant challenge in identifying a discrete cause for specific mental illnesses [1, 6]. The mechanisms by which the brain holds and processes information to generate emotions and behavioral outputs are largely unknown, and evaluation of gross brain morphology in mental illness is often normal [1]. The additional interactions between environmental cues and gene function that may underly psychiatric disease development are complex and difficult to ascertain given the challenges in obtaining brain tissue for testing and the absence of blood-based biomarkers of mental illness [1, 6].

Without reliable markers of disease and a strong understanding of the underlying pathophysiology of mental illness, classification systems for psychiatric conditions based on syndromes were created to help improve the diagnosis of patients with mental illness [1, 2]. Two related classification systems were developed by the American Psychiatric Association (Diagnostic and Statistical Manual of Mental Disorders) and the World Health Organization (International Classification of Disease) with the goal to facilitate clinical diagnostic criteria and increase reliability between medical providers [1]. These systems have improved clinical syndromic based diagnosis but have created heterogenicity with regards to identifying the underlying pathophysiology of disease due to multiple symptoms being grouped within a syndrome [1, 2]. This approach creates challenges in drug development in that effective treatments would need to cover multiple underlying pathologies to treat the syndrome, which ultimately leads to the potential for greater medication side effects and less targeted therapy [2].

An additional challenge in drug development to treat mental illness is that the disease cannot be easily modeled in animals for preclinical studies [6]. Most preclinical animal models for mental illness involve artificially stressing an animal, in a method that is dissimilar to stressors encountered by humans, to produce the desired behavior or mood changes to be studied [6]. For example, in order to induce anxiety in a mouse, the animal is placed in a water tank until an anxious phenotype is observed. Following this anxiety-producing event, the animal is exposed to a compound used to treat anxiety, and the resultant behavioral modification is analyzed. This approach to preclinical data acquisition to test the effect of a medication on a behavior leads to a significant degree of variability and often results in compounds that lack efficacy in humans, despite showing benefits in animal models [6]. Advancements of these compounds into human clinical trials often result in unnecessary human testing, risks of significant side effects, reduced efficacy, and increased cost through development of compounds that lack effectiveness [1].

7.1.2 Neurotransmitter Systems

Given the challenges associated with serendipitous drug discovery, pharmaceutical companies have moved to a rational therapeutic design approach by selecting a specific receptor target within a neurotransmitter pathway and a detailed indication for treatment based on an understanding of the role of the target within a neurologic circuit [7].This method focuses on the underlying pathophysiology of disease, thus reducing the risk of drug development by increasing the odds of creating a successful therapy. Targeted drug design additionally allows for a stepwise approach to drug development with greater reliability between preclinical and human studies and permits for the identification of specific populations of patients that will derive the most benefit from the new medication [5]. Implementation of a rational approach to drug design decreases time in development, overall expense, allows a mechanism for the detection of potential blood biomarkers for earlier identification of disease, and can identify patients that would derive the greatest benefit from the medication [5].

Identification of neural circuits that are conserved between animals and humans is key to designing preclinical studies with defined endpoints that are measured in similar methods between animals and humans [6, 8]. For example, mammalian brains show distinct changes in electrical activity during sleep that can be measured in noninvasive ways including an electroencephalogram (EEG) that measures the electrical activity of the brain and polysomnography (PSG) that measures eye movements and muscle activity [6]. These techniques can be used in both animals and humans to measure changes in sleep and wakefulness in additional to evaluating the effect of a medication on this neural circuit [6]. The conservation of these circuits between mammalian and human brains allows for results obtained in preclinical studies to be directly translated to human studies, thus increasing overall validity of the preclinical data [8].

Rational drug design in the field of psychiatry additionally must emphasize a single symptom or behavior, for example, insomnia, rather than a syndrome like major depressive disorder (MDD), which consists of multiple symptoms including insomnia, depressed mood, anhedonia, reduced energy, reduced appetite, guilt, hopelessness, irritability, etc. [8]. Symptoms are the direct output of brain function in a single neural pathway that are amenable to treatment with a compound utilizing a single mechanism of action, whereas syndromes are created constructs that affect multiple pathways with the potential to generate numerous unintended side effects when using a single medication for treatment [8, 9]. Targeting a specific symptom or behavior reduces uncertainty about the effect of the medication and allows for the development of high-throughput systems where multiple animals can be assessed simultaneously for medication dose and effect relationships, thus reducing the timeline for drug development and minimizing cost [6, 9]. This method also allows a means to evaluate unintended targets of the medication, for example, the effect of the medication on cytochrome P450 enzymes [9]. This approach allows for the identification of other important regulatory proteins including enzymes and receptors within a pathway that can be used as additional potential targets for therapy [9].

7.1.3 Genetic Basis for Drug Discovery

An understanding of genetics and the role of environmental interactions on gene function is an important component of the drug discovery process [1]. Suvorexant is the first CNS medication that was developed directly as a result of understanding genetic contributions to a neuronal circuit [7]. This medication was discovered by researchers who were searching fragments of human DNA for previously unidentified G-protein-coupled receptors, also known as orphan receptors [7]. Receptors that were identified through this process were then used to detect new neurotransmitters and gain a better understanding of the functions of the various components within the identified neural pathways [2, 7].

A pharmacogenetic approach that identifies ligands that turn "on and off" specific populations of cells or axonal terminals allows researchers to understand the contribution of a specific cell to a symptom or behavior [1]. A review of the human genome shows that roughly 7% of the approximately 21,000 genes in the human genome are involved in circadian function [10]. The orexin pathway, which regulates the circadian cycle of sleep and wakefulness, was identified as a result of this genetic approach to drug discovery [7]. By gaining an understanding of this pathway and the role orexin plays in the sleep-wake cycle, researchers have been able to develop several dual orexin receptor antagonists that are used in the treatment of primary insomnia [7, 11]. The use of genetic information and molecular pharmacology for the identification of novel receptors, enzymes, neurotransmitters, and neural circuits allows for the development of medications with novel molecular mechanisms of action and provides the potential to gain an in-depth understanding of the pathophysiology underlying specific psychiatric conditions [5].

7.1.4 Timeline from Discovery to Approval

The development of new medications is a long process with most neuropsychiatric drugs taking 12–18 years to reach clinical practice, which is twice as long as medications used for other indications including infectious disease treatments [3, 4]. The success rate of medications in development to treat neuropsychiatric conditions is 8.2%, which is significantly less than other therapies at 15% [3]. The time to gain regulatory approval from the FDA is also significantly longer for neurological drugs at 2 years compared to 1 year for most other clinical indications [3]. Most neurologic compounds used for therapy tend to fail in phase 3 clinical trials, which are later in the development process [3].

Suvorexant was novel in the CNS drug development process in that it went from discovery to approval in 8 years, which is half the time needed for most CNS-based therapies [7]. Merck, the pharmaceutical company responsible for the development of suvorexant, began a high-throughput screen to identify orexin receptor antagonists in 2006, and by 2014 suvorexant was approved by the FDA for the treatment of primary insomnia [6]. This rapid timeline for approval of a medication was made possible through rational drug design that utilized a genetic basis to identify the mechanism of action of a receptor within a neural pathway, a high-throughput approach to screen for drug targets using noninvasive mechanisms to measure clinical endpoints, conservations of neural pathways between animals and humans, and objective measurements of the effect of a medication on a specific symptom [6, 7].

7.2 Orexin Pathway

The discovery of the orexin neuropeptide occurred in 1998 by two independent researchers [11, 12]. DeLecea et al. identified two neurotransmitters, hypocretin-1 (more commonly known as orexin A) and hypocretin 2 (more commonly known as orexin B), that were derived from a pre-pro-peptide precursor located in the synaptic vesicles of neurons in the hypothalamus [11]. In an independent study, Sakurai et al. identified the hypocretin/orexin receptors, locations of their expression within the CNS, and identified appetite stimulating properties of these molecules in animals [11]. Orexin is produced by approximately 100,000 excitatory neurons localized in the hypothalamus with broad effects on neuromodulation in the brain through neuronal projections into the cerebral cortex, thalamus, hypothalamus, and brainstem [7, 11].

Early animal studies showed that mice that lacked the orexin 2 receptor gene had symptoms similar to humans with narcolepsy [12, 13]. This finding was confirmed in studies in dogs with mutations in orexin receptors that also possessed characteristics of narcolepsy [7, 12, 13]. Later studies in humans revealed an approximately 90% reduction in orexin neurons in patients with narcolepsy, with these patients having significant changes in the degree of sleepiness and sleep-wake cycle stability [7, 12, 13]. These findings demonstrated the critical role of orexin in maintaining arousal and for the stability of the awake state of the brain through modulation of neurotransmitters including serotonin, histamine, acetylcholine, and dopamine [7, 14]. Additional studies showed that the role of orexin in circadian function was conserved between rodents, dogs, primates, and humans with greater than 90% genetic sequence conservation between rodents and humans [7, 13].

7.2.1 Orexin Neurotransmitters and Receptors

The orexin neurotransmitter has two chemically distinct forms known as orexin A and orexin B that have only 46% sequence homology [12]. Orexin A and orexin B are produced in the lateral and posterior hypothalamic areas of the diencephalon from a common pre-pro-peptide precursor [12, 13, 15]. Orexin A is 33 amino acids in length with 2 disulfide bonds, and orexin B contains 28 amino acids and forms its secondary protein structure through hydrogen bonds of the alpha helices [12]. Orexin A and orexin B bind to two transmembrane G-protein-coupled receptors known as orexin receptor 1 and orexin receptor 2 [11–13, 15]. Both receptors signal through an increase in intracellular calcium levels and increased cyclic adenosine monophosphate (cAMP) levels within neurons [13]. Orexin A has been found to bind equally to both orexin receptors, but orexin B has been shown to bind with approximately tenfold greater affinity for the orexin 2 receptor [16].

The orexin 1 receptor has been shown in studies to be more selective for orexin A [11, 12]. This receptor is located in cholinergic neurons in the pedunculopontine and lateral dorsal tegmental nuclei [11, 13]. Orexin receptor 1 is the only orexin receptor that is located in the adrenergic neurons within the locus coeruleus [11, 15]. Both orexin receptor subtypes are found in the serotonin-releasing neurons of the dorsal raphe nucleus and the dopamine-releasing neurons of the ventral tegmental area [11, 13]. The orexin receptor 2 is the only orexin receptor in the histaminergic tuberomammillary nucleus [11, 13, 15].

7.2.2 Mechanism of Action

The most prevalent effect of the orexin signaling pathway is the maintenance of wakefulness through the continuous depolarizing effects in the wake-promoting nuclei in the brain [12]. Intracerebroventricular (ICV) administration, or direct administration of a drug into the CNS bypassing the blood brain barrier, of orexin A in mice increased wakefulness and reduced rapid eye movement (REM) and nonrapid eye movement (NREM) sleep [11]. Orexin A given to mice without either the orexin receptor 1 or orexin receptor 2 has reduced wakefulness and NREM sleep [11]. Mice without the orexin receptor 2 are the most effected with regards to wakefulness and sleep [11]. These findings suggest that both orexin receptors are important in the maintenance of wakefulness and NREM sleep, with orexin receptor 2 having the dominate role in this pathway [11, 15]. Additional studies have shown that orexin peptide expression changes in a circadian manner with a peak during the daytime and reduced levels during normal sleep [13]. Increased locomotor activity has also been noticed in preclinical studies with the exogenous administration of orexin peptides [13].

The sleep-wake cycle is a complex system with nuclei in the brain that reciprocally regulate neural pathways under a feedback mechanism to allow stable transitions between sleep and wakefulness [12]. The ascending reticular activating system (ARAS) promotes wakefulness and is involved in the stimulation of cholinergic neurons, monoaminergic cell bundles, and orexin nuclei in the hypothalamus [12]. Activation of the ventrolateral preoptic region (VLPO) releases inhibitory GABA and galanin that promote sleep [12]. Orexin-producing neurons innervate many of the nuclei that promote wakefulness in the ARAS including the locus coeruleus, lateral dorsal tegmentum, pedunculopontine tegmentum, dorsal raphe nucleus, and tuberomammillary nucleus [12].

Orexin has also been shown to have effects in memory, emotions, motivation, attention, autonomic control, feeding, and energy maintenance [12]. Studies in humans with narcolepsy who have reduced concentrations of orexin have a higher incidence of metabolic syndrome with an increased body mass index (BMI) compared to controls with normal orexin levels [11]. Studies in mice have shown that overexpression of orexin has been linked with increased feeding behavior [11].

7.2.3 Role of Orexin in CNS Diseases

Orexin has been shown to have a role in numerous neuropsychiatric conditions including Alzheimer's disease, Parkinson's disease, anxiety, stress, post-traumatic stress disorder (PTSD), substance abuse disorder, depression, and chronic pain [6]. Orexin knockout mice used in preclinical studies of Alzheimer's disease have been shown to have decreased amyloid beta (Aβ) plaque deposition, suggesting that orexin may have a role in the pathogenesis of Alzheimer's disease [11]. Human studies in patients with mild cognitive impairment or Alzheimer's disease with subjective reported sleep problems have elevated CSF orexin levels compared to patients without sleep disturbance or matched controls [12]. Additionally, reductions in sleep duration have been correlated with increased Aβ levels and increased plaque formation [6]. These findings suggest that antagonism of the orexin receptors may have a beneficial effect on cognition in patients with mild cognitive impairment or Alzheimer's disease through both reductions in Aβ deposition and improvements in sleep [12].

Orexin has been implicated to have a role in the progression of Parkinson's disease [6, 17]. Postmortem studies in humans have shown a correlation between the loss of orexin-producing neurons and the clinical stages of Parkinson's disease with a 23% loss in early stages of disease and a 62% loss in later disease stages [17]. Additional studies have reported a 25% loss of orexin A concentrations in the cerebrospinal fluid (CSF) and a 40% decrease in orexin A levels in the prefrontal cortex in patients diagnosed with Parkinson's disease [17]. The number of orexin secreting neurons has been shown to be significantly decreased in both animal models of Parkinson's disease and humans with the disease. Reductions in orexin within these models and patients have been linked to significant reductions in sleep and a decline in cognition [17]. The potential for use of orexin as a therapeutic target for Parkinson's disease is currently under further investigation.

Studies have indicated that orexin may play a key role in the regulation of mood disorders [12, 15]. Genetic variants of the orexin 1 receptor have been associated with the development of major depressive disorder (MDD) and the severity of depressive symptoms [15]. Seltorexant, which is a selective orexin receptor 2 antagonist, has been shown to improve mood in patients with MDD and insomnia when combined with antidepressant therapy [15]. Suvorexant, which is a dual orexin receptor antagonist, has been shown to improve quality of sleep and mood in patients with insomnia and coexisting mental illness [15]. Orexin has also been linked to anxiety and panic disorder with increased orexin levels measured in the CSF of humans with panic-related anxiety [15]. Genetic variations of the orexin receptor 1 have been associated with the development of panic disorder [15]. Additional studies have revealed that the orexin pathway is downregulated in chronic stress with low levels of orexin A in depressed patients and elevated orexin A in patients with panic-related anxiety [15].

The orexin neurotransmitter pathway additionally plays an important role in substance use including cocaine, alcohol, and opioid use [15, 16]. Addictive substances have been shown to act locally on orexin-producing neurons in the lateral hypothalamus and the ventral tegmental area with activation of the system promoting drug-seeking behavior [15]. The selective orexin receptor 1 antagonist SB334867 has been shown to decrease reward threshold, impulsive behavior, and cocaine and alcohol self-administration in preclinical animal studies [15]. These findings suggest that orexin receptor 1 antagonists may have a role in the treatment of substance use disorder [15, 16].

7.3 History of Dual Orexin Receptor Antagonists

Once the potential therapeutic benefits of modulation of the orexin pathway in treating insomnia were discovered, over 50 applications for patents were filed to identify orexin receptor antagonists between 1999 and 2007 [6]. Dual orexin receptor antagonists offered the potential for a targeted approach to the treatment of insomnia that blocked a pathway known to affect the sleep-wake cycle with limited side effects on other sleep promoting neurological circuits [12]. Prior to the discovery of orexin, medications used to treat insomnia included antihistamines, benzodiazepines, non-benzodiazepine receptor agonists, tricyclic antidepressants, and melatonin agonists, all of which has reduced efficacy and noted side effects [12]. The use of targeted therapy with dual orexin receptor antagonists significantly reduced side effects and improved overall patient safety.

7.3.1 Almorexant

Almorexant was the first dual orexin receptor agonist to proceed to phase 2 sleep disorder studies [12, 15]. This compound that was developed by Actelion and GlaxoSmithKline in 2007 showed enhanced total REM sleep time in orexin receptor 1 knockout mice [15]. Additional studies in animals showed that almorexant had low to moderate bioavailability, easily crossed the blood brain barrier, induced sleepiness, and reduced movement and muscle tone [12]. Almorexant was advanced into phase 3 clinical trials and was shown to reduce locomotor activity, increase sleep cataplexy, improve sleep efficiency, increase REM sleep, and decrease sleep initiation time [12]. The relatively long half-life of this compound at 13–19 hours was shown to have longer effects compared to other molecules in development [12, 16]. Despite the noted positive effects of almorexant on the symptoms of insomnia, the development of this compound was stopped in 2011 due to safety concerns related to abnormal liver enzyme concentrations noted in the clinical trials [12, 16].

7.3.2 SB-649868

SB-649868 was developed by GlaxoSmithKline as an orally administered dual orexin receptor antagonist in 2007 [12, 16]. This compound had a shorter half-life at 3–6 hours and was shown in preclinical studies in rats to increase NREM and REM sleep and reduce sleep latency without associated impairments in motor function [12, 16]. Clinical trials in healthy male volunteers showed increased total sleep time, increased REM sleep duration, reduced waking after sleep onset, reduced sleep latency, and improved sleep induction and maintenance [12, 15, 16]. SB-649868 was shown to be a significant inhibitor of CYP3A4 in vitro and was noted to have significant drug-drug interactions during clinical testing [16]. This compound was stopped in development due to an undisclosed preclinical toxicity [16].

7.3.3 *Lemborexant*

Lemborexant was developed by Eisai in 2011 and was created from a parent compound that was shown in preclinical rat studies to decrease wakefulness and promote NREM sleep with no noted effects on REM sleep [12, 18]. Phase 2 clinical trials using lemborexant showed improved sleep efficiency with shorter sleep latency and reduced waking after sleep onset [12, 15]. Phase 3 clinical trials that began in 2018 in patients with insomnia showed mild side effects of somnolence, headache, and sleep paralysis [12]. Lemborexant was FDA approved for the treatment of insomnia in 2019 [18].

7.3.4 *Filorexant*

Filorexant was a dual orexin receptor inhibitor developed by Merck as a potential treatment for episodic migraine headaches and diabetic neuropathy but was found to be ineffective for these indications [12]. Preclinical studies showed a dose-dependent decrease in locomotor activity, increased NREM and REM sleep, and reduced active wake time [12, 16]. Pharmacokinetic studies using this compound have shown increased bioavailability and more rapid binding to orexin receptors at reduced doses [12]. Phase 3 clinical trials showed improved sleep efficiency in non-elderly patients with insomnia and improvements in sleep onset and maintenance [12, 16]. Filorexant has a short half-life of 3–6 hours, and somnolence was noted at doses above 10 mg [12]. Filorexant was stopped in drug development in 2018 for undisclosed reasons [19].

7.4 Suvorexant

Suvorexant was the first orally available dual orexin receptor antagonist to receive FDA approval in August of 2014 for the treatment of insomnia [6, 11, 12, 14, 16, 20]. The discovery of this drug began with a high-throughput assay of over two million compounds within sample collections at Merck in 2005. The search of compounds focused on potential drugs that blocked orexin A signaling in recombinant human cells that overexpressed orexin receptors 1 and 2 [6]. Of the molecules identified from the initial screen, an initial compound (compound 1) was identified that possessed a seven membered diazepane ring with a modular structure and appropriate polarity that allowed for easy manipulation of the chemical structure and access through the blood brain barrier [6]. Suvorexant was created through a series of manipulations to the chemical structure of the compound to enhance binding affinity, improve pharmacokinetic properties, enhance brain penetration, improve efficacy, and create a favorable toxicology profile prior to entering clinical studies [6].

7.4.1 Mechanism of Action

Suvorexant works by selectively blocking the binding of orexin A and orexin B to the orexin receptors in the brain [14]. Suvorexant binding to the orexin 1 and orexin 2 receptors is 6000-fold more selective than over 170 known receptors and enzymes, with greater than 90% receptor occupancy [20]. This strong receptor binding modulates endogenous orexin signaling to rapidly induce sleep in a dose-dependent manner [20]. Suvorexant promotes REM and NREM sleep, reduces locomotor activity, increases REM sleep duration and total sleep time, reduces sleep latency time, and improves sleep maintenance [11, 12, 15]. Suvorexant acts in a dose dependent manner with a strong effect in the first night and improved overall sleep maintenance and onset of sleep over 3 months of nightly treatment [11, 15].

7.4.2 Clinical Trial Results

The clinical program that supported the FDA approval of suvorexant consisted of 36 clinical studies with 34 phase 1 studies and 4 phase 2/3 studies [6]. Phase 1 studies in rodents showed that suvorexant was well tolerated with peak plasma levels achieved at 2 hours post dose under fasting conditions with a drug half-life of 12 hours [16, 20]. The bioavailability of suvorexant is approximately 80% with steady state achieved in 3 days and elimination primarily occurring through CYP3A-mediated metabolism [6, 14, 20]. Preclinical animal safety studies failed to determine a LD50 dose, which is a dose at which 50% of the animals die as a result of administering the drug, indicating a wide safety margin for this medication [9]. The three phase 3 clinical trials utilizing a double blind, randomized, placebo control design with 3 month and 1 year data in adults and participants over age 65 showed improved sleep onset and maintenance with increased time spent in all sleep stages [6, 11]. The effects of the medication on insomnia treatment were maintained after 1 year of use [6, 11, 14, 12, 20].

The most common adverse effect was somnolence, which occurred within the first 2 weeks of therapy and was rated as mild to moderate [6, 12]). Additional reported adverse effects included fatigue, dry mouth, headaches, abnormal dreams, cough, and diarrhea [11, 12]. Additional studies in humans showed no suppression of respiratory drive, no impairments in next day driving performance, no effects on balance or memory, and no changes on cardiac conduction [9]. Studies reviewing the discontinuation of suvorexant after 1 year showed no evidence of rebound or withdrawal from the medication [20].

7.5 Targeted Drug Development

A targeted approach to drug development results in improved patient safety and reduced side effects. The main objective with a targeted approach to drug design is to reduce uncertainty regarding the effect of the drug on a patient [9]. Utilization of a high-throughput screening method reduces the chance of the effects of the medication on unintended targets, thus improving overall safety and reducing side effects [9]. This restricts the pharmacology of the drug to the primary mechanism underlying the medical condition. A key example of this approach is the development of suvorexant that specifically inhibits the orexin receptors in the treatment of insomnia [9].

Development of medications based on the underlying pathophysiology of a disease state reduces the risk to the patient and the pharmaceutical company, allows for a stepwise approach with greater fidelity between animal and human studies, and allows for the identification of specific populations of patients that will derive the most benefit from the medication [5]. This process was highlighted in this chapter through the development of suvorexant by Merck that utilized a high-throughput approach to compound screening, a validated target with known specific effects on a symptom, and a streamlined clinical strategy which allowed for FDA approval of the medication in only 8 years [6]. Future development of CNS therapies will likely switch to this approach of targeted design using a singular neurotransmitter system to improve overall outcomes and reduce the associated cost of drug development.

7.6 Conclusions

Treatment of mental illness has largely proceeded through serendipitous drug discovery with medications primarily being identified by chance. The discovery of new neurotransmitter systems within the brain has led to new specific drug targets and a more directed approach to the drug development process. Identification of the orexin neurotransmitter pathway and the role this system plays in the development and treatment of insomnia are classic examples of how a deeper understanding of the pathophysiology underlying medical conditions can be used to develop specific neural targets to mitigate disease. This approach to drug development maximizes efficacy and minimizes risk with reduced time to clinical utilization, improved patient safety, reduced side effects, and reduced overall cost.

References

1. Kaffman A, White JD, Wei L, Johnson FK, Krystal JH. Enhancing the utility of preclinical research in neuropsychiatry drug development. Methods Mol Biol. 2019;2011:3–22. https://doi.org/10.1007/978-1-4939-9554-7_1.
2. Preskorn SH. Drug development in psychiatry: the long and winding road from chance discovery to rational development. Handb Exp Pharmacol. 2019;250:307–24. https://doi.org/10.1007/164_2018_169.
3. Mohs RC, Greig NH. Drug discovery and development: Role of basic biological research. Alzheimer's Dement (N Y). 2017;3(4):651–7. Published 2017 Nov 11. https://doi.org/10.1016/j.trci.2017.10.005.

4. Macaluso M, Krams M, Preskorn S. Phase I Trials: From traditional to newer approaches part II. J Psychiatr Pract. 2011;17(4):277–84.
5. Preskorn SH. CNS drug development: lessons learned part 3: psychiatric and central nervous system drugs developed over the last decade—implications for the field. J Psychiatr Pract. 2017a;23:352–60.
6. Uslaner J, Herring W, Coleman P. The discovery of Suvorexant: lessons learned that can be applied to other CNS drug development efforts. ACS Pharmacol Transl Sci. 2020;3(1):161–8. https://doi.org/10.1021/acsptsci.9b00110.
7. Preskorn SH. CNS drug development: lessons from the development of ondansetron, aprepitant, ramelteon, varenicline, lorcaserin, and suvorexant. Part I. J Psychiatr Pract. 2014;20(6):460–5.
8. Preskorn SH. CNS drug development: lessons learned Part 2. Symptoms, not syndromes as targets consistent with the NIMH research domain approach. J Psychiart Pract. 2015;21:60–6.
9. Preskorn SH. CNS drug development, lessons learned, part 5: how preclinical and human safety studies inform the approval and subsequent use of a new drug—suvorexant as an example. J Psychiatr Pract. 2018;24(2):104–10.
10. Preskorn S. Psychiatric and central nervous system drugs developed over the last decade: what are the implications for the field? J Psychiatr Pract. 2017b;23(5):352–60.
11. Chow M, Cao M. The hypocretin/orexin system in sleep disorders: preclinical insights and clinical progress. Nat Sci Sleep. 2016;8:81–6. Published 2016 Mar 14. https://doi.org/10.2147/NSS.S76711.
12. Janto K, Roxanne Prichard J, Pusalavidyasagar S. An update on dual orexin receptor antagonists and their potential role in insomnia therapeutics. J Clin Sleep Med. 2018;14(8):1399–408.
13. Wallace D, Mangion I, and Coleman P. Discovery and chemical development of suvorexant – a dual orexin antagonist for sleep disorder. Comprehensive accounts of pharmaceutical research and development: from discovery to late-stage process development Vol 1. Chapter 1. 2016. P1–36.
14. Rhyne DN, Anderson SL. Suvorexant in insomnia: efficacy, safety and place in therapy. Ther Adv Drug Saf. 2015;6(5):189–95. https://doi.org/10.1177/2042098615595359.
15. Han Y, Yuan K, Zheng Y, Lu L. Orexin receptor antagonists as emerging treatments for psychiatric disorders. Neurosci Bull. 2020;36(4):432–48. https://doi.org/10.1007/s12264-019-00447-9.
16. Kuduk S, Winrow C, Coleman P. Orexin Receptor Antagonists in Development for Insomnia and CNS Disorders. 2013. Chapter 6. Ann Rep Med Chem. 48:73–87.
17. Liu C, Xue Y, Liu MF, Wang Y, Chen L. Orexin and Parkinson's disease: a protective neuropeptide with therapeutic potential. Neurochem Int 2020 Sep;138:104754. doi: https://doi.org/10.1016/j.neuint.2020.104754. Epub 2020 May 15.
18. Scott LJ. Lemborexant: first approval. Drugs. 2020 Mar;80(4):425–32. https://doi.org/10.1007/s40265-020-01276-1.
19. National Center for Biotechnology Information. PubChem Compound Summary for CID 25128145, Filorexant. 2021. Retrieved 5 Oct 2021 from https://pubchem.ncbi.nlm.nih.gov/compound/Filorexant.
20. Yang LPH. Suvorexant: first global approval. Drugs. 2014;74:1817–22. https://doi.org/10.1007/s40265-014-0294-5.
21. Preskorn SH. CNS drug development, lessons learned, part 4: the role of brain circuitry and genes—tasimelteon as an example. J Psychiatr Pract. 2017c;23(6):425–30.

Chapter 8
Reverse Engineering Drugs: Lorcaserin as an Example

Tiffany Schwasinger-Schmidt and Sheldon H. Preskorn

Abstract Novel central nervous system (CNS)-based therapies have been difficult to produce due to the complexity of the brain, limited knowledge of CNS-based disease development and associated pathways, difficulty in penetrating the blood brain barrier, and a lack of reliable biomarkers of disease. Reverse engineering in drug development allows the utilization of new knowledge of disease pathways and the use of innovative technology to develop medications with enhanced efficacy and reduced toxicities. Lorcaserin was developed as a specific $5HT_{2C}$ serotonin receptor agonist for the treatment of obesity with limited off-target effects at the $5HT_{2A}$ and $5HT_{2B}$ receptors. This receptor specificity limited the hallucinogenic and cardiovascular side effects noted with other serotonin receptor agonists. Reverse engineering approaches to drug development reduce the cost of producing new medications, identify specific populations of patients that will derive the most benefit from therapy, and produce novel therapies with greater efficacy and limited toxicity.

Keywords Lorcaserin · Reverse engineering in drug development · Serotonin · 5HT receptors · 5HTc agonists

T. Schwasinger-Schmidt (✉)
Department of Internal Medicine, University of Kansas School of Medicine-Wichita, Wichita, KS, USA
e-mail: tschwasinger-schmidt@kumc.edu

S. H. Preskorn
Department of Psychiatry and Behavioral Science, University of Kansas School of Medicine-Wichita, Wichita, KS, USA
e-mail: spreskon@kumc.edu

M. Macaluso et al. (eds.), *Drug Development in Psychiatry*, Advances in Neurobiology 30,
https://doi.org/10.1007/978-3-031-21054-9_8

8.1 Introduction

Mental illness and CNS disorders are exceedingly common and have a disproportionate effect on society [1]. In the United States, approximately 51.5 million people have a mental illness, 7.5 million have dementia (neurocognitive disorders), 6.8 million had a prior stroke, 1 million have Parkinson's disease, and 25 million have additional, more rare neurologic diseases including atrophic lateral sclerosis (ALS) and Huntington's disease [2, 3]. The estimated annual socioeconomic cost of CNS-based disorders exceeds 0.8 trillion dollars annually in direct healthcare costs, direct nonmedical costs, and lost productivity with the three most burdensome CNS disorders including stroke, dementia, and migraine headaches [3, 4]. The cost of healthcare associated with neurological disease is anticipated to double in the next 10 years, with the cost of dementia- and stroke-associated care alone projected to exceed 600 billion dollars by 2030 [3].

Despite the increasing prevalence of CNS disorders, development of new therapies with novel mechanisms of action for CNS diseases has been sparse over the past few years [5]. In general, approximately 11–15% of medications that enter development programs reach clinical practice [5–7]. Medications used in therapy for CNS disorders have a reduced probability of reaching the market at 7–8% and take an average of 12–13 years for approval compared to other clinical indications that take approximately 6–7 years for regulatory approval [5, 7]. The success rates in producing medications for CNS-based indications are low due to the inherent complexity of the central nervous system; increased CNS-based side effects including dizziness, nausea, and seizures; the presence of the blood-brain barrier; difficulties in getting medications to the specific target areas of the brain; and a paucity of biomarkers to identify and diagnose disease [5].

8.2 Overview of CNS Disorders and Drug Development

Innovation in pharmaceutical development is driven by three key overlapping factors in the race to produce a therapy with a novel mechanism of action [1]. The first key component in drug development includes the initial research to gain an understanding of the underlying pathophysiology of the condition being treated and identifying a target for therapy [1]. This approach reduces the risk in drug development, permits for a stepwise approach to identify the mechanism of action, and allows for identification of specific populations that will derive the most benefit from the medication [8]. The second component in drug development includes production of a safe and effective therapy with reliable quality that can be mass produced [1]. The medication must be effective at treating the desired indication with minimal side effects in order to advance to human clinical testing [8]. The final step includes regulatory (Federal Drug Administration (FDA) in the United States) approval and dissemination to bring the medication into clinical practice for general use [1].

The innovation aspects in drug development for CNS disorders begin with fundamental advancements in understanding the molecular- and cellular-based processes of disease to identify the underlying etiology [1]. Through this process, potential therapies are identified that can disrupt or block signaling pathways in the development of disease [1]. An effective approach to identifying these molecules must include the development of multiple compounds with diverse structures for a particular target. This approach allows for modification of the molecular composition or structure to eliminate toxicities and enhance binding to the target receptor, thus minimizing off-target effects [5]. These molecular agents include small molecules, proteins, or biologics that cover a broad portfolio of potential therapies [1, 5]. A possible limitation to this approach is that CNS-based disorders are inherently complex, and addressing a single target may not be sufficient to effectively treat the condition [5].

A successful product development approach must combine effective preclinical and clinical research that is translatable between species [5]. The development and use of predictive animal models with CNS disorders, produced in similar mechanisms to human disease development, with similar phenotypic clinical output must be utilized to produce reliable results with predictive pharmacological responses [5]. This approach needs to include high-throughput screening methods for multiple targets within the identified disease pathway, the ability to chemically alter different substances to enhance binding to the target receptor and reduce toxicity, and genomic-based approaches including pharmacogenomic, proteomic, and metabolomic tools to identify key polymorphisms or gene alterations in disease [5, 6]. This method of translational research allows for quicker, more cost-effective drug development processes that deliver innovative therapies through a more efficient and streamlined approach [1, 5].

Recommendations for the successful development of medications for CNS disorders include gaining an understanding of the underlying disease pathophysiology and genetics [5]. This allows for the selection of specific targets to modulate disease with an emphasis on in vitro systems that are physiologically relevant [5, 8]. Multiple targets within a pathway with diverse substrates allow for a greater success rate in drug discovery through diversification of the platform using small molecules, proteins, or biologics as substrates [1, 5]. Preclinical testing in animal models needs to examine the most relevant aspects of the disorder with a focus on clinically significant endpoints to produce the desired clinical outcomes [5]. An emphasis must be placed on understanding the pharmacogenetic, pharmacodynamic, pharmacokinetic, and overall metabolism of the compound to minimize off-target effects and resultant toxicities [5, 6]. The identification and use of clinical biomarkers to identify disease must also be a priority in the drug development process. The use of multiple modalities including imaging, biochemistry, and behavioral outcomes enhances the likelihood of the translation of research findings into clinically significant patient outcomes [1, 5].

8.3 Approaches to Drug Development

Numerous approaches to the drug development process have been utilized with each having unique strengths and challenges [5, 9]. One method that is readily utilized by pharmaceutical companies is a broad-based strategy with an emphasis on screening large groups of targets to select effective compounds that produce the desired clinical result [5]. An example of this approach included the discovery of suvorexant through the screening of G-protein-coupled receptors, which resulted in the identification and characterization of the orexin signaling pathway involved in sleep regulation, and is covered elsewhere in this text [9]. This screening process using large categories of known receptors or enzymes that modify signaling pathways within the CNS allows for the identification and testing of multiple compounds to advance the drug development process [5].

Another approach to drug development includes reverse engineering using previously approved medications to develop novel therapies with improved efficacy or reduced toxicity [5]. This process was used to develop lorcaserin, which is a weight loss medication with effects on the serotonin receptors in the brain. This medication had reduced toxicity compared to prior medications that modulate the same signaling pathway by targeting specific receptors and limiting off-target effects [9].

A final approach that is often used to develop novel medications focuses on basic research into the pathophysiology of a particular disease to gain a deeper understanding of the overall disease process [5, 8]. This approach is the most challenging, complex, time-consuming and is associated with increased risk, but it allows for the opportunity to alter the progression of a disease rather than just focus on improving the symptoms [5, 8]. This approach has been applied to the development of aducanumab, which is a medication used to treat Alzheimer's disease through blocking the deposition of amyloid beta proteins within the brain, thus limiting plaque formation [5].

8.4 Reverse Engineering

A reverse engineering approach to drug development utilizes a known proven mechanism to explore potential new targets to modulate disease [10]. This approach is associated with less risk in the production of a safe and effective substance to treat disease, but is more challenging in developing a medication that can be differentiated from other similar medications that are clinically available for treatment. The following section will describe some of the medications that were developed utilizing this approach to maximize drug potency, target multiple sites within a pathway, enhance prior knowledge of a mechanism of disease, and reduce toxicity.

Reverse engineering to maximize potency of a medication is clearly evidenced in the development of atorvastatin [10, 11]. This medication was the 5th statin to be approved by the FDA and was discovered 10 years after lovastatin, which was the first medication to be identified within this class [10, 11]. Atorvastatin has become the best-selling drug in the lipid-lowering medication class and has far greater clinical use than lovastatin [10]. Atorvastatin was developed utilizing a drug scaffold with resultant chemical modifications to maximize potency in reducing low density lipoprotein (LDL) levels compared to other statin medications available on the market [10, 11]. Different synthetic approaches were utilized in the development of atorvastatin with a focus on enhancing potency and reducing toxicity from the parent compound [10, 11].

Lapatinib and bendamustine were developed specifically to effect more than one biological target for cancer treatment [10]. Lapatinib was discovered through the optimization of a medication that targeted two receptors, epidermal growth factor receptor (EGFR) and human epidermal growth factor 2 (HER2), which are known to be commonly involved in breast cancer development [10, 12]. Knowledge of the homonology between EGFR and HER2 was used in combination with molecular modeling to develop lapatinib which targeted both receptors. This medication became widely used in women with breast cancer with tumors that overexpress HER2 [10, 12]. Bendamustine was developed to treat chronic lymphocytic leukemia (CLL) by inducing cytotoxicity of the cancer cells through apoptosis and mitotic catastrophe with greater overall response rates and longer progression free survival time compared to other available therapies [10, 13]. This mechanism of drug development focused on multiple targets to enhance cancer treatment by concentrating on different mechanisms that promote tumor growth [10].

Pregabalin was initially developed as a γ-aminobutyric acid (GABA) aminotransferase inhibitor for the treatment of epilepsy [10, 14]. Further review of the mechanism of action of pregabalin revealed that this compound worked by inhibiting glutamate release, which is similar to gabapentin that has been classically used in the treatment of neuropathic pain [10, 14]. By using this knowledge of the mechanism of action, pregabalin was tested on numerous neuropathic pain conditions and became the first medication approved to treat fibromyalgia [10]. Prior knowledge of the mechanism of action of a similar compound can be applied to other medications in development to improve treatment options for various medical conditions.

Dexfenfluramine was a $5HT_{2c}$ receptor agonist approved for the treatment of obesity in 1996 [10, 15]. This medication resulted in clinically significant weight loss but was withdrawn from the market in 1997 due to heart valve malformations associated with treatment [10]. This significant side effect was later determined to be due to the effects of dexfenfluramine as an agonist at the $5HT_{2B}$ receptor, which is expressed in cardiac tissue [10]. Lorcaserin was developed as a potent $5HT_2c$ agonist used to treat obesity with minimal activity at the $5HT_{2B}$ receptor, thus limiting the cardiac toxicity noted in the parent compound [10, 16]. Increased selectivity of lorcaserin toward the CNS-based $5HT_{2c}$ receptor allowed for a method to promote weight loss through central mechanisms with reduced toxicity due to limited off-target effects [10, 17]. By using reverse engineering approaches to drug development, side effects and unwanted toxicities can be removed through modernized preclinical assays [10].

8.5 History of Seratonin Receptors

Serotonin plays a significant role in numerous CNS disorders through the widespread innervation of serotonergic neurons and modulation of other neurotransmitter systems via the serotonin (5HT) receptors [18, 19]. The complexity of the serotonin signaling system has increased since the initial identification of the 5HT receptors in the 1980s [18]. Currently there are 7 types of serotonin receptors with 14 subtypes identified within the brain. Two of the serotonin receptors, $5HT_{1e}$ and $5HT_{5b}$, are classified only as gene products since they have not been linked to a functional outcome in vivo to date [18, 20, 21]. In humans, the serotonin receptors are coded by 17 genes with numerous genetic polymorphisms with changes in the receptors clearly linked to changes in behavioral output [18].

Selective modulation of specific serotonin receptors provides a targeted approach to drug development. The use of biased agonists that focus on specific 5HT receptors produces a desire effect with limited off-target side effects [18, 19]. This approach may offer therapeutic benefits over the currently available medications for CNS disorders by limiting adverse side effects and maximizing potency at the desired receptor within a specific part of the brain [18]. Given the diverse effects of serotonin on modulating neurotransmission within the brain, this approach has the potential to additionally mediate changes in neuronal architecture and neuronal migration [18, 19]. The different distribution patterns of the 5HT receptor subtypes in the brain additionally are associated with distinct CNS functions that may allow for the development of specific targeted CNS therapies [18].

$5HT_{1A}$ receptor agonists have multiple effects with the CNS including alterations in motor function, body temperature regulation, and activity of the neuroendocrine systems [18]. Agonists at this receptor have been shown to have antidepressant and anxiolytic effects that originate early in development, with the loss of these receptors early in life leading to anxiety and depression in later adulthood [18]. Buspirone is a $5HT_{1A}$ agonist that was approved in 1990 for the treatment of anxiety and is still widely used in clinical practice for augmentation with selective serotonin reuptake inhibitors (SSRIs) for anxiety treatment [18]. Vilazodone and vortioxetine used for the treatment of depression additionally have high affinity for the serotonin reuptake transporter and effects at the $5HT_{1A}$ receptors, among others [18, 22].

The $5HT_{2A}$ receptor is known to contribute to head twitching, drug discrimination, hyperthermia, and exploratory behaviors in animals [18]. This receptor is additionally responsible for the hallucinogenic effects of psychedelic drugs in humans, which has limited the utility of receptor agonists for this target due to pronounced side effects [18, 20]. Pimavanserin, which is a $5HT_{2A}$ receptor inverse agonist, has been used in the treatment of psychosis in Parkinson's disease and is currently being investigated for use in treatment-resistant depression as an augmentation therapy [18, 23].

Functions of the $5HT_{2C}$ receptor agonists include compulsive drug and food-seeking behavior, control of energy homeostasis, oral dyskinesia, wakefulness, and control of the seizure threshold [18]. Activation of the $5HT_{2C}$ receptors reduces the consumption of palatable foods by promoting satiety through effects on the hypothalamic nuclei, thus regulating energy homeostasis [18, 19]. In 1976, fenfluramine was approved by the FDA for weight loss and was in clinical use until 1997, when it was withdrawn from the market due to pulmonary hypertension and cardiac toxicity [18, 19]. Locarserin was approved by the FDA in 2012 for weight loss as a selective $5HT_{2C}$ receptor agonist but was withdrawn from the market in 2020 given concerns of increased cancer risk associated with medication use [15, 18, 19, 24].

The $5HT_3$ receptors have a well-established role in the control of nausea and vomiting [18]. Ondansetron was one of the initial therapies developed that targeted this specific receptor as a treatment for nausea and vomiting associated with chemotherapy [11]. Alostron is a $5HT_3$ receptor agonist that has been shown to reduce emotional responses in patients with irritable bowel syndrome (IBS) with this effect being mediated by reductions in limbic system activation as measured using PET scans of the brain [18]. Vortioxetine, which is indicated for the treatment of major depressive disorder, has atypical pharmacology at the $5HT_3$ receptor with an initial agonist action followed by long-term receptor inhibition [18, 22]. This medication has additionally been shown to improve cognition through $5HT_3$ receptors actions on cortical GABA interneurons [18].

Drugs that target specific serotonin receptors have the potential to mitigate specific neurologic symptoms based on the location and effect of the receptor within the CNS [18]. Biased agonists and specific receptor antagonists may be used to modulate behaviors and specific neurological symptoms with limited off-target effects [18]. By using current knowledge of the localization and effects of the different serotonin receptor subtypes, medications could be developed to specifically reduce impulsivity, prevent cognitive decline, and treat depression and anxiety [18].

8.6 $5HT_2$ Receptor Agonists

Fenfluramine was a nonspecific $5HT_2$ receptor agonist that was approved by the FDA for the treatment of obesity in 1973 [16, 25]. This medication was a derivative of amphetamine that was initially developed in the 1960s to promote weight loss and satiety through reduced food intake [19, 25]. The effects of fenfluramine on food intake are mediated by interactions with the $5HT_{1B}$ and $5HT_{2C}$ receptors in the hypothalamus [19].

One of the earliest clinical trials for fenfluramine included 60 middle aged women with no history of cardiovascular disease who were randomized to fenfluramine or placebo for 12 weeks [25]. Results obtained from the study showed a 9-pound weight loss in the medication arm of the study with a 0.4-pound weight gain in the placebo arm [25]. Additional studies indicated that fenfluramine use resulted in a 6–7-pound weight loss compared to placebo. Increased dropout rates at 9% for fenfluramine versus 2% for placebo were noted within the study [25]. Clinical trials conducted from the 1970s through the 1990s showed that fenfluramine improved blood sugar levels in patients with type II diabetes, improved blood pressure, and improved lipid levels [25]. Results obtained from additional studies using fenfluramine affirmed that 5HT receptor agonists had the potential to promote weight loss and improve metabolic parameters [25].

Given the success of fenfluramine for treating obesity, dexfenfluramine, which is the active isomer of fenfluramine, was created [25]. This drug was developed in a reverse engineering approach to improve efficacy and minimize toxicity [25]. In 1996, dexfenfluramine became the first long-term medication to treat obesity that was approved in the United States [16, 25]. Additional studies into fenfluramine and dexfenfluramine revealed an increased incidence of valvular heart disease and pulmonary hypertension, which resulted in the removal of both medications from the market in 1997 [15, 19, 21, 25]. Further investigation revealed that the cardiotoxicity associated with the medications was due to the agonistic effects on the $5HT_{2B}$ receptors that are located on the valvular cardiac interstitial cells and the smooth muscle cells in the pulmonary arteries [16, 25].

Concerns regarding the modest weight loss achieved with fenfluramine and dexfenfluramine led to the implementation of combination therapy to promote increased weight loss in patients with obesity [21, 25]. Fenfluramine and phentermine were combined and promoted for weight loss under the name of Fen-Phen, which was a popular weight loss option in the 1990s [21, 25]. A 34-week clinical trial of 121 overweight patients, defined as being 130–180% of their ideal total body weight, showed that subjects on the medication lost approximately 31 pounds compared to 10 pounds in patients on placebo [25]. Additional long-term studies showed continued weight loss up to 4 years on therapy. Further analysis of the data showed an increased attrition rate with only 1/3 of participants completing the study with most of the patients regaining the lost weight during the later stages of the trial [25]. Fen-Phen was ultimately removed from the market in 1997 due to concerns of cardiac valvular disease and pulmonary hypertension associated with the medication [25, 26].

8.7 Lorcaserin

The role of serotonin in regulating weight through modulation of food intake, increased satiety, and appetite suppression has been well established and has served as an active area of investigation for medications used to treat obesity [20, 24, 25]. Given the cardiac toxicity associated with agonists at the $5HT_{2B}$ receptor, a reverse engineering approach was used to develop a medication that specifically targets $5HT_{2C}$ receptors to help address weight loss without the associated negative cardiac effects [20, 26]. Locaserin is a novel, selective $5HT_{2C}$ receptor agonist developed by Arena Pharmaceuticals that regulates appetite and reduces food intake [9, 15, 17, 21, 26]. This molecule binds to the $5HT_{2C}$ receptor in the hypothalamus with approximately 100-fold affinity, with minimal effects at the $5HT_{2A}$ and $5HT_{2B}$ receptors, thus limiting the hallucinogenic and cardiovascular side effects noted with other serotonin receptor agonists [15, 16, 21].

Lorcaserin controls food intake through its effects on the $5HT_{2C}$ receptors on the pre-opiomelanocortin (POMC) neurons in the arcuate nucleus of the hypothalamus [21, 24]. Activation of the POMC neurons triggers the release of alpha melanocyte-stimulating hormone that acts on the melanocyte 4 receptors in the paraventricular nucleus of the hypothalamus to reduce appetite and thus reduce food intake [21, 24]. Additional studies have shown that lorcaserin reduces the incentive associated with food intake and improves impulse control, which reduces the motivational value of food [15]. Lorcaserin has no effect on dopamine or norepinephrine release and does not affect energy expenditure in patients on the medication [17].

Lorcaserin is rapidly absorbed in the gastrointestinal tract with a peak plasma concentration at 1.5–2 h with minimal delays noted with fatty food intake ([16, 17, 24]). This medication is distributed evenly to the central nervous system and in the cerebral spinal fluid and has a half life of 11 h [16, 17, 21, 24]. Lorcaserin is 70% bound to plasma proteins and is metabolized in the liver with 92% of the medication excreted in the urine [16, 21, 24]. Gender, ethnicity, age, and body mass index (BMI) have no effects on lorcaserin's pharmacokinetics, but renal and hepatic impairment increase the plasma concentration by 2- to 6-fold and the half-life by 5–9 h, respectively [15]. This medication inhibits CYP2D6 metabolism and cannot be removed through hemodialysis [16, 21].

Phase 1 testing for lorcaserin began in 2004 and has been studied in 18 clinical trials with greater than 8,500 participants on the medication [15, 26]. Lorcaserin was shown in the initial studies to have highly consistent and predictable pharmacokinetic properties with a proportional dose-dependent response and steady state reached within 5 days of initiating the drug [15]. The first clinical efficacy and safety study consisted of a 4-week placebo-controlled trial with multiple doses ranging from 1 mg to 15 mg daily in 352 individuals with obesity [15]. Lifestyle modifications were intentionally excluded from this study and results showed that all doses were well tolerated with significant weight loss noted only with the 15 mg per day dosage arm of the study [15].

Phase 2 clinical trails began in 2005 with a 12-week randomized, double blind, placebo-controlled parallel arm study with 469 participants with a body mass index of 30–45 kg/m^2 [15, 26]. Lorcaserin use resulted in significant weight loss with the greatest change noted in the 10 mg twice daily dosage arm of the study with noted improvements in waist circumference, fasting glucose, and total cholesterol [15, 25]. A separate phase 2 placebo-controlled clinical trial in overweight or obese subjects that received treatment for 56 days showed that lorcaserin reduced weight by decreasing appetite and food intake rather than increasing energy expenditures [15].

Three phase 3 clinical trials lead to the FDA approval of lorcaserin for the treatment of obesity in 2012 [21]. The Behavioral Modification and Lorcaserin for Overweight and Obesity Management (BLOOM) study was a multicenter, double blind, 2-year-long efficacy trial with 3,182 overweight or obese participants initiated in 2007 [15, 16, 24]. Results from this study showed that at 1 year of therapy 47.5% of participants in the lorcaserin group and 20.3% of participants on placebo lost at least 5% of their body weight with maintained weight loss in 67.9% of participants on lorcaserin at 2 years [16, 24]. The BLOOM-DM study was a randomized, placebo-controlled trial that enrolled 604 patients with type II diabetes on metformin, sulfonylureas, or combination therapy with an HbA1c of 7–10% and body mass index of 27–45%. Noted improvements in weight loss, fasting blood sugar, and HbA1c were observed in a dose-dependent manner in patients on the medication [15, 16, 24]. The Behavioral Modification and Lorcaserin Second Study for Obesity Management (BLOSSOM) study was a 1-year clinical trial with 4,008 participants that were obese or overweight. Results of this study showed statistically significant improvements in lipid parameters, blood sugar levels, and quality of life measurements, with half of the participants on the treatment arm achieving 5% weight loss and one quarter losing 10% of their initial total body weight [15]. In the clinical trials, lorcaserin was well tolerated with mild to moderate headaches that were self-limited, nausea, and dizziness reported [16, 26].

Lorcaserin was reviewed by the FDA in 2010 and was rejected by a vote of 9 to 5 given concerns regarding efficacy of the medication and potential toxicities including cardiac valve damage and breast cancer development associated with use of the drug [16, 17]. Lorcaserin was reviewed again by the FDA and approved for use in obesity treatment in June of 2012 with an extended-release formulation approved for use in 2016 [15, 21, 24]. Unfortunately, due to concerns regarding the increased risk of breast cancer development associated with lorcaserin use, the medication was voluntarily withdrawn from the market in the United States in February 2020 [15].

8.8 Reverse Engineering in Drug Discovery

Reverse engineering in drug development is associated with reduced risk by using a parent compound with a known mechanism of action and modifying specific components to allow for optimization of efficacy and reduce toxicity [10, 27]. This approach relies on increased knowledge of the pathophysiology of disease and advancements in technology to produce new medications [10]. Unwanted side effects of medications can be removed through targeted therapy that specifically enhances binding to a desired receptor and limits off-target effects of the medication [10]. This approach to drug design is promising in that it limits the cost of development, produces more efficacious medications, can be used to select specific patient populations that will have the greatest response to the medication, and eliminates unwanted side effects [10, 27].

8.9 Conclusions

CNS-based disorders have a disproportionate effect on society, and identification of novel therapies to treat these disorders has been challenging due to the complexity of the brain, difficulty in penetrating the blood brain barrier, and a lack of reliable biomarkers used to diagnose disease. A reverse engineering approach to drug development utilizing known mechanisms of action of a previously identified target allows for drug developers to alter the chemical structure or binding profile of a molecule or compound to maximize efficacy and minimize toxicity due to off-target effects. Lorcaserin is a prime example of reverse engineering in drug development by specifically targeting the $5HT_{2C}$ receptor to promote weight loss with limited effects at the $5HT_{2A}$ and $5HT_{2B}$ receptors, thus limiting the hallucinogenic and cardiovascular side effects noted with other serotonin receptor agonists. This approach to drug development minimizes cost and improves efficiency in the development process through the utilization of new knowledge of disease development pathways and innovative technology to produce safer and more effective treatment options for patients with disease.

References

1. Nutt DJ, Attridge J. CNS drug development in Europe—past progress and future challenges. Neurobiol Dis. 2014;61:6–20.
2. National Institute of Health. Mental Health Information. 2019. Accessed at https://www.nimh.nih.gov/health/statistics/mental-illness
3. Gooch CL, Pracht E, Borenstein AR. The burden of neurological disease in the United States: a summary report and call to action. Ann Neurol. 2017 Apr;81(4):479–84. https://doi.org/10.1002/ana.24897.
4. GBD 2017 US Neurological Disorders Collaborators. Burden of neurological disorders across the US From 1990–2017: a global burden of disease study. JAMA Neurol. 2021;78(2):165–76. https://doi.org/10.1001/jamaneurol.2020.4152.
5. Pangalos MN, Schechter LE, Hurko O. Drug development for CNS disorders: strategies for balancing risk and reducing attrition. Nat Rev Drug Discov. 2007;6(7):521–32.
6. Kaitin KI. Deconstructing the drug development process: the new face of innovation. Clin Pharmacol Ther. 2010;87(3):356–61. https://doi.org/10.1038/clpt.2009.293. Epub 2010 Feb 3. Erratum in: Clin Pharmacol Ther. 2011;89(1):148. PMID: 20130565; PMCID: PMC2953249.
7. Mohs RC, Greig NH. Drug discovery and development: role of basic biological research. Alzheimer's Dement (N Y). 2017;3(4):651–657. Published 2017 Nov 11. https://doi.org/10.1016/j.trci.2017.10.005.
8. Preskorn S. Psychiatric and central nervous system drugs developed over the last decade: what are the implications for the field? J Psychiatr Pract. 2017;23(5):352–60.
9. Preskorn SH. CNS drug development: lessons from the development of ondansetron, aprepitant, ramelteon, varenicline, lorcaserin, and suvorexant. Part I. J Psychiatr Pract®. 2014;20(6):460–5.
10. Zhao H, Guo Z. Medicinal chemistry strategies in follow-on drug discovery. Drug Discovery Today. 2009;14(9–10):516–22. ISSN 1359-6446. https://doi.org/10.1016/j.drudis.2009.02.008.
11. Roth BD. The discovery and development of atorvastatin, a potent novel Hypolipidemic agent. Progr Med Chem. 2002;40:1–22. https://doi.org/10.1016/S0079-6468(08)70080-8.
12. Cameron DA, Stein S. Drug insight: intracellular inhibitors of HER2--clinical development of lapatinib in breast cancer. Nat Clin Pract Oncol. 2008;5(9):512–20. https://doi.org/10.1038/ncponc1156. Epub 2008 Jul 1.
13. Gandhi V, Burger JA. Bendamustine in B-cell malignancies: the new 46-year-old kid on the block. Clin Cancer Res. 2009;15(24):7456–61. https://doi.org/10.1158/1078-0432.CCR-08-3041.

14. Gale JD, Houghton LA. Alpha 2 Delta (α(2)δ) Ligands, gabapentin and pregabalin: What is the evidence for potential use of these ligands in irritable bowel syndrome. Front Pharmacol. 2011;2:28. Published 2011 Jun 9. https://doi.org/10.3389/fphar.2011.00028.
15. Higgins GA, Fletcher PJ, Shanahan WR. Lorcaserin: a review of its preclinical and clinical pharmacology and therapeutic potential. Pharmacol Ther. 2020;205. https://doi.org/10.1016/j.pharmthera.2019.107417.
16. Bai B, Wang Y. The use of lorcaserin in the management of obesity: a critical appraisal. Drug Des Devel Ther. 2010;5:1–7. Published 2010 Dec 20. https://doi.org/10.2147/DDDT.S11945.
17. Miller LE. Lorcaserin for weight loss: insights into US Food and Drug Administration approval. J Acad Nutr Diet. 2013;113(1):25–30.
18. Sharp T, Barnes NM. Central 5-HT receptors and their function; present and future. Neuropharmacology. 2020;177:108155.
19. Fletcher PJ, Higgins GA. Serotonin and reward-related behavior: focus on 5-HT 2C receptors. In: 5-HT2C receptors in the pathophysiology of CNS disease. Humana Press, Totowa; 2011, pp. 293–324.
20. Greenway FL, et al. Safety and tolerability review of lorcaserin in clinical trials. Clin Obesity. 2016;6(5):285–95.
21. Kim GW, Lin JE, Blomain ES, Waldman SA. Anti obesity pharmacotherapy: new drugs and emerging targets. Clin Pharmacol Ther. 2014;95(1):53–66. https://doi.org/10.1038/clpt.2013.204.
22. Schwasinger-Schmidt TE, Macaluso M. Other antidepressants. In: Macaluso M, Preskorn S, editors. Antidepressants. Handbook of experimental pharmacology, vol. 250. Cham: Springer; 2018. https://doi.org/10.1007/164_2018_167.
23. Kianirad Y, Simuni T. Pimavanserin, a novel antipsychotic for management of Parkinson's disease psychosis. Expert Rev Clin Pharmacol. 2017;10(11):1161–1168. doi: https://doi.org/10.1080/17512433.2017.1369405. Epub 2017 Oct 17.
24. Brashier DB, Sharma AK, Dahiya N, Singh SK, Khadka A. Lorcaserin: a novel antiobesity drug. J Pharmacol Pharmacother. 2014;5(2):175–8. https://doi.org/10.4103/0976-500X.130158.
25. Bays HE. Lorcaserin and adiposopathy: 5-HT2c agonism as a treatment for 'sick fat' and metabolic disease. Expert Rev Cardiovasc Ther. 2009;7(11):1429–45.
26. Hopkins CR. ACS chemical neuroscience molecule spotlight on lorcaserin. ACS Chem Neurosci. 2010;1(11):718–9.
27. Caban A, Pisarczyk K, Kopacz K, Kapuśniak A, Toumi M, Rémuzat C, Kornfeld A. Filling the gap in CNS drug development: evaluation of the role of drug repurposing. J Mark Access Health Policy. 2017;5(1):1299833. https://doi.org/10.1080/20016689.2017.1299833. PMID: 28473889; PMCID: PMC5405562.

Chapter 9
Back to the Future of Neuropsychopharmacology

Anton Bespalov, Marcel van Gaalen, and Thomas Steckler

Abstract Disappointments in translating preclinical findings into clinical efficacy have triggered a number of changes in neuroscience drug discovery ranging from investments diverted to other therapeutic areas to reduced reliance on efficacy claims derived from preclinical models. In this chapter, we argue that there are several existing examples that teach us on what needs to be done to improve the success rate. We advocate the reverse engineering approach that shifts the focus from preclinical efforts to "model" human disease states to pharmacodynamic activity as a common denominator in the journey to translate clinically validated phenomena to preclinical level and then back to humans. Combined with the research rigor, openness, and transparency, this reverse engineering approach is well set to bring new effective and safe medications to patients in need.

Keywords Drug discovery · Translational research · Preclinical model · Reverse engineering · Research rigor

9.1 Breakthrough Discoveries in the Past: What Made Them Possible?

The history of psychopharmacology is for a good part a history of serendipity, that is, the discovery of drugs used for the treatment of mental disorders based on initial clinical observations of psychoactive drug effects in patients treated for other conditions, rather than by design [27, 45]. One example of such discoveries is

---

A. Bespalov (✉)
Exciva GmbH, Heidelberg, Germany
e-mail: anton.bespalov@exciva.com

M. van Gaalen
Evotec, Göttingen, Germany

T. Steckler
Janssen Pharmaceutica, Beerse, Belgium

M. Macaluso et al. (eds.), *Drug Development in Psychiatry*, Advances in Neurobiology 30,
https://doi.org/10.1007/978-3-031-21054-9_9

chlorpromazine, a drug originally used for its sedative properties in surgical patients and subsequently also in psychotic patients, where its potent antipsychotic properties were noted. Subsequently, the dopamine D_2 receptor blocking effects of chlorpromazine were discovered [37]. Another example is the drug iproniazid, originally used for treatment of tuberculosis and observed to have mood-elevating effects in tuberculotic patients. Because of its effects on mood, the drug was subsequently used as antidepressant and identified to act via inhibition of the enzyme monoamine oxidase [26, 58]. Likewise, the tricyclic drug imipramine, originally developed as an antihistamine, was serendipitously found to have antidepressant effects and to inhibit monoamine (serotonin and noradrenaline) reuptake [26].

Some psychiatric drugs were developed by design as well, but, in general, those new discoveries consisted of relatively small incremental steps rather than real breakthroughs. Notable examples are the selective serotonin reuptake inhibitors or serotonin-noradrenaline reuptake inhibitors, classes of extremely useful antidepressant drugs characterized by reduced side effects when compared to their predecessors, the nonselective tricyclic antidepressants. Nevertheless, they still targeted the same main molecular mechanism of action, i.e., serotonin and/or noradrenaline reuptake, and while side effects are reduced, their efficacy remains comparable [26].

An important recent development is the discovery of rapid antidepressant properties of the NMDA receptor channel blocker ketamine, even in hitherto treatment resistant depressed patients [4, 26]. It is important to note though that while there was a growing body of preclinical data in support of antidepressant effects of NMDA receptor antagonism prior to the first clinical study of ketamine, there were no preclinical studies that would predict rapid onset of action of ketamine in depression (also note that other NMDA receptor antagonists like memantine failed to show antidepressant effects in clinical trials, [60]).

Clearly, the discovery of ketamine is encouraging but unfortunately represents an exception, and it must be concluded that the successful development of efficacious drugs with truly novel mechanism of action and confirmed clinical efficacy for the treatment of psychiatric disorders has been extremely limited, despite tremendous efforts [17]. The attrition rate of CNS drugs is higher than that of non-CNS drugs [34], and the likelihood of approval in the field of psychiatry has been shown to be the lowest among non-oncology disease areas [61]. Part of the issue may be the poor translatability from animal models of psychiatric disorders to the patient [49].

9.2 The Schizophrenic Mouse 1.0: How It Was Done in the Past

Classical animal models of schizophrenia used for the development of novel antipsychotics focused on drug-induced, mechanistic models [31, 56]. For example, dopaminergic mechanisms were stimulated (e.g., by acute administration of amphetamine) and behavioral effects (e.g., amphetamine-induced hyperactivity) measured.

The ability of test compounds to reverse the behavioral effects (e.g., reversal of amphetamine-induced hyperactivity) was then taken as evidence for potential therapeutic efficacy in schizophrenia. The amphetamine model was based on the hypothesis that schizophrenia is a disorder characterized by dopaminergic dysfunction, which in turn was inferred from the ability of amphetamine to elicit psychotic features in people and that effective antipsychotics were known to block dopamine D_2 receptors. Indeed, clinical effective antipsychotics – all D_2 receptor antagonists – were able to reverse amphetamine-induced hyperactivity. However, chances were high to identify even more drugs with D_2 antagonistic properties and efficacy in schizophrenia or drugs with other mechanisms of action that also reduced locomotor activity but that did lack beneficial clinical effects. Another relatively frequently used drug-induced animal model in drug development is based on acute administration of another NMDA receptor channel blocker phencyclidine (PCP, [29]). PCP is also known to induce psychosis in man and hyperactivity in animals. Notably, acute PCP administration is associated with activation of dopamine neurotransmission, which in turn may affect locomotor activity and explain the efficacy of D_2 receptor antagonists in this model.

Also of note, those classical models focus on positive symptoms, i.e., hallucinations and delusions, while the negative and cognitive symptoms are not readily addressed. Preclinical research in the past also concentrated on the development of pharmacotherapy against positive symptoms, while negative and cognitive symptoms were widely neglected. This coincides with the main effects of classical antipsychotics on positive symptoms, while efficacy to treat negative and cognitive symptoms is limited [39]. But because of the difficulty to model hallucinations, delusions, etc., in animals, researchers reverted to what could be measured, e.g., locomotor activity, representing more side effects rather than primary symptoms and played with the neurotransmitter system where classical antipsychotics work.

The issues highlighted for animal models of schizophrenia are not unique, but similar shortcomings can be identified for a range of other classical animal models of psychiatric disorders other than schizophrenia, e.g., the forced swim test as a model to predict antidepressant activity of drugs [54] or the elevated plus maze as a test to detect novel anxiolytics [7]. In fact, the term "model" may be misleading. As the targeted pathophysiological processes underlying the psychiatric conditions are not modeled, a term "assay" may be more appropriate. This is not to say that these assays are without utility. Usually, they are excellent to detect drugs that act at the same mechanism of action as the one of the drug classes used to originally validate the test (e.g., dopamine-D_2 antagonism to reverse amphetamine-induced hyperactivity,serotonin/noradrenaline reuptake inhibition in forced swim and $GABA_A$-positive allosteric modulation in the elevated plus maze). However, there are many false-positive compounds reported in these assays that don't translate into the clinic, as well as false negatives, such as selective serotonin reuptake inhibitors that don't work in the elevated plus maze yet are considered first-line treatment in many anxiety disorders in man [10]. Thus, one may argue the assays have predictive validity, that is, the extent to which the outcome from the assay predicts clinical efficacy but mostly within the classical mechanism of action of psychiatric drugs used during the original characterization of the assays (which may not come as a surprise). Predictive validity for novel molecular targets may be poor. Some assays are also based on face validity, that is, the extent to which the assay appears to measure, at face value, what is altered in the disorder. Face validity of amphetamine-induced hyperlocomotion is poor as it is merely a proxy of the positive symptoms seen in schizophrenia. Face validity of the elevated plus maze is better as it is considered to measure anxiety-related behavior in animals. However, it must be noted that anxiety is a normal emotion, both in humans and nonhuman animals. Whether the anxiety-related behavior in a rodent exploring the maze reflects the pathological anxiety seen in anxiety patients remains questionable. This leads to the third type of validity, that is, construct validity. Construct validity is the extent to which the mechanism used to induce the phenotype in animals reflects the disease etiology in patients. Classical assays hardly addressed construct validity, which for a good part is due to the lack of knowledge about the disease etiology underlying psychiatric disorders.

There are additional issues associated with the abovementioned assays, such as the fact that drugs are usually tested after acute treatment, while therapeutic effects in patients are delayed and often require chronic treatment. Furthermore, there has been a striking negligence of pharmacokinetic/pharmacodynamic principles in the past: compounds have been reported to be effective without knowing whether they cross the blood-brain barrier at all and would even reach their target organ, i.e., the brain; time of compound testing neglected time-concentration relationships of the compound in the target organ; dose-response relationships were often missing; and whether the compound affected its molecular target (e.g., a G-protein-coupled receptor) in a meaningful way to elicit a pharmacodynamic response was often unknown.

Further, the area of CNS disorders to a large extent dealt with syndromal diagnoses in the past, with divergent symptoms and largely unknown pathophysiology [48]. Current diagnostic manuals allow diagnoses with high diversity of symptoms, for example, depression associated with or without inhibition or agitation, with or without significant weight loss or weight gain, with or without a decrease or increase in appetite, or with or without insomnia or hypersomnia [2]. Because of the lack of understanding of the underlying pathophysiology, hypotheses of disease pathology were often developed retrospectively, based on the pharmacological mechanism of action of drugs that were discovered, e.g., the dopamine hypothesis of schizophrenia [38], the 5-HT hypothesis of depression [43], and the GABA hypothesis of anxiety [32].

More novel approaches focused on, e.g., genetics (e.g., [56]), such as the involvement of the gene locus "disrupted in schizophrenia 1" (DISC1), and based on these admittedly very interesting findings, novel animal models were developed, such as mice carrying DISC1 mutations [30], but unfortunately those animal models did not lead to any breakthrough in the development of novel therapies either.

9.3 The Schizophrenic Mouse 2.0: Reverse Engineering Approaches

The term reverse engineering in drug development has been used to describe the process of deciphering the formulation parameters of a marketed drug in an attempt to develop generic formulations. In the context of this chapter, reverse engineering during the early stages of drug development is defined by the following:

1. In its strictest sense, the notion that pharmacological or nonpharmacological treatment used in the clinic, with a defined molecular mechanism of action, causes (adverse) effects that suggest utility of targeting that mechanism of action to develop a therapeutic for another disorder. Such novel therapeutic would be tested in animals for a pharmacodynamic readout that is predictive for the desired effect at the molecular target, i.e., a pharmacodynamic model would be developed.
2. The notion that a clinically effective intervention with action at multiple molecular targets has unwanted effects in patients. New molecules with selectivity for a specific molecular target would be developed with an aim to maintain efficacy while dialling out side effects. Again, novel therapeutics would be tested in a pharmacodynamic model to show an in vivo effect via interaction with the molecular target.
3. The discovery of a molecular mechanism/pathway or a functional readout being altered in a patient population and an attempt to develop new therapies that interact with the molecular mechanism/pathway or the functional readout. As before, the newly developed therapeutics would be tested in a pharmacodynamic model predictive for efficacy on the molecular mechanism/pathway or for the functional readout in the clinic.

It can be argued that at least scenarios two and three merely reflect back-translational strategies. But importantly, in all three scenarios, efficacy would be tested in a pharmacodynamic animal model, and no claims would be made about modelling a disease. This is an important distinction as the first generates a functional readout that should be predictive of an interaction with the (molecular or functional) target and is agnostic about the disease, while the latter, especially in the absence of knowledge about disease pathophysiology, makes claims based on resemblances with symptoms, i.e., based on face validity, that are based on prevailing but so far unproven disease hypotheses or that are completely unsubstantiated (Fig. 9.1).

An example of a reverse engineering approach is the development of a 5-HT_{2C} agonist for treatment of obesity. A commonly known side effect of antipsychotics is weight gain. Meta-analysis revealed that all antipsychotic drugs increase body weight, whereby atypical antipsychotics have the greatest potential to induce this side effect [1]. Atypical or second-generation antipsychotic drugs have antagonistic properties at the 5-HT_{2C} receptor. Given that 5-HT_{2C} receptor knockout mice develop overweight due to abnormal control of feeding [53], the 5-HT_{2C} receptor has been implicated in this adverse effect of atypical antipsychotics. Furthermore, human

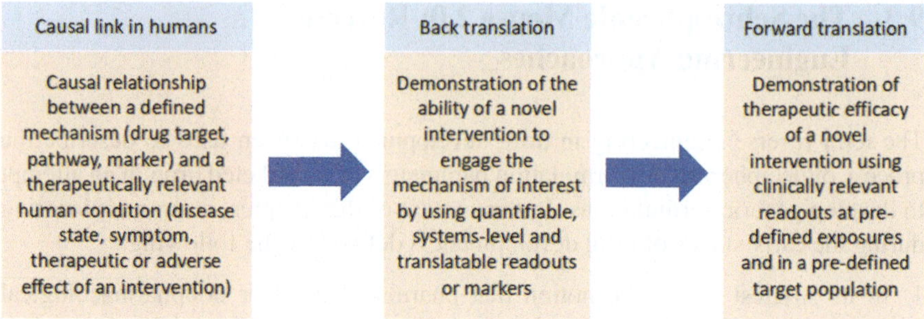

Fig. 9.1 The reverse engineering concept in drug discovery

genetic studies indicated that polymorphisms of the promoter region of the 5-HT_{2C} receptor gene associated with higher transcription levels are associated to resistance to obesity [59]. Two years later, it was shown that polymorphism of the 5-HT_{2C} receptor regulatory region affects treatment-induced weight gain in first-episode schizophrenic patients [44], further strengthening the hypothesis that 5-HT_{2C} receptor antagonism plays a role in weight gain. As reviewed by Garfield and Heisler [21], various 5-HT_{2C} receptor agonists reduce food intake and body weight in both rodents and humans, although selectivity for the various 5-HT_2 receptor subtype remains an issue for some of the compounds. The 5-HT_{2C} receptor agonist lorcaserin received approval by the FDA for the treatment of obesity in 2012 which made this approach a success story.

An example of translating side effects back to the preclinical setting with perhaps less face validity is the muscarinic M_1 receptor agonist approach for cognitive functioning. The cholinergic hypothesis of Alzheimer's disease postulates that learning and memory deficits result from the loss of cholinergic neurotransmission. These cognitive deficits can be partially recovered by increasing extracellular acetylcholine concentrations, by inhibiting acetylcholine-catabolizing cholinesterase [3]. Galantamine, donepezil, and rivastigmine are cholinesterase inhibitors that are approved treatments for Alzheimer's disease. These treatments show some, albeit limited, efficacy on cognitive function, in particular in the early stages of disease. However, cholinesterase inhibitors induce a range of adverse effects due to cholinergic stimulation both in the brain and in the periphery that limit clinical use (see, for example, [28]). Increasing extracellular acetylcholine concentrations stimulates nicotinic and muscarinic receptors, and several nicotinic and muscarinic receptor subtypes have been identified. It is tempting to develop agonist with selectivity for one of those receptor subtypes to improve the efficacy/side effect ratio.

Scopolamine, a nonselective muscarinic receptor antagonist, induces cognitive deficits in healthy volunteers. These effects in humans appear to mimic some of the cognitive deficits associated with Alzheimer's disease. There are also many studies showing scopolamine-induced cognitive deficits in rodents [12, 15]. To optimize the back-translation from clinical to preclinical setting, comparable tests were developed that are sensitive to muscarinic blockade for humans and animals, allowing cross-species comparison [24, 52]. This is important because many rodent cognitive tasks are difficult to translate to the clinic. Furthermore, it is crucial to investigate an AD-relevant cognitive domain. To complicate matters, there are five muscarinic receptor subtypes (M_1–M_5). Developing treatment for a specific subtype may result in a better therapeutic window. The M_1 receptor is highly expressed in areas such as the hippocampus and cortex [18]. Blocking the M_1 receptors results in cognitive impairment as well, and these effects have been argued to be more specific to cognitive function alone than the effects of nonspecific agonist such as scopolamine, which can induce sedative effects [9]. Many pharmaceutical companies started drug discovery programs targeting M_1 receptors, resulting in several M_1 receptor agonists that indeed improved cognition in Alzheimer's disease. Yet, the first-generation muscarinic agonists had modest selectivity for the M_1 receptors compared to other muscarinic receptors, which led to cholinergic adverse responses, predominantly within gastrointestinal and cardiovascular systems [18]. Nevertheless, the development of next-generation agonists is still continuing (see, for example, [11]), and when selectivity over other muscarinic receptors can be achieved, the reverse engineering strategy and the associated use of animal models may be successful.

Another member of the 5-HT receptor family, the 5-HT_{2A} receptor, has been suggested as being responsible for unique antipsychotic properties of clozapine. This hypothesis has led to development of a series of antipsychotics with potent antagonism at 5-HT_{2A} receptors in addition to a conventional dopamine D_2 antagonistic activity [42]. The logical next step was to ask whether selective 5-HT_{2A} antagonists have antipsychotic effects, which could lead to antipsychotic drugs devoid of dopamine D2 antagonism. Would selective 5-HT_{2A} receptor antagonists be effective in the same disease states that respond to treatment to D_2 receptor antagonists (e.g., schizophrenia)?

In animal models of psychosis, pharmacological challenges with psychostimulants as well as 5-HT_{2A} receptor agonists are commonly used. Models which use NMDA receptor antagonists (such as phencyclidine, ketamine, or MK-801) or drugs that increase extracellular dopamine concentrations (such as amphetamine) are often considered by some as more appropriate models of psychosis compared to models that use serotonergic hallucinogens (mescaline, psilocybin, or LSD). However, serotonergic hallucinogens have been described to induce anxiety, difficulty in thinking and concentration, alterations in body image, marked alterations in sensory perception (especially visual perception), kaleidoscopic visual "pseudohallucinations," true hallucinations, and occasionally loss of insight in man [57], and several of these phenomena are seen in schizophrenic patients as well. Furthermore, the hallucinogen psilocybin induces phenomena that could be observed in healthy volunteers which are conceptualized both as positive and negative symptoms of schizophrenic psychosis, whereas effects of the stimulant amphetamine were limited to increase of general activity with increased vigilance, drive, and talkativeness [23]. In a later study, Gouzoulis-Mayfrank et al. [22] compared the effect of the hallucinogen N,N-dimethyltryptamine (DMT) with ketamine in humans. In that study, they found that the phenomena resembling positive symptoms of schizophrenia were stronger after DMT, while ketamine induced stronger effects on phenomena resembling negative symptoms of schizophrenia. Exposure to the hallucinogenic drug LSD can also lead to psychotic symptoms which can be treated with antipsychotics [36]. These effects of hallucinogens in humans and the similarity to symptoms in schizophrenia further strengthening the hypothesis that 5-HT_{2A} antagonist may be beneficial for treatment of patients with psychotic disorders.

Hallucinations cannot be readily measured in animals. In fact, it is not known if rodents hallucinate at all. Therefore, methods based on hallucinogenic 5-HT_{2A} agonist-induced behaviors (e.g., head twitches) may have good predictive validity for target occupancy but no face or construct validity. For the reverse engineering approach though, this is not an obstacle. Clinical evidence suggests involvement of 5-HT_{2A} receptors in psychosis, preclinical methods are robust enough to identify drug candidates with the desired activity at 5-HT_{2A} receptors, and translational tools are available for confirming interaction of drug candidates with 5-HT_{2A} receptor in humans. Thus, the there is a solid foundation enabling a definitive answer to the question as to whether 5-HT_{2A} receptor blockade alone has therapeutically relevant antipsychotic effects.

Pimavanserin (ACP-103, Acadia Pharmaceuticals) is a potent 5-HT_{2A} receptor inverse agonist that readily penetrates the CNS and blocks central 5-HT_{2A} agonist-induced behaviors in animals [20, 55]. While other members of this class (volinanserin, eplivanserin) have shown limited efficacy in schizophrenia patients, pimavanserin is now the first and only medication approved by the US Food and Drug Administration (FDA) for the treatment of hallucinations and delusions associated with Parkinson's disease psychosis (PDP). This example illustrates that an important limitation of the reverse engineering approach is that there are certain decisions related to selection of a therapeutic indication or target patient (sub)population that may be driven by additional clinical, commercial, or other considerations.

A related example of reversed engineering that receives currently much attention is the story of 5-HT_{2A} agonism in depression. In recent years, evidence is accumulating that 5-HT_{2A} receptor agonists may have beneficial effects in patients with major depression [33]. This led to interest in the pharmaceutical industry to start drug discovery programs for 5-HT_{2A} agonists. It is, however, of note that binding of psychedelics to these receptors activates different signal transduction pathways, the canonical pathway and a parallel, G-protein-independent pathway mediated by β-arrestins. The bias to these pathways differs for the various psychedelics, which makes it challenging to pick the right balance for new compounds. Whether the 5-HT_{2A}-related animal models that helped developing pimavanserin are also useful of the agonist approach is not clear yet.

Here, we presented several successful examples of reverse engineering approaches. Yet, there will undoubtedly also be non-successful programs, published or unpublished. The predictive validity of animal models of human neuropsychiatric disorders has been questioned. The examples listed above indicate that the use of animal models may be more straightforward.

The probability for success using animal models that are not relevant for clinical symptoms (such as hallucinations) and their underlying pathophysiology (as this is still poorly understood) is limited. The readouts obtained from pharmacodynamic animal models that focus on the molecular target, whether a compound engages with that target and affects the right pathways and physiological responses mediated by those pathways (be it a behavior, an electrophysiological response, or else), rather than aiming at modelling the complexity of human disease states, may be more promising. But even here, care should be taken when interpreting the specificity of the readout. For example, most, if not all, drugs will reduce body weight or have sedative properties when dosed high enough, which is not necessary target mediated.

9.4 Redefining the Use of Animal Models in Neuropsychiatric Drug Discovery

Looking retrospectively into neuropsychopharmacology research of the 1970s and 1980s, one will recognize the excitement and high hopes triggered by the clinical success of many mechanistically related drugs that followed the initial discoveries of chlorpromazine and imipramine. Animal models reliably detected efficacy of these drugs and the new to be developed drugs were therefore expected to be effective in the same models too.

As failures in the clinic started to mount in the 1990s, positive effects generated by preclinical models have come into focus. The following is best described by the change curve model originally developed by Elisabeth Kübler-Ross to describe how terminally ill patients cope with their impending deaths. This model has been repeatedly modified to apply to various situations where people deal with loss.

The first dozens (!) of clinical-stage failures in neuropsychiatric drug development have hardly been noticed. Or, more correct to say, implications of these failures for the way animal models were built and used have been denied (Fig. 9.2, "denial" stage). Instead, a number of corrective measures have been tried ranging from a closer attention to PK/PD relationship to repeated structural changes in the drug discovery organizations. Some of these changes cause nothing more than a smile today (e.g., an open-office concept as a way to stimulate the innovation; [5]), but it all led to a realization that the old paradigm based on detecting human disease-like signs and symptoms in a laboratory animal was not the right way.

This realization has triggered a frustrative action with most companies leaving or reducing the efforts in the neuroscience area (and psychiatry in particular), departments closed, and scientists laid off (Fig. 9.2, "frustration" stage). After several years, the field got used to the ideas that animal models may be of little use in neuropsychiatric drug discovery research, and such thoughts have fueled a more general criticism of the use of laboratory animals in biomedical research (Fig. 9.2, "depression" stage).

Who was left out in this discussion? The patient. Just like 50 years ago, there is a large unmet medical and social need to develop new safe and effective therapies

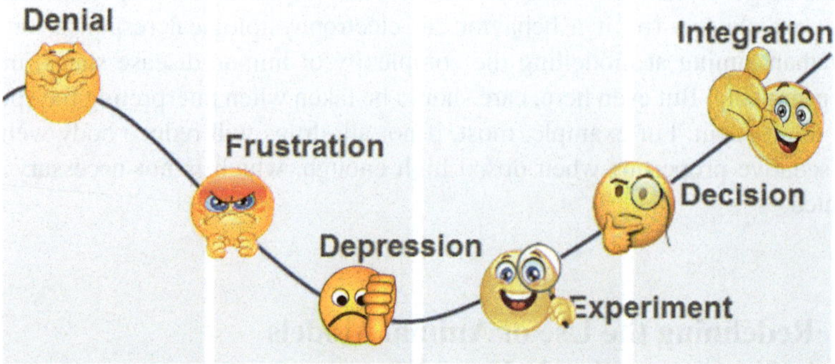

Fig. 9.2 The Kübler-Ross stages of changing use of animal models in neuropsychiatry drug discovery

for most neuropsychiatric indications. With that in mind, a large number of initiatives have been established to critically appraise the use of animal models in the past (Fig. 9.2, "experiment" stage). This is where we are today, ready to re-evaluate the value of animal models and to start using them with all the lessons learned taken into account (Fig. 9.2, "decision" and "integration" stages). What are those lessons learned?

9.4.1 *Lesson 1: Do Not Expect a Mouse with Schizophrenia*

As summarized in the sections above, for a vast majority of mental disorders, it is not reasonable to expect laboratory animals to express behaviors or other systems-level signs that are equivalent or are sufficiently close to human disease signs and symptoms.

The future use of animal models in psychiatry drug discovery research will most likely be limited to the following two scenarios (Table 9.1).

First, several efforts are under way to define basic dimensions of functioning that are trans-diagnostic and for which multiple levels of information (genetics, imaging, electrophysiology, behavior) can be integrated. The aim of these efforts is to understand the brain circuits underlying these dimensions, understand the homologous circuits in animals and humans (if they exist), and eventually build models to address those circuits with a novel therapy. One of the most well-known examples is the Research Domain Criteria (RDoC) established in 2019 by the US National Institute of Mental Health (https://www.nimh.nih.gov/research/research-funded-by-nimh/rdoc). Another prominent example is the IMI-funded PRISM consortium that aims to develop a quantitative biological approach to the understanding and

Table 9.1 Past and current scenarios for using animal models in psychiatric drug discovery

Animal model use	Objective	Use in psychiatric drug discovery research
Disease model	Expression of behaviors and other systems-level signs similar to human disease symptoms	Very limited (with the possible exception of certain disease states where etiological factors are better understood such as substance and alcohol use disorders)
Disease-relevant pharmacodynamic biomarker	Expression of behaviors and other systems-level signs and markers mechanistically similar to impairments observed in human disease	RDoC is an example of a research framework for basic dimensions of functioning that integrates all levels of information and can identify specific dimensions (constructs, subconstructs) to target in a drug discovery program
Target-specific pharmacodynamic biomarker	Evidence of target engagement that: (i) can be translated from humans to animals and vice versa, and (ii) can be used to establish safety margins in nonclinical development	As the blood-brain barrier can make the target access for drugs challenging, such biomarkers are particularly valuable for CNS drug discovery

treatment of neuropsychiatric diseases (https://prism-project.eu/en/about-prism/project-summary/).

Importantly, dimensions addressed by such initiatives are truly trans-diagnostic and, in many cases, are not even explicitly recognized or mentioned by current diagnostic systems. Therefore, primary value of such dimensions is in being: (i) intimately related to mechanisms of dysfunction in a human disease, (ii) amenable to study in animals, and, therefore, (iii) able to support preclinical-to-clinical translation of pharmacodynamic activity of novel therapy.

For example, RDoC's positive valence systems subconstruct "effort" is one of the key domains of goal-directed behavior affected in subjects with apathy. Since decades, dopamine and mesocorticolimbic dopaminergic system have been implicated in various aspects of goal-directed behavior. Reduced dopamine tone (e.g., using tetrabenazine in animal studies or in Parkinson's disease patients) shifts performance in effort-based tasks toward low-effort alternatives [50]. Dopaminergic stimulants such as methylphenidate and amphetamine produce shifts toward high-effort choices in animals as well as in human experimental pharmacology studies [51]. Clinical studies confirm clinical effectiveness of drugs such as methylphenidate in management of apathy [41, 47].

Second, one of the main uses of preclinical data on pharmacodynamic activity of a novel drug is to support determination of a safety window. Without this information, human dose selection may only be guided by evidence of target occupancy (with the arbitrarily defined quantitative goal) and/or by limits imposed by toxicities observed in nonclinical safety studies. Needless to explain, a decision on desired target occupancy should be educated by the doses and tissue concentrations associated with the pharmacodynamic activity of interest. Similarly, following solely nonclinical safety information may lead to testing novel drugs at doses that are safe but are too low to target the receptors and processes thought to be relevant for therapeutic efficacy.

9.4.2 Lesson 2: Understand Drug-Target Interactions

During early stages of drug discovery, there is a substantial amount of attention devoted to understanding the molecular- and cellular-level aspects of interactions of drugs with their receptor targets. This focus does not always continue as the programs advance and effects of drugs are explored at systems level in animals and later in humans.

Targets pursued for neuropsychiatry indications present an additional challenge since target accessibility can be hindered by the blood-brain barrier (BBB). Therefore, conducting a clinical PoC with a dose that results in plasma concentrations that are projected from preclinical studies and used to model likely target occupancy may not be predictive of centrally mediated effects. A case study is given by Cook et al. [14], where the positive allosteric mGluR2 modulator AZD8529 failed to show the expected effects in a clinical phase 2 study in schizophrenia patients. Among several factors thought to contribute to the trial failure were variable exposure observed with the compound and the failure to develop a PET ligand or other biomarkers to measure target engagement.

How often have drug development projects in neuropsychiatry not been supported by robust PK/PD models and target engagement translational strategies? Optimally, this question should be addressed both by accessing exposure, target binding, and evaluating the biological consequence of target binding by "functional pharmacological activity." A group of scientists has searched the Thomson Reuters Cortellis database to identify all drug development projects in the field of schizophrenia between 1994 and 2014 [8]. After excluding all marketed drugs and combinations thereof, agents with unidentifiable mechanism (e.g., herbal products), and D_2 dopamine receptor antagonists, there were a total of 72 novel drugs subjected to clinical phase II PoC studies (representing 34 drug targets). The publicly and commercially available information resources as well as unpublished data from internal projects were used to identify drugs tested in phase II at the doses supported by PET and target engagement biomarkers. For some targets (e.g., 5-HT_{2A} receptors), good-quality PET ligands are available, and therefore dose selection could be validated (prospectively or retrospectively). For other targets, there was a robust pharmacodynamic response that could be used in clinical phase I studies to guide dose selection (e.g., reduction in EEG alpha-power for T-type calcium channel blockers, [16]; sensory gating as a pharmacodynamic biomarker for α_7 nicotinic receptor agonists, [25]). However, overall, evidence of biomarker-driven dose selection could not be found for 80.5% of drugs tested in clinical phase II schizophrenia trials (representing 70.6% of targets and mechanisms that were supposed to be tested).

Solutions to improve study design and to support selection of dosing paradigms in preclinical studies requiring brain penetration have already been proposed. For example, Kleinman and Ehlers [35] have explicitly listed the compound specific data that should be generated and reported to justify dose selection: (i) expected or measured plasma exposure of the study drug in the preclinical species during the study, (ii) expected or measured target organ exposure of the study drugs in the preclinical species during the study, (iii) expected or measured free fraction (unbound by protein) of the study drugs in the target organ of the preclinical species during the study, and (iv) expected or measured potency of the study drug against the hypothesized activity in vitro.

9.4.3 Lesson 3: Be Confident in the Data

When a clinical hase II trial fails to meet its primary predicted endpoints, the preclinical data upon which the prediction was made is one area in the development chain that is often called into question. There has been much discussion about the issue of reproducibility of preclinical data which has become a major concern for funding agencies such as the NIH and, ultimately, the taxpayers [13].

Over the past several years, there have been numerous discussions about reproducibility of data in various fields of science, and the research community has not been united regarding the extent of the issue or the urgency to act upon it. There are, however, two aspects of this discussion that are rarely debated. Both are particularly relevant for drug discovery and even more important for areas where success rates have been unacceptably low.

First, internal validity of published studies supporting therapeutic use of novel drugs is generally rather low. In preclinical research, studies are typically underpowered, are conducted without proper blinding and randomization, and in the absence of pre-defined inclusion / exclusion criteria, hypotheses, or data analysis plans. There is no evidence that research conducted by biopharmaceutical companies to support clinical development candidates follows more rigorous practices compared to academic research.

Lack of pre-specified hypotheses and analysis plans is particularly threatening when confidence in the results is critical for decision-making. For example, studies are more likely to be repeated if results are negative and therefore not in support of the original hypothesis. In such cases, researchers are often more likely to critically review study conditions and vigorously look for factors that may account for negative results. This seems to be a common practice dictated by the need to justify a replication study against the time and resource pressure discussed above (Fig. 9.3).

Second, every laboratory has its preferred protocols, preferred suppliers of tools and reagents, and preferred source and type of study subjects. This is dictated by prior experience, budgetary, logistics, and other reasons. And this uniqueness of the research environment in every laboratory makes comparison between results generated across labs very important. Seen from the perspective of drugs being developed

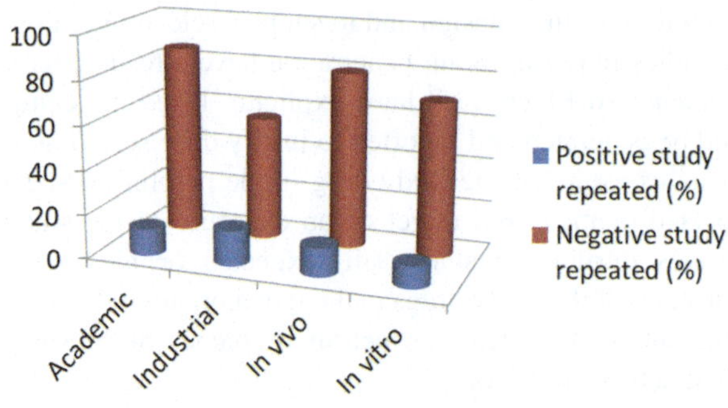

Fig. 9.3 Both industrial and academic neuroscientists have been asked to answer questions on whether or not they were repeating studies if results were positive or negative (i.e., confirming or disproving the hypothesis). Results are based on survey of 42 scientists

for highly heterogeneous patient populations, data generated by someone's lab make sense only when they can also be obtained under distinct conditions of other labs. In other words, the broader can the preclinical efficacy claims be generalized, the higher is the likelihood of detecting efficacy signals in clinical studies. And, in contrast, efficacy that is seen only under unique highly standardized conditions may be detrimental. One example of research strategy explicitly focused on generalizability is the use of purpose-bred animals with maximal genetic heterogeneity [40].

Generalizability of preclinical data is not just about laboratory conditions and animal strains, age, and sex. There is an aspect of preclinical data sustainability that may be especially relevant for neuroscience where traditionally a significant proportion of efficacy studies are conducted with acute application of the test compounds. Indeed, in experimental psychopharmacology, there is a remarkable paucity of studies with repeated drug administration [6]. It is largely unknown how often compounds' efficacy has been confirmed in [unpublished] studies with chronic administration prior to initiation of a clinical trial to exclude potential tolerance development.

9.4.4 Lesson 4: Adopt Transparent and Open Science Practices

Discovery and development of a novel therapy is the result of a collective effort of a very large number of individuals and organizations. This effort is built from numerous sequential advances where every new experiment utilizes and further extends the knowledge generated by the prior ones. Given the complexity of modern research environments, it is no wonder that accurate and complete reporting of materials and methods has become very important.

In a recently completed Reproducibility Project: Cancer Biology, a total of 193 experiments from 53 papers were selected for replication. One particularly disturbing outcome of this project was that "none of the 193 experiments were described in sufficient detail for the project team to design protocols to repeat them" [46]. Without having done or being involved in a similar effort, we dare to say that the completeness and transparency of reporting in the neuroscience field is not much different from cancer biology.

Different research areas also share similar issues with the reporting of completed studies in general. It has been repeatedly demonstrated that results of a significant proportion of studies are never written up and published [19]. What is omitted may of course be failed studies of technically poor quality, and this would be of little or no concern. However, existing evidence points at a substantial publication bias in favor of studies reporting positive results, while studies with negative, null, or neutral results stay largely unpublished. Such bias is highly problematic for a proper assessment of the drugs, drug targets, and models.

Drug discovery has always been seen as a highly competitive enterprise with little tolerance toward sharing of information or resources. Over the past several years, there were already several remarkable examples of fruitful cooperation between industry and academic setting precedents of precompetitive collaboration. Adoption of transparent research practices and adherence to the principles of open sciences should further facilitate the most efficient use of information and resources and is the basis for future advances in psychopharmacology.

Taken together, neuropsychiatric drug discovery has been a challenging field. A number of disappointing results at the advanced clinical development stages have significantly reduced the enthusiasm of those who provide financial and nonfinancial resources to drive innovation in this field. In this chapter, we argue that there are several previous success stories that teach us on what needs to be done to improve the success rate. In fact, the reverse engineering approach does not reveal anything fundamentally novel. It only shifts the focus from preclinical efforts to "model" human disease states to pharmacodynamic activity as a common denominator in the journey to translate clinically validated phenomena to preclinical level and then back to humans. Combined with the research rigor, openness, and transparency, this reverse engineering approach is well set to bring new effective and safe medications to the patients in need.

Conflicts of Interest AB has financial interests in and serves as managing director and CSO/CMO of Exciva GmbH, a pharmaceutical company focused on treating behavioral and psychological symptoms of dementia, and managing director of PAASP GmbH (Heidelberg, Germany), a company focused on developing and providing assessment and accreditation of scientific practice. MvG owns stock of AbbVie, uniQure, and Charles River Laboratories. TS is an employee of J&J and holds stock from J&J.

References

1. Allison DB, Mentore JL, Heo M, Chandler LP, Cappelleri JC, Infante MC, Weiden PJ. Antipsychotic-induced weight gain: a comprehensive research synthesis. Am J Psychiatry. 1999;156:1686–96.

2. American Psychiatric Association. Diagnostic and statistical manual of mental disorders (DSM-5). 5th ed. American Psychiatric Association; 2020.
3. Bartus RT, Dean RL, Beer B, Lippa AS. The cholinergic hypothesis of geriatric memory dysfunction. Science. 1982;217:408–14.
4. Berman RM, Cappiello A, Anand A, Oren DA, Heninger GR, Charney DS, Krystal JH. Antidepressant effects of ketamine in depressed patients. Biol Psychiatry. 2000;47:351–4.
5. Bernstein E, Waber B. The truth about open offices. Harv Bus Rev. 2019;97:83.
6. Bespalov A, Müller R, Relo AL, Hudzik T. Drug tolerance: a known unknown in translational neuroscience. Trends Pharmacol Sci. 2016b;37:364–78.
7. Bespalov A, Steckler T. Pharmacology of anxiety or pharmacology of elevated plus maze? Biol Psychiatry. 2021;89:e73.
8. Bespalov A, Steckler T, Altevogt B, Koustova E, Skolnick P, Deaver D, Millan MJ, Bastlund JF, Doller D, Witkin J, Moser P, O'Donnell P, Ebert U, Geyer MA, Prinssen E, Ballard T, Macleod M. Failed trials for central nervous system disorders do not necessarily invalidate preclinical models and drug targets. Nat Rev Drug Discov. 2016a;15:516.
9. Blokland A, Sambeth A, Prickaerts J, Riedel WJ. Why an M1 antagonist could be a more selective model for memory impairment than scopolamine. Front Neurol. 2016;7:10–3.
10. Borsini F, Podhorna J, Marazziti D. Do animal models of anxiety predict anxiolytic-like effects of antidepressants? Psychopharmacology. 2002;163:121–41.
11. Brown AJH, Bradley SJ, Marshall FH, Brown GA, Bennett KA, Brown J, Cansfield JE, Cross DM, de Graaf C, Hudson BD, et al. From structure to clinic: design of a muscarinic M1 receptor agonist with potential to treatment of Alzheimer's disease. Cell. 2021;184:5886–901.
12. Buccafusco JJ. The revival of scopolamine reversal for the assessment of cognition-enhancing drugs. In: Buccafusco JJ, editor. Methods of behavior analysis in neuroscience. Boca Raton, FL: CRC Press/Taylor & Francis; 2009.
13. Collins FS, Tabak LA. Policy: NIH plans to enhance reproducibility. Nature. 2014;505:612–3.
14. Cook D, Brown D, Alexander R, March R, Morgan P, Satterthwaite G, Pangalos MN. Lessons learned from the fate of AstraZeneca's drug pipeline: a five-dimensional framework. Nat Rev Drug Discov. 2014;13:419–31.
15. Ebert U, Kirch W. Scopolamine model of dementia: electroencephalogram findings and cognitive performance. Eur J Clin Investig. 1998;28:944–9.
16. Egan MF, Zhao X, Smith A, Troyer MD, Uebele VN, Pidkorytov V, Cox K, Murphy M, Snavely D, Lines C, Michelson D. Randomized controlled study of the T-type calcium channel antagonist MK-8998 for the treatment of acute psychosis in patients with schizophrenia. Hum Psychopharmacol. 2013;28:124–33.
17. Erhel F, Scanff A, Naudet F. The evidence base for psychotropic drugs approved by the European Medicines Agency: a meta-assessment of all European Public Assessment Reports. Epidemiol Psychiatr Sci. 2020;29:e120.
18. Felder CC, Goldsmith PJ, Jackson K, Sanger HE, Evans DA, Mogg AJ, Broad LM. Current status of muscarinic M1 and M4 receptors as drug targets for neurodegenerative diseases. Neuropharmacology. 2018;136:449–58.
19. Franco A, Malhotra N, Simonovits G. Social science. Publication bias in the social sciences: unlocking the file drawer. Science. 2014;345:1502–5.
20. Gardell LR, Vanover KE, Pounds L, Johnson RW, Barido R, Anderson GT, Veinbergs I, Dyssegaard A, Brunmark P, Tabatabaei A, Davis RE, Brann MR, Hacksell U, Bonhaus DW. ACP-103, a 5-hydroxytryptamine 2A receptor inverse agonist, improves the antipsychotic efficacy and side-effect profile of haloperidol and risperidone in experimental models. JPET. 2007;322:862–70.
21. Garfield AS, Heisler LK. Pharmacological targeting of the serotonergic system for the treatment of obesity. J Physiol. 2009;587:49–60.
22. Gouzoulis-Mayfrank E, Heekeren K, Neukirch A, Stoll M, Stock C, Obradovic M, Kovar K-A. Psychological effects of (S)-ketamine and N,N-dimethyltryptamine (DMT): a double-blind, cross-over study in healthy volunteers. Pharmacopsychiatry. 2005;38:301–11.

23. Gouzoulis-Mayfrank E, Thelen B, Habermeyer E, Kunert HJ, Kovar KA, Lindenblatt H, Hermle L, Spitzer M, Sass H. Psychopathological, neuroendocrine and autonomic effects of 3, 4-methylenedioxyethylamphetamine (MDE), psilocybin and d-methamphetamine in healthy volunteers Results of an experimental double-blind placebo-controlled study. Psychopharmacology. 1999;142:41–50.
24. Harel BT, Pietrzak RH, Snyder PJ, Maruff P. Effect of cholinergic neurotransmission modulation on visual spatial paired associate learning in healthy human adults. Psychopharmacology. 2013;228:673–83.
25. Hashimoto K. Targeting of α7 nicotinic acetylcholine receptors in the treatment of schizophrenia and the use of auditory sensory gating as a translational biomarker. Curr Pharm Des. 2015;21:3797–806.
26. Hillhouse TM, Porter JH. A brief history of the development of antidepressant drugs: from monoamines to glutamate. Exp Clin Psychopharmacol. 2015;23:1–21.
27. Howland RH. Serendipity and psychopharmacology. J Psychosoc Nurs Ment Health Serv. 2010;48:9–12.
28. Inglis F. The tolerability and safety of cholinesterase inhibitors in the treatment of dementia. Int J Clin Pract Suppl. 2012;127:45–63.
29. Jentsch JD, Roth RH. The neuropsychopharmacology of phencyclidine: from NMDA receptor hypofunction to the dopamine hypothesis of schizophrenia. Neuropsychopharmacology. 1999;20:201–25.
30. Johnstone M, Thomson PA, Hall J, McIntosh AM, Lawrie SM, Porteous DJ. DISC1 in schizophrenia: genetic mouse models and human genomic imaging. Schizophr Bull. 2011;37:14–20.
31. Jones CA, Watson DJG, Fone KFC. Animal models of schizophrenia. Br J Pharmacol. 2011;164:1162–94.
32. Kalueff A, Nutt DJ. Role of GABA in memory and anxiety. Depress Anxiety. 1996/1997;4:100–10.
33. Kelmendi B, Kaye AP, Pittenger C, Kwan AC. Psychedelics. Curr Biol. 2021;32:R55–71.
34. Kesselheim AS, Hwang TJ, Franklin JM. Two decades of new drug development for central nervous system disorders. Nat Rev Drug Disc. 2015;14:815–6.
35. Kleiman RJ, Ehlers MD. Data gaps limit the translational potential of preclinical research. Sci Transl Med. 2016;8:320ps1.
36. Leikin JB, Krantz AJ, Zell-Kanter M, Barkin RL, Hryhorczuk DO. Med Toxicol Adverse Drug Exp. 1989;4:324–50.
37. Lopez-Munoz F, Alamo C, Cuenca E, Shen WW, Clervoy P, Rubio G. History of the discovery and clinical introduction of chlorpromazine. Ann Clin Psychiatry. 2005;17:113–35.
38. Meltzer HY, Stahl SM. The dopamine hypothesis of schizophrenia: a review. Schizophrenia Bull. 1976;2:19–76.
39. Millan MJ, Fone K, Steckler T, Horan WP. Negative symptoms of schizophrenia: clinical characteristics, pathophysiological substrates, experimental models and prospects for improved treatment. Eur Neuropsychopharmacol. 2014;24:645–92.
40. Miller RA, Harrison DE, Astle CM, Floyd RA, Flurkey K, Hensley KL, Javors MA, Leeuwenburgh C, Nelson JF, Ongini E, Nadon NL, Warner HR, Strong R. An aging interventions testing program: study design and interim report. Aging Cell. 2007;6:565–75.
41. Mintzer J, Lanctôt KL, Scherer RW, Rosenberg PB, Herrmann N, van Dyck CH, Padala PR, Brawman-Mintzer O, Porsteinsson AP, Lerner AJ, Craft S, Levey AI, Burke W, Perin J, Shade D, ADMET 2 Research Group. Effect of methylphenidate on apathy in patients with Alzheimer disease: the ADMET 2 randomized clinical trial. JAMA Neurol. 2021;78:1324–32.
42. Möller HJ. Risperidone: a review. Expert Opin Pharmacother. 2005;6:803–18.
43. Ögren SO, Fuxe K, Agnati LF, Gustafssons JA, Jonsson G, Holm AC. Reevaluation of the indoleamine hypothesis of depression. Evidence for a reduction of functional activity of central 5-HT systems by antidepressant drugs. J Neural Transm. 1979;46:85–103.
44. Reynolds GP, Zhang ZJ, Zhang XB. Association of antipsychotic drug-induced weight gain with a 5-HT2C receptor gene polymorphism. Lancet. 2002;359:2086–7.

45. Robinson E. Psychopharmacology: from serendipitous discoveries to rationale design, but what next? Brain Neurosci Adv. 2018;2:1–11.
46. Rodgers P, Collings A. What have we learned? elife. 2021;10:e75830.
47. Rosenberg PB, Lanctôt KL, Drye LT, Herrmann N, Scherer RW, Bachman DL, Mintzer JE, Investigators ADMET. Safety and efficacy of methylphenidate for apathy in Alzheimer's disease: a randomized, placebo-controlled trial. J Clin Psychiatry. 2013;74:810–6.
48. Rush AJ, Ibrahim HM. Speculations on the future of psychiatric diagnosis. J Nerv Ment Dis. 2018;206:481–7.
49. Stanford SC. Some reasons why preclinical studies of psychiatric disorders fail to translate: what can be rescued from the misunderstanding and misuse of animal 'models'? Alt Lab Animals. 2020;48:106–15.
50. Salamone JD, Yohn SE, López-Cruz L, San Miguel N, Correa M. Activational and effort-related aspects of motivation: neural mechanisms and implications for psychopathology. Brain. 2016;139:1325–47.
51. Soder HE, Cooper JA, Lopez-Gamundi P, Hoots JK, Nunez C, Lawlor VM, Lane SD, Treadway MT, Wardle MC. Dose-response effects of d-amphetamine on effort-based decision-making and reinforcement learning. Neuropsychopharmacology. 2021;46:1078–85.
52. Talpos JC, Aerts N, Fellini L, Steckler T. A touch-screen based paired-associates learning (PAL) task for the rat may provide a translatable pharmacological model of human cognitive impairment. Pharmacol Biochem Behav. 2014;122:97–106.
53. Tecott LH, Sun LM, Akana SF, Strack AM, Lowenstein DH, Dallman MF, Julius D. Eating disorder and epilepsy in mice lacking 5-HT2C serotonin receptors. Nature. 1995;374:542–6.
54. Trunnell ER, Carvalho C. The forced swim test has poor accuracy for identifying novel antidepressants. Drug Disc Today. 2021;26:2898–904.
55. Vanover KE, Weiner DM, Makhay M, Veinbergs I, Gardell LR, Lameh J, Del Tredici AL, Piu F, Schiffer HH, Ott TR, Burstein ES, Uldam AK, Thygesen MB, et al. Pharmacological and behavioral profile of N-(4-fluorophenylmethyl)-N-(1-methylpiperidin-4-yl)-N′-(4-(2-methylpropyloxy) phenylmethyl) carbamide (2R, 3R)-dihydroxybutanedioate (2: 1) (ACP-103), a novel 5-hydroxytryptamine2A receptor inverse agonist. J Pharmacol Exp Ther. 2006;317:910–8.
56. Winship IR, Dursun SM, Baker GB, Balista PA, Kandratavicius L, Maia-de-Oliveira JP, Hallak J, Howland JG. An overview of animal models related to schizophrenia. Can J Psychiatr. 2019;64:5–17.
57. Wolbach AB Jr, Miner EJ, Isbell H. Comparison of psilocin with psilocybin, mescaline and LSD-25. Psychopharmacologia. 1962;3:219–23.
58. Yanez M, Padin JF, Arranz-Tagarro JA, Camina M, Laguna R. History and therapeutic use of MAO-A inhibitors: a historical perspective of MAO-A inhibitors as antidepressant drug. Curr Topics Med Chem. 2012;12:2275–82.
59. Yuan X, Yamada K, Ishiyama-Shigemoto S, Koyama W, Nonaka K. Identification of polymorphic loci in the promoter region of the serotonin 5-HT(2c) receptor gene and their association with obesity and type II diabetes. Diabetologia. 2000;43:373–6.
60. Zarate CA Jr, Singh JB, Quiroz JA, De Jesus G, Denicoff KK, Luckenbaugh DA, Manji HK, Charney DS. A double-blind, placebo-controlled study of memantine in the treatment of major depression. Am J Psychiatry. 2006;163:153–5.
61. Zhu T. Challenges of psychiatry drug development and the role of human pharmacology models in early development—a drug developer's perspective. Front Psych. 2021;11:562660.

Chapter 10
Targeted Treatments for Fragile X Syndrome

Devon Johnson, Courtney Clark, and Randi Hagerman

Abstract The histories of targeted treatment trials in fragile X syndrome (FXS) are reviewed in animal studies and human trials. Advances in understanding the neurobiology of FXS have identified a number of pathways that are dysregulated in the absence of FMRP and are therefore pathways that can be targeted with new medication. The utilization of quantitative outcome measures to assess efficacy in multiple studies has improved the quality of more recent trials. Current treatment trials including the use of cannabidiol (CBD) topically and metformin orally have positive preliminary data, and both of these medications are available clinically. The use of the phosphodiesterase inhibitor (PDE4D), BPN1440, which raised the level of cAMP that is low in FXS has very promising results for improving cognition in adult males who underwent a controlled trial. There are many more targeted treatments that will undergo trials in FXS, so the future looks bright for new treatments.

Keywords Fragile X syndrome · Premutation · Treatments · Medications · FXTAS · Minocycline · AFQ056 · Metformin · Arbaclofen

D. Johnson (✉) · C. Clark
MIND Institute, University of California Davis Health, Sacramento, CA, USA

R. Hagerman
MIND Institute, University of California Davis Health, Sacramento, CA, USA

Department of Pediatrics, University of California Davis Health, Sacramento, CA, USA

M. Macaluso et al. (eds.), *Drug Development in Psychiatry*, Advances in Neurobiology 30,
https://doi.org/10.1007/978-3-031-21054-9_10

10.1 Introduction: Overview of Fragile X Spectrum Disorders Including the Full Mutation and FXS and Premutation Disorders

Fragile X syndrome (FXS) is a genetic disorder characterized by a trinucleotide repeat, CGG, at the 5′ end of the fragile X mental retardation 1 gene (*FMR1*) at Xq27.3. FXS is the most common inherited form of intellectual disability (ID) and the most common single-gene cause of autism spectrum disorder (ASD), so the *FMR1* DNA test is typically ordered whenever a patient presents with ID or ASD, and the cause is unknown. Those with FXS will have a full mutation (>200 CGG repeats) in the *FMR1* gene, causing methylation and silencing of the gene so that little or no *FMR1* mRNA is produced. Therefore, little or no *FMR1* protein (FMRP) is made. It is the absence or deficiency of FMRP that causes FXS.

FMRP deficiency leads to ID in over 85% of males with FXS and 30% of females. Those that are higher functioning are typically mosaic by size, meaning some cells have the premutation and some cells have the full mutation. One can also be mosaic by methylation differences, where some cells have the full mutation and are unmethylated, but others are methylated. These types of mosaicism lead to more FMRP production and a higher IQ [1].

Approximately 50–60% of boys with FXS have autism spectrum disorder [2]. Additional behavior problems include ADHD [3], anxiety [4], tantrums, and sometimes aggression [5]. Girls are less affected by the full mutation because they have two X chromosomes and the normal X is producing FMRP. Approximately 30% of women with the full mutation will have a normal IQ, 30% will have a borderline IQ, and 30% will have ID. The cognitive abilities of females with the full mutation correlate with their activation ratio (AR), which is the percentage of cells with the normal X as the active X. The higher the AR, the more FMRP is produced and the higher the IQ [6].

Individuals with the premutation are called carriers, and they are common in the general population. Approximately 1 in 200 females and 1 in 400 males is a carrier, whereas those with FXS are less common, and approximately 1 in 5000 has FXS [7]. Individuals with the premutation are usually unaffected intellectually, but approximately 20% of women have early menopause before age 40 called fragile X-associated primary ovarian insufficiency (FXPOI), approximately 40% of men and 16% of females develop a neurodegenerative syndrome called the fragile X-associated tremor ataxia syndrome (FXTAS), and approximately 50% develop one or more fragile X-associated neuropsychiatric disorders (FXAND) including neuropsychiatric disorders such as anxiety, depression, chronic fatigue, chronic pain syndrome, and insomnia.

Figure 10.1 [8]* shows the differential effects of CGG repeats in the normal range (<55), premutation range (55–200), and full mutation range (>200). Excessive CGG repeats in the premutation range lead to elevated mRNA and toxicity that

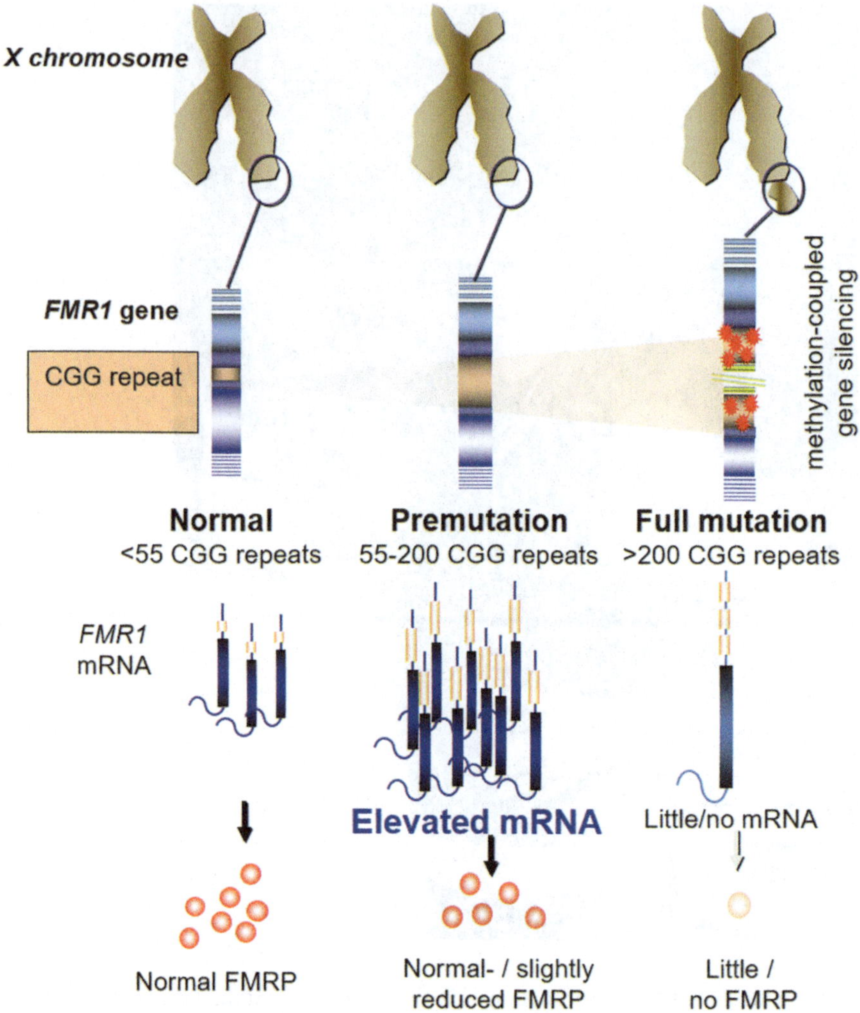

Fig. 10.1 Model of CGG repeats and FMR1 mRNA level

sequesters proteins important for neuronal function, oxidation, and the mitochondria. Despite elevated mRNA, the FMRP is normal or paradoxically reduced secondary to less efficient translation [9]. The illustration shows >200 CGG repeats results in methylation and silencing of the gene, creating little/no mRNA and little/no FMRP.

*Edited and used with permission from [8]. First steps into developing EEG as outcome measure for targeted treatments trials in fragile X. Poster presented at 2010 MRNET conference, Erlangen, Germany.

There have been robust preclinical findings from animal models for pathways involved in FXS.

Treatments targeting these pathways have largely been effective in animal models but have had limited success in their subsequent translation into human subjects. In this chapter we review pharmacology and pathways related to FXS, current treatments for FXS including targeted treatments that are and are not FDA approved, and ultimately the lessons learned for future research.

FXS drug development is an important example of preclinical research on a monogenic condition that inspired an industry but had multiple high-profile failures translating animal models to human studies. It stands as a significant learning opportunity for improving future research in FXS, single-gene causes of autism, and other monogenic disorders.

10.2 Animal Models Guiding Targeted Treatments

10.2.1 KO Mouse Model and Drosophila *Model for FXS*

Animal models have been used to discover signaling pathways and pharmaceutical targets in FXS. The use of mouse, drosophila, rat, and zebra fish models led to development of pharmacologic treatments from the bottom up: from gene discovery to pathophysiology and preparation for application in humans.

Experimental research on how loss of FMRP in mouse models changed systems led to finding multiple phenotypic targets that could be measured to assess the efficacy of pharmaceuticals including protein synthesis, dendritic spine density, seizures, brain changes, among others. In *Fmr1* knockout mice (most common model used), many pharmaceuticals have been used to rescue the FXS phenotype.

FMR1 has a functional homologue in the mouse, and *Fmr1* knockout mice have absent FMRP production, leading to a phenotype comparable to symptoms in human patients. They exhibit macroorchidism, seizure susceptibility, learning deficits, social impairment, and hyperactivity [10]. In addition, they show anomalies in dendritic spine density similar to affected human brains [11]. These models can be crossed with other mutant lines, such as *Grm5* gene mutant mice leading to reduced expression of mGluR5 [12], to see if changes correct diverse FXS phenotypes. A *Drosophila* model for FXS was developed based on loss-of-function mutants of *dfmr1* which is a single homologue of the *FMR1* gene [13]. Studies showed parallels to humans related to circadian rhythms, synaptic branching, and memory. Other FXS animal models include the rat, zebra fish, zebra finch, roundworm, and primate model.

10.2.2 Downside of Animal Models

Although vital to FXS research, animal models do not have the complexity of cognition and emotional features present in humans. We've found difficulties in translating any preclinical findings to a clinical model due to differences in development, pathophysiology related to CGG repeats and methylation, variability in outcomes, and dissimilarity in extrapolating behavior. No single model is reflective of the wide spectrum of symptoms seen in human FXS patients. From an evolutionary perspective, FMRP is well-conserved across species, but neurodevelopment is distinct in each species with different windows of plasticity obscuring optimal times for treating humans. Humans with FXS have >200 CGG repeats leading to methylation and silencing at *FMR1*, however all animal models (except primate model) differ from humans in terms of methylation and trinucleotide repeat numbers [14].

There exists high variability and small effect size in cognition when studying the mouse model. Variability depends on the laboratory as deficits are found in some mice, but normal performance in others [15]. Outcome measures differ between humans and animals; in most studies, behavioral traits and not cognition have been primarily measured in humans. Significant error is introduced when extrapolating how social behavior in one animal model translates to human behavior. For example, the phenotype of the *Drosophila* model exhibits decreased courtship behavior; how does this relate to behavior in humans [16]? Self-grooming behavior in mice is dissimilar to the human behavior it is seeking to mimic. Behavior is a major reason for referral for FXS and of great interest as an outcome measure, despite the difficulties translating complex behaviors from model to human. Overall *Fmr1* mouse model has not been effective or sufficient to predict preclinical to clinical results.

10.3 FMRP Deficits and Pathways that Are Dysregulated in the Absence of FMRP

Significant research over the past two decades has characterized multiple pathways related to the pathophysiology of FXS. These changes are caused by the decrease or absence of FMRP, an RNA-binding protein that binds mRNAs, transports them to the synapse, and regulates their translation and therefore the synthesis of many proteins. FMRP is expressed ubiquitously throughout the brain [17] and is involved in neurological processes critical for neurodevelopment, most notably related to neuron long-term potentiation (LTP) and synaptic plasticity. FMRP acts as a translational repressor which binds to ribosomes and mRNA [10], and its absence leads to loss of inhibitory control over protein and receptor expression.

Research has identified multiple neurotransmitter systems, important protein pathways, and channelopathies described below. Specific neurotransmitter and neuromodulator systems known to be affected in fragile X syndrome and targeted by therapeutics involve the glutamatergic, GABAergic, and endocannabinoid systems primarily. FMRP loss also dysregulates important protein pathways including matrix metalloproteinase9 (MMP9), eukaryotic translation initiation factor 4E (eIF4E), linked to the mammalian target of rapamycin (mTOR) pathway, cyclic adenosine monophosphate (cAMP) production, and mitogen-activated protein kinase/extracellular signal-regulated kinase (MAP/ERK) [18]. FMRP loss also leads to dysfunction of numerous ion channels, changing the intrinsic excitability of neuron synaptic transmission and action potentials [19]. Each of these levels underlies the downstream effects of the FXS phenotype and helps to generate a framework for targeting by therapeutics.

Glutamate

Glutamate is the primary excitatory neurotransmitter and a key compound in cellular metabolism. It exerts its actions via ionotropic receptors and G-protein-coupled metabotropic receptors that are responsible for basal excitatory synaptic transmission and multiple forms of synaptic plasticity. *Fmr1* knockout mice were found to have excessive protein synthesis in the several pathways including those connected to group I metabotropic glutamate receptors (mGluRs) [20].

Genetic reduction of the mGluR5 pathway by 50% in mutant mice corrected multiple phenotypes representing different FXS symptoms. For example, it reduced seizures, accelerated body growth, rescued hippocampal proteins and memory, increased dendritic spine density, and reversed visual changes [21]. Pharmaceutical suppression of mGluR5 also rescued mouse model phenotypes of seizures, anxiety, and behavior [22, 23] . Pharmaceutical suppression of mGluR5 also rescued phenotypes in the *Drosophila* model, including courtship, memory defects, and synaptic plasticity [24]. The summation of these rescued phenotypes validated the theory that many FXS symptoms are secondary to exaggerated mGluR5 activation. This paved the way for human trials targeting mGluR5 including fenobam, mavoglurant (AFQ056), and basimglurant. However, recent studies have shown that the amygdala has lowered levels of mGluR5 activation in the rat model of FXS compared to the upregulation of mGluR5 in the hippocampus, so different areas of the brain may respond differently to targeted treatments for FXS [25].

GABA

The GABA system is the major inhibitory neurotransmitter and exerts its actions through the ionotropic GABA-A receptor and metabotropic GABA-B receptor. GABAergic inhibition tightly regulates glutamatergic transmission. In addition to the exaggerated glutamatergic response, the GABAergic system is also dysregulated in FXS. FMRP has been shown to bind mRNAs of GABA-B receptor [26] and GABA-A subunit δ and α1 [27], which presents as targets for intervention.

Mouse and *Drosophila* models of FXS show a deficit in GABA-A receptors [28, 29]. PET scans on human brains with FXS find 10% fewer GABA-A receptors on average [30]. GABA-B receptors have been found to upregulate FMRP expression in neurons [31]. Brain areas that correlate with behavior such as prefrontal cortex, amygdala, striatum, and hippocampus also show decreased GABA signaling and inhibition in mice models [32–34]. Pharmacologic treatment of both mouse [35–37] and *Drosophila* [38] models with GABA-A and GABA-B agonists has been effective at reversing FXS phenotype. This research set the stage for human trial using arbaclofen, acamprosate, ganaxolone, and gaboxadol.

Endocannabinoid System
Endocannabinoids function as retrograde synaptic messengers. Activation of CB1 exerts negative feedback on presynaptic neurons, inhibiting presynaptic neurotransmitter release, resulting in differing effects depending on which neurotransmitter is affected [39]. Of significance, mutations in FMRP, mGlu-R, and NMDA-R disrupt endocannabinoid presynaptic regulation [40]. Activation of mGluR5 enhances endocannabinoid retrograde signaling to suppress inhibitory release in brain tissues [41] and is increased in FXS. The endocannabinoid system also exerts a neurodevelopmental regulatory role partly through activation of mTORc1 signaling [42].

Fmr1 knockout mice have impaired endogenous endocannabinoid production in multiple brain regions [43]. mGlu5R-driven endocannabinoid signaling in the striatum is controlled by FMRP [44]. Long-term depression by endogenous endocannabinoid 2-arachidonoyl-glycerol (2-AG) at excitatory synapses of striatum and prefrontal cortex is absent in knockout mice [45]. Targeting the cannabinoid system therapeutically in mouse models has remained complex. Genetic blockade of CB1 normalized multiple traits in FXS mouse model. CB1 and CB2 receptors are linked to anxiety behavior in the KO mouse model [43]. Pharmacologic inhibition and genetic inhibition of CB1 normalized multiple behavioral phenotypes in the mouse model. CB2 inhibition increased anxiety in the mouse model, normalizing the anxiolytic phenotype. However, increasing endocannabinoid ligands 2-AG corrects behavioral phenotype in mouse model [46]. A transdermal cannabidiol preparation (Zynerba) is currently being studied in humans.

MAP/ERK and mTOR
mTOR is a metabolic sensor and regulator of cell growth and is controlled via mGluR. FMRP regulates translation of multiple Gq-linked receptors which activate to produce phospholipase C and signaling through ERK- and mTOR-dependent pathways, resulting in loss of FMRP inhibition at the ribosome and leading to enhanced production of multiple proteins. MAP/ERK functions as a cell growth and proliferation pathway, activated by growth factors. Both mTOR and MAP/ERK pathways are elevated in *Fmr1* knockout mice [47].

Pharmacologic treatment with mTOR inhibitor temsirolimus reversed cognitive impairment in animal model [43]. Chronic treatment of mouse model with mTOR1 inhibitor, rapamycin, did not reverse behavioral phenotypes however and had adverse effects [48]. Lovastatin, a drug typically used to treat cholesterol, reduces ERK signaling and corrected the fragile X phenotype by preventing learning deficiencies with early administration in rat model [49]. Human randomized controlled trial (RCT) using lovastatin in combination with normal language interventions did show a benefit, but this was equivalent to the benefit seen with a parent-implemented language intervention alone [50]. Metformin inhibits mTOR1 in addition to affecting other relevant pathways and is discussed in greater detail under the targeted treatments section.

MMP9

MMP9 is an extracellular protein involved in connective tissue, reproduction, cell migration, and memory found in overabundance in humans with FXS and *Fmr1* KO mice. In humans elevated MMP-9 is thought to contribute to neuropsychiatric changes and epilepsy [51], as well as connective tissue problems that can manifest in changes to the cardiovascular system, genitourinary system, muscles, and ligaments [52].

MMP-9 deficiency suppresses elevations in mTOR, elf4E, and Akt in mouse models [53], exerting effect on cell growth pathways. Genetic removal of MMP9 rescues the dendritic spine abnormalities, behavior abnormalities, mGluR5-dependent LTD, and macroorchidism seen in the KO mouse [53]. Pharmacologic inhibition of MMP-9 via minocycline corrected anxiety and cognitive deficits in mouse models [54]. Human studies with minocycline followed and were helpful as described below.

Cyclic AMP (cAMP) Alteration

Cyclic AMP -CREB is a learning and memory pathway. When FMRP is overexpressed genetically, cAMP levels increase [55]. cAMP levels are decreased in brains of the mouse and *Drosophila* FXS models [56]. Production of cAMP is also decreased in various human FXS cell lines including platelets [57], lymphoblastoid cells and fibroblasts [58], and neuronal human progenitor cells [56]. cAMP is degraded by phosphodiesterase activity, and PDE-4 is the most abundant cAMP-specific phosphodiesterase in the brains of mammals. Acute and chronic pharmaceutical inhibition of PDE-4, which increases cAMP, rescued memory and brain defects in the mouse and *Drosophila* models [59].

Wnt/ß-Catenin Signaling Pathway

The Wnt/ß-catenin signaling pathway is an overlapping dysfunctional pathway in both FXS and idiopathic ASD [60]. This pathway regulates the balance between proliferation and differentiation in cortical neural precursor cells during development, determining the number of neurons and their size within each brain region [61]. ß-catenin, an evolutionarily conserved transcription factor, is the effector molecule of the Wnt pathway regulating cell fate. In the absence of Wnt signaling, ß-catenin is degraded by the ß-catenin "destruction complex" composed of proteins including GSK-3. *Fmr1* KO mice have abnormally elevated "catenin destruction complex" protein (GSK-3β) [62], resulting in reduced ß-catenin and defective WNT signaling, altering neurogenesis. GSK3 influences social and anxiety-related behaviors in *Fmr1* KO mice [63]. Pharmacologic activation of Wnt using lithium and GSK3 inhibitors resulted in reversal of neurobehavioral phenotype in mouse models [62, 64]. Multiple medication's used for symptomatic treatment of FXS affect this pathway through Wnt activation including valproic acid, lithium, and SSRIs [65]. Methylphenidate and amphetamine also modulate GSK3 affecting pathway [66].

Channelopathies

Many ion channels are dysfunctional in *Fmr1* knockout mouse models, including action potential-related Na+ & K+, neurotransmitter-related-Ca2+ & K+, dendritic function-related-K+, Ca2+, and hyperpolarization-activated cyclic nucleotide-gated (HCN) channels, and glutamatergic/GABAergic receptors [19]. Data suggests there is increased excitability at the network level, driven by synaptic mechanisms and cellular excitability.

An example is the large conductance voltage and Ca2+-activated K+ (BK) channels directly regulated by FMRP and facilitates neuron excitability, action potential duration, and neurotransmitter release. It is notable in that it has a larger conductance compared to other channels. FMRP loss in KO mice and stem cell-derived neurons of humans reduce BK channel activity. Genetic upregulation of BK activity in mouse models corrected numerous behavioral phenotypes [67] as well as pharmaceutical intervention with BKCa modulator flindokalner (BMS-204352) [68]. No human studies have yet been carried out in those with FXS.

10.4 Symptomatic Treatments for FXS

There are no FDA-approved treatments for FXS, and treatment relies heavily on symptomatic treatment of FXS using multiple pharmaceutical types. Because few RCTs have been conducted for symptom management in individuals with FXS specifically, symptomatic treatment is often based on ASD-related research; standard behavioral treatment is often used off-label and includes stimulants for ADHD symptoms, antidepressants for anxiety and behavior, antipsychotics for aggression and sleep disturbances, and anticonvulsants for behavior and seizure disorders. Ongoing research continues to find potential mechanisms of action specific to FMRP-related pathways for many of these long-prescribed medications.

10.4.1 Stimulants and Alpha Agonists

Stimulants including methylphenidate and dextroamphetamine are typically used for the treatment of inattention and hyperactivity in children with FXS over 5 years old and are the most common medications prescribed [69]. This is a common symptom cluster, and approximately 59–70% of boys meet full DSM criteria for ADHD [70, 71]. An observational study including 12 boys with FXS treated with stimulants found improvements on academic measures higher than those reported for children with other intellectual disabilities [72]. The only published study evaluating behavior was a double-blind placebo-controlled trial of 15 individuals lasting 1 week that evaluated effectiveness of methylphenidate and dextroamphetamine, measured through parent and teacher behavioral questionnaires [73]. They found improvements in socialization skills and attention span. For children under 5 years old, stimulants are typically avoided as they can lead to irritability and behavioral problems.

A study by Hagerman et al. found children with FXS are uniquely affected by stimulants. They looked at electrodermal responses (EDRs) as a measure of sympathetic activity, the theory being that individuals with FXS have an overarousal to sensation that can be measured via increased sweating. Children with FXS and ADHD (15 males, 4 females) treated with stimulants had a significant decrease in EDRs compared to age- and IQ-matched control patients with ADHD (12 males 4 females), suggesting stimulants uniquely dampen sympathetic arousal in FXS through enhancement of inhibitory systems [74].

For children under 5, nonstimulant medications including alpha-adrenergic agonists clonidine and guanfacine are used for hyperactivity and attention. An observational study surveying parents of 35 children taking clonidine reported 63% of parents found clonidine was "very helpful" for their child, 6% found no effect, and 11% reported worse behavior [75]. The most common uses were for hyperactivity, aggression, anxiety, and sleep disturbances. The alpha-agonist guanfacine also improves ADHD symptoms [76, 77]. It is particularly recommended for children with FXS under 5 as it is less sedating than clonidine and has a longer half-life allowing BID dosing [78].

10.4.2 Antidepressants

Antidepressants such as selective serotonin reuptake inhibitors (SSRIs) are the primary treatment used for anxiety and emotional or behavioral problems. They are also beneficial for the treatment of compulsive behaviors, fixations, tolerability of environmental stimuli [69], and even language and cognition in children. About 40% of individuals with FXS over age 5 are prescribed SSRIs [79]. There is reason to consider SSRIs to be targeted treatment as serotonin is dysregulated in FXS [80]. Children with ASD have disrupted serotonin synthesis [81], decreased 5-HT receptor binding [82], and decreased 5-HT precursors [83], and adults with diets lacking precursor had worse symptoms [84]. A metabolic study found abnormalities in tryptophan metabolism among cell lines from patients with FXS and non-syndromic ASD [83]. SSRIs have been found to stimulate hippocampal neurogenesis [85] and stimulate brain-derived neurotrophic factor (BDNF). The use of SSRIs has been suggested as helpful for both ASD and related FXS (2006) [86].

Case report evidence suggests the use of sertraline in combination with other modalities can improve cognition and behavior in children with FXS [87]. Two controlled trials have now used low-dose sertraline in children with FXS and ASD as young as 2 years old, finding significant improvements only among FXS children. In 2016 a double-blind placebo-controlled RCT found improvements in expressive language development in young children with FXS given low-dose sertraline [88]. The trial employed a parallel two-arm design using sertraline 2.5 mg or 5.0 mg vs. placebo for 52 children aged 2–6 years and measured language via the Mullen Scales of Early Learning (MSEL), expressive language (EL), and Clinical Global Impression Scale. Significant improvements were also seen in motor and visual perception. The second trial involving 58 children with non-syndromic ASD did not show significant improvements [89]. The differential response is thought to be from increased anxiety among those with FXS compared to idiopathic ASD and thus benefiting more from SSRIs [90].

Anxiety is a prominent feature in FXS and often starts by age 2–4 years old. SSRI usage increases with age into adulthood [91]. Patient surveys found that SSRIs were helpful for anxiety or aggression in more than 70% of patients [92]. Approximately 20% of patients also experienced activation or hyperarousal as a side effect. This side effect can be managed with a decrease in dosage or discontinuation. Other side effects include sleep difficulties, nausea, diarrhea, dizziness, and weight change.

10.4.3 Antipsychotics

Antipsychotics including aripiprazole and risperidone are used for severe behavioral disturbances that pose a significant threat, including self-injurious behavior and aggression. Aripiprazole and risperidone have been shown to be effective in decreasing aggression in autism spectrum disorder. Three RCTs looking at risperidone in children with ASD over age 5 showed it significantly decreases tantrums, aggression, and self-injurious behavior [93, 94], and one showed communication improvements [95]. In FXS, aripiprazole was found to decrease irritable behavior by at least 25% and improve social responsiveness and hyperactivity among the 12 participants (11 male, 1 female) over 12 weeks [96]. Among boys with FXS, a retrospective analysis measured 33% of the 21 boys improved via the Clinical Global Impressions-Improvement (CGI-I) scale [97].

Side effects limit use and depend primarily on the unique receptor profile of each antipsychotic. These can include extrapyramidal symptoms (worse with typical antipsychotics), sedation, weight gain, constipation, and gynecomastia. Atypical antipsychotics are generally chosen in individuals with FXS due to fewer extrapyramidal side effects, although recent Ding et al. study suggested trifluoperazine as a potential future candidate based on transcriptome-based computation of ideal antipsychotic profiles [98].

10.4.4 Mood Stabilizers

Valproic acid and lithium are mood stabilizers that are used for behavior. Valproate has a dual function as an anticonvulsant for those with seizure disorders as well. Valproate has been shown to enhance the Wnt signaling pathway via inhibition of destruction of the protein complex GSK-3, providing a possible mechanism of action specific to FXS [99]. Lithium acts as an agonist at the Wnt signaling pathway, also inhibiting GSK-3 [100]. Lithium reverses phenotypes in *Drosophila* and mouse models (*dfxr* and *FMR1* knockout, respectively) effectively reducing abnormally increased mGluR-activated translation [101]. In humans, an open-label trial of lithium was conducted to evaluate safety and efficacy for 15 individuals with fragile X syndrome (age 6–23) titrated to blood levels of 0.8–1.2 mEq/L. They found significant improvements in behavior, but about half experienced side effects of polyuria/polydipsia ($n = 7$) or elevated TSH ($n = 4$) [102]. A controlled trial has yet to be carried out.

10.5 mGluR5: The Failed Translation of Preclinical Success

10.5.1 Target Supported by Theory

Efforts to correct the FXS mouse model phenotype predominantly targeted the rescue of synaptic plasticity, thought to be the primary deficit with loss of FMRP resulting in intellectual disability. The first two neurotransmitter pathways targeted were mGuR and GABA signaling, as both are involved in local translation of proteins necessary for plasticity and dysfunctional in FXS models. Significant preclinical evidence supported the theory that loss of synaptic plasticity was secondary to loss of FMRP repressor function leading to increased Glu-related protein synthesis. Following this theory, correction with mGlu antagonists or GABA agonists would be beneficial to those with FXS.

These were targeted treatments designed from the bottom up. The protein FMRP attenuates protein synthesis by inhibiting translation of many messages. Stimulation of mGlu results in de novo protein synthesis. *Fmr1* knockout mice have increased levels of mGlu-protein-dependent processes such as hippocampal LTP. And finally, genetic reduction of mGlu in some mouse and *Drosophila* models reverses the phenotype [40]. Scientists and clinicians were excited to translate these findings into humans. A pathophysiological mechanism was found, and drugs that worked on this pathway cured animal models of FXS. However, despite the hope and excitement, the results of the following human trials were disappointing, leading those in the field to take a hard look at why and how translational science can be improved for the future.

10.5.2 mGluR5 Human Trials

To tell the story of the mGluR5 antagonist human research, we can start with fenobam, the first mGluR5 antagonist to be evaluated for individuals with FXS with an open-label trial showing trend of improvement [103]. Following this exciting first study, mavoglurant (AFQ056), a selective mGluR5 antagonist used in preclinical studies, was chosen and evaluated for safety and behavior measures. Two well-powered phase II/III open-label studies, in adolescents (12–19 years, n = 119) and adults (18–45 years, n = 148) evaluating safety and behavior as a secondary endpoint, were conducted, but both were terminated early due to lack of efficacy in their core studies. A gradual behavioral improvement trend was measured via ABC-C and CBI-I greater than the placebo arm; however, trials were open label with no control group present, a significant patient dropout had occurred due to perceived lack of efficacy, and it was not deemed appropriate to test whether improvements were statistically significant [104]. In addition, mavoglurant was tested in two well-powered phase IIb RCTs lasting 3 months in adolescents (12–17 years, n = 139) and adults (18–45 years, n = 175) using ABC, CGI-I, and ET as outcome measures but found lack of efficacy in both trials for the primary outcomes [105].

Basimglurant, a negative allosteric modulator at mGlu5, was then tested after promising preclinical results and found to be well-tolerated in a low-powered 6-week study. Basimglurant went forward into two phase II RCTs for children (age 5–13) and adolescent/adults (age 14–50) but showed no improvement over placebo, and the Roche group terminated the development of basimglurant for FXS [106]. The results of mavoglurant and basimglurant suggest short-term mGluR5 antagonism is not the correct target for behavioral improvement. A major limitation of these studies is that cognition and language were not investigated, and their focus was on parent-reported behavioral outcome measures which are subject to placebo effect.

While human trials suggest mGlu5 antagonism is not as effective as hoped, GABA agonism has also been brought into human trials targeting multiple GABA receptors with drugs including arbaclofen, ganaxolone, gaboxadol, and acamprosate. These studies have met with mixed results.

10.6 Targeted Treatments Not Yet FDA Approved

Arbaclofen

Arbaclofen is an R-enantiomer of baclofen that acts as a potent and selective GABA-B receptor agonist. Arbaclofen differs from the racemic baclofen in terms of metabolism, activity, and side effects and undergoes renal elimination. Anecdotal clinical evidence supported the use of baclofen for behavior in FXS and ASD. Studies using FXS mouse models found baclofen protected against audiogenic seizures [107], and evidence that baclofen would suppress LTP plasticity in humans provided support that this drug should be studied in a clinical trial.

A phase II trial was then conducted. This trial lasted 1 month, included 63 participants (8 females) aged 6–40 years old, and looked at irritability as measured through the ABC [108]. Arbaclofen was well tolerated and showed improvement over placebo in post hoc analyses for the more socially impaired subgroup.

Following these results, two phase III studies were done on children and adolescents/adults looking at social avoidance instead of irritability [109]. The subjects of the studies were aged 5–11 and 12–50, respectively. The primary endpoints were measured via social avoidance subscale of the ABC-C; the secondary endpoints included other behavior scores of the ABC, Vineland-II, and CGI-S, CGI-I. A total of 119 subjects (26 females) completed the adult/adolescent study and 159 subjects (25 females) in the child study. Neither study improved social avoidance in FXS. However, the children receiving the highest dose of arbaclofen showed improvements in irritability ($p = 0.03$) and decreased parental stress (parental stress index, $p = 0.03$) with an effect size similar to SSRIs, but the adolescent/adult showed no improvements on any metric. Limitations cited for the study included lack of full enrollment and potential exaggeration of symptoms by families in order to meet the severity listed in the inclusion criteria.

Arbaclofen is overall well-tolerated at doses used for individuals with FXS with some side effects limiting use. In clinical trials, behavior was not improved compared to placebo. In post hoc, improvements were seen in full-methylation patients. The subsequent phase III trials had mixed results but suffered from similar problems with outcome measures as mGlu5 studies.

Ganaxolone

Ganaxolone is a synthetic analog of allopregnanolone, a natural neurosteroid, and is a GABA-A receptor positive allosteric modulator. Neuroactive steroids (NASs) persistently increase tonic inhibition [37] via GABA-A modulation. Ganaxolone was tested in a phase II double-blind RCT in ages 6–17 years, lasting 1.5 months with 59 participants [110]. Primary outcome measure included the Clinical Global Impression-Improvement (CGI-I) and secondary measures measuring anxiety with the PARS-R, ABC-C, ADAMS, and VAS. Overall they found a lack of efficacy. However, in post hoc analysis, subgroups with $IQ \leq 45$ and those with high baseline anxiety showed trended reductions in anxiety/hyperactivity, suggesting ganaxolone may benefit the most affected and anxious individuals.

Gaboxadol

Gaboxadol (OV101) is a δ-subunit-specific GABA-A positive modulator similar to acamprosate and ganaxolone. It was tested in 23 participants (13 adolescents, 10 adults) in a 12-week phase IIa RCT and was found to be safe and well-tolerated, with a positive initial signal trend for improvement based on caregiver/clinician assessments, but larger RCTs are needed to confirm results [111].

Trofinetide

Trofinetide (NNZ-2566) is a novel synthetic analog of insulin-like growth factors (IGF-1). Trofinetide has neuroprotective and anti-inflammatory properties and was developed to attenuate apoptotic neuronal death following TBI and stroke [112]. It has shown promising results in Rett syndrome with family/caregiver assessments and is now studied for treatment of FXS. An exploratory phase 2, multicenter, double-blind RCT was conducted in 72 adolescent and adult males with FXS over 28 days, and trofinetide was shown to be well-tolerated and safe, with primary side effects related to diarrhea [113]. Preliminary efficacy at higher doses was observed in caregiver/clinician assessments, suggesting potential for meaningful improvements in core symptoms.

Phosphodiesterase (PDE4D) Inhibitors

BPN14770 is a phosphodiesterase inhibitor studied in a recent phase 2 study. We learned that PDE4D is a key modulator of cAMP in the brain with the discovery of a rare missense mutation of PDE4D that caused a neurodevelopmental disorder [114], followed by PET imaging showing altered expression in the cortex and hippocampus [115]. BNP14770 is a selective inhibitor that in *Fmr1* knockout mice ameliorated behavioral phenotypes. The suspected mechanism is rewiring of excitatory cortical connections on dendritic spines.

The trial conducted was a single-center, phase 2, randomized, double-blind, placebo-controlled, two-period crossover study using 25 mg of BPN14770 BID vs. placebo to assess safety and efficacy [116]. Thirty adult participants were enrolled (average age mid-30s). All were male with a median IQ of 42.6 ranging from 24.6 to 66.2.

The study objective was to obtain preliminary assessment of the efficacy and safety of BPN14770 in adult males with FXS. Primary outcome was safety and tolerability, and secondary outcomes involved cognitive performance and then subjective caregiver and physician rating scales.

Medication was tolerated well by the 30 adult participants without any dropping out, and adherence was good. Cognition improvement was measured by the National Institutes of Health Toolbox Cognition Battery (NIH-TCB) and Test of Attentional Performance for Children (KiTAP). The drug was found to improve language function on both tests and the perceptions of caregivers/physicians. It also found overall improvements in daily functioning. These effects also persisted for at least 12 weeks after the last dose. Biomarkers collected were not significant but were directionally consistent with improvement in cortical circuitry and reduction in cognitive deficits seen. Further studies should include children to potentially improve neurodevelopment and women, who are differentially affected depending on the proportion of X-chromosome activation ratio. Overall BPN14770 increases cAMP levels in the brain through inhibition of PDE4D, overcoming one of the pathway deficits thought to be inherent to FXS.

10.7 Targeted Treatments Available Currently

10.7.1 Minocycline

Minocycline is a tetracycline antibiotic approved for the use of infections and treatment of severe acne. In patients with FXS, minocycline is thought to inhibit matrix metalloproteinaise-9 activity (MMP-9), which is normally overactive due to the absence of FMRP shown in the *fmr1* knockout mice model [117]. In a 2010 open-label trial of minocycline in 20 individuals with FXS, it was found to be well tolerated and had significant functional benefits, especially the ABC-C subscale and the CGI. A side effect of second-generation agents like minocycline can be diarrhea, and this was documented as the primary side effect during the length of the study [118]. A double-blind RCT of minocycline over 3 months also demonstrated some behavioral benefits in children with FXS, so it is sometimes used in clinic. However, the side effect of darkening of teeth if used in patients under 8 years or darkening of the skin can be a problematic side effect [119].

10.7.2 Metformin

Metformin is a biguanide AMP-activated protein kinase (AMPK) agonist that lowers MMP9 expression and targets ELF4E phosphorylation, PI3K/Akt/mTOR, and MAP/ERK signaling pathways approved by the FDA to treat type 2 diabetes. Preliminary studies including one with nine children between ages 2 and 7 treated clinically demonstrated improvements in language development and behavior in most patients [120]. Another 7 individuals with FXS were treated clinically (one with T2DM, three with Prader-Willi phenotype, two with obesity, and one with only FXS) and found improvements in irritability, hyperactivity, social responsiveness, and language as measured by family [121]. Both studies recommended clinical trials.

Two double-blind clinical trials in both the United States and Canada followed subjects 6 to 25 years old with the primary outcome of analyzing language deficits and expressive language sampling (ELS). Psychiatry, behavior, and eating were studied over 4 months. Results of the study will help determine the usefulness of metformin as a targeted drug for FXS (NCT03479476, NCT03862950).

10.7.3 Cannabidiol (CBD)

Cannabidiol is a cannabinoid with low affinity to CB1 and CB2 and partial agonist at 5ht-1 and has activity at μ- and *δ-opioid receptors* via allosteric modulator. It is FDA approved for treatment of seizures in Lennox-Gastaut syndrome, Dravet syndrome, and tuberous sclerosis complex. Based on clinical data in ASD and theory, case reports, and early clinical data in FXS, cannabidiol is a promising candidate currently studied primarily for mood and behavioral support. Data supporting the potential therapeutic role for CBD in ASD initially came from studies of children/adolescents treated for seizures and had improvements in autistic traits [122]. In patients with FXS, evidence from case series reported improvements in sleep, anxiety, language, and sensory processing for three patients with FXS taking oral doses of CBD between 32.0 and 63.9 mg daily, with two experiencing a reemergence of anxiety following cessation of treatment [123].

An open-label trial assessed CBD for safety, tolerability, and initial efficacy of transdermal CBD gel in 20 children and adolescents (aged 6–17 years) with FXS. They titrated doses from 50 mg to 250 mg BID maximum over 12 weeks.

They found a significant improvement in the Anxiety, Depression, and Mood Scale (ADAMS), behavior (ABC), quality of life (PedsQL), and CGI with most adverse events termed mild [124]. These translate to decreased anxiety, social avoidance, irritability, severity of symptoms, and increased quality of life.

After a successful phase 1/2 trial, Zynerba CBD gel moved on to a multicenter, phase 3 double-blind, placebo-controlled RCT in male and female children and adolescents with FXS (aged 3–17 years). This trial, called CONNECT-FX, demonstrated significant improvements in the primary outcome measure of social avoidance from the ABC only in the 80% of patients that had a full mutation that was >90% methylated [125]. These patients with FXS are the most affected although the overall group of 212 patients did not show a significant benefit. Therefore, a second phase 3 trial called RECONNECT, enrolling approximately 160 patients with 100% methylation and 40 with partial FMR1 methylation, is now taking place at multiple centers.

10.8 Lessons Learned

10.8.1 Easier to Cure the Mouse than the Human

Preclinical success in treatment of animal models, particularly with excitatory and inhibitory domains, has surprisingly led to many negative results when tested in human trials. These negative results suggest the mouse model may be overpredictive or not representative and insufficient alone to justify RCTs in humans. Using multiple genetically distinct animal models such as the rat or zebra fish models is likely to decrease error. Developing newer models of FXS, particularly those which capture repeat expansion and methylation, should be developed. Humans with FXS have a wide diversity of symptomatology secondary to genetic and environmental diversity as well as differences in methylation status; it follows that ideally animal models could reflect human diversity in methylation status. While it is not possible for animal and human studies to use the same outcome measures, overlap with some quantitative tests using the same parameters should be employed, such as EEG or imaging studies.

10.8.2 How to Avoid a Placebo Effect

A meta-analysis of FXS clinical trials between 2006 and 2018 found clinical improvements among trial participants taking the placebo to be common [126]. The placebo effect was strongest among caregiver-rated efficacy endpoints and stronger among adolescents and adults than children. They found no placebo effect among objective performance-rated measures suggesting objective performance-based outcome measures should be used. Several mechanisms for this include "placebo by proxy" induced by family members or clinicians, expectancy, and implicit learning [127]. Placebo effect size is also higher in open-label studies of individuals with ID and a genetic diagnosis. The use of open-label studies should be reduced to safety data or biological marker collection and replaced with placebo-controlled studies.

10.8.3 Quantitative Outcome Measures Are a Necessity

Past failure has led to important lessons for future clinical trials.

The Aberrant Behavior Checklist (ABC-C) has been widely applied as a key endpoint in multiple trials of intellectual disability described here as it correlates with behaviors and deficits in FXS. Its relevance to FXS has been questioned because of its dependence on parental report [128].

Caregiver-based measures are subjective and thus reduce the reliability of their use in study designs. Their use has been cited as a reason previous studies may have failed to show an effect [129]. The ADAMS, ABC, SRS, and VAS assessments are recorded by caregivers and thus could reflect this bias but are still indispensable tools when evaluating clinical manifestations of FXS. Clinician-based measures like the CGI-S, CGI-I, RBANS, and VABS-II could offer reduced bias.

Development of better endpoints is crucial and should reflect meaningful improvements in quality of life. There has been a search to correlate the many behaviors and cognitive deficits seen in FXS to objective and quantitative measures that can be used to measure changes with intervention.

The EEG and event-related potentials (ERP) allow researchers to look at information processing. The measure looks at the summation of electrical activity from groups of neurons in response to stimuli. These stimuli can be compared with typically developing individuals to characterize how sensory processing is different. Attention and memory formation can also be measured with this method [130]. Similar EEG findings are observed in humans [131] and rodent models [132, 133] of FXS and are correlated with behavioral sensitivity biomarkers.

fMRI is an objective imaging measure that has also been used to differentiate FXS and TD as well as ASD. fMRI creates brain maps that give information on structural mechanisms and functional aspects of the brain and give insight into connectivity networks. It has significant advantages in research as fMRI has widespread availability, is noninvasive, and has good spatial resolution [134].

PET scans allow for use of radioactive tracers to measure changes in metabolic processes as well as blood flow and other activities. Examples of its use in FXS include quantification of GABA-A receptors in the brains of patients with FXS [30], imaging of mGlu5 receptor expression in humans [135]. Limitations for these modalities are the difficulty in obtaining reliable measurements due to behavioral noncompliance.

10.8.4 Measuring Cognition with the NIH Toolbox

The ultimate goal of targeted controlled trials and treatments for FXS is improvement in ID, and validated measures for tracking treatment responses are necessary. Data on the sensitivity of interventions for the treatment of behaviors seen in FXS is currently limited in many of the past double-blind controlled clinical trials [136]. Previous cognitive tools have been postulated to not have been sensitive enough to track improvements for moderate to severe ID as males with FXS can have average IQ in the 40s. Following the limited success in translating successful preclinical glutamate and GABA-based pharmaceuticals to humans, calls were put forth to use the National Institutes of Health Toolbox Cognitive Battery (NIH-TCB) or NIH Toolbox. The toolbox has been extensively validated, including in the population with FXS, and works well for children, adolescents, and through adulthood. The toolbox has excellent test-retest stability [137] and has the benefit of minimizing floor and ceiling effects to allow a much better capture of cognition. Since its introduction, the cognitive benefit of BPN14770 phosphodiesterase inhibitor was measured using NIH-TCB along with KiTAP.

10.8.5 New Language Outcome Measures

Compared to individuals of normal development, those with FXS experience extreme difficulty with nearly all communication including receptive, written, and above all expressive language. Due to the presence of profound language deficits, novel ways to capture and measure language have been needed for evaluating improvements in clinical trials. Expressive language capturing techniques are required to accurately evaluate a baseline and improvements in individuals with FXS [138]. Expressive language sampling (ELS) is found to have good compliance and strong test-retest reliability across ranges of age and IQ, suggesting validity for measuring vocabulary, syntax, and unintelligibility particularly in those aged 6 to 23 years with FXS [139]. Language outcomes are popular selections for primary outcomes of current and past clinical trial studies, and the use of new language measures such as ELS in the clinical trial of metformin will support capture of improvements possibly missed in previous trials.

10.8.6 Multimodality Treatment Can Be Synergistic

Cannot limit to one drug The logistics of studying a single medication in the FXS syndrome population would be both problematic and ill-advised. Individuals with FXS can range from mildly to severely impaired and often have remarkable medical histories. These individuals are often taking medications for long-standing chronic health problems expected in those with FXS, including but not limited to AEDs for seizures and mood stabilization, psychostimulants for concentration and executive function, and antidepressants and antipsychotics for behavior and mood. Recruitment for studies would also suffer if there involved a requirement to cease medications that individuals already find helpful.

Synergy benefit In a perfect world, one medication or therapy would balance each of the multiple biological pathways altered by loss of FMRP in appropriate amounts. Many of the drugs studied here are targeting individual pathways affected, and layering these drugs to correct for multiple pathway disturbances is likely the most effective method of seeing improvement.

Difficulty The downsides of dealing with multimodal synergistic treatments are that they are difficult to study together. Polypharmacy and drug combinations should be studied together preclinically [18]. Combining drugs can affect their clearance and thus increase or decrease drug exposure.

In human studies, a sufficient number of participants are needed to power studies; dividing participants into different groups based on the types and numbers of medications they have taken is challenging. A proposed method of increasing power in lieu of larger trials is using a series of N-of-1 trials, which are crossover trials conducted in a single patient at a time, allowing for more flexibility with recruitment.

10.8.7 Earlier Treatments Can Build a Better Brain

Neurodevelopmental research on autism spectrum disorders [140], attention deficit disorders [141], and FXS [142] have found that the clinical symptomatology of individuals changes over the course of their lifetime. The protein FMRP has cumulative effects during neurodevelopment, and earlier intervention can mean larger disease-modifying effects.

There exist windows of plasticity or critical periods of brain development, where intervention should target for optimal outcomes. However, exact windows of plasticity are difficult to elucidate, and there are few clinical studies in children less than 5 years old. More data-driven studies are needed to give guidance on optimal periods of development to target with therapeutics [143].

The correct targets change over time with dynamic sensitivity during developmental periods. For example, inhibition of mGluR5 during critical period is sufficient to provide a persistent improvement in FXS mouse models, but chronic administration at other times produces treatment resistance and will no longer inhibit seizures, decrease hypersensitivity, and correct protein synthesis abnormalities [144]. Gene therapies also have windows to restore balance, and hearing in KO mice require immature inner ear target to be effective. For this example, the human fetus hears by gestational week 19, and this therapeutic window would close early second trimester [145].

In the case report, early intervention combined with targeted treatment promoted cognitive and behavioral improvements in two young children with fragile X syndrome [87]. The ethical question, however, is when you start a treatment earlier not only can it have larger positive effects but can also have larger detrimental effects. And with the current record for translational science in FXS, how do you decide which treatments should begin at earlier ages and possibly have larger positive effects?

10.9 Summary

Preclinical studies on pathophysiology and subsequent amelioration of symptoms in animal models pointed to developing therapeutics targeting the glutamatergic and GABAergic systems. Overall, human trials suggest mGluR5 inhibitors (basimglurant and mavoglurant) are ineffective at improving behavior. GABA-A agonists, however, such as arbaclofen, do have some efficacy at improving behavior in some subgroups.

Current management is primarily symptom-based and includes stimulants, SSRIs, mood stabilizers, and antipsychotics. The current trial of metformin may change the routine treatment of FXS if efficacy is demonstrated because this medication is available clinically. There is much hope that the PDE4D inhibitor, BPN1440, will continue to show cognitive improvements in childhood and adulthood and the results of the current trials are awaited with enthusiasm. In the near future gene therapy will remarkably change the prognosis of FXS.

10.10 Definitions/Assessments

Activation ratio – The fraction of normal FMR1 alleles on the active X chromosome.
Intellectual disability – Disability originating before age 18 in intellectual functioning (IQ generally <70) related to learning and problem-solving as well as adaptive functioning related to communication and living.
Diagnostic and Statistical Manual of Mental Disorders (DSM) – Used for the classification of mental disorders by standard criteria.
A*berrant Behavior Checklist-Community (ABC-C)* – Scale designed to measure behavior and psychiatric symptoms of individuals with intellectual disability across domains, irritability, social withdrawal, stereotypic behavior, inappropriate speech.
Anxiety, Depression, and Mood Scale (ADAMS) – Scale measuring manic/hyperactive behavior, depressed mood, social avoidance, general anxiety, and compulsive behavior for ages 10+.

Clinical Global Impressions Scale (CGI) – A three-item observer-rated scale that measures severity (CGI-S), improvement (CGI-I), and efficacy (CGI-E) over the previous 7 days based on symptoms, behavior, and function.
Vineland-II – Survey assessment used for measuring daily functioning and adaptive behavior including communication, daily living skills, socialization, and motor skills.
Visual Analogue Scale (VAS) – Scale measuring a characteristic that ranges on a continuum. VAS pain measures pain intensity on a scale of 010 with corresponding faces ranging from smiling (no pain) to tearful (worst pain).
Parental Stress Index (PSI) – Designed to measure magnitude of stress in a parent-child system, ranging from 36-item screening to 120-questions on domains including child and characteristics and situational and demographic life stress.
Social Responsiveness Scale (SRS) – Measures social ability of children age 2 years, 5 months to 18 years, primarily used with individuals with ASD.
The Children's Sleep Habits Questionnaire (CSHQ) – Parent-reported screening survey designed to assess sleep problems in children aged 4–10 years.
Repeatable Battery for the Assessment of Neuropsychological Status (RBANS) – Neuropsychiatric screening battery used to measure cognitive decline or improvement on five cognitive domains aged 12 to 90 years.

Acknowledgements We acknowledge the support of the Azrieli Foundation and also the MIND Institute IDDRC grant HD103526 in support of our research on those with fragile X syndrome.

Conflicts Dr. Hagerman has received funding from the Azrieli Foundation and Zynerba for treatment trials in those with fragile X syndrome.

References

1. Loesch DZ, Huggins RM, Hagerman RJ. Phenotypic variation and FMRP levels in fragile X. Ment Retard Dev Disabil Res Rev. 2004;10:31–41.
2. Thurman AJ, McDuffie A, Kover ST, Hagerman RJ, Abbeduto L. Autism symptomatology in boys with fragile X syndrome: a cross sectional developmental trajectories comparison with nonsyndromic autism spectrum disorder. J Autism Dev Disord. 2015;45:2816–32.
3. Cornish K, Steele A, Monteiro CRC, Karmiloff-Smith A, Scerif G. Attention deficits predict phenotypic outcomes in syndrome-specific and domain-specific ways. Front Psychol. 2012;3:227.
4. Cordeiro L, Ballinger E, Hagerman R, Hessl D. Clinical assessment of DSM-IV anxiety disorders in fragile X syndrome: prevalence and characterization. J Neurodev Disord. 2011;3:57–67.
5. Hess KL, Morrier MJ, Heflin LJ, Ivey ML. Autism treatment survey: services received by children with autism spectrum disorders in public school classrooms. J Autism Dev Disord. 2008;38:961–71.
6. Hagerman RJ, et al. Fragile X syndrome. Nat Rev Dis Prim. 2017;3:17065.
7. Tassone F, et al. FMR1 CGG allele size and prevalence ascertained through newborn screening in the United States. Genome Med. 2012;4:100.
8. Schneider A. MRNET Conference. Mol Syndromol. 2010;1:136–57.

9. Ludwig AL, et al. CNS expression of murine fragile X protein (FMRP) as a function of CGG-repeat size. Hum Mol Genet. 2014;23:3228–38.
10. The Dutch-Belgian Fragile X Consorthium et al. Fmr1 knockout mice: a model to study fragile X mental retardation. Cell. 1994;78:23–33.
11. Bakker CE, Oostra BA. Understanding fragile X syndrome: insights from animal models. Cytogenet Genome Res. 2003;100:111–23.
12. Lu Y-M, et al. Mice lacking metabotropic glutamate receptor 5 show impaired learning and reduced CA1 Long-term potentiation (LTP) but normal CA3 LTP. J Neurosci. 1997;17:5196–205.
13. Wan L, Dockendorff TC, Jongens TA, Dreyfuss G. Characterization of dFMR1, a Drosophila melanogaster homolog of the fragile X mental retardation protein. Mol Cell Biol. 2000;20:8536–47.
14. Kulkarni P, Sevilimedu A. The known unknowns: missing pieces in in vivo models of fragile X syndrome. J Rare Dis Res Treat. 2020;5:1–9.
15. Errijgers V, Kooy RF. Genetic modifiers in mice: the example of the fragile X mouse model. Cytogenet Genome Res. 2004;105:448–54.
16. Dockendorff TC, et al. Drosophila lacking dfmr1 activity show defects in circadian output and fail to maintain courtship interest. Neuron. 2002;34:973–84.
17. Gholizadeh S, Halder SK, Hampson DR. Expression of fragile X mental retardation protein in neurons and glia of the developing and adult mouse brain. Brain Res. 2015;1596:22–30.
18. Berry-Kravis EM, et al. Drug development for neurodevelopmental disorders: lessons learned from fragile X syndrome. Nat Rev Drug Discov. 2018;17:280–99.
19. Deng P-Y, Klyachko VA. Channelopathies in fragile X syndrome. Nat Rev Neurosci. 2021;22:275–89.
20. Huber KM, Gallagher SM, Warren ST, Bear MF. Altered synaptic plasticity in a mouse model of fragile X mental retardation. Proc Natl Acad Sci. 2002;99:7746–50.
21. Dölen G, et al. Correction of fragile X syndrome in mice. Neuron. 2007;56:955–62.
22. Burket JA, Herndon AL, Winebarger EE, Jacome LF, Deutsch SI. Complex effects of mGluR5 antagonism on sociability and stereotypic behaviors in mice: possible implications for the pharmacotherapy of autism spectrum disorders. Brain Res Bull. 2011;86:152–8.
23. Thomas AM, Bui N, Perkins JR, Yuva-Paylor LA, Paylor R. Group I metabotropic glutamate receptor antagonists alter select behaviors in a mouse model for fragile X syndrome. Psychopharmacology. 2012;219(47–58):47.
24. McBride SMJ, et al. Pharmacological rescue of synaptic plasticity, courtship behavior, and mushroom body defects in a drosophila model of fragile X syndrome. Neuron. 2005;45:753–64.
25. Fernandes G, et al. Correction of amygdalar dysfunction in a rat model of fragile X syndrome. Cell Rep. 2021;37:109805.
26. Wolfe SA, et al. FMRP regulates an ethanol-dependent shift in GABABR function and expression with rapid antidepressant properties. Nat Commun. 2016;7:12867.
27. Braat S, et al. The GABA A receptor is an FMRP target with therapeutic potential in fragile X syndrome. Cell Cycle. 2015;14:2985–95.
28. Braat S, Kooy RF. Insights into GABAAergic system deficits in fragile X syndrome lead to clinical trials. Neuropharmacology. 2015;88:48–54.
29. Reyes ST, et al. GABA measurement in a neonatal fragile X syndrome mouse model using 1H-magnetic resonance spectroscopy and mass spectrometry. Front Mol Neurosci. 2020;13:612685.
30. D'Hulst C, et al. Positron emission tomography (PET) quantification of GABAA receptors in the brain of fragile X patients. PLoS One. 2015;10:e0131486.
31. Zhang W, et al. GABAB receptor upregulates fragile X mental retardation protein expression in neurons. Sci Rep. 2015;5:10468.
32. Kramvis I, et al. Dysregulated prefrontal cortex inhibition in prepubescent and adolescent fragile X mouse model. Front Mol Neurosci. 2020;13:88.

33. Selby L, Zhang C, Sun Q-Q. Major defects in neocortical GABAergic inhibitory circuits in mice lacking the fragile X mental retardation protein. Neurosci Lett. 2007;412:227–32.
34. Centonze D, et al. Abnormal striatal GABA transmission in the mouse model for the fragile X syndrome. Biol Psychiatry. 2008;63:963–73.
35. Heulens I, D'Hulst C, Van Dam D, De Deyn PP, Kooy RF. Pharmacological treatment of fragile X syndrome with GABAergic drugs in a knockout mouse model. Behav Brain Res. 2012;229:244–9.
36. Olmos-Serrano JL, Corbin JG, Burns MP. The GABAA receptor agonist THIP ameliorates specific behavioral deficits in the mouse model of fragile X syndrome. Dev Neurosci. 2011;33:395–403.
37. Modgil A, et al. Neuroactive steroids reverse tonic inhibitory deficits in fragile X syndrome mouse model. Front Mol Neurosci. 2019;12:15.
38. Chang S, et al. Identification of small molecules rescuing fragile X syndrome phenotypes in Drosophila. Nat Chem Biol. 2008;4:256–63.
39. Khalsa JH. Medical and health consequences of Marijuana. In: Marijuana and the Cannabinoids. Humana Press, pp. 237–252. https://doi.org/10.1007/978-1-59259-947-9_10.
40. Fyke W, Velinov M. FMR1 and autism, an intriguing connection revisited. Genes (Basel). 2021;12:1218.
41. Varma N, Carlson GC, Ledent C, Alger BE. Metabotropic glutamate receptors drive the endocannabinoid system in hippocampus. J Neurosci. 2001;21:RC188.
42. García-Rincón D, et al. Contribution of altered endocannabinoid system to overactive mTORC1 signaling in focal cortical dysplasia. Front Pharmacol. 2019;9:1508.
43. Busquets-Garcia A, et al. Targeting the endocannabinoid system in the treatment of fragile X syndrome. Nat Med. 2013;19:603–7.
44. Maccarrone M, et al. Abnormal mGlu 5 receptor/endocannabinoid coupling in mice lacking FMRP and BC1 RNA. Neuropsychopharmacology. 2010;35:1500–9.
45. Jung K-M, et al. Uncoupling of the endocannabinoid signalling complex in a mouse model of fragile X syndrome. Nat Commun. 2012;3:1080.
46. Pirbhoy PS, et al. Increased 2-arachidonoyl-sn-glycerol levels normalize cortical responses to sound and improve behaviors in Fmr1 KO mice. J Neurodev Disord. 2021;13:47.
47. Osterweil EK, Krueger DD, Reinhold K, Bear MF. Hypersensitivity to mGluR5 and ERK1/2 leads to excessive protein synthesis in the hippocampus of a mouse model of fragile X syndrome. J Neurosci. 2010;30:15616–27.
48. Saré RM, et al. Negative effects of chronic rapamycin treatment on behavior in a mouse model of fragile X syndrome. Front Mol Neurosci. 2018;10:452.
49. Asiminas A, et al. Sustained correction of associative learning deficits after brief, early treatment in a rat model of fragile X syndrome. Sci Transl Med. 2019;11:eaao0498.
50. Thurman AJ, et al. Controlled trial of lovastatin combined with an open-label treatment of a parent-implemented language intervention in youth with fragile X syndrome. J Neurodev Disord. 2020;12:12.
51. Konopka A, et al. Matrix metalloproteinase-9 (MMP-9) in human intractable epilepsy caused by focal cortical dysplasia. Epilepsy Res. 2013;104:45–58.
52. Ramírez-Cheyne JA, et al. Fragile X syndrome and connective tissue dysregulation. Clin Genet. 2019;95:262–7.
53. Sidhu H, Dansie LE, Hickmott PW, Ethell DW, Ethell IM. Genetic removal of matrix metalloproteinase 9 rescues the symptoms of fragile X syndrome in a mouse model. J Neurosci. 2014;34:9867–79.
54. Bilousova TV, et al. Minocycline promotes dendritic spine maturation and improves behavioural performance in the fragile X mouse model. J Med Genet. 2008;46:94–102.
55. Berry-Kravis E, Ciurlionis R. Overexpression of fragile X gene (FMR-1) transcripts increases cAMP production in neural cells. J Neurosci Res. 1998;51:41–8.
56. Kelley DJ, et al. The cyclic AMP cascade is altered in the fragile X nervous system. PLoS One. 2007;2:e931.

57. Berry-Kravis E, Huttenlocher PR. Cyclic AMP metabolism in fragile X syndrome. Ann Neurol. 1992;31:22–6.
58. Berry-Kravis E, Hicar M, Ciurlionis R. Reduced cyclic AMP production in fragile X syndrome: cytogenetic and molecular correlations. Pediatr Res. 1995;38:638–43.
59. Choi CH, et al. PDE-4 inhibition rescues aberrant synaptic plasticity in drosophila and mouse models of fragile X syndrome. J Neurosci. 2015;35:396–408.
60. Salcedo-Arellano MJ, et al. Overlapping molecular pathways leading to autism Spectrum disorders, fragile X syndrome, and targeted treatments. Neurotherapeutics. 2021;18:265–83.
61. Kalkman HO. A review of the evidence for the canonical Wnt pathway in autism spectrum disorders. Mol Autism. 2012;3:10.
62. Franklin AV, et al. Glycogen synthase Kinase-3 inhibitors reverse deficits in long-term potentiation and cognition in fragile X mice. Biol Psychiatry. 2014;75:198–206.
63. Mines MA, Yuskaitis CJ, King MK, Beurel E, Jope RS. GSK3 influences social preference and anxiety-related behaviors during social interaction in a mouse model of fragile X syndrome and autism. PLoS One. 2010;5:e9706.
64. Yuskaitis CJ, et al. Lithium ameliorates altered glycogen synthase kinase-3 and behavior in a mouse model of fragile X syndrome. Biochem Pharmacol. 2010;79:632–46.
65. Baranova J, et al. Autism spectrum disorder: signaling pathways and prospective therapeutic targets. Cell Mol Neurobiol. 2021;41:619–49.
66. Mines MA, Jope RS. Brain region differences in regulation of Akt and GSK3 by chronic stimulant administration in mice. Cell Signal. 2012;24:1398–405.
67. Deng P-Y, Klyachko VA. Genetic upregulation of BK channel activity normalizes multiple synaptic and circuit defects in a mouse model of fragile X syndrome. J Physiol. 2016;594:83–97.
68. Hébert B, et al. Rescue of fragile X syndrome phenotypes in Fmr1KO mice by a BKCa channel opener molecule. Orphanet J Rare Dis. 2014;9:124.
69. Berry-Kravis E, Potanos K. Psychopharmacology in fragile X syndrome?Present and future. Ment Retard Dev Disabil Res Rev. 2004;10:42–8.
70. Sullivan K, et al. ADHD symptoms in children with FXS. Am J Med Genet Part A. 2006;140A:2275–88.
71. Baumgardner TL, Reiss AL, Freund LS, Abrams MT. Specification of the neurobehavioral phenotype in males with fragile X syndrome. Pediatrics. 1995;95:744–52.
72. Roberts JE, et al. Treatment effects of stimulant medication in young boys with fragile X syndrome. J Neurodev Disord. 2011;3:175–84.
73. Hagerman RJ, Murphy MA, Wittenberger MD. A controlled trial of stimulant medication in children with the fragile X syndrome. Am J Med Genet. 1988;30:377–92.
74. Hagerman RJ, et al. Influence of stimulants on electrodermal studies in fragile X syndrome. Microsc Res Tech. 2002;57:168–73.
75. Hagerman RJ, Riddle JE, Roberts LS, Breese K, Fulton M. Survey of the efficacy of clonidine in fragile X syndrome. Dev Brain Dysfunct. 1995;8:336–44.
76. Scahill L, et al. A placebo-controlled study of Guanfacine in the treatment of children with tic disorders and attention deficit hyperactivity disorder. Am J Psychiatry. 2001;158:1067–74.
77. Joseph A, et al. Comparative efficacy and safety of attention-deficit/hyperactivity disorder pharmacotherapies, including guanfacine extended release: a mixed treatment comparison. Eur Child Adolesc Psychiatry. 2017;26:875–97.
78. Hagerman RJ, et al. Advances in the treatment of fragile X syndrome. Pediatrics. 2009;123:378–90.
79. Valdovinos MG, Parsa RA, Alexander ML. Results of a nation-wide survey evaluating psychotropic medication use in fragile X syndrome. J Dev Phys Disabil. 2009;21:23–37.
80. Hanson AC, Hagerman RJ. Serotonin dysregulation in fragile X syndrome: implications for treatment. Intractable Rare Dis Res. 2014;3:110–7.
81. Chugani DC, et al. Developmental changes in brain serotonin synthesis capacity in autistic and nonautistic children. Ann Neurol. 1999;45:287–95.

82. Oblak A, Gibbs TT, Blatt GJ. Reduced serotonin receptor subtypes in a limbic and a neocortical region in autism. Autism Res. 2013;6:571–83.
83. Boccuto L, et al. Decreased tryptophan metabolism in patients with autism spectrum disorders. Mol Autism. 2013;4:16.
84. McDougle C. Effects of tryptophan depletion in drug-free adults with autistic disorder. Arch Gen Psychiatry. 1996;53:993.
85. Song N-N, et al. Reducing central serotonin in adulthood promotes hippocampal neurogenesis. Sci Rep. 2016;6:20338.
86. Zhou C, et al. Meta-analyses of comparative efficacy of antidepressant medications on peripheral BDNF concentration in patients with depression. PLoS One. 2017;12:e0172270.
87. Winarni TI, Schneider A, Borodyanskara M, Hagerman RJ. Early intervention combined with targeted treatment promotes cognitive and behavioral improvements in young children with fragile X syndrome. Case Rep Genet. 2012;2012:1–4.
88. Greiss Hess L, et al. A randomized, double-blind, placebo-controlled trial of low-dose sertraline in young children with fragile X syndrome. J Dev Behav Pediatr. 2016;37:619–28.
89. Potter LA, et al. A randomized controlled trial of sertraline in young children with autism spectrum disorder. Front Psych. 2019;10:810.
90. Rajaratnam A, et al. Review of autism profiles and response to sertraline in fragile X syndrome-associated autism vs. non-syndromic autism; next steps for targeted treatment. Front Neurol. 2020;11:581429.
91. Amaria RN, Billeisen LL, Hagerman RJ. Medication use in fragile X syndrome. Ment Heal Asp Dev Disabil. 2001;4:143–7.
92. Hagerman RJ, et al. A survey of fluoxetine therapy in fragile X syndrome. Dev Brain Dysfunct. 1994;7:155–64.
93. McCracken JT, et al. Risperidone in children with autism and serious behavioral problems. N Engl J Med. 2002;347:314–21.
94. Shea S, et al. Risperidone in the treatment of disruptive behavioral symptoms in children with autistic and other pervasive developmental disorders. Pediatrics. 2004;114:e634–41.
95. Nagaraj R, Singhi P, Malhi P. Risperidone in children with autism: randomized, placebo-controlled, double-blind study. J Child Neurol. 2006;21:450–5.
96. Erickson CA, et al. A prospective open-label study of aripiprazole in fragile X syndrome. Psychopharmacology. 2011;216:85–90.
97. Dominick KC, et al. Risperidone treatment for irritability in fragile X syndrome. J Child Adolesc Psychopharmacol. 2018;28:274–8.
98. Ding Q, et al. Transcriptome signature analysis repurposes trifluoperazine for the treatment of fragile X syndrome in mouse model. Commun Biol. 2020;3:127.
99. Wang L, Liu Y, Li S, Long Z-Y, Wu Y-M. Wnt signaling pathway participates in valproic acid-induced neuronal differentiation of neural stem cells. Int J Clin Exp Pathol. 2015;8:578–85.
100. Xia M-Y, et al. Activation of Wnt/β-catenin signaling by lithium chloride attenuates <scp>d</scp> –galactose-induced neurodegeneration in the auditory cortex of a rat model of aging. FEBS Open Bio. 2017;7:759–76.
101. Liu Z-H, Huang T, Smith CB. Lithium reverses increased rates of cerebral protein synthesis in a mouse model of fragile X syndrome. Neurobiol Dis. 2012;45:1145–52.
102. Berry-Kravis E, et al. Open-label treatment trial of lithium to target the underlying defect in fragile X syndrome. J Dev Behav Pediatr. 2008;29:293–302.
103. Berry-Kravis E, et al. A pilot open label, single dose trial of fenobam in adults with fragile X syndrome. J Med Genet. 2009;46:266–71.
104. Hagerman R, et al. Mavoglurant in fragile X syndrome: results of two open-label, extension trials in adults and adolescents. Sci Rep. 2018;8:16970.
105. Berry-Kravis E, et al. Mavoglurant in fragile X syndrome: results of two randomized, double-blind, placebo-controlled trials. Sci Transl Med. 2016;8:321ra5.

106. Youssef EA, et al. Effect of the mGluR5-NAM basimglurant on behavior in adolescents and adults with fragile X syndrome in a randomized, double-blind, placebo-controlled trial: FragXis phase 2 results. Neuropsychopharmacology. 2018;43:503–12.
107. Pacey LKK, Heximer SP, Hampson DR. Increased GABA B receptor-mediated signaling reduces the susceptibility of fragile X knockout mice to audiogenic seizures. Mol Pharmacol. 2009;76:18–24.
108. Berry-Kravis EM, et al. Effects of STX209 (arbaclofen) on neurobehavioral function in children and adults with fragile X syndrome: a randomized, controlled, phase 2 trial. Sci Transl Med. 2012;4:152ra127.
109. Berry-Kravis E, et al. Arbaclofen in fragile X syndrome: results of phase 3 trials. J Neurodev Disord. 2017;9:3.
110. Ligsay A, et al. A randomized double-blind, placebo-controlled trial of ganaxolone in children and adolescents with fragile X syndrome. J Neurodev Disord. 2017;9:26.
111. Budimirovic DB, et al. Gaboxadol in fragile X syndrome: a 12-week randomized, double-blind, parallel-group, phase 2a study. Front Pharmacol. 2021;12:2762.
112. Bickerdike MJ, et al. NNZ-2566: a Gly–Pro–Glu analogue with neuroprotective efficacy in a rat model of acute focal stroke. J Neurol Sci. 2009;278:85–90.
113. Berry-Kravis E, et al. A double-blind, randomized, placebo-controlled clinical study of trofinetide in the treatment of fragile X syndrome. Pediatr Neurol. 2020;110:30–41.
114. Lee H, et al. Exome sequencing identifies PDE4D mutations in acrodysostosis. Am J Hum Genet. 2012;90:746–51.
115. Wakabayashi Y, et al. Discovery, radiolabeling, and evaluation of subtype-selective inhibitors for positron emission tomography imaging of brain phosphodiesterase-4D. ACS Chem Neurosci. 2020;11:1311–23.
116. Berry-Kravis EM, et al. Inhibition of phosphodiesterase-4D in adults with fragile X syndrome: a randomized, placebo-controlled, phase 2 clinical trial. Nat Med. 2021;27:862–70.
117. Dziembowska M, et al. High MMP-9 activity levels in fragile X syndrome are lowered by minocycline. Am J Med Genet Part A. 2013;161:1897–903.
118. Paribello C, et al. Open-label add-on treatment trial of minocycline in fragile X syndrome. BMC Neurol. 2010;10:91.
119. Leigh MJS, et al. A randomized double-blind, placebo-controlled trial of minocycline in children and adolescents with fragile X syndrome. J Dev Behav Pediatr. 2013;34:147–55.
120. Biag HMB, et al. Metformin treatment in young children with fragile X syndrome. Mol Genet Genomic Med. 2019;7:e956.
121. Dy ABC, et al. Metformin as targeted treatment in fragile X syndrome. Clin Genet. 2018;93:216–22.
122. Fleury-Teixeira P, Caixeta FV, Ramires da Silva LC, Brasil-Neto JP, Malcher-Lopes R. Effects of CBD-enriched cannabis sativa extract on autism spectrum disorder symptoms: an observational study of 18 participants undergoing compassionate use. Front Neurol. 2019;10:1145.
123. Tartaglia N, Bonn-Miller M, Hagerman R. Treatment of fragile X syndrome with Cannabidiol: a case series study and brief review of the literature. Cannabis Cannabinoid Res. 2019;4:3–9.
124. Heussler H, et al. A phase 1/2, open-label assessment of the safety, tolerability, and efficacy of transdermal cannabidiol (ZYN002) for the treatment of pediatric fragile X syndrome. J Neurodev Disord. 2019;11:16.
125. Palumbo J, et al. Reconnect (ZYN2-CL-033): design of a phase 3 trial of ZYN002 Cannabidiol transdermal gel in children and adolescents with fragile x syndrome based upon learnings from connect-FX (ZYN2-CL-016). J Am Acad Child Adolesc Psychiatry. 2021;60:S258.
126. Luu S, et al. Response to placebo in fragile X syndrome clinical trials: an initial analysis. Brain Sci. 2020;10:629.
127. Curie A, et al. Placebo responses in genetically determined intellectual disability: a meta-analysis. PLoS One. 2015;10:e0133316.
128. Duy PQ, Budimirovic DB. Fragile X syndrome: lessons learned from the most translated neurodevelopmental disorder in clinical trials. Transl Neurosci. 2017;8:7–8.

129. Díaz-Caneja C, et al. A white paper on a neurodevelopmental framework for drug discovery in autism and other neurodevelopmental disorders. Eur Neuropsychopharmacol. 2021;48:49–88.
130. Knoth IS, Lippé S. Event-related potential alterations in fragile X syndrome. Front Hum Neurosci. 2012;6:264.
131. Wang J, et al. A resting EEG study of neocortical hyperexcitability and altered functional connectivity in fragile X syndrome. J Neurodev Disord. 2017;9:11.
132. Jonak CR, Lovelace JW, Ethell IM, Razak KA, Binder DK. Multielectrode array analysis of EEG biomarkers in a mouse model of fragile X syndrome. Neurobiol Dis. 2020;138:104794.
133. Kozono N, Okamura A, Honda S, Matsumoto M, Mihara T. Gamma power abnormalities in a Fmr1-targeted transgenic rat model of fragile X syndrome. Sci Rep. 2020;10:18799.
134. Glover GH. Overview of functional magnetic resonance imaging. Neurosurg Clin N Am. 2011;22:133–9.
135. Mody M, et al. In vivo imaging of mGlu5 receptor expression in humans with fragile X syndrome towards development of a potential biomarker. Sci Rep. 2021;11:15897.
136. Budimirovic DB, et al. Updated report on tools to measure outcomes of clinical trials in fragile X syndrome. J Neurodev Disord. 2017;9:14.
137. Hessl D, et al. The NIH toolbox cognitive battery for intellectual disabilities: three preliminary studies and future directions. J Neurodev Disord. 2016;8:35.
138. Berry-Kravis E, et al. Development of an expressive language sampling procedure in fragile X syndrome. J Dev Behav Pediatr. 2013;34:245–51.
139. Abbeduto L, et al. Expressive language sampling as a source of outcome measures for treatment studies in fragile X syndrome: feasibility, practice effects, test-retest reliability, and construct validity. J Neurodev Disord. 2020;12:10.
140. Taylor JL, Seltzer MM. Changes in the autism behavioral phenotype during the transition to adulthood. J Autism Dev Disord. 2010;40:1431–46.
141. Wilens TE, Spencer TJ. Understanding attention-deficit/hyperactivity disorder from childhood to adulthood. Postgrad Med. 2010;122:97–109.
142. Lee M, Martin GE, Berry-Kravis E, Losh M. A developmental, longitudinal investigation of autism phenotypic profiles in fragile X syndrome. J Neurodev Disord. 2016;8:47.
143. Razak KA, Dominick KC, Erickson CA. Developmental studies in fragile X syndrome. J Neurodev Disord. 2020;12:13.
144. Stoppel DC, McCamphill PK, Senter RK, Heynen AJ, Bear MF. mGluR5 negative modulators for fragile X: treatment resistance and persistence. Front Psych. 2021;12:718953.
145. Wang L, Kempton JB, Brigande JV. Gene therapy in mouse models of deafness and balance dysfunction. Front Mol Neurosci. 2018;11:300.

Chapter 11
The Difficult Path to the Discovery of Novel Treatments in Psychiatric Disorders

Valentin K. Gribkoff and Leonard K. Kaczmarek

Abstract CNS diseases, including psychiatric disorders, represent a significant opportunity for the discovery and development of new drugs and therapeutic treatments with the potential to have a significant impact on human health. CNS diseases, however, present particular challenges to therapeutic discovery efforts, and psychiatric diseases/disorders may be among the most difficult. With specific exceptions such as psychostimulants for ADHD, a large number of psychiatric patients are resistant to existing treatments. In addition, clinicians have no way of knowing which psychiatric patients will respond to which drugs. By definition, psychiatric diagnoses are syndromal in nature; determinations of efficacy are often self-reported, and drug discovery is largely model-based. While such models of psychiatric disease are amenable to screening for new drugs, whether cellular or whole-animal based, they have only modest face validity and, more importantly, predictive validity. Multiple academic, pharmaceutical industry, and government agencies are dedicated to the translation of new findings about the neurobiology of major psychiatric disorders into the discovery and advancement of novel therapies. The collaboration of these agencies provide a pathway for developing new therapeutics. These efforts will be greatly helped by recent advances in understanding the genetic bases of psychiatric disorders, the ongoing search for diagnostic and therapy-responsive biomarkers, and the validation of new animal models.

Keywords Psychiatric disorders · Drug discovery · Diagnostic biomarkers · Combination therapy

---

V. K. Gribkoff (✉)
Department of Internal Medicine, Section on Endocrinology, Yale University School of Medicine, New Haven, CT, USA
e-mail: valentin.gribkoff@yale.edu

L. K. Kaczmarek (✉)
Department of Pharmacology, Yale University School of Medicine, New Haven, CT, USA

Department of Cellular and Molecular Physiology, Yale University School of Medicine, New Haven, CT, USA
e-mail: leonard.kaczmarek@yale.edu

M. Macaluso et al. (eds.), *Drug Development in Psychiatry*, Advances in Neurobiology 30,
https://doi.org/10.1007/978-3-031-21054-9_11

11.1 Is There a Problem with the Discovery of New Therapeutics for Psychiatric Disorders?

Simply put, the answer is yes; there is a serious problem in this field, characterized by significant unmet medical need and a paucity of new and novel drugs at or near the approval stage. There are many reasons for this, which we will touch on below, all made more difficult by decades-old nature-versus-nurture arguments about disease causality. While there has been increasing and compelling evidence that major psychiatric disorders have a genetic linkage, expression of which can be influenced to a greater or lesser degree by environmental factors, there is little agreement on how to translate this complex data set into testable hypotheses and novel therapeutic approaches, and there is widespread recognition that there is a problem [1–6]. For these and other reasons, there is significant reluctance on the part of the pharmaceutical industry to venture into new territory in the search for new medications. Much of the recent history of psychiatric drug discovery has been characterized by attempts to exploit known and tractable targets, regardless of their shortcomings, in an attempt to reduce the burden of unwanted side effects, increase patient compliance, and establish a foothold in the market.

11.1.1 Why Are Most Drugs for Psychiatric Disorders Similar?

While there is hope for a greater understanding of the neurobiology of psychiatric illness, there are also factors that make this quite challenging. We will use the example of schizophrenia as our primary example throughout this review. Most in the field would now agree that successful treatment for schizophrenia must take into account the cellular etiology of what clinically presents as chronic and serious behavioral disturbance. For decades, treatment options have relied upon the initial and largely serendipitous (wholly accidental) discovery of a drug. Serendipity, when paired with skilled observation and study [7, 8], later developed into classes of similar drugs that could dampen the dysfunctional behavioral effects and, to a much lesser extent, improve the disordered thinking associated with the illness. In fact, their discovery helped propel forward the first and prevailing comprehensive theories about the neurobiological causes of schizophrenia [9–11]. The early finding that affective disorders and schizophrenia involved disruptions in biogenic amine-mediated neurotransmission or neuromodulation resulted in many reasonably effective drugs [12]. Attempts to discover new, more effective, and more specific treatments have, however, not been very successful, and the overwhelming numbers of undertreated or untreated patients have contributed to significant societal problems, including homelessness. Nevertheless, new potential targets for drugs have been identified, and there are attempts underway to shift the focus of new discovery efforts towards these new targets.

While there are a number of new treatments, primarily small molecule drugs, in the pipeline for serious psychiatric disorders (we will discuss some of these below), there are a number of issues that contribute to a lack of new treatments that incorporate the knowledge gained from advances in the neurobiology, specifically new insights into the genetic causes of most of the major psychiatric disorders. We are not psychiatrists, and this chapter does not attempt to add to the growing body of evidence concerning the polygenic nature of major psychiatric disorders, nor will it offer insights into specific new treatment approaches, although some will be offered as examples. Instead, as neurobiologists, with specific expertise in neurophysiology, neuropharmacology, and central nervous system (CNS) drug discovery [13], we will briefly discuss the process of drug discovery and development. We will point out some of the difficulties in drug discovery in general, discuss the inherent obstacles to discovery of central nervous system (CNS) drugs in particular, and use the psychiatric disorders to illustrate the very difficult case history of CNS disease targets.

We will approach the issue of the neurobiological causation of the major psychiatric disorders as settled science, although the precise mechanisms and specific systems involved in illness causation, and the degree to which environmental factors can influence etiology, may not be well understood in many cases [2, 14–21]. Just as with many cancers and other diseases, we believe that disorders such as schizophrenia, most forms of autism, attention deficit-hyperactivity disorder (ADHD), bipolar disorder, and major depression result from a combination of complex genetic factors and environmental/developmental influences. In this case, they are ultimately expressed as behavioral dysfunction resulting in significant impact on an individual's quality of life and ability to function well in their specific milieu. While drugs have been approved over the years to treat many of these disorders, their effectiveness is often questionable, and their significant side effects, in many cases, contribute to difficult prescription paradigms and often low levels of patient compliance. New therapeutic directions are badly needed, but progress has been slow and halting.

11.2 Drug Discovery

11.2.1 Defining Drugs and the Limitations of Therapeutic Benefit

Drugs are essentially anything introduced into the body with the express purpose of treating a disease or condition. This does not mean they are effective or approved; this is simply their stated purpose. This is why we have regulatory agencies, such as the Food and Drug Administration (FDA) in the USA, the Medicines and Healthcare Products Regulatory Agency (MHRA) in the UK, and the European Medicines Agency (EMA) in the EU, and their many counterparts in the rest of the world. It is their purview to examine the evidence for safety and efficacy of new drugs, to grant marketing approval for new medications and therapies, and to review, on an ongoing basis, the evidence of their safety and efficacy in real-world patient populations after approval.

Historically, drugs have primarily been small synthetic or naturally derived molecules, including such biologically derived molecules as hormones and plant and fungal alkaloids. Over the past several decades, these classes have been enriched with antibodies, other immunotherapies, and synthetic peptides and proteins. Most recently, synthetic genetic materials, including messenger RNA (mRNA), specific oligonucleotides, and DNA sequences for direct and indirect gene therapy, have been included in this mix. Soon, gene editing, still in its therapeutic infancy, may be added. As a mechanism for the potential correction of identified and important mutations, it brings its own substantial set of ethical dilemmas.

Regardless of the specific molecule or approach, most drug therapies are not designed to "cure" a disease but rather to decrease the deleterious effects, treat symptoms, and possibly extend disease-free periods in a patient's life. In most cases, at least up to the present, there have been few instances where an actual alleviation of a disease would even be possible; notable exceptions are anti-infective therapy, where the desire and target is the total eradication of the disease vector, and anticancer therapy, where in some cases early and aggressive treatment can effectively eliminate all traces of tumor in the patient. Most other drugs are intended to treat symptoms of the disease, as is the case for anti-epileptic drugs (AEDs), which by raising seizure thresholds can reduce the likelihood of additional seizure episodes. Probably the most classic example of this, and one with classical roots and current social consequences, is the treatment of pain by opiates. The effects of opium on pain have been known for millennia, as have the consequences of using it, and since the development of derivatives in the nineteenth century, opiates have had an important place in pain treatment. To this day (and this is a major failure thus far in the advancement of the field), opiates remain the single most effective class of drugs at treating severe pain, both acute and chronic, despite their obvious drawbacks and our dramatic increases in knowledge about the etiology of pain and the identification of possible new targets [22–25]. Opiates, however, do not directly address the cause of pain, only the perception of pain, and equally notable, pain is not a disease but rather the result of other disease, injury, or inflammation.

Non-steroidal anti-inflammatory drugs (NSAIDS) may not treat pain as well as opiates nor do they provide more than temporary symptomatic relief (unless inflammation is causal rather than a secondary effect), but they actually target some of the root causes of pain with fewer of the behavioral side effects. In contrast, there are many examples that are similar to the treatment of pain with opiates, including most of the drugs currently used to treat major psychiatric diseases or conditions. Frankly, this is true of most CNS drugs, including AEDs. In general, psychiatric drugs rely upon the temporary manipulation of the levels of neurochemicals thought to be in either excess or deficient in a condition, with the hoped-for end result being a normalization of disordered thought and behavior. As such they are not generally thought to have significant effects on the disease mechanisms themselves. There has never been the expectation that a course of drug treatment for a major psychiatric disorder, once established, would produce a "cure"; the hope has always been that the effects of the disease would be blunted sufficiently so that the patient would be able to function while continuing therapy, that their quality of life would be improved, and they would have the lowest level of deleterious side effects. The expectation in many such scenarios is that successful drug therapy is therapy for life, or at least until other factors, such as aging, may reduce the severity of some symptoms, as is frequently observed with positive symptoms in schizophrenia [26]. In a cynical vein, this is ideal from the perspective of the pharmaceutical industry, regardless of target, because establishment of such long-term if not lifelong drug therapy guarantees a long-term customer for any approved and established medication. The actual situation in psychiatric drug therapy is that while some patients are helped immensely by existing drug therapies, many are not, and many more do not comply with required drug regimens or are afflicted with debilitating side effects that the patients themselves consider to be as bad or worse than the disease. While new targets have been suggested in the literature, this has only recently started to be successfully exploited.

11.2.2 The Process of Drug Discovery

While the specifics of all drug discovery are dependent on the target and the disease, the basic structure of modern campaigns to find new drugs is similar regardless of target. Prior to a rigorous understanding of the biological bases of disease, therapeutic effects of drugs for any disease were discovered by extensions from folk medicine accompanied by careful observation or, as occurred frequently, purely by chance. In many cases drugs were tested for many potential uses, before eventually settling on a disease or condition where the compound had demonstrable efficacy (something that still occurs today frequently enough to have its own category of "drug repurposing"). This latter pathway was followed in the case of the discovery of the first antipsychotic medication, the phenothiazine derivative chlorpromazine (Thorazine) [9, 27], which was initially examined for its potential as an antimalarial drug, an antihistamine, an augmenter of anesthesia, and eventually as an antipsychotic.

Drug discovery often still depends on blind luck and persistence augmented with skilled observation. Buspirone, a serotonin 1A (5HT1A) receptor partial agonist (among its other receptor affinities) that was being developed by the Bristol-Myers Company as an antipsychotic, went on to be approved as an antianxiety treatment without sedative side effects [28, 29]. One of the authors (VKG) early in their career was a research scientist at the Bristol-Myers Company CNS Drug Discovery Division (later the Bristol-Myers Squibb Company, BMS), working with the group of researchers who first synthesized, discovered, and developed buspirone. The story of its discovery as an antianxiety drug is another example of serendipity versus targeted discovery. While feeding monkeys in a preclinical trial of buspirone, it was noted that monkeys dosed with the drug did not display their usual levels of anxiety and agitation while interacting with veterinary staff during feeding, a finding that was correlated with results in other animal models of anxiety [29]. While there are many such stories in the history of drug discovery, the reliance on chance or luck is not the way such campaigns are, or should be, structured today. We are all prepared to take advantage of serendipity but should strive to discover new drugs through better understanding of disease biology, genetic linkage, and with the computational and medicinal chemistry skill necessary to maximize these.

There is another lesson to be learned from such stories. The standard treatments for anxiety, primarily benzodiazepines, had sedative and even mildly euphoric effects (with accompanying abuse potential), while buspirone was relatively free of these effects. Much of the assessment of the severity of anxiety and its amelioration through pharmacotherapy is based on personal reporting [30]. Lacking obvious evidence of drug effect, patients were often convinced that the drug was not working, even though standard measures of anxiety were reduced. Similar problems exist in other areas of CNS drug discovery, including pain management.

The discovery of a new drug that targets a previously treatment-resistant disease and that is approved and advanced into humans can trigger a natural scramble to exploit these findings and create new and better versions of the same drug, based on the presumed mechanism of action (MOA) or the compound class/pharmacophore. This is the way that drug discovery for psychiatric disorders has largely progressed, with the development of "me-too drugs." While this approach is not unique, it has perhaps been more prevalent in this field than in most other areas of medicine [1, 9]. In other words, most drugs with some effect on one or more major psychiatric disorders (or cluster of symptoms) were discovered initially by accident or an observation-based hunch. The subsequent improvements resulted from the demonstration of a possible (and the word *possible* is key) MOA of these drugs, development of in vitro assays based on the putative molecular targets of these compounds, and, where possible, testing in animal models thought to have predictive value for drug efficacy in humans. While this is not unusual in the history of the hunt for effective drugs in any area, there are some unique aspects of psychiatric drug discovery that make this approach less effective than in most other areas.

Modern successful drug discovery campaigns rely on hypothetical MOAs that identify one or more molecular targets believed to be involved in the etiology of the disease or whose manipulation may modulate other targets to produce positive results in the disease phenotype. The key for success here is the quality of the evidence that the selected molecular target, whether a G-protein-coupled receptor (GPCR), an enzyme, an ion channel, or other molecular entity, such as a gene or its regulatory mechanisms, is accurately associated with disease etiology or progression.

While the discovery of anti-infective drugs is far from a trivial endeavor, it represents an ideal model for such an MOA-based approach [31–33]. Knowledge of the biology of the infection vector(s) results in the identification of drug targets. These may, for example, involve mechanisms of cell attachment, cell entry of the organism, or its genetic material. In turn, target identification drives a search for molecules that can block some key feature of the target's actions. Molecules can be designed in silico given sufficient structural knowledge of the drug's target and/or by physical screening campaigns using large molecular libraries representing great structural diversity. Initial in vitro assays can be constructed that test the interaction of the compounds of interest with the molecular target, and cellular assays can be designed to test those compounds that pass the initial in vitro assays against the infection vector. Once highly potent compounds that act on the target and produce the desired effects on the bacterium, virus, or parasitic organism have been identified, the major focus shifts to the safety of the compound (and the safety of affecting the particular target) [31–33]. While nothing is certain in drug discovery, one can more readily assume that you have a good chance of escaping side effects associated with a *specific* target when that target is unique or structurally very different in the invasive organism than when the target is widely expressed and functionally critical in human physiology.

With apologies to those seeking to discover new anti-infective agents, whether classical drugs, vaccines, or other forms of therapeutic intervention, because all of what we just described is difficult and always riddled with opportunity for frustration and failure, this is about as straightforward a drug discovery process as one can imagine. Again, if you can identify a unique feature of the biology of the invasive agent, you have a much higher probability of eventually finding a therapeutic approach that targets this unique biology and that controls or kills the invasive agent with few or no significant effects on the host. Given this, then why have the number of new antibiotics not kept up with disease resistance? The drop-off in the search for new antibiotics is driven by lack of investment [34], partly because, in the absence of drug-resistant strains of invasive agents, drugs that cheaply and quickly cure a disease lack as much market potential as other types of drugs. Specifically, after a brief course of treatment, the patient no longer requires the drug. The economics of drug discovery are a reality in every area and are among the most important factors in the pace of discovery in a particular therapeutic area. Larger-scale new efforts to discover effective anti-infective drugs have, however, been initiated, driven by the discovery of new sources of wide-spread infections and drug-resistant strains of infective agents. To be successful, these often require significant contributions of public funds in their development. In contrast, the search for new approaches to psychiatric drug discovery is about as far from this linear model as you can get, largely because of our lack of detailed understanding of the etiology and underlying neurobiology of these disorders. Nevertheless, it is still driven by the same market forces that shape all drug discovery programs.

11.2.3 The Economics of Drug Discovery and Development

Drug discovery, which is largely preclinical, and drug development, which is largely a multistaged clinical program ideally leading to regulatory approval for the manufacture and sale of a new drug, are expensive and time-consuming endeavors. Recent estimates for the costs of new drug approvals vary from about $1.3 billion to ~$2.6 billion, factoring in cost contributors such as failures and the money needed to finance the discovery and development campaigns [13, 35]. This varies, of course, depending on the size and complexity of the clinical trials required for regulatory approval, the difficulty of the target in the initial discovery campaign, the cost of goods (active pharmaceutical ingredients or API), the size and complexity of the companies undertaking the project, and the difficulties in API formulation, among many other factors. It generally does not include the subsequent post-marketing costs of manufacturing, distribution, and advertising. Failure rates are always quite high and significantly higher in some therapeutic areas than others. This can be due to off-target pharmacology, lack of efficacy in predictive models (if such models exist), poor handling of the drug, and the like. Clinical failures result from many factors but in general include poor efficacy in the disease, as has occurred in the area of neurodegenerative disorders such as Alzheimer's disease [36–39], poor human pharmacokinetics or toxicology, and unacceptable levels of side effects that affect safety and patient compliance.

It should be noted that all of the factors that increase the time of development and cost of bringing a drug to the patient (the market) have to be assessed in the light of a limited time frame of market exclusivity. This is determined by patent expiration, including attacks on seemingly valid patents, essentially attempts to invalidate patents to open them up to early generic competition. It is also affected by any extensions to market exclusivity that can be obtained based on the disease targeted (e.g., orphan disease status) or on recoupment of time spent in the approval process (e.g., Hatch-Waxman extensions). Given the expense and complexity, it should come as no surprise that pharmaceutical/biotechnology companies weigh these factors carefully in deciding which therapeutic areas they wish to pursue and which diseases and conditions, and their associated targets (established or hypothetical), they select for drug discovery campaigns. In the cold light of day, these are business decisions. In the modern era, small companies have been formed in the biotechnology/biopharmaceutical arena to approach new and riskier diseases and targets. The high failure rates of candidate drugs, particularly those that are not exploiting proven MOAs, make obtaining the funding for these high-risk ventures more difficult and competitive. The result of all of this, of course, is expensive drugs, particularly in the US, where regulation of drug prices is minimal, allowing the generation of larger offsetting profits. The economic inertia for approaching difficult diseases and conditions, using novel and unproven approaches, is therefore quite substantial.

These economic factors have the result that some areas of significant medical need are largely ignored by those with the resources required to attempt novel drug discovery and that many other areas, such as psychiatric diseases, are served largely by the near constant, if low-level, infusion of new chemical entities that are effectively "me-too" drugs. Such drugs *may* represent a significant improvement in therapy, perhaps by demonstrating better general efficacy or side-effect profiles, or superior efficacy in a specific subset of a patient population. They are unlikely, however, to represent game-changing new ways of treating patients, particularly those who are either refractory or resistant to existing medications.

11.3 The Unique Problems of CNS Drug Discovery in General and Psychiatric Drug Discovery in Particular

11.3.1 Why Is CNS Drug Discovery Difficult?

CNS drug discovery presents greater difficulty than comparable efforts in most other organ systems [13]. While brain penetration definitely remains on the list of reasons for this because of the blood-brain barrier (BBB) [40], there are a number of additional reasons that are substantially more important. With small molecule drugs, if you can prove in principle that engagement with a particular CNS target will prove efficacious, then it is likely that a directed medicinal chemistry effort will eventually be able to create a molecule that can get into the CNS at adequate levels. This has been less true with other classes of drugs, such as peptides, but this too is changing. It makes the effort more expensive and fraught with greater risk, which can narrow the field of those willing to undertake (and underwrite) that risk, but it can be accomplished. Moreover, significant recent work has helped to improve brain penetration (and models of the BBB) and thereby improve candidate selection [40–42]. The other problems, however, can prove much more difficult. We have written about this previously [13] but will briefly review these issues here.

Almost any disease or condition thought of as primarily a peripheral (to the CNS) problem can have a CNS counterpart. These include infection, cancers, cardiovascular disease, and the like. There are, however, a large number of conditions that are primarily CNS in origin and which are very difficult to approach from a therapeutic standpoint. Even those CNS diseases that have a non-CNS counterpart, such as infections and cancers, have an added layer of difficulty, not only because of the BBB but also because of the limited ability of neurons to regenerate in the adult. This is compounded by the prolonged development of the CNS and the sheer complexity of the interconnections in the CNS. Any agent with a propensity for damaging developing neurons or disrupting the complex network of neuronal interconnections is problematic and may result in failure of the candidate or even an entire class of targets. Use of any CNS active agent in a patient whose CNS has not yet fully developed, which we now understand to mean well into adolescence and beyond, must be undertaken very carefully. Drug effects, and more importantly side effects, may be different in these patients and may result in limitations in use. For example, the use of some antidepressant therapeutics in teenagers and young adults has required "black-box" warnings of increased potential for suicide [43, 44]. Additionally, many CNS drugs have as their primary target one or more receptors or alter the activity of an ion channel. Because many systems are likely to use the same neurotransmitters or ion channels for normal neuronal functions not impacted by the disease, the likelihood of serious or at least dose-limiting side effects that reduce the potential for efficacy is quite high. Drugs that target these proteins must be cleverly designed.

Some diseases that are largely or wholly unique to the nervous system can be surprisingly difficult to approach, despite the fact that we know (or think we know) a great deal about them and that *seemingly* valid cellular and animal models exist. These are a subset of neurological and neuro-cardiovascular disorders, including acute ischemic stroke and traumatic brain injury, seizure disorders, and some of the neurodegenerative disorders. For example, while a number of effective drugs do exist to treat some seizure disorders, a large percentage of patients are completely refractory to drug therapy, and another large percentage of persons with epilepsy are at risk of breakthrough seizures, requiring constant monitoring of circulating levels of several anti-epileptic drugs (AEDs) [45–47].

In the case of neurodegenerative disorders such as Alzheimer's disease (AD), motor neuron diseases such as amyotrophic lateral sclerosis (ALS), and Parkinson's disease (PD), recent years have seen the development of sophisticated models of disease etiology, largely based on genetic linkage studies in family clusters [13, 48–50]. From these findings, a number of hypotheses have been developed, therapeutic targets have been identified, and animal models of genetically linked disease have been created to aid in drug discovery and development. Despite the recent and controversial approval of an AD therapy [51], all of the drugs based on the central hypothesis that some form of errant handling and/or deposition of β-amyloid, or more recently, aberrant tau protein, underlies the progressive pathology of AD have thus far been found to be ineffective. These include drugs that indeed did reduce the target in the CNS [36–39, 52]. Much the same is true of drugs targeting ALS and other neurodegenerative disease. Why is this the case? To a large degree, above and beyond the shortcomings of any specific drug (its potency, efficacy, and ability to reach and interact with its target), much of the reason for failure lies in the likelihood that such diseases have multiple causal parameters, particularly in the large majority of so-called "sporadic" or "idiopathic" cases where no genetic linkage has yet been demonstrated. The logical lapse is the assumption that such a disease can be treated successfully with a drug that targets only a single receptor, enzyme, or protein identified in a small population of genetically linked cases. The animal models created to test putative therapeutics are usually based on identified familial mutations, and therefore their efficacy is dependent upon a particular underlying pathogenesis [48–50, 53–55]. If the disease is the result of many factors, both environmental (including toxicological) and a more complex set of genetic linkages, this approach is likely to fail. The failure results from the assumption that a single drug must treat all, or at least a significant proportion, of the patient population (more on this below).

The current situation is that the approved drugs for many neurodegenerative disorders are symptomatic treatments, such as acetylcholinesterase inhibitors and a single NMDA antagonist for AD and dopamine receptor ligands for PD, and a paucity of drugs that demonstrably affect disease trajectory. Multiple sclerosis is an exception, where treatments targeting a form of the disease are much more effective, and there is now hope that future treatments will actually stop disease progression [56].

Given the expense, in terms of both time and money, of the development of a new drug, it is no surprise that the hope is that a relatively large cohort of patients can be treated with this single agent. A drug that proved effective against a particular genetic variant in AD as demonstrated in an animal model may very well prove effective in treating those afflicted with that particular disease-causing mutation. By and large, however, drugs of this class have not been tested, because specific mutations are present only in small patient numbers and, given the societal burden of AD, the desire on the part of the medical establishment and pharmaceutical companies is to treat the wider population. The lack of a known mechanistic link between the genetic mutation and the cause of the majority of neurodegenerative disease cases was largely overlooked until a few years ago, and this attitude persists in some form or other to this day. Drugs developed against such targets have been tested against large cross sections of patients with AD or ALS, and, for the most part, the result has been failure. Following a complete lack of success, or results that were significantly below stated goals, there is a flurry of attempts to explain the lack of apparent efficacy. In the case of AD, this has often taken the form of discussions of the need to intervene prior to the onset of frank symptoms. This is probably true; there is usually a need to intervene at the earliest possible time point in any disease process, perhaps even before the appearance of any recognizable symptoms [36, 57, 58]. Where this has been tested explicitly in AD, however, the results have been hopeful but, at best, modest and often unconvincing [51, 59–62]. Clearly what is needed, and this is an object lesson for psychiatric drug discovery as well, is an appreciation of the heterogeneous nature of CNS disease. It may prove more fruitful in the long run to obtain small victories by targeting specific subsets of each disease, such as patients with specific identified mutations or those presenting with some discrete biomarker that can be linked to the disease or therapeutic intervention. With the knowledge obtained in these efforts, combination therapy approaches that treat larger subsets of the disease population could be attempted. Alternatively, the economics of creating precision drugs for treating smaller subsets of patients could be rewarded, perhaps through pricing regulations or patent life extensions.

Genetic linkage with small, identified cohorts of patients is an approach that has already been initiated in certain forms of epilepsy, for example, where gain-of-function and loss-of-function mutations in specific potassium ion channels have been associated with a number of severe forms of neonatal seizure disorders as well as generalized epilepsy. Understanding the role of those ion channels in neuronal excitability and understanding how the mutations in these proteins promote seizures in neuronal networks have resulted in a number of current efforts to treat severe genetically linked seizure and developmental disorders. These include migrating malignant partial seizures of infancy (MMPSI), also termed EIMFS (epilepsy of infancy with migrating focal seizures) linked to KCNT1, Slack, channel mutations [63–66]; Liang-Wang syndrome (LIWAS) linked to KCNMA1, BK channel mutations [67–69]; and BFNE (benign familial neonatal epilepsy), mediated by Kv7, KCNQ, and potassium channels [66, 70–73]. Despite the fact that the population frequency of these disorders is limited and that there are some uncertainties concerning how gain-of-function mutations in some of these mutated potassium channels result in hyperactivity [74], the hope is that successful treatment of these disorders will both provide therapeutic benefit for these patients and provide information and therapies useful in the treatment of a larger number of patients with epilepsy. This has resulted in significant drug discovery efforts at all of these targets, although as of this writing it has not resulted in an approved therapy. The field of psychiatric drug discovery could potentially be helped by this approach, where a similar model of partitioning of patients into groups with the greatest genetic and symptomatic similarity may be useful.

As stated above, when a drug candidate is identified, it must be put through a series of steps, involving in vitro assays as well as animal assays, which provide evidence of safety, the distribution and metabolism of the drug, and, if possible, some measure of efficacy. Some of these steps are strictly enforced by regulatory agencies, for example, pharmacokinetic profiling and toxicity. Efficacy is not so strictly regulated and for good reasons. In many diseases, efficacy in animal models is controversial at best, and this is the case in many psychiatric disorders. Nevertheless, before a drug is given to human subjects, particularly those with the target disease, some in vivo evidence of efficacy supporting the risk of giving a human a new and previously untried molecule is expected if not required in statute. This will almost certainly be expected from those providing funds for expensive clinical studies. This has proven difficult in CNS diseases, even when animal models or conditions appear to mimic human counterparts, such as ischemic stroke. As has been said previously [13], while many animal models of thromboembolic stroke exist and can be made to appear to have significant face validity, they have proven to have little to no predictive validity in terms of drug efficacy in humans. There are many reasons for this but three key ones are (*i*) the genetic uniformity of the animal subjects, at least with most test species such as rodents; (*ii*) the age of the animals at which the testing is carried out (young animals are usually used versus older humans with actual acute ischemic strokes); and (*iii*) at least in previous campaigns, a lack of basic understanding of stroke evolution as well as marketing decisions that led to dosing regimens that were destined to fail.

Where animal models show the most promise is when a specific gene is identified with a specific condition, and genetic models, usually mice, are created to express the disease-related mutation. This is, of course, what has been done in many neurodegenerative diseases, including ALS and AD, but it also requires that we understand the role of the gene product in the function of the CNS in that animal (and its relationship to its function in humans) and whether the single gene is the only factor in disease expression. Mutating a gene in an animal may not result in a phenotype recognizable as having any immediate relationship with the disease target in humans. It also means that we accept that only subsets of patients may be amenable to therapy by a drug with a particular MOA and that combination therapies may be necessary. This is probably not a significant problem for psychiatric disorders, as prescribing psychiatrists are used to the need to individualize pharmacotherapy for their patients. Based on genetic linkage studies, and a deeper understanding of CNS structure and function, there is reason to be optimistic about the development of new animal models for CNS diseases, including psychiatric disorders.

11.3.2 The Complicated Landscape of Psychiatric Drug Discovery

Psychiatric drug discovery is perhaps even more complicated than the already complex field of CNS drug discovery in general. While strict diagnostic criteria can be applied to most neurological disorders and are increasingly helped by imaging studies and the identification of diagnostic and predictive biomarkers, this has proven elusive in psychiatric disorders. The very definitions and diagnostic criteria of psychiatric disorders have frequently changed (perhaps less so now than previously), and diagnoses are largely based on behavioral outcomes or personal report [75]. For a number of reasons, most major forms of psychiatric disorders are syndromes, with an individual patient's condition placed on a spectrum. In addition, there is considerable overlap in the symptoms of many of these conditions. A further complication is that, although the finding of a genetic linkage to, for example, schizophrenia (and other major psychoses) remains among the most convincing piece of evidence for the neuronal bases of these conditions, such linkages are very complex, and full expression often results from a combination of genetic and environmental factors. To our knowledge, no single mutation, resulting in the identification of a single drug discovery target, has been definitively linked to a predominant form of any major psychiatric disorder, including schizophrenia [20, 76–80].

Clear genetic linkages have, however, been found for conditions such as Fragile-X syndrome, which is one of the most prevalent developmental disorders leading to intellectual impairment and autistic behavior. Fragile-X syndrome is a trinucleotide repeat disorder that produces silencing of the *FMR1* gene and the loss of its product fragile X mental retardation protein (FMRP). FMRP is an mRNA-binding protein that can regulate translation of a subset of neuronal transcripts [81–83] but also has other "noncanonical functions" that include direct binding to several plasma membrane ion channels to alter neuronal firing and neurotransmitter release [84–86]. Thus, loss of FMRP results in widespread developmentally-linked changes in neuronal (particularly synaptic) morphology and excitability [87–89]. The knowledge that the target for disease intervention was a single gene has resulted in numerous directed efforts to discover successful therapeutics. Regardless of the complexity of the genetic association landscape for psychiatric diseases, some associations may prove more tractable and more frequent or may lead to greater insights concerning novel targets.

While the complex polygenic nature of the relationship between mutations and psychiatric disease has not resulted in a significant epiphany in terms of identification and validation of discrete drug targets, at least not yet [19, 77, 80, 90], it is to be hoped that this situation is changing for the better. For example, in the past two decades, mutations of the regulator of G-protein signaling 4 (RGS4) gene have been among the most robust and frequently associated genetic linkages with schizophrenia. A member of a large gene family, RGS4, is widely but specifically distributed in the CNS, where it is thought to regulate neurotransmission mediated by a wide range of G-protein-coupled receptors, including those most closely (and previously) associated with schizophrenia spectrum disorders [79, 91–96]. The evidence for its relevance in schizophrenia is significant and as constant as it gets in this area. It could, therefore, potentially represent a single drug target that, if modulated appropriately, may confer normalization of function to an array of aberrant neurotransmitter systems in this complex disease. Like TAAR1, which will be discussed later, it may represent a novel target, associated with the disease, that either alone or in combination with other drugs targeting other proteins may prove effective across a spectrum of symptoms. While modulators of RGS4 have been discovered, these findings have not yet translated into a specific molecular species for schizophrenia spectrum disorders [97, 98]. However, because RGS4 exerts its influence on G-protein-coupled receptors by increasing GTP hydrolysis, which in turn inhibits G-protein-coupled receptor function, such modulators have been proposed as potential drugs for Parkinson's disease, where an inhibitor of RGS4 could play a role in amplifying dopaminergic neurotransmission [97]. Thus, this is an exciting direction, and one with a clear path from genetic linkage analyses in a psychiatric disorder, but it is still in the early stages of development for these indications.

One aspect that is perhaps unique to psychiatric disorders is that, because these are conditions defined by their expression in aberrant human behavior, disordered thoughts, hallucinations, delusions, or affect, no readily identifiable animal models of these diseases exist. These cannot be assayed in animals, making this a difficult problem. Animal models do exist, of course, and have been used for decades in the hunt for new antipsychotic therapies [2, 99–102]; they are mostly based on the behavioral effects of current drug treatments or the effects of drugs acting on specific neurotransmitter receptor populations, often dopamine. They have been critical in discovering new medications over the years. Because, however, these medications were discovered and refined to detect effects attributable to known MOAs, they have good predictive validity only in that respect. They are of much less value as actual models of disease that can be used to find novel treatments. Sensory-gating disruptions, which are present in many patients with schizophrenia, and can be easily induced and detected in animals, may be useful in animal models of schizophrenia spectrum disorders. We will discuss this below.

To illustrate these problems in more detail, and to illustrate both current approaches that are attempting to mitigate them, and any tangible results, we will focus the remainder of this chapter on major psychoses, in particular on schizophrenia. We will briefly discuss the disease, or constellation of conditions, that come under this category of psychiatric illness or condition, the history of therapeutic intervention, the current state of drug discovery and allied efforts, and speculate about future approaches that may prove useful.

11.3.3 Schizophrenia Spectrum as a Disorder and a Drug Target

Schizophrenia and other major psychoses comprise a schizophrenia spectrum, according to the Diagnostic and Statistical Manual of Mental Disorders (DSM-5) [103]. As a cluster of disorders, they are characterized by the presence of one or more symptoms, all of which have substantial formal definitions but which sufficiently speak for themselves for the purposes of this review. These symptoms include delusions, of which there are many categories and degrees of severity, hallucinations, disorganized thinking (as reflected in speech), abnormal motor behavior up to and including catatonia. This set of symptoms are referred to as the positive symptoms. A second set of symptoms that are considered separately are the negative and cognitive symptoms. These include diminished emotional expression, decreases in self-motivated purposeful activities (avolition) and others such as diminished speech (alogia), and diminished ability to perceive or remember pleasurable stimuli or experiences (anhedonia) and asociality, in addition to cognitive dysfunction. While most antipsychotic medications have some degree of success at managing the so-called positive symptoms of schizophrenia and related disorders, the negative and cognitive symptoms, which often predate diagnosis, have proven much more difficult to approach therapeutically [3, 9, 104, 105].

The major psychoses are certainly not orphan diseases. Schizophrenia is estimated to have a lifetime prevalence in the USA of approximately 1.0%, with a great deal of variability due to differences in which populations are measured and means of measurement [106–108]. The prevalence in the rest of the world has a similar range. The incidence, i.e., the number of individuals newly diagnosed as having schizophrenia during a particular year, is also highly variable but is probably in the range of 1–1.5 per 10,000 persons in the US. While the prevalence is relatively low, this nevertheless means that the number of people with schizophrenia, or schizophrenia spectrum disorders, may be more than 1.5 million at any given time in the US. This relatively low prevalence belies the importance of the disease both in terms of human suffering and societal cost. For example, the prevalence in the homeless community is quite high [109], and the disease contributes to significant morbidity and low productivity even in those maintaining sufficient function to have some level of employment.

Drug treatment for schizophrenia began in the 1950s with the approval of the typical or first-generation antipsychotic chlorpromazine [9, 27]. Its predominant pharmacology, and that of subsequent antipsychotics such as haloperidol (Haldol) [9, 110, 111], is dopamine receptor antagonism, specifically dopamine D2 receptor antagonism, although all of these drugs interact to one degree or another with numerous other receptors. Over the course of the next several decades, this common pharmacological pathway resulted in a hypothetical construct, the dopamine hypothesis, placing aberrant (overabundant) dopamine, and more generally biogenic amine neurotransmitter function at the center of schizophrenia spectrum disorders and many other psychiatric conditions [6, 10, 11]. While these "first-generation" antipsychotic medications are effective in many patients for treating positive symptoms of the illness, such as hallucinations and delusions, they are relatively ineffective or transient in their effects on negative symptoms and cognitive difficulties. They also have a marked propensity for producing extrapyramidal motor effects such as tardive dyskinesia, believed to result from high dopamine D2 receptor occupancy [9, 112, 113], and a host of other side effects that are both serious and which have significant effects on patient adherence.

The development of "atypical" or second-generation antipsychotics, the first of which was clozapine [9, 18, 114–117], attempted to address side effects, treatment-resistant schizophrenia, and the lack of effects of typical antipsychotics on negative symptoms and cognitive dysfunctions. While clozapine interacts with dopamine receptors, it also has significant and potent interactions with other neurotransmitter receptors, such as serotonin receptors (5HT). Each of the approximately 20 approved antipsychotics, typical and atypical, has relatively good efficacy on positive symptoms and more variable and less dramatic effects on negative symptoms and cognitive dysfunction. They all have as their primary target either dopamine receptors or a combination of dopamine receptors and receptors for other biogenic amines. Not a single approved antipsychotic specifically targeted a novel mechanism. In addition, not a single one of these drugs is specific for only a single receptor within the classes of receptors accepted as being concordant with schizophrenia therapy.

The reason for the homogeneity in action of the current antipsychotics is straightforward. If all of the preclinical screens and models were created to discover drugs that produce effects similar to other antipsychotics, then you would be unlikely to discover a drug targeting a novel mechanism. All of these compounds were selected to move forward in the discovery process because of their receptor binding and efficacy profiles, and so all represent "tweaks" of the same general theme. Moving away from this paradigm will require the identification of novel possible targets for schizophrenia spectrum disorders. There has been significant progress in this regard, resulting to some degree from genetic linkage studies [19, 20, 77, 79, 90, 91, 95, 118, 119]. The number of potential targets has increased dramatically in recent years and also increased in complexity. Even in the area of neurotransmitters, there now exists significant evidence that a number of neurotransmitter systems other than dopamine are altered in schizophrenia spectrum disorders. These include glutamate and its receptors, acetylcholine and cholinergic receptors, and other biogenic amines such as serotonin, which has been long implicated in both schizophrenia spectrum disorders and major depression [6, 17, 120, 121]. They also include the opiate μ receptor and TAAR1 (trace amine-associated receptors 1), and they will be discussed further in a later section.

11.4 How Can We Move Psychiatric Drug Discovery Forward?

Drug discovery and development are difficult, time-consuming, and very expensive. Failure rates are particularly high when targeting diseases in new ways. CNS drugs have added layers of difficulty, and psychiatric drugs represent perhaps the most vexing of all if the hope is to find something effective and safe beyond current therapy and outside existing target confines. To significantly alter the current trajectory of psychiatric drug discovery in general, and for schizophrenia spectrum disorders in particular, the specific aspects of neurobiological activity that become dysfunctional at the onset of the illness, including but not limited to genetic factors, will have to be determined [1, 3, 5, 103, 122, 123]. Together with this, the discovery and exploitation of objective measures that have a high degree of correlation with the disease, in other words, biomarkers, will be required. This is particularly true in psychiatry, where clinical diagnosis and treatment outcomes rely on quantification of largely subjective measures, including self-report. Validated biomarkers have become one of the most sought-after components of a drug discovery campaign. While potential biomarkers have been proposed for schizophrenia and related disorders, psychiatry remains the only field of medicine where there are no useful and routine objective clinical indicators of illness presence or severity. There is currently no blood test or imaging paradigm that a physician can order to determine the presence or severity of schizophrenia in a patient or to determine if a drug regimen has been effective; such a measure or measures would be a great stride forward in all aspects of the treatment of this and other psychiatric diseases.

11.4.1 Biomarkers in Psychiatry

A biomarker can be many things, but generally is thought of as comprising a characteristic that can be measured, that reflects an underlying biological processes, that changes when this process leads to or indicates a disease or dysfunction, and that reverses with effective treatment, including drug treatment [124]. Within the catchphrase of biomarker are a number of categories. These including diagnostic biomarkers, which indicate disease presence (or the likelihood of future emergence and disease progression, such as prognostic or susceptibility/risk biomarkers) and predictive biomarkers, which indicate the likelihood of therapeutic success (theranostic biomarkers). While biomarkers with some correlation to a disease can be proposed, their utility lies in their validation, a complex process involving at least a high degree of correlation with clinical assessments and endpoints. To label a potential biomarker as a surrogate of a disease or condition is even more complex, and can usually only be accomplished after it has been shown that the characteristic varies in the same way, and is correlated with outcome to the same degree, when measured following several unrelated therapeutic approaches. This is a high burden for psychiatric illness.

The magnitude of response to a brief but intense sensory stimulus that evokes a startle response has been proposed as a biomarker for schizophrenia. The acoustic startle response and startle responses to other sensory modalities are behavioral phenomenon associated with sensorimotor gating, and these responses have well-defined neuroanatomical and neurophysiological circuits [125, 126]. Startle responses are present in many species, indicating their likely evolutionary advantage. They are also subject to significant alteration as a function of previous experience in both humans and animals. For example, the response of a subject to a stimulus that normally evokes a startle can be markedly blunted if they are given a pre-pulse of significantly lower magnitude (i.e., below a threshold for a startle) just before the startle stimulus itself. This effect, often measured using the eyeblink response in humans, is termed pre-pulse inhibition (PPI). The opposite can also happen. If the priming stimulus and the second stimulus are in the same sensory modality and separated by a sufficient interval between presentations, a facilitation of the response to the second stimulus is often observed, even if that second stimulus is not sufficient to produce startle when presented alone. This is termed pre-pulse facilitation (PPF). Repeated presentations of a startling stimulus also result in habituation, a reduction in the response. These phenomena are very similar in humans and species such as rats [102, 127–129].

Exaggerated responses to environmental stimuli have long been noted in individuals with schizophrenia. Nearly four decades ago, it was determined that schizophrenia was often correlated with a reduced level of PPI [130] and could also often be associated with a reduced potential for habituation. Elevated PPF has also been observed in subjects with schizophrenia. While the presence of reduced PPI or enhanced PPF has not been universally demonstrated or accepted as a diagnostic marker, perhaps because of differences in stimulation protocols and measurement tools, it is widely accepted that sensory-motor gating deficits exist in at least many patients with schizophrenia. It may even be observed in normal volunteers without schizophrenia but who carry schizophrenia risk alleles, nonschizophrenic siblings of people with schizophrenia, and individuals with schizophrenia in a period before full expression of the illness [129–132]. Its variability over time and between patients, and its presence in many subjects without schizophrenia, precludes its utility as a diagnostic criterion in human disease. Nevertheless, it has been shown to be a useful construct for creating animal models of disease (genetic models or models produced via pharmacological manipulation) with a fairly clear relationship to a well-characterized phenomenon observed in many human subjects with the disorder. In many respects sensory-motor deficits represent the first and most widely researched biomarker in the schizophrenia spectrum disorders. Even though it is a behavioral characteristic, rather than a molecular or anatomical biomarker in the usual sense, it nevertheless allows for the objective measurement of a characteristic that is altered in many (but certainly not all) subjects with schizophrenia. Like all biomarkers in a heterogeneous disease or disease spectrum such as schizophrenia, its prominence in some subjects and absence or near absence in others may have value in approaching schizophrenia therapeutically as a disease composed of subgroups, a fact that is strongly indicated by the polygenic nature of known genetic correlates. Animal models of PPI impairment are used widely in models employed in drug discovery campaigns for new antipsychotics [102, 128, 133, 134].

A number of biomarker classes have been proposed for major psychoses, reflecting hypothetical or demonstrated disease-related alterations in regional neuroanatomy [14, 20, 135], neurotransmitter systems [6, 14, 120, 136–138], and other measures such as neuroendocrine factors and levels of inflammation [139–141]. All of these have their proponents as well as significant problems associated with their validation. To be useful they have to allow for their use without unreasonable risk to a patient and therefore be relatively noninvasive. Imaging of the CNS can be done noninvasively, and there may be neuroanatomical measures that can be observed and quantified using neuroimaging techniques. To be useful in the discovery of novel treatments for schizophrenia spectrum disorders, as opposed to their validation as some form of diagnostic biomarker, they must also be able to be used in animal models, along with some form of behavioral anomaly thought to reflect a "disease phenotype."

Although not yet validated in psychiatric illness, there is perhaps the greatest near-term promise for the use of imaging techniques [20, 80, 135, 142–148], particularly when coupled with large-scale population studies of genetic risk in schizophrenia [19, 20]. Neuroimaging falls into several categories, including structural imaging, neurochemical imaging, and imaging relying on metabolic differences between discrete regions either at baseline or as a result of stimulation. Evidence exists that techniques such as diffusion-weighted magnetic resonance imaging (MRI) to examine white matter [148], T1-weighted MRI to examine areas with higher densities of cell somata [145], and fMRI (functional MRI) BOLD (blood-oxygen-level-dependent) signals to examine cellular energetics in discrete brain regions [142, 149, 150] can reveal distinct signatures in the brains of patients with schizophrenia. Machine learning algorithms have been used for analyses of these data sets to increase the power to detect subtle and highly variable characteristics, including widespread reductions in cortical gray matter volumes [135]. Great progress has been made in this area, but the variability observed in the patient populations has thus far precluded the use of these as biomarkers for diagnosis, although they may have greater utility in assessing the effects of antipsychotics in the CNS [151, 152]. The hope is that they will have significant utility in drug discovery and development efforts in the near future, including in all areas of psychiatry and in other areas of CNS drug discovery [139, 147, 153]. This will not be trivial, but the recent recognition of the utility of large research consortia in both the validation or discovery of potential biomarkers and in their use in drug discovery will greatly increase the chances of success [19, 20, 118].

11.5 New Approaches in Drug Discovery for Schizophrenia

While this final section will discuss some new approaches and candidate drugs for schizophrenia, it is not intended in any way as an exhaustive listing of these ventures. While we continue to use schizophrenia as our example, much of what we have said throughout applies to all areas of psychiatric drug discovery. While we agree with nearly everyone that has written on this subject that things have been pretty bleak [1, 3, 5, 9, 103, 122, 123, 154–156], recent developments suggest that the situation is getting better. Perhaps most critical to this shift has been the recognition and exploitation of recent knowledge concerning neurotransmitter systems other than the dopamine and serotonin receptor pharmacology that has dominated the field of antipsychotic drug discover for decades. As stated earlier, evidence has accumulated implicating a wide array of central transmitter systems in the pathology associated with major psychiatric illness, including schizophrenia spectrum disorders. A thorough discussion of all of these potential targets is beyond our current scope, but we will discuss examples of recently approved antipsychotics, or antipsychotics in advanced clinical trials, that are believed to involve one of these receptors at least as part of their mechanism of action.

Glutamic acid and related compounds are the most widespread excitatory neurotransmitters in the CNS, with several classes of receptor mediating their functional consequences on neuronal transmission and function [136, 138, 157, 158]. These include (*i*) the ionotropic α-amino-3-hydroxy-5-methyl-4-isoxazolepropionic acid (AMPA) and kainate receptors, which mediate fast excitatory neurotransmission; (*ii*) the ionotropic but condition-dependent *N*-methyl-D-aspartate (NMDA) receptors, which are modulated by ketamine and phencyclidine (PCP) and other psychoactive drugs; and (*iii*) the metabotropic glutamate (mGluR) receptors. This latter group are G-protein-coupled receptors that influence cellular function indirectly (either postsynaptically or presynaptically), by modulation of other transmitter systems, ion channels, and other cellular functions. There is evidence for both hyper- and hypofunction of glutamate receptors in schizophrenia [17, 121, 136, 159], and one early therapeutic approach was the prolongation of the duration of action of synaptic glutamate by positive allosteric modulators, such as the ampakines (e.g., CX516) [160]. Recent efforts have concentrated on mGluR and NMDA ligands.

TAAR1 is a member of a new class of G-protein-coupled receptors discovered in 2001 by its sequence homology with biogenic amine receptors. It responds to a number of non-biogenic amines present in trace amounts in the CNS and elsewhere (e.g., *p*-tyramine, 2-phenylethylamine, and tryptamine) [161, 162]. These receptors are widely distributed in the CNS and are believed to have a modulatory influence on other neurotransmitter systems, including the biogenic amines, and to be involved in the actions of a number of psychoactive drugs. Of particular interest, TAAR genes, including TAAR1, are found in the region of chromosome 6 (q23), an area rich in psychosis-related mutations [16, 19, 77, 93, 118, 119, 163].

Newly approved antipsychotics, including those approved within the last year, are not entirely novel, although their innovation generally relies on either the fact that they avoid high receptor occupancy antagonism of dopamine D2 receptors or that they have additional novel pharmacology (in addition to dopamine D2 or serotonin 5HT2 receptor antagonism). The pharmacology of the drugs we discuss below shows that, either as stand-alone drugs or in combination with an atypical antipsychotic, they interact with one of the receptor classes discussed above. Perhaps because of this, they may provide additional benefit in treating negative symptoms and cognition and/or avoid significant side effects, such as weight gain or extrapyramidal symptoms associated with "classical" antipsychotics. The three drugs are (1) a combination of the atypical antipsychotic olanzapine and the μ-opioid antagonist samidorphan (approved in 2021), (2) lumateperone (approved in 2019), and (3) SEP-363856 (SEP856) which appears to be quite novel and is currently in advanced clinical trials.

The olanzapine/samidorphan combination drug for schizophrenia is important, less for its degree of efficacy per se than for the fact that it represents the sort of approach likely to be more successful in treating multiple symptom parameters or, in this case, reducing specific side effects. Its first component, olanzapine, is an efficacious atypical antipsychotic approved for both schizophrenia and bipolar 1 disorder [164], but it has a very significant drawback in that patients exhibit pronounced weight gain and increases in Type II diabetes while on olanzapine therapy [165, 166]. Addition of samidorphan, a μ-opioid antagonist, was found to significantly reduce this effect of olanzapine while having no discernable effect on its antipsychotic profile [167]. While this is a minor advance in some respects, taking an effective but flawed atypical antipsychotic and reducing one of the major side effects suggest that combination therapies can target specific disease or therapeutic outcomes. These may be disease modalities, such as drug combinations that are a mixture of drugs with effects on positive symptoms and drugs that have greater effects on negative symptoms and cognition, or they may target one or more of the dose- or compliance-limiting side effects, as was accomplished with the olanzapine/samidorphan combination product. Polypharmacy, approaching precision medicine, may be the only way to approach a disease spectrum presenting with a range of phenotypes, which are presumably mediated by the polygenic nature of the underlying neurobiological pathology [18–20, 33, 77, 78, 80, 90, 93].

Another recently approved, effective, and relatively novel antipsychotic is lumateperone [168–171]. Unlike other first generation and atypical antipsychotics, lumateperone is a partial agonist at presynaptic dopamine (D2) receptors and an antagonist at postsynaptic receptors, a combination thought to inhibit dopaminergic neurotransmission without the need for high dopamine receptor occupancy which can lead to dopamine receptor upregulation and tardive dyskinesia [112, 113]. It is also a serotonin receptor (5HT2) antagonist, a serotonin reuptake inhibitor, and a modulator of glutamate neurotransmission via dopamine D1 receptor antagonism. This latter function is interesting because glutamate hypofunction has long been hypothesized to be a contributing factor to at least some of the symptomatology of schizophrenia [6, 121] and was the logic behind the earlier unsuccessful development of an ampakine compound as an antipsychotic [134, 160, 172, 173]. Lumateperone is effective and has less potential for extrapyramidal side effects, but its greatest import may be that it is the first approved antipsychotic to have glutamate modulation, albeit indirect, as a feature of its target.

SEP-363856 (SEP856) is our final example and is the only one of the three that has not been approved. As of this writing (according to clinicaltrials.gov), it continues in phase 3 clinical trials for schizophrenia, and preliminary efficacy from phase 2 has been positive, with particular emphasis on efficacy in negative symptoms [174]. SEP-363856 may be unique because direct modulation of dopamine does not appear to be a feature of its pharmacology. While a number of efforts have attempted to investigate non-D2 ligands in schizophrenia, prior results in clinical settings have overwhelmingly been failures [155]. SEP-363856 is a serotonin 5HT1A autoreceptor antagonist and, prominently, a TAAR1 agonist. If ultimately successful, it may represent a groundbreaking new direction in the search for antipsychotic pharmacotherapies [133, 174–176]. Importantly, while its effect alone is being studied on schizophrenia, it is also being investigated in combination with other atypical antipsychotics. These combinations may result in a reduction in extrapyramidal side effects, because of lower doses of the D2-preferring drug, as well as in a broader range of symptoms that are positively affected by the combination(s).

11.6 Conclusions

The history of the search for new psychiatric therapeutics has been interesting but often frustrating [3, 5, 9, 27, 103, 122, 123]. The early discovery of classes of compounds that had significant efficacy in at least a proportion of the populations suffering from major disorders such as schizophrenia spectrum disorders, depression, and anxiety made it appear as if these diseases would be readily amenable to pharmacotherapy. The importance of these early medications cannot be overstated; they helped patients, at least a large number of patients, immensely, and helped to drive the search for disease mechanisms in an area that was encumbered by vague, nearly mystical, and often misogynistic theories of disease etiology. While much remains to be understood in all these illnesses, a great deal has been learned about the underlying neurobiology and the genetic associations/causations and their environmental interactions.

Unfortunately, until quite recently drug discovery has been in a serious rut with respect to almost all of these conditions. Most second-generation therapies were, at best, modest improvements on previous compounds in terms of pharmacological profile, and all suffered to varying degrees from serious potential side effects. In schizophrenia spectrum disorders, both typical and atypical antipsychotics were effective against positive symptoms but were relatively ineffective on negative and cognitive symptoms. Antipsychotics and antidepressants carry warnings of potentially fatal consequences in certain age-related populations. The use of antipsychotics can actually lead to the death of the aged and those suffering from dementia, while in children, adolescents, and young adults, the risk of death by suicide can be increased with antidepressant use, although there is some controversy about this in the field. It has proven difficult to separate whole classes of drugs from these and other side effects because, quite simply, they are the result of compound class pharmacology.

There is widespread recognition of the need for new approaches. New therapeutics are being advanced, albeit slowly and cautiously. These have come from recognition of new neuronal targets, which have been identified from results of analyses of complex polygenic associations in large patient populations, and a clearer understanding of the biology of neuronal systems that control behavior. There is good reason to be more optimistic in this decade than during the previous 30–40 years. One positive outcome, triggered by frustrations of the past, has been the emergence of academic and industrial consortia, which can pool large data sets for genetic analyses and clinical data analysis [19, 20, 77, 118, 135, 154], as well as a recognition of an increased role for government agencies in both funding and coordinating such efforts [1, 154]. We also have many new tools available for the analysis of large data sets and new ways to approach therapy. We are no longer limited, at least in theory, to classical small molecule drugs for treating these disorders. As more is understood about how candidate genes affect behavior and mental illness, tools for regulating gene expression, and even for gene editing, now exist where no such approach was even contemplated in prior decades.

While this chapter was concerned with discussing the difficulties in finding and developing new therapeutics for psychiatric illness in general, with schizophrenia spectrum disorders as a specific example, a central goal was to convey the difficulties of discovering any drug. We have always found it to be borderline miraculous, when viewed from both a background of the basic neurobiology of disease and decades of drug discovery experience, that it is possible that an effective treatment can be found (eventually) for almost anything that can go wrong with any system. That this has proven more difficult in many CNS disorders, and more difficult yet in psychiatric indications, is a direct result of the redundant and complex nature of neuronal systems and of the difficulties in defining and treating human disease when there is a significant subjective component to its diagnosis and assessment of the clinical effects of candidate drugs. In psychiatric illness, in particular, lack of effective diagnostic biomarkers of the disease and of treatment outcomes and the difficulty in creating model systems that have sufficient face and predictive validity have compounded the difficulty. In the pharmaceutical industry, success breeds success (or at least the incentives to try), and recent successes with novel approaches, such as those discussed above, should improve the near horizon for psychiatric drug discovery.

References

1. Conn PJ, Roth BL. Opportunities and challenges of psychiatric drug discovery: roles for scientists in academic, industry, and government settings. Neuropsychopharmacology. 2008;33(9):2048–60.
2. Kesby JP, Eyles DW, McGrath JJ, Scott JG. Dopamine, psychosis and schizophrenia: the widening gap between basic and clinical neuroscience. Transl Psychiatry. 2018;8(1):30.
3. Keshavan MS, Lawler AN, Nasrallah HA, Tandon R. New drug developments in psychosis: challenges, opportunities and strategies. Prog Neurobiol. 2017;152:3–20.
4. Owens DC, Johnstone EC. The development of antipsychotic drugs. Brain Neurosci Adv. 2018;2:1–6.
5. van der Doef TF, Zaragoza Domingo S, Jacobs GE, Drevets WC, Marston HM, Nathan PJ, et al. New approaches in psychiatric drug development. Eur Neuropsychopharmacol. 2018;28(9):983–93.
6. Yang AC, Tsai S-J. New targets for schizophrenia treatment beyond the dopamine hypothesis. Int J Mol Sci. 2017;18(8):1689.
7. Ban TA. The role of serendipity in drug discovery. Dialogues Clin Neurosci. 2006;8(3):335–44.
8. Baumeister AA, Hawkins MF, López-Muñoz F. Toward standardized usage of the word serendipity in the historiography of psychopharmacology. J Hist Neurosci. 2010;19(3):253–70.
9. Carpenter WT Jr, Davis JM. Another view of the history of antipsychotic drug discovery and development. Mol Psychiatry. 2012;17:1168–73.
10. Sayed Y, Garrison JM. The dopamine hypothesis of schizophrenia and the antagonistic action of neuroleptic drugs--a review. Psychopharmacol Bull. 1983;19(2):283–8.
11. Seeman P. Dopamine receptors and the dopamine hypothesis of schizophrenia. Synapse (New York, NY). 1987;1(2):133–52.
12. Potter WZ. Psychotherapeutic drugs and biogenic amines. Current concepts and therapeutic implications. Drugs. 1984;28(2):127–43.
13. Gribkoff VK, Kaczmarek LK. The need for new approaches in CNS drug discovery: why drugs have failed, and what can be done to improve outcomes. Neuropharmacology. 2017;120:11–9.
14. Fallon JH, Opole IO, Potkin SG. The neuroanatomy of schizophrenia: circuitry and neurotransmitter systems. Clin Neurosci Res. 2003;3(1):77–107.
15. Kahn RS, Sommer IE. The neurobiology and treatment of first-episode schizophrenia. Mol Psychiatry. 2015;20(1):84–97.
16. Khavari B, Cairns MJ. Epigenomic dysregulation in schizophrenia: in search of disease etiology and biomarkers. Cells. 2020;9(8):1837.
17. McCutcheon RA, Krystal JH, Howes OD. Dopamine and glutamate in schizophrenia: biology, symptoms and treatment. World Psychiatry. 2020;19(1):15–33.
18. Nucifora FC Jr, Woznica E, Lee BJ, Cascella N, Sawa A. Treatment resistant schizophrenia: clinical, biological, and therapeutic perspectives. Neurobiol Dis. 2019;131:104257.
19. Ripke S, Neale BM, Corvin A, Walters JTR, Farh K-H, Holmans PA, et al. Biological insights from 108 schizophrenia-associated genetic loci. Nature. 2014;511(7510):421–7.
20. van Erp TGM, Walton E, Hibar DP, Schmaal L, Jiang W, Glahn DC, et al. Cortical brain abnormalities in 4474 individuals with schizophrenia and 5098 control subjects via the enhancing neuro imaging genetics through meta analysis (ENIGMA) consortium. Biol Psychiatry. 2018;84(9):644–54.
21. Wojtalik JA, Smith MJ, Keshavan MS, Eack SM. A systematic and meta-analytic review of neural correlates of functional outcome in schizophrenia. Schizophr Bull. 2017;43(6):1329–47.
22. Frood A. High-dose opiates could crack chronic pain. Nature. 2012. https://doi.org/10.1038/nature.2012.9796.
23. Obeng S, Hiranita T, León F, McMahon LR, McCurdy CR. Novel approaches, drug candidates, and targets in pain drug discovery. J Med Chem. 2021;64(10):6523–48.

24. Rosenblum A, Marsch LA, Joseph H, Portenoy RK. Opioids and the treatment of chronic pain: controversies, current status, and future directions. Exp Clin Psychopharmacol. 2008;16(5):405–16.
25. Woolf CJ. Capturing novel non-opioid pain targets. Biol Psychiatry. 2020;87(1):74–81.
26. Jeste DV, Maglione JE. Treating older adults with schizophrenia: challenges and opportunities. Schizophr Bull. 2013;39(5):966–8.
27. Ban TA. Fifty years chlorpromazine: a historical perspective. Neuropsychiatr Dis Treat. 2007;3(4):495–500.
28. New JS. The discovery and development of buspirone: a new approach to the treatment of anxiety. Med Res Rev. 1990;10(3):283–326.
29. Riblet LA, Eison AS, Eison MS, Taylor DP, Temple DL, VanderMaelen CP. Neuropharmacology of buspirone. Psychopathology. 1984;17(Suppl 3):69–78.
30. Rose M, Devine J. Assessment of patient-reported symptoms of anxiety. Dialogues Clin Neurosci. 2014;16(2):197–211.
31. Aulner N, Danckaert A, Ihm J, Shum D, Shorte SL. Next-generation phenotypic screening in early drug discovery for infectious diseases. Trends Parasitol. 2019;35(7):559–70.
32. Katsuno K, Burrows JN, Duncan K, van Huijsduijnen RH, Kaneko T, Kita K, et al. Hit and lead criteria in drug discovery for infectious diseases of the developing world. Nat Rev Drug Discov. 2015;14(11):751–8.
33. Van Voorhis WC, Hol WG, Myler PJ, Stewart LJ. The role of medical structural genomics in discovering new drugs for infectious diseases. PLoS Comput Biol. 2009;5(10):e1000530.
34. Plackett B. No money for new drugs. Nature. 2020;586:S50–S2.
35. DiMasi JA, Grabowski HG, Hansen RW. Innovation in the pharmaceutical industry: new estimates of R&D costs. J Health Econ. 2016;47:20–33.
36. Cummings JL, Morstorf T, Zhong K. Alzheimer's disease drug-development pipeline: few candidates, frequent failures. Alzheimers Res Ther. 2014;6(4):37.
37. Berk C, Sabbagh MN. Successes and failures for drugs in late-stage development for Alzheimer's disease. Drugs Aging. 2013;30(10):783–92.
38. Husna Ibrahim N, Yahaya MF, Mohamed W, Teoh SL, Hui CK, Kumar J. Pharmacotherapy of Alzheimer's disease: seeking clarity in a time of uncertainty. Front Pharmacol. 2020;11:261.
39. Mehta D, Jackson R, Paul G, Shi J, Sabbagh M. Why do trials for Alzheimer's disease drugs keep failing? A discontinued drug perspective for 2010–2015. Expert Opin Investig Drugs. 2017;26(6):735–9.
40. Begley DJ. Delivery of therapeutic agents to the central nervous system: the problems and the possibilities. Pharmacol Ther. 2004;104(1):29–45.
41. Hajal C, Le Roi B, Kamm RD, Maoz BM. Biology and models of the blood–brain barrier. Annu Rev Biomed Eng. 2021;23(1):359–84.
42. Wilhelm I, Krizbai IA. In vitro models of the blood–brain barrier for the study of drug delivery to the brain. Mol Pharm. 2014;11(7):1949–63.
43. Morrison J, Schwartz TL. Adolescent angst or true intent? Suicidal behavior, risk, and neurobiological mechanisms in depressed children and teenagers taking antidepressants. Int J Emerg Ment Health. 2014;16(1):247–50.
44. Henry A, Kisicki MD, Varley C. Efficacy and safety of antidepressant drug treatment in children and adolescents. Mol Psychiatry. 2012;17(12):1186–93.
45. Campos MSA, Ayres LR, Morelo MRS, Carizio FAM, Pereira LRL. Comparative efficacy of antiepileptic drugs for patients with generalized epileptic seizures: systematic review and network meta-analyses. Int J Clin Pharm. 2018;40(3):589–98.
46. Johannessen Landmark C, Johannessen SI, Patsalos PN. Therapeutic drug monitoring of antiepileptic drugs: current status and future prospects. Expert Opin Drug Metab Toxicol. 2020;16(3):227–38.
47. Yang S, Wang B, Han X. Models for predicting treatment efficacy of antiepileptic drugs and prognosis of treatment withdrawal in epilepsy patients. Acta Epileptologica. 2021;3(1):1.

48. Dawson TM, Golde TE, Lagier-Tourenne C. Animal models of neurodegenerative diseases. Nat Neurosci. 2018;21(10):1370–9.
49. Gitler AD, Dhillon P, Shorter J. Neurodegenerative disease: models, mechanisms, and a new hope. Dis Model Mech. 2017;10(5):499–502.
50. Slanzi A, Iannoto G, Rossi B, Zenaro E, Constantin G. In vitro models of neurodegenerative diseases. Front Cell Dev Biol. 2020;8:328.
51. Rubin R. Recently approved Alzheimer drug raises questions that might never be answered. JAMA. 2021;326(6):469–72.
52. Mullard A. Failure of first anti-tau antibody in Alzheimer disease highlights risks of history repeating. Nat Rev Drug Discov. 2021;20(1):3–5.
53. Dawson TM, Ko HS, Dawson VL. Genetic animal models of Parkinson's disease. Neuron. 2010;66(5):646–61.
54. de Boer AS, Koszka K, Kiskinis E, Suzuki N, Davis-Dusenbery Brandi N, Eggan K. Genetic validation of a therapeutic target in a mouse model of ALS. Sci Transl Med. 2014;6(248):248ra104.
55. Drummond E, Wisniewski T. Alzheimer's disease: experimental models and reality. Acta Neuropathol. 2017;133(2):155–75.
56. Hauser SL, Cree BAC. Treatment of multiple sclerosis: a review. Am J Med. 2020;133(12):1380–90.e2.
57. Crous-Bou M, Minguillón C, Gramunt N, Molinuevo JL. Alzheimer's disease prevention: from risk factors to early intervention. Alzheimers Res Ther. 2017;9(1):71.
58. Grossberg GT, Tong G, Burke AD, Tariot PN. Present algorithms and future treatments for Alzheimer's disease. J Alzheimers Dis. 2019;67(4):1157–71.
59. Knopman DS, Jones DT, Greicius MD. Failure to demonstrate efficacy of aducanumab: an analysis of the EMERGE and ENGAGE trials as reported by Biogen, December 2019. Alzheimers Dement. 2021;17(4):696–701.
60. Schneider L. A resurrection of aducanumab for Alzheimer's disease. Lancet Neurol. 2020;19(2):111–2.
61. Selkoe DJ. Alzheimer disease and aducanumab: adjusting our approach. Nat Rev Neurol. 2019;15(7):365–6.
62. Sevigny J, Chiao P, Bussière T, Weinreb PH, Williams L, Maier M, et al. The antibody aducanumab reduces Aβ plaques in Alzheimer's disease. Nature. 2016;537(7618):50–6.
63. Heron SE, Smith KR, Bahlo M, Nobili L, Kahana E, Licchetta L, et al. Missense mutations in the sodium-gated potassium channel gene KCNT1 cause severe autosomal dominant nocturnal frontal lobe epilepsy. Nat Genet. 2012;44(11):1188–90.
64. Møller RS, Heron SE, Larsen LH, Lim CX, Ricos MG, Bayly MA, et al. Mutations in KCNT1 cause a spectrum of focal epilepsies. Epilepsia. 2015;56(9):e114–20.
65. Quraishi IH, Stern S, Mangan KP, Zhang Y, Ali SR, Mercier MR, et al. An epilepsy-associated KCNT1 mutation enhances excitability of human iPSC-derived neurons by increasing Slack K(Na) currents. J Neurosci Off J Soc Neurosci. 2019;39(37):7438–49.
66. Zhou P, He N, Zhang JW, Lin ZJ, Wang J, Yan LM, et al. Novel mutations and phenotypes of epilepsy-associated genes in epileptic encephalopathies. Genes Brain Behav. 2018;17(8):e12456.
67. Lee US, Cui J. {beta} subunit-specific modulations of BK channel function by a mutation associated with epilepsy and dyskinesia. J Physiol. 2009;587(Pt 7):1481–98.
68. Liang L, Li X, Moutton S, Schrier Vergano SA, Cogné B, Saint-Martin A, et al. De novo loss-of-function KCNMA1 variants are associated with a new multiple malformation syndrome and a broad spectrum of developmental and neurological phenotypes. Hum Mol Genet. 2019;28(17):2937–51.
69. Bailey CS, Moldenhauer HJ, Park SM, Keros S, Meredith AL. KCNMA1-linked channelopathy. J Gen Physiol. 2019;151(10):1173–89.

70. Maljevic S, Lerche H. Chapter 2 – potassium channel genes and benign familial neonatal epilepsy. In: Steinlein OK, editor. Progress in brain research, vol. 213. Amsterdam: Elsevier; 2014. p. 17–53.
71. Gribkoff VK. The therapeutic potential of neuronal K V 7 (KCNQ) channel modulators: an update. Expert Opin Ther Targets. 2008;12(5):565–81.
72. Maljevic S, Wuttke TV, Seebohm G, Lerche H. KV7 channelopathies. Pflugers Arch. 2010;460(2):277–88.
73. Yang W-P, Levesque PC, Little WA, Conder ML, Ramakrishnan P, Neubauer MG, et al. Functional expression of two KvLQT1-related potassium channels responsible for an inherited idiopathic epilepsy*. J Biol Chem. 1998;273(31):19419–23.
74. Niday Z, Tzingounis AV. Potassium channel gain of function in epilepsy: an unresolved paradox. Neuroscientist. 2018;24(4):368–80.
75. Association AP. Diagnostic and statistical manual of mental disorders. 5th ed. Washington, DC: American Psychiatric Association; 2013.
76. Dick DM, Riley B, Kendler KS. Nature and nurture in neuropsychiatric genetics: where do we stand? Dialogues Clin Neurosci. 2010;12(1):7–23.
77. Liu C, Kanazawa T, Tian Y, Mohamed Saini S, Mancuso S, Mostaid MS, et al. The schizophrenia genetics knowledgebase: a comprehensive update of findings from candidate gene studies. Transl Psychiatry. 2019;9(1):205.
78. McCutcheon RA, Reis Marques T, Howes OD. Schizophrenia—an overview. JAMA Psychiat. 2020;77(2):201–10.
79. Talkowski ME, Seltman H, Bassett AS, Brzustowicz LM, Chen X, Chowdari KV, et al. Evaluation of a susceptibility gene for schizophrenia: genotype based meta-analysis of RGS4 polymorphisms from thirteen independent samples. Biol Psychiatry. 2006;60(2):152–62.
80. Turner JA, Smyth P, Macciardi F, Fallon JH, Kennedy JL, Potkin SG. Imaging phenotypes and genotypes in schizophrenia. Neuroinformatics. 2006;4(1):21–49.
81. Darnell JC, Van Driesche SJ, Zhang C, Hung KY, Mele A, Fraser CE, et al. FMRP stalls ribosomal translocation on mRNAs linked to synaptic function and autism. Cell. 2011;146(2):247–61.
82. De Rubeis S, Bagni C. Fragile X mental retardation protein control of neuronal mRNA metabolism: insights into mRNA stability. Mol Cell Neurosci. 2010;43(1):43–50.
83. Richter JD, Bassell GJ, Klann E. Dysregulation and restoration of translational homeostasis in fragile X syndrome. Nat Rev Neurosci. 2015;16(10):595–605.
84. Brown MR, Kronengold J, Gazula VR, Chen Y, Strumbos JG, Sigworth FJ, et al. Fragile X mental retardation protein controls gating of the sodium-activated potassium channel Slack. Nat Neurosci. 2010;13(7):819–21.
85. Deng PY, Klyachko VA. Channelopathies in fragile X syndrome. Nat Rev Neurosci. 2021;22(5):275–89.
86. Zhang Y, Brown MR, Hyland C, Chen Y, Kronengold J, Fleming MR, et al. Regulation of neuronal excitability by interaction of fragile X mental retardation protein with Slack potassium channels. J Neurosci Off J Soc Neurosci. 2012;32(44):15318–27.
87. Bagni C, Zukin RS. A synaptic perspective of fragile X syndrome and autism spectrum disorders. Neuron. 2019;101(6):1070–88.
88. Crawford DC, Acuña JM, Sherman SL. FMR1 and the fragile X syndrome: human genome epidemiology review. Genet Med. 2001;3(5):359–71.
89. Hagerman RJ, Berry-Kravis E, Hazlett HC, Bailey DB Jr, Moine H, Kooy RF, et al. Fragile X syndrome. Nat Rev Dis Primers. 2017;3:17065.
90. Tsuang MT, Bar JL, Stone WS, Faraone SV. Gene-environment interactions in mental disorders. World Psychiatry. 2004;3(2):73–83.
91. Chowdari KV, Mirnics K, Semwal P, Wood J, Lawrence E, Bhatia T, et al. Association and linkage analyses of RGS4 polymorphisms in schizophrenia. Hum Mol Genet. 2002;11(12):1373–80.

92. Huang MW, Lin YJ, Chang CW, Lei FJ, Ho EP, Liu RS, et al. RGS4 deficit in prefrontal cortex contributes to the behaviors related to schizophrenia via system x(c)(-)-mediated glutamatergic dysfunction in mice. Theranostics. 2018;8(17):4781–94.
93. Levitt P, Ebert P, Mirnics K, Nimgaonkar VL, Lewis DA. Making the case for a candidate vulnerability gene in schizophrenia: convergent evidence for regulator of G-protein signaling 4 (RGS4). Biol Psychiatry. 2006;60(6):534–7.
94. Mirnics K, Middleton FA, Stanwood GD, Lewis DA, Levitt P. Disease-specific changes in regulator of G-protein signaling 4 (RGS4) expression in schizophrenia. Mol Psychiatry. 2001;6(3):293–301.
95. Schwarz E. A gene-based review of RGS4 as a putative risk gene for psychiatric illness. Am J Med Genet B Neuropsychiatr Genet. 2018;177(2):267–73.
96. Xu FL, Yao J, Wu X, Xia X, Xing JX, Xuan JF, et al. Association analysis between SNPs in the promoter region of RGS4 and schizophrenia in the Northern Chinese Han population. Neuropsychiatr Dis Treat. 2020;16:985–92.
97. Blazer LL, Storaska AJ, Jutkiewicz EM, Turner EM, Calcagno M, Wade SM, et al. Selectivity and anti-Parkinson's potential of thiadiazolidinone RGS4 inhibitors. ACS Chem Neurosci. 2015;6(6):911–9.
98. Roof RA, Sobczyk-Kojiro K, Turbiak AJ, Roman DL, Pogozheva ID, Blazer LL, et al. Novel peptide ligands of RGS4 from a focused one-bead, one-compound library. Chem Biol Drug Des. 2008;72(2):111–9.
99. Burrows EL, Hannan AJ. Cognitive endophenotypes, gene-environment interactions and experience-dependent plasticity in animal models of schizophrenia. Biol Psychol. 2016;116:82–9.
100. Jones CA, Watson DJG, Fone KCF. Animal models of schizophrenia. Br J Pharmacol. 2011;164(4):1162–94.
101. Koszła O, Targowska-Duda KM, Kędzierska E, Kaczor AA. In vitro and in vivo models for the investigation of potential drugs against schizophrenia. Biomol Ther. 2020;10(1):160.
102. Winship IR, Dursun SM, Baker GB, Balista PA, Kandratavicius L, Maia-de-Oliveira JP, et al. An overview of animal models related to schizophrenia. Can J Psychiatr. 2019;64(1):5–17.
103. Psychiatric drug discovery on the couch. Nat Rev Drug Discov. 2007;6(3):171.
104. Buckley PF, Stahl SM. Pharmacological treatment of negative symptoms of schizophrenia: therapeutic opportunity or cul-de-sac? Acta Psychiatr Scand. 2007;115(2):93–100.
105. Kantrowitz JT. Managing negative symptoms of schizophrenia: how far have we come? CNS Drugs. 2017;31(5):373–88.
106. Kahn RS, Sommer IE, Murray RM, Meyer-Lindenberg A, Weinberger DR, Cannon TD, et al. Schizophrenia. Nat Rev Dis Primers. 2015;1:15067.
107. McGrath J, Saha S, Chant D, Welham J. Schizophrenia: a concise overview of incidence, prevalence, and mortality. Epidemiol Rev. 2008;30:67–76.
108. Saha S, Chant D, Welham J, McGrath J. A systematic review of the prevalence of schizophrenia. PLoS Med. 2005;2(5):e141.
109. Ayano G, Tesfaw G, Shumet S. The prevalence of schizophrenia and other psychotic disorders among homeless people: a systematic review and meta-analysis. BMC Psychiatry. 2019;19(1):370.
110. Review: haloperidol is effective for schizophrenia but increases parkinsonism, dystonia, and akathisia. Evid Based Ment Health. 2001;4(4):112.
111. Beresford R, Ward A. Haloperidol decanoate. A preliminary review of its pharmacodynamic and pharmacokinetic properties and therapeutic use in psychosis. Drugs. 1987;33(1):31–49.
112. Ali Z, Roque A, El-Mallakh RS. A unifying theory for the pathoetiologic mechanism of tardive dyskinesia. Med Hypotheses. 2020;140:109682.
113. Caroff SN. Recent advances in the pharmacology of tardive dyskinesia. Clin Psychopharmacol Neurosci. 2020;18(4):493–506.
114. Crilly J. The history of clozapine and its emergence in the US market: a review and analysis. Hist Psychiatry. 2007;18(1):39–60.

115. Khokhar JY, Henricks AM, Sullivan EDK, Green AI. Unique effects of clozapine: a pharmacological perspective. Adv Pharmacol (San Diego, Calif). 2018;82:137–62.
116. Nucifora FC Jr, Mihaljevic M, Lee BJ, Sawa A. Clozapine as a model for antipsychotic development. Neurotherapeutics. 2017;14(3):750–61.
117. Siskind D, McCartney L, Goldschlager R, Kisely S. Clozapine v. first- and second-generation antipsychotics in treatment-refractory schizophrenia: systematic review and meta-analysis. Br J Psychiatry J Ment Sci. 2016;209(5):385–92.
118. Chen J, Calhoun VD, Lin D, Perrone-Bizzozero NI, Bustillo JR, Pearlson GD, et al. Shared genetic risk of schizophrenia and gray matter reduction in 6p22.1. Schizophr Bull. 2019;45(1):222–32.
119. Gratten J, Wray NR, Keller MC, Visscher PM. Large-scale genomics unveils the genetic architecture of psychiatric disorders. Nat Neurosci. 2014;17(6):782–90.
120. Scarr E, Gibbons A, Neo J, Udawela M, Dean B. Cholinergic connectivity: it's implications for psychiatric disorders. Front Cell Neurosci. 2013;7:55.
121. Rubio MD, Drummond JB, Meador-Woodruff JH. Glutamate receptor abnormalities in schizophrenia: implications for innovative treatments. Biomol Ther (Seoul). 2012;20(1):1–18.
122. Agid Y, Buzsáki G, Diamond DM, Frackowiak R, Giedd J, Girault J-A, et al. How can drug discovery for psychiatric disorders be improved? Nat Rev Drug Discov. 2007;6(3):189–201.
123. Brady LS, Potter WZ, Gordon JA. Redirecting the revolution: new developments in drug development for psychiatry. Expert Opin Drug Discovery. 2019;14(12):1213–9.
124. Califf RM. Biomarker definitions and their applications. Exp Biol Med. 2018;243(3):213–21.
125. Gómez-Nieto R, Hormigo S, López DE. Prepulse inhibition of the auditory startle reflex assessment as a hallmark of brainstem sensorimotor gating mechanisms. Brain Sci. 2020;10(9):639.
126. Jones CK, Shannon HE. Lesions of the laterodorsal tegmental nucleus disrupt prepulse inhibition of the acoustic startle reflex. Pharmacol Biochem Behav. 2004;78(2):229–37.
127. Mena A, Ruiz-Salas JC, Puentes A, Dorado I, Ruiz-Veguilla M, De la Casa LG. Reduced prepulse inhibition as a biomarker of schizophrenia. Front Behav Neurosci. 2016;10:202.
128. Swerdlow NR, Weber M, Qu Y, Light GA, Braff DL. Realistic expectations of prepulse inhibition in translational models for schizophrenia research. Psychopharmacology. 2008;199(3):331–88.
129. Wynn JK, Dawson ME, Schell AM, McGee M, Salveson D, Green MF. Prepulse facilitation and prepulse inhibition in schizophrenia patients and their unaffected siblings. Biol Psychiatry. 2004;55(5):518–23.
130. Geyer MA, Braff DL. Startle habituation and sensorimotor gating in schizophrenia and related animal models. Schizophr Bull. 1987;13(4):643–68.
131. Quednow BB, Ettinger U, Mössner R, Rujescu D, Giegling I, Collier DA, et al. The schizophrenia risk allele C of the TCF4 rs9960767 polymorphism disrupts sensorimotor gating in schizophrenia spectrum and healthy volunteers. J Neurosci Off J Soc Neurosci. 2011;31(18):6684–91.
132. Quednow BB, Frommann I, Berning J, Kühn KU, Maier W, Wagner M. Impaired sensorimotor gating of the acoustic startle response in the prodrome of schizophrenia. Biol Psychiatry. 2008;64(9):766–73.
133. Koblan KS, Kent J, Hopkins SC, Krystal JH, Cheng H, Goldman R, et al. A non-D2-receptor-binding drug for the treatment of schizophrenia. N Engl J Med. 2020;382(16):1497–506.
134. Lipina T, Weiss K, Roder J. The ampakine CX546 restores the prepulse inhibition and latent inhibition deficits in mGluR5-deficient mice. Neuropsychopharmacology. 2007;32(4):745–56.
135. Rozycki M, Satterthwaite TD, Koutsouleris N, Erus G, Doshi J, Wolf DH, et al. Multisite machine learning analysis provides a robust structural imaging signature of schizophrenia detectable across diverse patient populations and within individuals. Schizophr Bull. 2018;44(5):1035–44.
136. Crupi R, Impellizzeri D, Cuzzocrea S. Role of metabotropic glutamate receptors in neurological disorders. Front Mol Neurosci. 2019;12:20.

137. du Bois TM, Deng C, Huang X-F. Membrane phospholipid composition, alterations in neurotransmitter systems and schizophrenia. Prog Neuro-Psychopharmacol Biol Psychiatry. 2005;29(6):878–88.
138. Willard SS, Koochekpour S. Glutamate, glutamate receptors, and downstream signaling pathways. Int J Biol Sci. 2013;9(9):948–59.
139. Perkovic MN, Erjavec GN, Strac DS, Uzun S, Kozumplik O, Pivac N. Theranostic biomarkers for schizophrenia. Int J Mol Sci. 2017;18(4):733.
140. Kirkpatrick B, Miller BJ. Inflammation and schizophrenia. Schizophr Bull. 2013;39(6):1174–9.
141. Müller N, Weidinger E, Leitner B, Schwarz MJ. The role of inflammation in schizophrenia. Front Neurosci. 2015;9:372.
142. González-Vivas C, García-Martí G, Soldevila-Matías P, Sanz-Requena R, Aguilar EJ, Castro-Bleda MJ, et al. First-episode psychotic patients showed longitudinal brain changes using fMRI with an emotional auditory paradigm. Front Psych. 2020;11:1374.
143. Keshavan MS, Collin G, Guimond S, Kelly S, Prasad KM, Lizano P. Neuroimaging in schizophrenia. Neuroimaging Clin N Am. 2020;30(1):73–83.
144. Kraguljac NV, McDonald WM, Widge AS, Rodriguez CI, Tohen M, Nemeroff CB. Neuroimaging biomarkers in schizophrenia. Am J Psychiatr. 2021;178(6):509–21.
145. Oh J, Oh B-L, Lee K-U, Chae J-H, Yun K. Identifying schizophrenia using structural MRI with a deep learning algorithm. Front Psych. 2020;11:16.
146. Specht K. Current challenges in translational and clinical fMRI and future directions. Front Psych. 2020;10:924.
147. Suhara T, Chaki S, Kimura H, Furusawa M, Matsumoto M, Ogura H, et al. Strategies for utilizing neuroimaging biomarkers in CNS drug discovery and development: CINP/JSNP Working Group report. Int J Neuropsychopharmacol. 2017;20(4):285–94.
148. Wheeler AL, Voineskos AN. A review of structural neuroimaging in schizophrenia: from connectivity to connectomics. Front Hum Neurosci. 2014;8:653.
149. Kim MA, Tura E, Potkin SG, Fallon JH, Manoach DS, Calhoun VD, et al. Working memory circuitry in schizophrenia shows widespread cortical inefficiency and compensation. Schizophr Res. 2010;117(1):42–51.
150. Yang B, Chen Y, Shao QM, Yu R, Li WB, Guo GQ, et al. Schizophrenia classification using fMRI data based on a multiple feature image capsule network ensemble. IEEE Access. 2019;7:109956–68.
151. Abbott CC, Jaramillo A, Wilcox CE, Hamilton DA. Antipsychotic drug effects in schizophrenia: a review of longitudinal FMRI investigations and neural interpretations. Curr Med Chem. 2013;20(3):428–37.
152. Towlson EK, Vértes PE, Müller-Sedgwick U, Ahnert SE. Brain networks reveal the effects of antipsychotic drugs on schizophrenia patients and controls. Front Psych. 2019;10:611.
153. Borsook D, Becerra L, Fava M. Use of functional imaging across clinical phases in CNS drug development. Transl Psychiatry. 2013;3(7):e282.
154. Brady LS, Winsky L, Goodman W, Oliveri ME, Stover E. NIMH initiatives to facilitate collaborations among industry, academia, and government for the discovery and clinical testing of novel models and drugs for psychiatric disorders. Neuropsychopharmacology. 2009;34(1):229–43.
155. Girgis RR, Zoghbi AW, Javitt DC, Lieberman JA. The past and future of novel, non-dopamine-2 receptor therapeutics for schizophrenia: a critical and comprehensive review. J Psychiatr Res. 2019;108:57–83.
156. Markou A, Chiamulera C, Geyer MA, Tricklebank M, Steckler T. Removing obstacles in neuroscience drug discovery: the future path for animal models. Neuropsychopharmacology. 2009;34(1):74–89.
157. Traynelis SF, Wollmuth LP, McBain CJ, Menniti FS, Vance KM, Ogden KK, et al. Glutamate receptor ion channels: structure, regulation, and function. Pharmacol Rev. 2010;62(3):405–96.

158. Paoletti P, Bellone C, Zhou Q. NMDA receptor subunit diversity: impact on receptor properties, synaptic plasticity and disease. Nat Rev Neurosci. 2013;14(6):383–400.
159. Lee G, Zhou Y. NMDAR hypofunction animal models of schizophrenia. Front Mol Neurosci. 2019;12:185.
160. Goff DC, Leahy L, Berman I, Posever T, Herz L, Leon AC, et al. A placebo-controlled pilot study of the ampakine CX516 added to clozapine in schizophrenia. J Clin Psychopharmacol. 2001;21(5):484–7.
161. Miller GM. The emerging role of trace amine-associated receptor 1 in the functional regulation of monoamine transporters and dopaminergic activity. J Neurochem. 2011;116(2):164–76.
162. Rutigliano G, Accorroni A, Zucchi R. The case for TAAR1 as a modulator of central nervous system function. Front Pharmacol. 2018;8:987.
163. Lencz T, Yu J, Khan RR, Flaherty E, Carmi S, Lam M, et al. Novel ultra-rare exonic variants identified in a founder population implicate cadherins in schizophrenia. Neuron. 2021;109(9):1465–78.e4.
164. Duggan L, Fenton M, Rathbone J, Dardennes R, El-Dosoky A, Indran S. Olanzapine for schizophrenia. Cochrane Database Syst Rev. 2005;(2):CD001359.
165. Huang J, Hei G-R, Yang Y, Liu C-C, Xiao J-M, Long Y-J, et al. Increased appetite plays a key role in olanzapine-induced weight gain in first-episode schizophrenia patients. Front Pharmacol. 2020;11:739.
166. Mathews J, Newcomer JW, Mathews JR, Fales CL, Pierce KJ, Akers BK, et al. Neural correlates of weight gain with olanzapine. Arch Gen Psychiatry. 2012;69(12):1226–37.
167. Correll CU, Newcomer JW, Silverman B, DiPetrillo L, Graham C, Jiang Y, et al. Effects of olanzapine combined with samidorphan on weight gain in schizophrenia: a 24-week phase 3 study. Am J Psychiatry. 2020;177(12):1168–78.
168. Correll CU, Davis RE, Weingart M, Saillard J, O'Gorman C, Kane JM, et al. Efficacy and safety of Lumateperone for treatment of schizophrenia: a randomized clinical trial. JAMA Psychiat. 2020;77(4):349–58.
169. Greenwood J, Acharya RB, Marcellus V, Rey JA. Lumateperone: a novel antipsychotic for schizophrenia. Ann Pharmacother. 2021;55(1):98–104.
170. Snyder GL, Vanover KE, Davis RE, Li P, Fienberg A, Mates S. Chapter eleven – a review of the pharmacology and clinical profile of lumateperone for the treatment of schizophrenia. In: Zorn SH, editor. Advances in pharmacology, vol. 90. Cambridge, MA: Academic Press; 2021. p. 253–76.
171. Snyder GL, Vanover KE, Zhu H, Miller DB, O'Callaghan JP, Tomesch J, et al. Functional profile of a novel modulator of serotonin, dopamine, and glutamate neurotransmission. Psychopharmacology. 2015;232(3):605–21.
172. Lynch G. Glutamate-based therapeutic approaches: ampakines. Curr Opin Pharmacol. 2006;6(1):82–8.
173. Lynch G, Gall C. Mechanism based approaches for rescuing and enhancing cognition. Front Neurosci. 2013;7:143.
174. Correll CU, Koblan KS, Hopkins SC, Kent J, Cheng H, Goldman R, et al. Safety and effectiveness of SEP-363856 in schizophrenia: results of a 6-month, open-label extension study. CNS Spectr. 2021;26(2):148–9.
175. Begni V, Sanson A, Luoni A, Sensini F, Grayson B, Munni S, et al. Towards novel treatments for schizophrenia: molecular and behavioural signatures of the psychotropic agent SEP-363856. Int J Mol Sci. 2021;22(8):4119.
176. Dedic N, Jones PG, Hopkins SC, Lew R, Shao L, Campbell JE, et al. SEP-363856, a novel psychotropic agent with a unique, non-D(2) receptor mechanism of action. J Pharmacol Exp Ther. 2019;371(1):1–14.

158. Barbon R, Bellone C, Zhu Q. NMDA receptor subunit diversity: impact on receptor properties, synaptic plasticity and disease. Nat Rev Neurosci. 2013;14(6):383–400.
159. Lee G, Zhou Y. NMDAR hypofunction animal models of schizophrenia. Front Mol Neurosci. 2019;12:185.
160. Olds DC, [illegible] et al. [illegible] controlled pilot study of the impact of [illegible] in [illegible] schizophrenia. J Clin Psychopharmacol. 20[illegible].
161. Millan MJ. The neurobiology and control of anxious states. [illegible] of [illegible] 2003;[illegible]
162. [illegible] The receptor [illegible] of [illegible] system function. Curr Pharmacol. [illegible]
163. Lee G, [illegible] Novel [illegible] schizophrenia. Neuron. [illegible]
164. [illegible]
165. [illegible]
166. [illegible]
167. [illegible]
168. [illegible]
169. [illegible]
170. [illegible]
171. [illegible] 2018;[illegible]
172. [illegible] 2018;[illegible]
173. Lin [illegible] 2013;[illegible]
174. [illegible] placebo-controlled [illegible] study. CNS Spectr. 20[illegible]
175. [illegible] novel treatment [illegible] psychedelic agent [illegible] Ther. 2019;[illegible]

Chapter 12
Biomarkers in Psychiatric Drug Development: From Precision Medicine to Novel Therapeutics

Rudy Lozano Carreon, Ana Maria Rivas-Grajales, Nicholas Murphy, Sanjay J. Mathew, and Manish K. Jha

Abstract Burden of psychiatric disorders is compounded by their wide prevalence as well as the limited efficacy of currently available treatments and the current approaches for prescribing these treatments. The selection of treatments continues to be subjective and often results in a trial-and-error approach. Emerging research suggests that biological markers (or biomarkers) can be used to develop precision medicine approaches for psychiatric disorders. Furthermore, the biomarkers also promise to elucidate the underlying pathophysiological mechanisms which in turn can be used to develop novel therapeutic treatments. In this chapter we have focused on mood disorders and reviewed studies on electroencephalography (EEG),

R. L. Carreon
Menninger Department of Psychiatry and Behavioral Sciences, Baylor College of Medicine, Houston, TX, USA

Michael E. Debakey VA Medical Center, Houston, TX, USA

A. M. Rivas-Grajales
Department of Psychiatry, Boston University Chobanian & Avedisian School of Medicine, Boston, MA, USA

N. Murphy
Menninger Department of Psychiatry and Behavioral Sciences, Baylor College of Medicine, Houston, TX, USA

The Menninger Clinic, Houston, TX, USA

S. J. Mathew
Menninger Department of Psychiatry and Behavioral Sciences, Baylor College of Medicine, Houston, TX, USA

Michael E. Debakey VA Medical Center, Houston, TX, USA

The Menninger Clinic, Houston, TX, USA

M. K. Jha (✉)
Center for Depression Research and Clinical Care, O'Donnell Brain Institute, UT Southwestern Medical Center, Dallas, TX, USA
e-mail: manish.jha@utsouthwestern.edu

M. Macaluso et al. (eds.), *Drug Development in Psychiatry*, Advances in Neurobiology 30,
https://doi.org/10.1007/978-3-031-21054-9_12

magnetic resonance imaging (MRI), and blood-based biomarkers that can guide selection of one treatment versus another (treatment-selection biomarker) as well as biomarkers that can guide the development of novel therapeutics. These studies suggest that the use of objective physiological data is poised to alter the landscape of psychiatric diagnosis and treatment. However, practical and economic barriers remain as major hurdles. The key to finding such translational diagnostic and therapeutic biomarkers is a better understanding of the underlying pathophysiology, and despite the tremendous advances in neuroscience, it is clear there remains much left to be elucidated.

Keywords Biomarker · Psychiatric drug development · Precision medicine · Antidepressant · Neuroimaging

12.1 Introduction

Psychiatric disorders are widely prevalent and are some of the leading causes of disability in the world [1]. Burden related to psychiatric disorders has been compounded by the limited efficacy of currently available treatments and the current approaches for prescribing these treatments. As an example, major depressive disorder (MDD) is estimated to affect one in five adults in the United States during their lifetime [2]. While there are numerous treatment options available for MDD, including psychotherapy and antidepressant medications, the selection of treatment still continues to be based on subjective measures (such as overall depression severity) [3] and often results in a trial-and-error approach [4]. While clinical features of psychiatric disorders, such as symptom severity [5, 6], functional impairments [7–9], and quality of life [10, 11], have been proven to be helpful in prognosticating the response to treatment, their utility in guiding the selection of one treatment versus another has been limited [12]. Recent reports and ongoing studies are attempting to bridge these knowledge gaps by focusing on biological markers (biomarkers).

The importance of studying the biomarkers of psychiatric illnesses comes from the much-needed ability to provide psychiatry with nuanced, objective data that can help shed light into the underlying pathophysiologic processes that underpin psychiatric disorders. This may in turn serve as a way of improving psychiatric nosology by allowing for the subtyping of heterogeneous clinical phenotypes that share a syndrome-based diagnosis. As an extension of this, biomarkers could also serve as a guide for novel drug developments and may be the way of actualizing a personalized medicine approach to the treatment of psychiatric disorders. As for what exactly constitutes a biomarker, in 2015 the Food and Drug Administration (FDA) and National Institute of Health (NIH) created a collaborative Biomarker Working Group which defined a biomarker as "a defined characteristic that is measured as an indicator of normal biological processes, pathogenic processes or response to an exposure or intervention" [13]. In this chapter we will cover the electroencephalography (EEG), magnetic resonance imaging (MRI), and blood-based biomarkers that can help precision psychiatry by guiding selection of one treatment versus another (treatment-selection biomarker) as well as biomarkers that can guide the development of novel therapeutics while focusing on mood disorders to limit the scope.

A. *Electroencephalography (EEG) Biomarkers*

Integral to the investigation of candidate psychiatric biomarkers is an increased understanding of neurophysiology and the need for identification of "normal" neural parameters from which physiologic deviances can be described. Potential neurophysiologic biomarkers can be studied via a wide variety of techniques that provide spatial, temporal, and circuit-based insights into clinical pathology and are discussed here as diagnostic and treatment-selection biomarkers as well as neurophysiologic biomarkers for novel drug development.

Diagnostic and Treatment-Selection Biomarkers

Physiological findings commonly tend to be transdiagnostic, thereby underscoring the inherent limitations to the current syndrome-based classification system of the DSM-5 [14] and making it difficult for objective neurophysiologic observations to serve as a specific diagnostic biomarker. Therefore, an ideal diagnostic biomarker would have a specificity high enough that it could be used as an objective determinant of a disease-specific state. One such potential diagnostic biomarker can be the analysis of gamma oscillations which has been used as a method to distinguish unipolar depression from a bipolar disorder, with findings suggesting that these disorders feature unique signal patterns even when they have phenotypically similar affective states (i.e., active unipolar depressive episode compared to active bipolar depressive episode) [15].

Successful subtyping within DSM-5 diagnoses can provide clinical outcome and treatment response prediction data. An exemplar of this concept can be seen in a study using machine learning analysis of resting-state EEG data to identify two power envelope connectivity-based subtypes of patients for both MDD and PTSD, subtypes that were subsequently validated using functional magnetic resonance imaging (fMRI) connectivity correlates. These subtypes then underwent first-line clinical treatment of MDD and PTSD with results finding that those with the largest functional connectivity deviances were less likely to respond to antidepressant medication and psychotherapy, respectively [16].

EEG Biomarkers to Identify Novel Treatment Targets

The identification of either high-sensitivity or high-specific biomarkers can provide researchers insight into the pathophysiologic disturbances that are underlying a given disorder which can then be used as a guide to help identify novel targets of pharmacotherapeutic intervention. Studies into the influences behind mismatch negativity (MMN) reveal that it correlates with the overall status of N-Methyl-D-aspartate (NMDA) receptor functioning [17]. This is of particular note given the recent resurgence in research working to identify the mechanisms behind the antidepressant effects of ketamine. Given that NMDA antagonism is responsible for the anesthetic and psychomimetic effects of ketamine, the search for the mechanism behind the rapid and long lasting antidepressant effects seen at subanesthetic doses has brought the focus toward the glutamatergic system and, more broadly, neural network functional connectivity as related to glutamatergic regulation of synaptic plasticity [18–20]. With the glutamatergic system featuring a tripartite of ionotropic receptors, this has subsequently led researchers into the investigation of the kainate and α-amino-3-hydroxy-5-methyl-4-isoxazolepropionic acid (AMPA) receptors with preclinical studies supporting the concept that these receptors have influences on mood and could thus serve as targets for drug development [21, 22]. A notably impactful study done by Zhou et al. [23] revealed that AMPA receptor antagonism could prevent ketamine-induced increases in mammalian target of rapamycin (mTOR) and brain-derived neurotrophic factor (BDNF), which are thought to play a crucial role in the antidepressant effects of ketamine.

In addition to the previously mentioned studies of disease-specific bipolar disorder and unipolar depression gamma oscillation variations, there have been other studies done that show an increase in gamma oscillations and gamma power in those experiencing antidepressant response with ketamine [24]. Due to its influences on fast excitatory neurotransmission, gamma oscillations, long-range synchronization, long-term potentiation, and BDNF-induced neurogenesis, the AMPA receptor presents an area of high interest in neurologic/psychiatric drug development [21, 25]. In a successful demonstration of the experimental medicine approach, a recent study demonstrated the NMDA receptor engagement by lanicemine, a selective low-trapping NMDA antagonist, with changes in gamma oscillations in patients with post-traumatic stress disorder [26].

B. *Magnetic Resonance Imaging (MRI) Biomarkers*

Magnetic resonance imaging (MRI) has greatly advanced our understanding of abnormalities in brain structure and function in psychiatric disorders. MRI has also the potential to guide treatment selection and to make the process of drug development more efficient. Currently, indications for a specific treatment are based on clinical characteristics such as disease subtype and severity, prior treatment failures and tolerability, comorbid conditions, and patient's preferences. However, data to support specific recommendations are limited. This scarcity of evidence has encouraged efforts to evaluate predictive biomarkers that could be applied at the level of the individual patient [27].

MRI Treatment-Selection Biomarkers

As treatments are theorized to act by diverging neural mechanisms, individual differences in brain structure and function may explain variations in clinical response. As such, the identification of this variability using MRI may improve the precision of treatment selection [28]. In this context, it has been proposed that a treatment-selection biomarker should predict response to a specific treatment and predict non-response to an alternative treatment, to be clinically meaningful [29].

The Predictors of Remission in Depression to Individual and Combined Treatments (PReDICT) study was a randomized controlled trial which was specifically designed to identify biological, psychological, and clinical factors that may differentially impact treatment outcomes to three interventions, cognitive behavioral therapy (CBT), duloxetine, and escitalopram [30]. One-hundred twenty-two patients with major depressive disorder underwent resting-state fMRI at baseline and 12 weeks after treatment. They found that resting-state connectivity of the subcallosal cingulate cortex was differentially associated with response to CBT and pharmacological treatment. Specifically, positive connectivity was associated with remission with CBT and treatment failure with medication, while negative connectivity was associated with remission with medication and treatment failure with CBT. The authors concluded that these biomarkers may represent brain states that are differentially responsive to treatments with divergent mechanism of action, and this information could be used to personalize treatments [31].

A widely use biomarker to measure therapeutic response is change in ventral striatal activity and connectivity. This biomarker was incorporated in the Establishing Moderators and Biosignatures of Antidepressant Response in Clinical Care (EMBARC) trial, which was a placebo-controlled trial designed to explore clinical and biological markers of response to sertraline [32]. They found that abnormal patterns of ventral striatal activity before treatment were associated with better response to sertraline compared to placebo, which suggests that reward-related ventral activity could be used as an objective neural marker to guide antidepressant treatment choice [33]. Similarly, in a report from the Canadian Biomarker Integration Network in Depression (CAN-BIND) project, changes in ventral striatal functional connectivity were associated with improvement of depressive symptoms after 8 weeks of treatment with escitalopram [34].

MRI Biomarkers to Facilitate Development of Novel Treatments

Neuroimaging biomarkers are also being used to understand the antidepressant effects of ketamine in order to facilitate the development of novel and robust rapidly acting antidepressants [35]. Global brain connectivity (GBC) is a biomarker that allows the evaluation of large-scale functional connectivity networks and has been used to investigate normal and abnormal brain states [36, 37]. The antidepressant effect of ketamine has been shown to normalize aberrant global brain connectivity in open-label [38] and placebo-controlled studies [39, 40]. However, a study failed to replicate these results, which was attributed to differences in scan interval [41]. This led to the suggestion that GBR may be an immediate marker of neuronal reactivity to ketamine, with decreased sensitivity 2 days following treatment [41].

A major limitation in the development of new treatments in disorders of the central nervous system (CNS), is the lack of appropriate biomarkers to evaluate treatment effects in early-stage clinical trials to confirm target engagement, which has resulted in the withdrawal of some large pharmaceutical companies from CNS drug development [42]. Biomarkers are assumed to be closer to the biological mechanism of actions of the drug, and their inclusion in clinical trials is expected to reduce nonspecific effects and potentially allow the development of more effective therapies. Krystal et al. 2020 [43] presented the first application of this approach in a randomized controlled trial of the selective kappa opioid receptor antagonist aticaprant (JNJ-67953964). Using functional MRI, they showed that treatment with aticaprant was associated with a significant increase in ventral striatal activation during a monetary incentive delayed task in patients with anhedonia across the spectrum of mood and anxiety disorders compared to placebo [43]. Using a similar approach, Costi et al. 2021 [18] conducted a proof-of-concept placebo-controlled trial to test the effects of ezogabine, an anticonvulsant with antidepressant properties, on striatal activation. While there were no significant group differences which were attributed to the small sample size, they showed that participants in the ezogabine group had a numerical increase in ventral striatum activation following treatment compared to placebo, supporting the use of ezogabine in future clinical trials [44].

C. *Blood-Based Biomarkers*

A psychiatrically relevant example of blood-based biomarker for safety monitoring is absolute neutrophil count (ANC) which serves to facilitate the safe utilization of clozapine for treatment-resistant schizophrenia. More recent studies of blood-based biomarkers, ranging from genomic (including pharmacogenomic), epigenomics, transcriptomics, proteomics, to metabolomics marker, have interrogated a wide variety of biological processes. These markers are discussed here in the context of their utility in guiding treatments and in developing novel treatments.

Blood-Based Treatment-Selection Biomarkers

Combinatorial pharmacogenetic test kits that rely on common polymorphisms in genes encoding for cytochrome P450 (CYP) enzymes, which can affect the metabolism of psychiatric medications, are one of the commercially available treatment-selection blood-based biomarkers [45]. The main utility of these kits has been in predicting side effects with insufficient data to support the widespread use of these kits [46]. C-reactive protein (CRP), a nonspecific marker of inflammation, has been shown to be helpful in selecting between serotonergic versus non-serotonergic antidepressants. In two separate reports, Uher et al. [47] and Jha et al. [48] found that depressed patients with CRP levels less than 1 mg/L prior to treatment initiation experienced significantly greater reduction in depression severity with escitalopram as compared to nortriptyline and combination of escitalopram and bupropion, respectively. Conversely, depressed patients with CRP levels ≥1 mg/L responded significantly better to nortriptyline and bupropion-escitalopram as compared to escitalopram [47, 48]. In a more recent report, the association between elevated CRP and poorer antidepressant outcome appeared to be significant in females but not in males [49], suggesting the need for future studies to prospectively validate these findings.

While CRP is a clinically pragmatic treatment-selection biomarker, it can be affected by a multitude of acute and chronic factors, including obesity which in turn has been shown to serve as a treatment-selection biomarker [50, 51]. Thus, there is a need to identify more specific biomarkers, such as interleukin 17 (IL-17), elevated levels of which in the Combining Medications to Enhance Depression Outcomes (CO-MED) trial were associated with greater reduction in depression severity with bupropion-escitalopram combination but not with escitalopram alone or combination of venlafaxine and mirtazapine [52]. Further reports from the CO-MED trial suggest that inflammatory markers may be associated with anhedonia [53] and prognosticate improvements in anhedonia [54, 55].

Blood-Based Biomarkers for Novel Drug Development
While the brain is usually considered as an immune privileged organ, emerging evidence implicates immune system dysfunction in the pathophysiology of psychiatric disorders. Initial evidence strongly suggesting this link were derived from the observation of patients who received cytokines as treatment for their medical conditions, such as hepatitis C or malignancies with over a third of these patients developing major depression [56]. Meta-analytic evidence further suggests elevated levels of specific immune markers, such as interleukin 6 (IL-6) [57] which can be targeted with monoclonal antibodies [58]. Recent reports of patients with psoriasis and depression have also raised the antidepressant potential of monoclonal antibodies targeting Il-17-mediated immune response [59, 60]. Infliximab is the only specific antibody to be tested in patients with psychiatric disorder in the absence of an overt autoimmune disease. In their initial study, Raison et al. found no significant difference overall between infliximab and placebo [61]. However, in a post hoc analysis, they found that those with elevated levels of inflammation, as indicated by CRP levels of 5 mg/L, were associated with greater improvement with infliximab versus placebo [61]. However, McIntyre et al. did not find a significant antidepressant effect of infliximab in depressed patients with bipolar disorder who were selected based on markers of inflammation, either biochemical or phenotypic [62]. In their own post hoc analysis, they found a significant effect of childhood maltreatment where history of childhood physical abuse was associated with higher response rates with infliximab versus placebo [62]. Major barriers in developing monoclonal antibodies against inflammatory cytokines include our limited understanding of the specific mechanisms that link immune dysregulation to syndromic features of psychiatric disorders as well as lack of commercially available tests for these immune markers. In fact, a recently completed phase 2 clinical trial (NCT02473289) of sirukumab, a monoclonal antibody against IL-6, augmentation in depressed patients had to utilize CRP levels $\geq$3 mg/L. Publicly available results of this trial (https://clinicaltrials.gov/ct2/show/results/NCT02473289) suggest that improvement in depression with sirukumab augmentation did not differ from those of placebo.

12.2 Conclusion

In conclusion, the expanding body of literature suggests that the use of objective physiological data will alter the landscape of psychiatric diagnosis and treatment. However, practical and economic barriers remain as major hurdles. Therefore, the FDA-NIH Biomarker Working Group included terminology for study endpoints including the concepts of validated surrogate endpoints as well as reasonably likely surrogate endpoints. These surrogate endpoints may be the key to practical and economically feasible biomarkers of neuropsychiatric illnesses. A prototypical illustration of a practical neurophysiologic biomarker is polysomnography, which has made its way into common clinical practice due to its ability to successfully and reliably differentiate between overlapping clinical phenotypes of obstructive sleep apnea (OSA) and major depressive disorder (MDD) which determines whether the most appropriate next step of care is an antidepressant or CPAP device [63]. The key to finding such translational diagnostic and therapeutic interventions is intrinsically connected with a better understanding of the underlying pathophysiology, and as much as we have learned within the realm of neuroscience, it is clear there remains much left to be elucidated.

References

1. GBD 2019 Diseases and Injuries Collaborators. Global burden of 369 diseases and injuries in 204 countries and territories, 1990-2019: a systematic analysis for the Global Burden of Disease Study 2019. Lancet. 2020;396:1204–22.
2. Hasin DS, Sarvet AL, Meyers JL, Saha TD, Ruan WJ, Stohl M, Grant BF. Epidemiology of adult DSM-5 major depressive disorder and its specifiers in the United States. JAMA Psychiat. 2018;75:336–46.
3. Park LT, Zarate CA Jr. Depression in the primary care setting. N Engl J Med. 2019;380:559–68.
4. Jha MK, Trivedi MH. Personalized antidepressant selection and pathway to novel treatments: clinical utility of targeting inflammation. Int J Mol Sci. 2018;19:233.
5. Jha MK, Minhajuddin A, Chin Fatt C, Kircanski K, Stringaris A, Leibenluft E, Trivedi MH. Association between irritability and suicidal ideation in three clinical trials of adults with major depressive disorder. Neuropsychopharmacology. 2020;45:2147–54.
6. Jha MK, Minhajuddin A, South C, Rush AJ, Trivedi MH. Irritability and its clinical utility in major depressive disorder: prediction of individual-level acute-phase outcomes using early changes in irritability and depression severity. Am J Psychiatry. 2019;176:358–66.
7. Jha MK, Minhajuddin A, Greer TL, Carmody T, Rush AJ, Trivedi MH. Early improvement in psychosocial function predicts longer-term symptomatic remission in depressed patients. PLoS One. 2016;11:e0167901.
8. Jha MK, Teer RB, Minhajuddin A, Greer TL, Rush AJ, Trivedi MH. Daily activity level improvement with antidepressant medications predicts long-term clinical outcomes in outpatients with major depressive disorder. Neuropsychiatr Dis Treat. 2017;13:803–13.
9. Jha MK, Minhajuddin A, Greer TL, Carmody T, Rush AJ, Trivedi MH. Early improvement in work productivity predicts future clinical course in depressed outpatients: findings from the CO-MED trial. Am J Psychiatry. 2016;173:1196–204.
10. Jha MK, Greer TL, Grannemann BD, Carmody T, Rush AJ, Trivedi MH. Early normalization of Quality of Life predicts later remission in depression: findings from the CO-MED trial. J Affect Disord. 2016;206:17–22.
11. Vittengl JR, Jha MK, Minhajuddin A, Thase ME, Jarrett RB. Quality of life after response to acute-phase cognitive therapy for recurrent depression. J Affect Disord. 2021;278:218–25.
12. Trivedi MH. Modeling predictors, moderators and mediators of treatment outcome and resistance in depression. Biol Psychiatry. 2013;74:2–4.
13. Group F-NBW. BEST (Biomarkers, EndpointS, and other Tools) Resource [Internet], 2016.
14. Association AP, Association AP. Diagnostic and statistical manual of mental disorders: DSM-5. Arlington; 2013.
15. Fitzgerald PJ, Watson BO. Gamma oscillations as a biomarker for major depression: an emerging topic. Transl Psychiatry. 2018;8:177.

16. Zhang Y, Wu W, Toll RT, Naparstek S, Maron-Katz A, Watts M, Gordon J, Jeong J, Astolfi L, Shpigel E, Longwell P, Sarhadi K, El-Said D, Li Y, Cooper C, Chin-Fatt C, Arns M, Goodkind MS, Trivedi MH, Marmar CR, Etkin A. Identification of psychiatric disorder subtypes from functional connectivity patterns in resting-state electroencephalography. Nat Biomed Eng. 2021;5:309–23.
17. Nagai T, Kirihara K, Tada M, Koshiyama D, Koike S, Suga M, Araki T, Hashimoto K, Kasai K. Reduced mismatch negativity is associated with increased plasma level of glutamate in first-episode psychosis. Sci Rep. 2017;7:2258.
18. Li BJ, Friston K, Mody M, Wang HN, Lu HB, Hu DW. A brain network model for depression: from symptom understanding to disease intervention. CNS Neurosci Ther. 2018;24:1004–19.
19. Yao N, Skiteva O, Zhang X, Svenningsson P, Chergui K. Ketamine and its metabolite (2R,6R)-hydroxynorketamine induce lasting alterations in glutamatergic synaptic plasticity in the mesolimbic circuit. Mol Psychiatry. 2018;23:2066–77.
20. Zanos P, Moaddel R, Morris PJ, Riggs LM, Highland JN, Georgiou P, Pereira EFR, Albuquerque EX, Thomas CJ, Zarate CA Jr, Gould TD. Ketamine and Ketamine metabolite pharmacology: insights into therapeutic mechanisms. Pharmacol Rev. 2018;70:621–60.
21. Alt A, Nisenbaum ES, Bleakman D, Witkin JM. A role for AMPA receptors in mood disorders. Biochem Pharmacol. 2006;71:1273–88.
22. Catches JS, Xu J, Contractor A. Genetic ablation of the GluK4 kainate receptor subunit causes anxiolytic and antidepressant-like behavior in mice. Behav Brain Res. 2012;228:406–14.
23. Zhou W, Wang N, Yang C, Li XM, Zhou ZQ, Yang JJ. Ketamine-induced antidepressant effects are associated with AMPA receptors-mediated upregulation of mTOR and BDNF in rat hippocampus and prefrontal cortex. Eur Psychiatry. 2014;29:419–23.
24. Gilbert JR, Zarate CA Jr. Electrophysiological biomarkers of antidepressant response to ketamine in treatment-resistant depression: gamma power and long-term potentiation. Pharmacol Biochem Behav. 2020;189:172856.
25. Fuchs EC, Doheny H, Faulkner H, Caputi A, Traub RD, Bibbig A, Kopell N, Whittington MA, Monyer H. Genetically altered AMPA-type glutamate receptor kinetics in interneurons disrupt long-range synchrony of gamma oscillation. Proc Natl Acad Sci U S A. 2001;98:3571–6.
26. Ramakrishnan N, Lijffijt M, Green CE, Balderston NL, Murphy N, Grillon C, Iqbal T, Vo-Le B, O'Brien B, Murrough JW, Swann AC, Mathew SJ. Neurophysiological and clinical effects of the NMDA receptor antagonist lanicemine (BHV-5500) in PTSD: a randomized, double-blind, placebo-controlled trial. Depress Anxiety. 2021;38:1108–19.
27. Collins FS, Varmus H. A new initiative on precision medicine. N Engl J Med. 2015;372:793–5.
28. Woo CW, Chang LJ, Lindquist MA, Wager TD. Building better biomarkers: brain models in translational neuroimaging. Nat Neurosci. 2017;20:365–77.
29. McGrath CL, Kelley ME, Holtzheimer PE, Dunlop BW, Craighead WE, Franco AR, Craddock RC, Mayberg HS. Toward a neuroimaging treatment selection biomarker for major depressive disorder. JAMA Psychiat. 2013;70:821–9.
30. Dunlop BW, Binder EB, Cubells JF, Goodman MM, Kelley ME, Kinkead B, Kutner M, Nemeroff CB, Newport DJ, Owens MJ, Pace TW, Ritchie JC, Rivera VA, Westen D, Craighead WE, Mayberg HS. Predictors of remission in depression to individual and combined treatments (PReDICT): study protocol for a randomized controlled trial. Trials. 2012;13:106.
31. Dunlop BW, Rajendra JK, Craighead WE, Kelley ME, McGrath CL, Choi KS, Kinkead B, Nemeroff CB, Mayberg HS. Functional connectivity of the subcallosal cingulate cortex and differential outcomes to treatment with cognitive-behavioral therapy or antidepressant medication for major depressive disorder. Am J Psychiatry. 2017;174:533–45.
32. Trivedi MH, McGrath PJ, Fava M, Parsey RV, Kurian BT, Phillips ML, Oquendo MA, Bruder G, Pizzagalli D, Toups M, Cooper C, Adams P, Weyandt S, Morris DW, Grannemann BD, Ogden RT, Buckner R, McInnis M, Kraemer HC, Petkova E, Carmody TJ, Weissman MM. Establishing moderators and biosignatures of antidepressant response in clinical care (EMBARC): rationale and design. J Psychiatr Res. 2016;78:11–23.

33. Greenberg T, Fournier JC, Stiffler R, Chase HW, Almeida JR, Aslam H, Deckersbach T, Cooper C, Toups MS, Carmody T, Kurian B, Peltier S, Adams P, McInnis MG, Oquendo MA, Fava M, Parsey R, McGrath PJ, Weissman M, Trivedi M, Phillips ML. Reward related ventral striatal activity and differential response to sertraline versus placebo in depressed individuals. Mol Psychiatry. 2020;25:1526–36.
34. Dunlop K, Rizvi SJ, Kennedy SH, Hassel S, Strother SC, Harris JK, Zamyadi M, Arnott SR, Davis AD, Mansouri F, Schulze L, Ceniti AK, Lam RW, Milev R, Rotzinger S, Foster JA, Frey BN, Parikh SV, Soares CN, Uher R, Turecki G, MacQueen GM, Downar J. Clinical, behavioral, and neural measures of reward processing correlate with escitalopram response in depression: a Canadian Biomarker Integration Network in Depression (CAN-BIND-1) report. Neuropsychopharmacology. 2020;45:1390–7.
35. Abdallah CG, Sanacora G, Duman RS, Krystal JH. Ketamine and rapid-acting antidepressants: a window into a new neurobiology for mood disorder therapeutics. Annu Rev Med. 2015;66:509–23.
36. Cole MW, Yarkoni T, Repovs G, Anticevic A, Braver TS. Global connectivity of prefrontal cortex predicts cognitive control and intelligence. J Neurosci. 2012;32:8988–99.
37. Anticevic A, Brumbaugh MS, Winkler AM, Lombardo LE, Barrett J, Corlett PR, Kober H, Gruber J, Repovs G, Cole MW, Krystal JH, Pearlson GD, Glahn DC. Global prefrontal and fronto-amygdala dysconnectivity in bipolar I disorder with psychosis history. Biol Psychiatry. 2013;73:565–73.
38. Abdallah CG, Averill LA, Collins KA, Geha P, Schwartz J, Averill C, DeWilde KE, Wong E, Anticevic A, Tang CY, Iosifescu DV, Charney DS, Murrough JW. Ketamine treatment and global brain connectivity in major depression. Neuropsychopharmacology. 2017;42:1210–9.
39. Abdallah CG, Averill CL, Salas R, Averill LA, Baldwin PR, Krystal JH, Mathew SJ, Mathalon DH. Prefrontal connectivity and glutamate transmission: relevance to depression pathophysiology and ketamine treatment. Biol Psychiatry Cogn Neurosci Neuroimaging. 2017;2:566–74.
40. Abdallah CG, Dutta A, Averill CL, McKie S, Akiki TJ, Averill LA, Deakin JFW. Ketamine, but Not the NMDAR Antagonist Lanicemine, increases prefrontal global connectivity in depressed patients. Chronic Stress (Thousand Oaks). 2018;2:2470547018796102.
41. Kraus C, Mkrtchian A, Kadriu B, Nugent AC, Zarate CA Jr, Evans JW. Evaluating global brain connectivity as an imaging marker for depression: influence of preprocessing strategies and placebo-controlled ketamine treatment. Neuropsychopharmacology. 2020;45:982–9.
42. Borsook D, Becerra L, Fava M. Use of functional imaging across clinical phases in CNS drug development. Transl Psychiatry. 2013;3:e282.
43. Krystal AD, Pizzagalli DA, Smoski M, Mathew SJ, Nurnberger J Jr, Lisanby SH, Iosifescu D, Murrough JW, Yang H, Weiner RD, Calabrese JR, Sanacora G, Hermes G, Keefe RSE, Song A, Goodman W, Szabo ST, Whitton AE, Gao K, Potter WZ. A randomized proof-of-mechanism trial applying the 'fast-fail' approach to evaluating κ-opioid antagonism as a treatment for anhedonia. Nat Med. 2020;26:760–8.
44. Costi S, Morris LS, Kirkwood KA, Hoch M, Corniquel M, Vo-Le B, Iqbal T, Chadha N, Pizzagalli DA, Whitton A, Bevilacqua L, Jha MK, Ursu S, Swann AC, Collins KA, Salas R, Bagiella E, Parides MK, Stern ER, Iosifescu DV, Han MH, Mathew SJ, Murrough JW. Impact of the KCNQ2/3 channel opener Ezogabine on reward circuit activity and clinical symptoms in depression: results from a randomized controlled trial. Am J Psychiatry. 2021;178:437–46.
45. Jha MK, Trivedi MH. Pharmacogenomics and biomarkers of depression. Handb Exp Pharmacol. 2019;250:101–13.
46. Zeier Z, Carpenter LL, Kalin NH, Rodriguez CI, McDonald WM, Widge AS, Nemeroff CB. Clinical implementation of pharmacogenetic decision support tools for antidepressant drug prescribing. Am J Psychiatry. 2018;175:873–86.
47. Uher R, Tansey KE, Dew T, Maier W, Mors O, Hauser J, Dernovsek MZ, Henigsberg N, Souery D, Farmer A, McGuffin P. An inflammatory biomarker as a differential predictor of outcome of depression treatment with escitalopram and nortriptyline. Am J Psychiatry. 2014;171:1278–86.

48. Jha MK, Minhajuddin A, Gadad BS, Greer T, Grannemann B, Soyombo A, Mayes TL, Rush AJ, Trivedi MH. Can C-reactive protein inform antidepressant medication selection in depressed outpatients? Findings from the CO-MED trial. Psychoneuroendocrinology. 2017;78:105–13.
49. Jha MK, Minhajuddin A, Chin-Fatt C, Greer TL, Carmody TJ, Trivedi MH. Sex differences in the association of baseline c-reactive protein (CRP) and acute-phase treatment outcomes in major depressive disorder: findings from the EMBARC study. J Psychiatr Res. 2019;113:165–71.
50. Jha MK, Wakhlu S, Dronamraju N, Minhajuddin A, Greer TL, Trivedi MH. Validating pretreatment body mass index as moderator of antidepressant treatment outcomes: findings from CO-MED trial. J Affect Disord. 2018;234:34–7.
51. Green E, Goldstein-Piekarski AN, Schatzberg AF, Rush AJ, Ma J, Williams L. Personalizing antidepressant choice by sex, body mass index, and symptom profile: an iSPOT-D report. Personal Med Psychiatry. 2017;1:65–73.
52. Jha MK, Minhajuddin A, Gadad BS, Greer TL, Mayes TL, Trivedi MH. Interleukin 17 selectively predicts better outcomes with bupropion-SSRI combination: novel T cell biomarker for antidepressant medication selection. Brain Behav Immun. 2017;66:103–10.
53. Jha MK, Miller AH, Minhajuddin A, Trivedi MH. Association of T and non-T cell cytokines with anhedonia: role of gender differences. Psychoneuroendocrinology. 2018;95:1–7.
54. Jha MK, Minhajuddin A, Gadad BS, Trivedi MH. Platelet-derived growth factor as an antidepressant treatment selection biomarker: higher levels selectively predict better outcomes with bupropion-SSRI combination. Int J Neuropsychopharmacol. 2017;20:919–27.
55. Jha MK, Minhajuddin A, Gadad BS, Chin Fatt C, Trivedi MH. Higher S100B levels predict persistently elevated Anhedonia with Escitalopram monotherapy versus antidepressant combinations: findings from CO-MED trial. Pharmaceuticals (Basel). 2019:12, 184.
56. Raison CL, Borisov AS, Broadwell SD, Capuron L, Woolwine BJ, Jacobson IM, Nemeroff CB, Miller AH. Depression during pegylated interferon-alpha plus ribavirin therapy: prevalence and prediction. J Clin Psychiatry. 2005;66:41–8.
57. Haapakoski R, Mathieu J, Ebmeier KP, Alenius H, Kivimäki M. Cumulative meta-analysis of interleukins 6 and 1β, tumour necrosis factor α and C-reactive protein in patients with major depressive disorder. Brain Behav Immun. 2015;49:206–15.
58. Kappelmann N, Lewis G, Dantzer R, Jones PB, Khandaker GM. Antidepressant activity of anti-cytokine treatment: a systematic review and meta-analysis of clinical trials of chronic inflammatory conditions. Mol Psychiatry. 2018;23:335–43.
59. Papp KA, Reich K, Paul C, Blauvelt A, Baran W, Bolduc C, Toth D, Langley RG, Cather J, Gottlieb AB, Thaçi D, Krueger JG, Russell CB, Milmont CE, Li J, Klekotka PA, Kricorian G, Nirula A. A prospective phase III, randomized, double-blind, placebo-controlled study of brodalumab in patients with moderate-to-severe plaque psoriasis. Br J Dermatol. 2016;175:273–86.
60. Griffiths CEM, Fava M, Miller AH, Russell J, Ball SG, Xu W, Acharya N, Rapaport MH. Impact of Ixekizumab treatment on depressive symptoms and systemic inflammation in patients with moderate-to-severe psoriasis: an integrated analysis of three phase 3 clinical studies. Psychother Psychosom. 2017;86:260–7.
61. Raison CL, Rutherford RE, Woolwine BJ, Shuo C, Schettler P, Drake DF, Haroon E, Miller AH. A randomized controlled trial of the tumor necrosis factor antagonist infliximab for treatment-resistant depression: the role of baseline inflammatory biomarkers. JAMA Psychiat. 2013;70:31–41.
62. McIntyre RS, Subramaniapillai M, Lee Y, Pan Z, Carmona NE, Shekotikhina M, Rosenblat JD, Brietzke E, Soczynska JK, Cosgrove VE, Miller S, Fischer EG, Kramer NE, Dunlap K, Suppes T, Mansur RB. Efficacy of adjunctive Infliximab vs Placebo in the treatment of adults with bipolar I/II depression: a randomized clinical trial. JAMA Psychiat. 2019;76:783–90.
63. Jehan S, Auguste E, Pandi-Perumal SR, Kalinowski J, Myers AK, Zizi F, Rajanna MG, Jean-Louis G, McFarlane SI. Depression, obstructive sleep apnea and psychosocial health. Sleep Med Disord. 2017;1:0012.

Chapter 13
The Role of fMRI in Drug Development: An Update

Owen Carmichael

Abstract Functional magnetic resonance imaging (fMRI) of the brain is a technology that holds great potential for increasing the efficiency of drug development for the central nervous system (CNS). In preclinical studies and both early- and late-phase human trials, fMRI has the potential to improve cross-species translation of drug effects, help to de-risk compounds early in development, and contribute to the portfolio of evidence for a compound's efficacy and mechanism of action. However, to date, the utilization of fMRI in the CNS drug development process has been limited. The purpose of this chapter is to explore this mismatch between potential and utilization. This chapter provides introductory material related to fMRI and drug development, describes what is required of fMRI measurements for them to be useful in a drug development setting, lists current capabilities of fMRI in this setting and challenges faced in its utilization, and ends with directions for future development of capabilities in this arena. This chapter is the 5-year update of material from a previously published workshop summary (Carmichael et al., Drug Discov Today 23(2):333–348, 2018).

Keywords Functional magnetic resonance imaging · Drug development · Central nervous system · Biomarkers

13.1 Introduction

Functional magnetic resonance imaging (fMRI) of the brain is frequently discussed in review articles on central nervous system (CNS) drug development as a potentially useful tool to enhance the drug development process. Over the past few years, both general articles on drug development [1–3] and those specific to pain [4–6],

---

O. Carmichael (✉)
Pennington Biomedical Research Center, Baton Rouge, LA, USA
e-mail: owen.carmichael@pbrc.edu

M. Macaluso et al. (eds.), *Drug Development in Psychiatry*, Advances in Neurobiology 30,
https://doi.org/10.1007/978-3-031-21054-9_13

schizophrenia [7], multiple sclerosis [8], and neurodevelopmental disorders [9] have called for greater incorporation of fMRI measurements into clinical trials of novel therapies. Yet, while a recent systematic review revealed that a large number (nearly 1400) of clinical trials had been registered on ClinicalTrials.gov with fMRI as an outcome measure, and many of these (over 400) were drug trials, only 7% of these studies were completed and published their results. In addition, regulatory agencies such as the Food and Drug Administration in the United States and the European Medicines Agency in the European Union have not, to date, recognized any fMRI measurements as qualified biomarkers for drug trials in any clinical area. Thus, while fMRI appears to have promising uses in the drug development process, that usefulness has not yet translated into high impact applications.

The goal of this chapter is to provide an update on the current status of fMRI as an emerging drug development tool. It will cover the rationale for its use, the requirements it must meet, its current capabilities, challenges that limit its use, and a set of activities that are proposed to meet the challenges. As such, it synthesizes material found in recent review articles on theoretical use cases for fMRI and/or biomarkers more generally in the drug development process, but not the technical capabilities and challenges of fMRI [1–9]. It connects this material to reviews of best practices in fMRI data collection and analysis, which do not cover drug development applications [10, 11]. It constitutes a 5-year update of a previously published workshop summary [12].

The remainder of this section provides broader context surrounding functional magnetic resonance imaging (fMRI) in drug development and shapes the remaining discussion of the state of the art, challenges in deploying it in clinical trials, and activities that could help to address the challenges.

13.1.1 fMRI Definitions

Data Acquisition Paradigms for fMRI

This chapter is concerned with fMRI of the brain, with data predominantly provided either by blood oxygenation level dependent (BOLD) [13, 14] or arterial spin labeling (ASL) perfusion MRI [15] sequences. Other dynamic MRI techniques (such as dynamic contrast-enhanced imaging) or dynamic neuroimaging techniques outside of MRI (such as positron emission tomography) are not covered. We consider the three main experimental settings within which fMRI data is collected. First, *task-based fMRI* uses sensory or cognitive stimuli to provoke responses from the brain regions or circuits involved in processing the stimuli. Second, *resting state fMRI* (rsfMRI) is used to examine fMRI data collected during ostensible times of rest—when no predesigned stimulus is presented to the individual [16]. Third, *pharmacological MRI* (phMRI) records fMRI signals collected before, during, and after the administration of pharmacological agents [17].

Data Analysis Paradigms for fMRI

BOLD data is typically analyzed with one of three goals in mind. The first goal, usually referred to as *subtraction paradigm* or *activation analysis*, is to assess relationships between BOLD signal amplitude and experimental conditions—for example, differences in BOLD signal amplitude between differing conditions such as the presence vs. absence of a certain stimulus. Often, the assessment of these relationships is performed separately at each spatial location in the brain (a so-called *mass univariate* analysis), and attempts are made post hoc to aggregate findings across locations. The second goal, usually referred to as a *functional connectivity* analysis, is to assess temporal relationships between BOLD signal characteristics at multiple spatial locations; for example, correlations between BOLD signals at a pair of spatial locations may be assessed [16]. Note that these temporal relationships may be assessed without regard to changing stimuli or pharmacological agents (or in the nominal absence of them, as in resting state fMRI), or differences in these spatiotemporal relationships between conditions can be assessed. The third goal of fMRI analysis is to assess quantitative signal characteristics of the local BOLD signal, especially its frequency characteristics such as the amplitude of low-frequency fluctuations of ALFF [18].

13.1.2 Drug Development Definitions

The drug development process starts with identification of a biological target hypothesized to be implicated in a disease process. Thousands of molecules may then be tested for their chemical properties and ability to bind to the target molecule *in vitro* [19, 20]. Of those, tens of molecules are tested in preclinical animal models of the disease. In addition to toxicity, molecules are tested for their pharmacokinetics (PK), bioavailability at the target organ, *in vivo* target engagement, biological or chemical response that can be directly linked to the molecular action in the organism (pharmacodynamics, PD), and efficacy in the animal model [21, 22]. This process builds confidence that the handful of molecules with the best *in vitro* and *in vivo* profiles will also be safe, engage the intended target, and potentially treat the disease in human patients. Activities then shift to human clinical trials, where the process can include four different phases. Phase 0 studies are employed to test scientific hypotheses or novel imaging methods in the absence of therapy or to evaluate novel therapeutic strategies at presumed subclinical ("micro") doses [23–25]. In Phase 1, tens of individuals are enrolled to demonstrate that the drug is tolerable and safe at multiple doses including those anticipated to evoke an efficacious clinical response [26–29]. PK and PD responses are often assessed in Phase 1 in order to provide better informed dose selection or design of subsequent Phase 2 trials. In Phase 2, roughly hundreds of participants are tested at a single or few doses to compare therapeutic responses against those of a similar cohort treated with placebo or control therapy. Safety assessments are made to assess less common drug side effects. In Phase 3, usually hundreds to thousands of participants are tested at multiple sites, typically at a single dose, to confirm safety and efficacy profiles suggested in earlier phase trials. Throughout this process, there is potential for imaging assays such as fMRI to make an impact on mechanistic evaluation of drugs and differentiation between treatment responders and nonresponders.

Definition of "Biomarker"

The FDA biomarkers, endpoints, and other tools (BEST) glossary of biomarker terms defined a biomarker as "A defined characteristic that is measured as an indicator of normal biological processes, pathogenic processes, or biological responses to an exposure or intervention, including therapeutic interventions. A biomarker is not a measure of how an individual feels, functions, or survives" [30]. Because this definition is broad enough to encompass any biological measurement that is relevant to health and disease, taxonomies have been articulated to clarify distinct use cases. The BEST taxonomy includes diagnostic, monitoring, predictive, prognostic, pharmacodynamic/response, safety, and susceptibility/risk biomarkers. These biomarkers respectively cover applications in detecting the presence of pathological processes, assessing change in them or in their response to a treatment over time, identifying individuals likely to react in one way or another to treatment, predicting changes in the clinical status of patients, assessing biological changes caused by exposure to a treatment, assessing likelihood of an adverse event following exposure to a treatment, and identifying currently clinically healthy individuals likely to develop a clinical condition over time [4, 31]. An additional category, the surrogate endpoint, is a health- or disease-related biological measurement that is targeted by treatment trials because it is believed to provide information about eventual clinical efficacy of the treatment. The EMA developed a similar taxonomy, including BEST-like prognostic, diagnostic, predictive, pharmacodynamic, safety, and surrogate endpoint categories, while omitting the susceptibility and monitoring categories and adding an enrichment category for measurements that identify individuals most likely to respond to a proposed treatment (European Medicines Agency: Guideline on the clinical investigation of medicines for the treatment of Alzheimer's disease). An alternative taxonomy that is more specific to CNS drug development, developed by Collegium Internationale Neuro-Psychopharmacologicum (CINP) and the Japanese Society of Neuropsychopharmacology (JSNP), articulates five tiers of biomarkers, each representing distinct and successive questions on the pathway of development of a novel agent [2]. The first three tiers quantify brain exposure to the agent, brain engagement of its specific target, and brain functional changes resulting from exposure to the agent. The fourth tier amounts to the EMA enrichment tier, and the fifth amounts to the surrogate endpoint category.

13.1.3 Theoretical Schema for Utilizing fMRI for Biomarkers in Drug Development

Investigators within several different clinical domains (neuropsychiatric disorders, pain, schizophrenia, multiple sclerosis, and neurodevelopmental disorders) have articulated how they believe fMRI may be best situated to effectively contribute to the drug development effort within that domain [3, 6–9]. Using the EMA and BEST taxonomies, fMRI is viewed within each of these domains as promising as a pharmacodynamic/response biomarker, with specific domains seeing a potential role as a diagnostic [6, 9], prognostic [6], predictive [6, 7], stratification [7, 9], and safety [9] biomarker. The FDA's Center for Drug Evaluation and Research (CDER) also articulated a possible role for fMRI as a new approach methodology (NAM) for assessing safety [1], and a review of the existing resting state fMRI literature has noted that to date, its main biomarker applications have been in pharmacodynamics and response [5]. The rest of this section elaborates on these fMRI biomarker applications at the aforementioned stages of the drug development pipeline.

Preclinical Phase

Preclinical studies use transgenic or inducible rodent models of disease, and *in vitro* studies that serve to characterize a novel molecule's pharmacologic properties and predicted clinical effect [32–35]. However, many of these approaches lack predictive validity to complex human neuropsychiatric disorders. This problem has motivated the search for alternative measurements in animal models that, alongside existing in vitro and animal model techniques, help to inform human studies via homologies between animal and human measurements. Because fMRI measurements in human studies relevant to drug development are prevalent, rodent fMRI is increasingly under consideration as a source of homologous measurements [36]. While less common than human fMRI studies, small animal phMRI studies are being performed in such clinical domains as Fragile X syndrome [37], depression [38], and drugs of abuse [39]. Methodological challenges that limit homologies between human and rodent fMRI, such as the need for optimal and standardized anesthesia protocols in the rodents, are also being explored [40].

Early-Phase Human Studies

In early-phase clinical studies, fMRI is typically used as a pharmacodynamic or response biomarker: detecting a functional CNS effect of pharmacological treatment in brain regions appropriate to the compound's mechanism and/or target population [17, 41–43]. Although it is not technically a marker of target engagement (*i.e.*, of pharmacological agent binding to a target site), an fMRI signal can provide indirect evidence of target engagement if a biologically plausible link can be established between the fMRI response and the molecular target [44–46]. Dose-response and exposure-response relationships established using fMRI are of particular value to guide dose selection for later phases [47–49]. Failure to observe such fMRI signals in early-phase studies can contribute to an overall body of knowledge about the agent, the sum total of which is used to decide whether to proceed to later-phase studies or change direction. One example of this approach, emerging from the pain literature, is the "N-of-1+Imaging study," which adds deep mechanistic phenotyping (including neuroimaging) to small early-phase pain therapeutic studies, which usually lean heavily on subjective ratings of pain reduction as their primary readout [50]; plausible changes in fMRI signals would add confidence to positive findings in the subjective pain data. The NIMH FAST-FAIL paradigm, emerging from the neuropsychiatric literature, is a more rigid and transparent approach along these lines—in early-phase studies, test agents are required to modify prespecified biomarkers (such as fMRI markers) in the specific fashion laid out in preregistered trial protocols, and if they don't, the agents are not allowed to progress to later trial phases [51, 52]. Alcohol use disorder (AUD) is one domain in which positive early-phase fMRI findings for one agent (varenicline) contributed to the decision to further develop the agent, which went on to show clinical efficacy to reduce alcohol cravings [53]; on the other hand, null early-phase fMRI findings for another AUD agent (pexacerfont) contributed to the decision to shift attention away from that agent and toward a competitor (verucerfont) [54]. Finally, resting state fMRI markers have been suggested as potentially useful in early-phase trials of Alzheimer therapeutics [55].

Late-Phase Human Studies

Later phases (phases 2 and 3) involve large patient studies at multiple clinical sites designed to identify or confirm clinical efficacy [56]. The emphasis for fMRI in these types of studies is more likely to be on demonstrating normalization of a disease-related fMRI signal, at one or very few dose levels. Most such studies aspire for submission to regulators as part of a new drug approval. Consequently, these studies might include fMRI to provide a more objective demonstration of disease modification, thereby increasing the evidence base for a regulatory submission. A 2021 systematic review [57] suggested that 109 phase 3 or phase 4 drug interventions with fMRI readouts were registered on ClinicalTrials.gov, or 21.6% of all registered drug interventions with fMRI readouts.

13.1.4 Current Regulatory Status of fMRI Biomarkers

The United States Food and Drug Administration (FDA) and the European Medicines Agency (EMA) are regulatory agencies that approve commercialization and human use of new drugs [58, 59]. Approval is based on review of evidence provided by drug sponsors on safety and efficacy of the new drug in treating specific disease indications [60, 61]. Both agencies have acknowledged that for the process of drug development to continue to thrive, novel technologies that facilitate the drug development process have to be continuously developed [62–64]. That is why both agencies have proposed a formal process for qualification of technologies such as fMRI for specific fit-for-purpose uses in drug development. Interested parties may submit a request for qualification of a biomarker if they believe there is a need that can be met by the biomarker in a specific context of use and enough data to support its use in that context. The agencies will review the application and either issue an opinion on whether they agree with the strength of the argument for the biomarker use and context or provide advice about what may be additionally required to issue a qualification opinion. Once qualified for use in a predefined context, the agencies will accept biomarker data as evidence, within the context of use, for new drug safety and efficacy.

To date, only small imaging-based biomarkers have been qualified by EMA or FDA. In 2012, EMA-qualified amyloid PET is an enrichment biomarker for Alzheimer's disease. Total kidney volume measured by MRI and other imaging techniques has been qualified as a prognostic biomarker for enrichment of clinical trials in autosomal dominant polycystic kidney disease by both agencies, but change in kidney volume has not been qualified as a PD biomarker [65]. EMA qualified dopamine transporter neuroimaging as an enrichment biomarker in early-stage Parkinson's disease trials. Several other biomarkers are currently in the process of development toward qualification in specific contexts of use, having received favorable feedback, specific questions, and suggestions for further development from the agencies. At this time, only one such biomarker, under consideration with the EMA, utilizes fMRI: a battery of brain MRI measurements that includes two task fMRI paradigms and a resting state acquisition, with a goal of enriching trials in autism spectrum disorder. Other imaging-based markers include CT-based tumor volume change in cancer; MRI- or CT-based anatomic features of bone for knee osteoarthritis; MRI-based iron-corrected T1 relaxation times, proton density fat fraction, and elastography of the liver for NASH and NAFLD; MRI-based elastography of the bowel and terminal ileum for Crohn's disease; and DXA-based bone mineral density change for hip and nonvertebral fracture.

The burden of proof for qualification of a biomarker such as fMRI is high and has not yet been fully standardized by the FDA and EMA. Typically, the agencies require data from more than a few small trials in which the biomarker has demonstrated value in the specified context of use. Also required are characterizations of the precision and reproducibility of the biomarker. Despite scientific interest in fMRI, there remain few industry-sponsored trials with sufficiently rigorous fMRI data for regulatory agencies to consider when reviewing an application for a new therapeutic. To explain why biomarker qualifications of fMRI have been limited to date, the following sections review what is required for fMRI to demonstrate value and current challenges that limit the ability of fMRI to be useful in this setting.

13.2 What Is Required of Any fMRI Biomarker?

13.2.1 Reproducibility and Modification by the Pharmacological Agent

As with any assay designed to assess an intervention, the reproducibility of an fMRI paradigm forms part of its initial characterization and validation as fit for purpose. In addition, the readouts should respond to pharmacological manipulation: a Phase 1 fMRI study, for example, would be expected to establish dose-response and exposure-response relationships between the selected readouts and the administered compound, in order to inform dose selection for subsequent patient trials [45, 48, 66–68]. Both reproducibility and responsiveness are important: a paradigm that is highly reproducible but impervious to pharmacological manipulation will not be useful, for example. Ideally, evidence of pharmacological modulation should be presented with suitable comparator compounds [69, 70].

The current status of fMRI markers with regard to reproducibility and modification by pharmacological agents is covered in Sects. 13.4.3, 13.4.4, and 13.4.5.

13.2.2 Well-Defined Measurement Characteristics

To ensure reliability, sensitivity, specificity, and accuracy of collected fMRI data, a quantitative, industry-standard method for assessing the fMRI measurement process is required. For structural MRI, the NIST/ISMRM system phantom provides such a method, including standardized MRI readouts to which scanners can be calibrated, as well as international standards for those readouts [71]. These readouts include contrast, resolution, and accuracy of distance and volume measurements in the image space. The NIST/RSNA QIBA ADC phantom similarly provides standardized readouts and international standards for diffusion MRI sequences [72]. Combining this approach with dynamic assessments such as temporal SNR [73] would be required to allow scanners worldwide to be quantitatively evaluated for their fMRI performance. The most widely used and validated techniques could not only be standardized but also tied to a quantitative gold standard recognized by global regulatory bodies. This level of industry-standard measurement quantification is required for fMRI to be a viable technique in late-stage clinical trials, especially multisite ones. Professional bodies have begun to address this measurement standardization issue, including the Functional Biomedical Informatics Research Network (fBIRN) and the QIBA FMRI Biomarker Committee.

13.2.3 Prespecification of Acquisition and Analysis Steps

In the context of drug development, fMRI methodology should be held to the same standard as other clinical endpoints—namely, methods must be prespecified and fixed for the duration of the study. This prespecification should include a thorough description of task design and implementation, image acquisition and quality control, data preprocessing, ROI definition, model estimation, and endpoint calculation [10, 11]. With such prescribed methodology in hand, an fMRI experiment can be reduced to a binary outcome more suitable to inform drug development decisions. To note, in drug development, primary end points and hypotheses are typically based on prespecified ROIs [74, 75], and power calculations are performed accordingly to avoid underpowering.

13.2.4 Real-World Applicability (Diverse Centers, Diverse Technologists)

Significant logistical requirements must be met when deploying fMRI to large numbers of heterogeneous, nonacademic imaging facilities is required for the typical late-stage clinical trial. Several consortia [76–80] have advanced the state of the art in techniques for meeting these requirements. For multicenter fMRI at the level of dozens to hundreds of imaging facilities, methods must be "turn-key" and able to accommodate all major manufacturers, models, and even field strengths. High-performance MRI scanners are not always located near participant recruitment centers, necessitating pragmatic decisions in the planning phase of a multicenter fMRI study. The most basic decision is whether fMRI acquisition can involve 1.5T or 3T systems. If acquisition must occur at multiple field strengths to facilitate recruitment, analytic end points emphasizing within-subject outcomes are most amenable, though ways to account for variance across magnets exist [81, 82]. The decision to include 1.5T systems also has implications for acquisition parameters and how these are standardized across sites. Clinical imaging facilities are so diverse in their equipment (vendor, model, software release, coil, and gradient configuration), and technologist expertise is so variable that perfect standardization of all acquisition parameters across all sites is not feasible. Instead, parameters most likely to impact endpoint derivation should be identified and fixed across sites, while other less critical settings must be allowed to vary between sites. Protocol optimization thus involves achieving certain fixed parameters and adjusting others to maximize performance at that site. At minimum, factors that should be consistent across sites are the type of pulse sequence, in-plane voxel size, slice thickness and spacing, temporal resolution and number of observations, flip angle, coverage, and behavioral conditions. Suggestions for this parameter set and the impact of different choices have been proposed [83]. A field map (including reconstruction of both magnitude and phase information) should be acquired to allow for distortion correction of fMRI images collected using EPI techniques [84]. Multichannel coils are preferred due to their increased signal to noise ratio [85]. Fat suppression and parallel imaging should be used if available to increase tissue contrast and reduce acquisition times [86, 87].

At the current time, task fMRI and phfMRI paradigms require a variety of infrastructure that can be expensive and specialized, thus unavailable at many imaging centers. However, the prerequisites for measuring whole-brain resting state functional connectivity in a valid fashion are present at the vast majority of clinical imaging facilities worldwide. Indeed, published data suggests that carefully controlled, prespecified, auditable resting state fMRI study can yield a rich connectome amenable to informing drug development, once logistical hurdles are overcome [88]. Unfortunately, due to broad diversity across sites, leading-edge acquisition techniques such as multiband and multiple-TE fMRI [89] are not presently realistic in this setting, nor are hardware peripherals needed for task-based fMRI typically available.

13.2.5 Rigorous Quality Control

A rigorous QC procedure must be in place to ensure the site acquires analyzable data for the duration of the trial. Errors left undetected quickly become systematic and could render worthless all scans from a site. This QC not only need to detect errors at the site but also must be tracked and fully auditable, especially in the context of regulatory approval [10]. Again, suggestions for what such a QC process should entail have been made [90], but at minimum should include DICOM header checks of protocol compliance, tests of dynamic range and temporal SNR, artifact inspection, adequacy of FOV placement, and head motion (now commonly reduced to a single vector, Framewise Displacement [91]).

13.3 What Can fMRI Biomarkers Do Currently?

Several review articles in recent years have summarized the established and developing uses for fMRI biomarkers within specific clinical domains including alcohol use disorder [92], multiple sclerosis [93, 94], opioid addiction [95], and epilepsy [96]. This section synthesizes those reviews together with individual drug trials utilizing fMRI.

13.3.1 Change in Response to Acute and Chronic Administration of Certain Drugs

A large number of published studies have demonstrated that a wide range of fMRI methods and paradigms are sensitive to changes following both acute (*i.e.*, after a single dose) or chronic (*i.e.*, multiple dose) pharmacological treatment. The vast majority of these have been academic studies using marketed drugs whose efficacy and effective doses have already been established.

Review articles from differing clinical domains suggest that knowledge about fMRI sensitivity to drug administration varies by domain. The effects of multiple sclerosis drugs on fMRI signals appear to be minimally studied, although there is growing recognition that fMRI could potentially be useful for assessing brain repair mechanisms [93, 94]. Similarly, there is little knowledge in this area within opioid use disorder, despite some initial evidence from the treatment of heroin addiction [95]. Meanwhile, the acute effects of pain therapeutics on fMRI signals have been heavily studied using a variety of compounds [97–107], and the fMRI effects of ketamine in healthy controls and patients with neuropsychiatric disorders have been investigated thoroughly [70, 108–116]. While pain studies have largely focused on intended effects on pain-related brain functions, in epilepsy, the main focus of phfMRI has been on predicting unintended, adverse side effects [96]. Additional drug classes that induce phMRI signals include antidepressants [117, 118], antipsychotics [119], cognitive enhancers [120], drugs of abuse [121, 122], calcium channel blockers [123, 124], cyclooxygenase-2 (COX-2) inhibitors [125], muscarinic acetylcholine receptor modulators [126–128], and therapies traditionally thought to impact solely immune system activity [129].

The largest subset of studies over the past 5 years was performed in healthy controls, to better understand drug effects in the absence of disease. Clinical populations under study span a wide variety of clinical conditions, including neuropsychiatric, developmental, and addictive disorders along with epilepsy, cognitive impairment, and migraine. All of the major fMRI measurement methods—phfMRI, task fMRI, and resting state—are represented, as are single-arm, parallel arm, and crossover designs. The majority assess acute effects of drug exposure on fMRI signals, most of the rest assess chronic effects of drug exposure, and a small number [130] assess the effects of chronic drug exposure on acute fMRI responses to ketamine. The latter can be thought of as using phfMRI as a standardized "challenge paradigm," analogous to the oral glucose tolerance test or the VO2max aerobic fitness test.

13.3.2 Identify Converging Mechanisms of Drug Response Across Drugs

Pain provides a clear case in which fMRI has identified pharmacodynamic effects in humans [97] and animals [131] on aberrant fMRI signals that are shared across compounds that differ wildly in mechanism (*e.g.*, opioids, nonsteroidal antiflammatories, and even tetrahydrocannabinol) [97]. In particular, a large meta-analysis has suggested that a variety of acute painful stimuli induce aberrant fMRI signals in the secondary somatosensory cortex, insula, cingulate cortex, and thalamus [107], and studies in fibromyalgia [132], osteoarthritis pain [133], complex regional pain syndrome [134], and models of neuropathic pain [135] have suggested that a variety of pharmacological strategies suppress these signals, especially those in the insula and cingulate cortex. These modulated fMRI signals may have clinical relevance—signals derived from the insula and inferior parietal lobe have been shown to predict pregabalin treatment response toward experimentally evoked or clinical pain states [132]. Nonetheless, the causal or correlational relationship between fMRI end points and subjective accounts of pain is not ubiquitous within the pain neuroimaging literature [136]. A similar investigation into convergent mechanisms underlying treatment responses is at an earlier stage in mood disorder-related cognitive impairment [137]. In this area, convergent treatment effects have been observed in the dorsolateral prefrontal cortex and default mode network, although synthesizing these results into a coherent model has been complex.

13.3.3 Support Translation Between Preclinical and Clinical Studies

In drug development, the clinical biomarker plan for a candidate therapeutic is formulated while the compound is still being optimized in preclinical testing (discovery phase). It is highly advantageous to be able to demonstrate an effect of the compound in preclinical species (typically rodents) on the same or similar biomarker as that being considered for use in the clinical phases. Unfortunately, few cross-species studies have been published to date demonstrating analogous effects of novel compounds in both preclinical models and human participants. Recent review articles have highlighted the importance of such cross-species studies [9, 138, 139]. NMDAR agonists have garnered the most attention in this regard [140, 141]. In one study, the NMDAR antagonist phencyclidine (PCP) elicited a strong phfMRI response in the rat brain involving the prefrontal and cingulate cortices and the thalamus, and to a lesser extent the hippocampus [115]. Rat 2-deoxy-glucose (2DG) studies with another NMDAR antagonist (ketamine) have shown similar results [142]. A very similar activation pattern is observed in healthy human volunteers given ketamine [69, 70, 108]. Further studies have directly compared ketamine responses in humans and rodents [141]. In a second example, the phMRI response to intravenous buprenorphine is concordant in many regions in rats and healthy humans [143], although deactivation of some regions is noted in rats but not humans. A recent review used contrasting case studies to point out that for any given drug, strong homologies across species are far from guaranteed: atomoxetine appears to generate similar phfMRI responses across species while selective serotonin reuptake inhibitors (SSRIs) do not [139].

Outside of phfMRI designs, very little data on cross-species homologies in drug modulation of fMRI signals has been reported. Data showing similar modulation of resting state functional connectivity between humans and rodents is in its infancy [144], and cross-species task fMRI data in the pharmacological modulation setting is minimal, due likely to the added difficulty of imposing task conditions in a compelling way across species. A small number of recent studies have continued to expand our understanding of drug effects on rodent models, with drugs of interest including ketamine [145], raclopride [146], Xiao Yao San [147], and telmisartan [38]. Meanwhile, the body of studies that assess across species homologies without drug modulation (beyond small-animal anesthetics) continues to grow [148–150].

13.4 What Are the Challenges in Developing fMRI Biomarkers?

In addition to the requirements of the drug development process and the capabilities of fMRI in this setting, there is a set of challenges that must be addressed to increase the utility of fMRI in clinical trials. The questions include technical ones about how fMRI studies should be performed, as well as biological ones about the inferences that should be made about the functioning of the brain based on fMRI data. Some recent review articles have focused on individual challenges such as poor replication of effects at the individual level (Sect. 13.4.4), while others have discussed a subset of the challenges listed below [151].

13.4.1 Lack of Agreed-Upon Concise Readouts from fMRI Exams

Although image-based representations of group-level analyses are commonly seen as the primary fMRI output of interest in a research setting, predefined numeric summary values from each scan are needed to use fMRI as a biomarker in drug development studies. There are several motivations for summary values. (1) The drastically lower dimensionality allows results of the fMRI experiment to be imported into standard databases, combined with other data such as pharmacokinetics, and analyzed by accredited statisticians. These databases are auditable and operate under strict revision and access control, safeguards that assure a high level of data integrity. (2) Defining summary readouts as primary end points before conducting the study requires the practitioner to prescribe specific and simple *a priori* hypotheses, an exercise that can highlight uncertainty in the anticipated outcome and prevent false positive rate inflation [152]. (3) Summary values can reduce the multiple comparison burden and thus avoid ill-defined choices among correction schemes. (4) End points extracted from voxelwise analyses may be subject to bias due to circularity [153], and determining the neuroanatomical localization of an effect in a set of voxels can be cumbersome or ill-defined.

However, while there is a clear need for low-dimensional fMRI summary measures, there is no widespread agreement on optimal approaches. Graph-theoretical summaries [154] and factor analytic methods such as independent components analysis (ICA) [155] are used extensively in exploratory research studies, but these methods have many variants and operating parameters. The more traditional approach of selecting a region of interest (ROI) and reducing all signals in a region to a single summary score [156] requires a choice of summary score (e.g., the mean, median, or mode). The optimal summary measure for any given clinical trial is unclear and may depend on the hypothesized action of the drug. If the treatment is hypothesized to strongly affect a single brain region, an ROI-based analysis might be effective, but if effective correlations among regions are hypothesized to be modified, a graph theoretical approach may be preferred. If the hypothesized effects cover a broad network of regions, a seed-based or ICA approach might capture the hypothesized effect. One reason for the lack of broad agreement on standardized summary measures for fMRI is that the range of task paradigms, pharmacological mechanisms, and imaging sites involved is large and few standardization studies have been published. Recently, with the rising deployment of artificial intelligence/machine learning (AI/ML) techniques throughout the biomedical sciences, AI/ML is increasingly thought of as an approach for deriving useful low-dimensional summary measures [157].

13.4.2 Poor Replication of Effects at the Individual Level

One key reason why fMRI was not considered fully developed for drug development applications is that the test-retest repeatability of many fMRI paradigms—that is, the ability of the paradigm to produce similar signals when applied to the same individual, under similar conditions, over time—was not uniformly high. For many years, the small amount of test-retest repeatability data that was published was highly mixed. For some task paradigms, including low level visual, auditory, motor, and eye tracking tasks [158, 159], painful stimuli [160], tasks in reward and working memory domains [161], tasks in stress, reward, and fear domains [162], face matching, reward processing, memory, and executive domains [163], published reproducibility covered a wide range between measurements, from poor to excellent. Other papers, including a meta-analysis of dozens of papers presenting test-retest data from many different paradigms, have suggested that on average test-retest repeatability is poor [164, 165]. Ketamine phfMRI had reported repeatability that was high in regions expected to respond to ketamine and fair to high elsewhere [69]. Resting state measures were reported to have repeatability that similarly covered a range, from fair to good [166, 167]. Even novel techniques for enhancing fMRI signal strength via contrast agents have shown a wide range of repeatabilities [168].

13.4.3 Poor Replication of Effects at the Group Level

While the published data on test-retest repeatability of individual fMRI measurements in individual research participants is mixed at best, group-level replication of fMRI findings—the finding of the same differences in fMRI measurement distribution between a pair of groups, across multiple samplings of those groups—may be even worse. For much of its history, fMRI research has emphasized mapping the linkage between brain activation and behavior, thus representing a signal detection problem [169]. Unfortunately attempts to distinguish signal from noise are fraught with confounds due to the high dimensionality of fMRI data, leading to a high probability of type I errors. The likelihood of false positives becomes even more dangerous when users do not understand the statistics underlying their empirical claims (Henson's "imager's fallacy" [170], or engage in circular selection ("voodoo correlation") [171] and non-independent analysis ("double dipping")) [153]. In relation to this, an important concern of underpowered fMRI studies was first pointed out in the commentary to "voodoo correlation" paper by [172] in the context of fMRI for human brain mapping. This was then later elaborated on in depth [173, 174]. When a study is underpowered, the power (probability) to detect a true effect is low. This results in consequences such as overestimation of the true effect (as only large observed effects pass the *p*-value threshold). This is also referred to also as winner's curse. Subsequently, a low reproducibility of follow-up studies ensues, as they find evidence of smaller or no effects, thus failing to reproduce findings of the initial study. As the field has evolved, more emphasis is placed on a clear reporting of methodology including experimental design, correction for multiple comparisons, ROI definition, and the statistical tests performed—ideally with sufficient detail to allow replication of the analysis [175, 176]. Yet, even the most clearly reported post hoc analysis is insufficient to inform a clinical trial where experimental power and reliability must be estimated in advance, with implications for study design and sample size [161].

13.4.4 A Replication Crisis?

This concerning body of test-retest and group-level replication findings have fueled the narrative that fMRI is in the midst of a "replication crisis." However, several authors have attempted to put these results in perspective. First, authors have noted that fMRI paradigms with low test-retest repeatability may be unsuitable for crossover study designs but could very well produce robust group-level differences that are replicable [177, 178]. For example, the same faces task that showed poor within-subject reproducibility [161] has been widely used in parallel-group designs, reproducibly showing exaggerated amygdala responses to negative emotional stimuli in conditions such as depression, and pharmacological attenuation of this response following drug treatment [117]. Second, the analytic methods we usually apply to fMRI were designed in the first place to identify group differences, not to provide highly repeatable individual-level readouts [178, 179]. Moving to alternative fMRI readouts, including those that separately model multiple components of repeatability [179], those that use machine-learning to derive useful multi-voxel activity patterns [178], and those that aggregate fMRI data across multiple tasks [178], could enhance test-retest repeatability. Third, the data acquisition methods we usually employ were not designed to carefully generate high test-retest repeatability data. As such, alternative acquisition approaches involving extended aggregation [178, 179], new pulse sequences such as multi-echo BOLD [179], and optimized stimulus design [179] could further improve test-retest repeatability. Finally, fMRI paradigms with poor test-retest repeatability may be the very same paradigms that respond robustly to the drug of interest. Clearly, both drug responses and repeatability/replication are important and need to be held in balance when designing novel studies.

13.4.5 Lack of Full Understanding of Molecular Modifiers of the fMRI Signal

PhMRI studies often ascribe treatment-related BOLD signal changes to changes in the functioning of neurons. However, such inferences must necessarily be physiologically agnostic because local changes in the BOLD signal are determined by at least three physiological effects—changes to cerebral blood volume (CBV), cerebral blood flow (CBF), and the cerebral metabolic rate of oxygen metabolism (CMRO2)—as well as glucose metabolism [180]. Past research shows an undeniable association between cerebral hemodynamics and activity of neuronal populations, although exact relationships are still a matter of debate [181–185]. There is a consensus that graded increases in neuronal metabolism result in graded increases in the BOLD response, while corresponding decreases in neuronal metabolism result in negative BOLD responses [186–189]. However, a precise understanding of the relative contributions of metabolic, vascular, and neural processes to the BOLD signal is still under development. For some pharmacological treatments, such a lack of physiological specificity may be a major limitation to the ability to make development decisions based on phMRI results. For example, investigators may have *a priori* hypotheses about which of the physiological changes are plausible for the treatment to elicit, based on earlier preclinical experiments; separate measurement of separate physiological effects could provide an important indicator of agreement with prior experiments. In addition, effects on specific physiological parameters may be seen as advantageous by developers. Emerging approaches could help to identify distinct physiological changes induced by the treatment. So-called "calibrated" BOLD, a T2*-based method, collects BOLD data during a controlled carbon dioxide challenge to "calibrate" the BOLD signal; together with simultaneous BOLD-ASL imaging sequences, this allows simultaneous estimation of CBF and CMRO2 [190–193]. Another method, a T2*- and T2-based method known as "quantitative BOLD" or "qBOLD" [194], provides similar information, but from a single scan that requires no hypercapnia challenge. T2-based methods, such as TRUST [195] and QUIXOTIC [196], are also promising [197]. The collection of BOLD and FDG PET data simultaneously, with FDG PET providing the measurement of tissue metabolic properties, is also emerging [198]. Similarly, the simultaneous collection of fMRI and neurotransmitter PET could help to clarify the contribution of molecular prerequisites for neuronal responses, to the hemodynamic response [199]. While these methods are promising, no systematic study has evaluated which ones add significant value to specific clinical trial designs or classes of compounds.

13.4.6 Lack of Full Understanding of Real-World Confounders/Best Practices for Participant Preparation, Etc.

In addition to fundamental questions about the biological drivers of the BOLD signal, the more practical question of how to best model and remove the effects of systemic physiological processes (including respiration and cardiac function) from the BOLD signal and even whether to do so at all when assessing drug effects is not fully understood despite some work in this area [200–202]. Similarly, while practical software exists for removing drift in the MRI signal caused by physical processes, a comprehensive understanding of that software and rigorous comparisons among them is still being developed [203]. The effects of numerous nuisance acquisition parameters on the BOLD signal is still under study, from specifics of rodent anesthesia administration [204–206] to the specifics of control conditions or tasks [207]. In addition, any extrinsic action taken by the individual that changes neuronal activity, blood volume, or perfusion is a potential confound of drug effects on fMRI. These include the use of concomitant psychoactive drugs or other drugs that affect perfusion, as well as any activities that may affect arousal, cognitive state, and blood flow, such as caffeine intake, sleep, and exercise [208]. However, it is unclear how aggressively each of these factors should be controlled to minimize confounding effects on fMRI data from clinical trials.

13.4.7 Lack of Understanding of Relationships Among Dose, fMRI Signals, and Clinical Outcomes

fMRI has relatively underexplored potential in early-phase drug discovery for informing dosing regimens. In principle, a consistent relationship between dose and the BOLD response could enable predictions about the expected brain response at higher or lower doses. However, there are very few examples of the use of this kind of pharmacometric fMRI modeling in the literature. For example, it is not clear whether it is more effective to model fMRI signals as a nonlinear function of plasma concentration of a compound (a pharmacokinetic approach; e.g., [209]) or as a mediator of the association between plasma concentration and PD effects [209–211]. Future work should evaluate whether important dose-related fMRI effects are present not only in the magnitude of the fMRI response but also in its temporal derivates and their dependency on spatial location, as suggested by early work on biphasic responses to dopamine and GABA manipulations [98, 212, 213]. Addressing such complexity with computationally tractable models is a continuing challenge [214].

Even if a plausible relationship between treatment dose and BOLD response can be obtained, a continuing challenge is the "inference gap": the uncertain relationship between the magnitude of a pharmacological effect seen in fMRI and probable clinical efficacy. In comparison, receptor occupancy (RO) PET imaging has matured to such a degree that goal occupancy ranges for some targets and molecular modes of action are established or can be estimated from animal studies. For example, striatal D2 dopamine RO of at least ~60% is associated with effective antipsychotic activity but occupancy greater than ~80% elicits undesirable side effects [45]. This goal range allows the company to confidently test a clinical hypothesis about the benefit of a treatment. The classic example of this is the neurokinin 1 (NK1) antagonist class of compounds, which turned out to be clinically ineffective in affective and pain indications, despite using doses that were known, based on RO-PET studies, to yield near 100% occupancy [215]. Without the occupancy data, it would not be possible to be sure whether the doses chosen were high enough to fully engage the target. Similarly, studies comparing the μ-opioid antagonists GSK1521498 and naltrexone were able to index phfMRI effects of those compounds against μ-opioid receptor occupancy in each participant [216–219]. Without the RO-PET data, dissociating on-target fMRI effects (i.e., those directly related to receptor binding in the spatial vicinity of the detected fMRI effect) from off-target or downstream effects (those distal to PET-reported receptor binding) is difficult [143, 215, 220–222]. The ability to plausibly bridge this inference gap would greatly enhance the value of fMRI for early-phase drug development. Combining RO-PET with fMRI is a step in this direction, by linking target engagement (RO-PET) to pharmacodynamic effects (fMRI) [216]. However, the ability to interpret the dose-response curve and relate it to probable clinical efficacy, especially for novel therapeutic targets, remains a challenge.

13.4.8 *Lack of Established Protocols for fMRI-Informed Participant Screening, Stratification, Trial Enrichment*

Most uses of fMRI in phase 3 clinical trials have centered on characterizing treatment effects. However, fMRI has the potential to identify subjects that should be enrolled in one arm of a trial or another. An example of this type of approach comes from the obesity world, in which fMRI can identify obese individuals whose brains are hyperresponsive, vs. weakly responsive, to images of food cues; the hyperresponsive individuals go on to respond more poorly to weight loss treatment, assumedly due to a poorer ability to maintain low food intake [223]. In clinical trials with weight loss as a primary outcome, the hyper-activators may be at higher risk of nonresponse and thus stratified to a higher dose regimen, or clinical trials with fMRI change as a primary outcome might enrich the sample for such hyper-activators, since these are the individuals most able to show a response to treatment: their fMRI activation is the farthest from normal and has the most room to normalize. Other recent attempts to use fMRI to predict treatment responses come from neuropsychiatry, where an rsfMRI-based neural signature of likely poor response to mainline therapy among PTSD patients has been identified [224], and from epilepsy, where an expert panel has determined that fMRI paradigms are useful to identify patients most likely to have poorer outcomes from resection surgery [225]. Another promising advance in this area is the identification of an rsfMRI marker that tracks with treatment response among migraine without aura patients [226]. Finally, one of the central goals of the large-scale EU-AIMS Longitudinal European Autism Project (LEAP) is to use fMRI to cluster individuals with autism spectrum disorder into subsets, including subsets that respond differently to different treatments [227]. Outside of these example areas, fMRI as a treatment response predictor appears to be underutilized.

13.5 How to Overcome the Challenges

Previous sections have made it clear that major obstacles must be overcome to enable broader use of fMRI in clinical trials. For several of these obstacles, specific research and development projects that could help to overcome them are obvious. But in addition, community-wide activities could further enhance the utility of fMRI in the drug development process. Each of these activities are centered on the ultimate goal of providing fMRI tools that are sensitive to drug-induced change, relative to repeatability, valid with respect to established clinical endpoints, and standardized across measurement platforms.

13.5.1 Form Public–Private Partnerships to Fund fMRI Method Development and Validation Studies

There is a general consensus that the following activities would enhance the utilization of fMRI in the drug development process: soliciting and reanalyzing data from trials with null findings, replicating findings from prior trials using differing treatments or imaging methods, enhancing the usability of research-grade fMRI processing software to make it easily adopted by the research community, and publishing well-designed and executed data sets to serve as publicly available gold standards for validation of novel imaging methods. However, each of these activities is difficult to obtain funding to do, both from funding agencies and industry sponsors. For example, despite isolated funding programs to enhance the usability of already developed neuroimaging software (see, for example, the NITRC program, www.nitrc.org), the NIH overwhelmingly focuses on providing funds for the development of novel neuroimaging techniques rather than maintenance and further enhancement of successful ones. Individual drug companies, meanwhile, are primarily focused on development of their own treatments rather than development of specific research tools (such as fMRI) that are used as part of the development pathway.

Because no specific entity makes a systematic effort to fund these activities, promising new fMRI techniques "die on the vine," novel studies recreate the mistakes or null findings of unpublished prior studies, and the viability of treatments remains unclear due to a lack of validation. Funding for such activities likely requires public–private partnerships including entities that jointly recognize that such activities have the potential to enhance the utility of fMRI in all clinical trials, including those sponsored by industry and funding agencies. Among the many partnerships operating currently, the Autism Biomarkers Consortium for Clinical Trials (ABC-CT, [228]) and Autism Spectrum Proof of Concept Initiative (ASPI, [229]) appear to be the largest efforts that at least partially involve developing fMRI biomarkers along these lines for a specific clinical domain; the NIH BRAIN Initiative continues to develop fMRI biomarkers more broadly [230], and the FDA Critical Path infrastructure has biomarker methodology development as part of its mission [231]. The need for partnerships that might involve fMRI biomarkers has also been identified in other clinical domains, including pain and opioid use disorder [232], as well as neuropsychiatric disorders [233]. Software partnerships could follow the NITRC model, funding software developers to make emerging neuroimaging technologies available on additional computing platforms, and able to interface with additional imaging data types and software systems. Such partnerships would need effective means to disseminate the resulting software and track the success of dissemination. Replication and validation partnerships could focus on identifying the most promising fMRI methods and findings and funding their replication using complementary measurement techniques or model systems. To date, there are few such replication studies [234]. The end result of these partnerships would be a broader set of validated software tools and greater knowledge about the generalizability of findings from one treatment to another.

13.5.2 Develop Infrastructure for Sharing Clinical Trial Data Without Exposing Sponsors or CROs to Legal Risks

The drug industry invests major resources in clinical trials that include fMRI, and there are numerous ways in which additional value could be extracted from that investment after primary analyses are completed. Subsequent analyses could be used to power novel studies, to understand the findings of similarly conducted studies, and to evaluate novel fMRI methods in general. However, any trial sponsor that makes such fMRI data public faces significant risks. A biased party could reanalyze the data to support a spurious claim about the trial. Dissemination of participant data also poses risks for confidentiality. Thus, there is a need to develop whatever infrastructure is needed to enable as much clinical trial data sharing as possible. Such data sharing could begin with FDA and EMA policies that require trial operators to submit all collected fMRI data to these regulatory bodies or an intermediary as a precondition to registering the trial. The data could be reidentified centrally, and various data characteristics could be provided only in the most general terms to avoid identification of the participant and trial. Regulatory bodies could, as a starting point, only release fMRI data from placebo arms to accelerate testing of methods for assessing longitudinal fMRI change. Importantly, informed consent forms would need to ensure that participants understand the implications of their consent to such data sharing. Data sharing efforts play a prominent role in the public–private consortia described above, and in recent years, specific efforts have sought to define data standards that individual trials should adhere to [235]. The need for greater work on enhanced data sharing across trial sponsors has been cited by at least one review article in recent years [31].

13.5.3 Establish an Ongoing, Regular Conference on fMRI in Clinical Trials

The previous sections should make it clear that there are significant unanswered questions about the role fMRI can and should play in clinical trials and a major need for ongoing research. But there is currently no conference venue designed for investigators to exchange information about the advancing state of the art in this area and for public–private partnerships aimed at overcoming structural issues to develop. None of the current conferences in neuropharmacology, neuroimaging, and neuroscience has the critical mass of focus on this topic to incorporate sessions on it in a well-reasoned way. A 1-day add-on meeting on fMRI in clinical trials attached to one of the major ongoing conferences or a stand-alone meeting would be beneficial. We hope the current exposition makes clear the interest that such a meeting would provoke from industry and academia and its high potential for self-sustaining growth.

13.5.4 Strengthen Publishing of All fMRI Validation Studies, Positive or Negative; Strengthen fMRI Method Reporting Standards

Functional MRI studies are time-consuming, labor-intensive, and expensive. Negative results in these studies are problematic usually as the study is not published and the data then languish—sometimes because the study was poorly executed, but often due to lack of power, an assumed statistical model that was incorrect, or some other problem [236]. But, due to this "file drawer" problem, it is difficult if not impossible to know how many other investigators have attempted to test the same hypothesis and found no result. Before funding an fMRI research study to test a hypothesis, it would be good to know whether unsuccessful studies have already been done on the subject or more likely whether imaging data exists, which could be reanalyzed to answer the question or refocus the study design.

There are a number of solutions to the file drawer problem. First is the publicly or collaboratively available neuroimaging data repository [237]. Within the United States, NIH-supported databases such as RDoC provide repositories for NIMH-supported research study data [238].

While public repositories address access to relevant unpublished data, they do not address the risk of non-publishability, relative to the high cost of doing the fMRI study. The neuroimaging community has realized that the data itself is worth publishing if it was collected well. Scientific Data (http://www.nature.com/sdata/) is one example of a journal whose purpose is to publish descriptions of valuable data sets, regardless of study results. It will be vital that such data set publications include enough detail on equipment, subject characteristics, and data acquisition parameters for other scientists to use the data to make decisions about the designs of their own studies.

Beyond disseminating and publishing the data, publishing negative results could be valuable for furthering the qualification of fMRI in specific contexts. However, a minimum of rigor should be placed in the presentation of negative results to enhance the community's confidence that the report does not constitute a "false negative." To that aim, the following should be observed when presenting negative results:

- In order to strengthen the case for publishing the negative results, the paper should present data and arguments that demonstrate that negative results are not due to (1) poor study design, (2) lack of standardization/harmonization of data acquisition protocol, (3) poor protocol compliance, (4) poor data quality, and (5) unique or nonstandard analyses. Following standardized method reporting checklists [239] would help to assure readers that all necessary methodological factors required to fully evaluate the validity of the null finding have been reported [240].
- Ideally, negative results should be published with public access to the original data sets.

- A publication discussion section should highlight possible reasons for the negative findings.
- Use of peer-reviewed preregistered reports (e.g., at Cortex, Drug, and Alcohol Dependence, others; https://osf.io/8mpji/wiki/home/) simultaneous with registration at ClinicalTrials.gov is encouraged. Publication of data analyzed via preregistered analytic plans is gradually becoming more common [241, 242], and review articles have pressed for greater use of preregistration [240].

The publication of null fMRI findings in drug trials is relatively rare [130], as are null findings with similar neurophysiology measurements such as MEG [243]. Once negative result publishing venues become more mature, funding agencies should build on their recent record of demanding more data sharing from its grantees; all grantees should be required to attempt publication of negative results in the event that it meets the quality standards outlined above.

References

1. Avila AM, Bebenek I, Bonzo JA, Bourcier T, Davis Bruno KL, Carlson DB, et al. An FDA/CDER perspective on nonclinical testing strategies: classical toxicology approaches and new approach methodologies (NAMs). Regul Toxicol Pharmacol. 2020;114:104662.
2. Suhara T, Chaki S, Kimura H, Furusawa M, Matsumoto M, Ogura H, et al. Strategies for utilizing neuroimaging biomarkers in CNS drug discovery and development: CINP/JSNP Working Group report. Int J Neuropsychopharmacol. 2017;20(4):285–94.
3. van der Doef TF, Zaragoza Domingo S, Jacobs GE, Drevets WC, Marston HM, Nathan PJ, et al. New approaches in psychiatric drug development. Eur Neuropsychopharmacol. 2018;28(9):983–93.
4. Davis KD, Aghaeepour N, Ahn AH, Angst MS, Borsook D, Brenton A, et al. Discovery and validation of biomarkers to aid the development of safe and effective pain therapeutics: challenges and opportunities. Nat Rev Neurol. 2020;16(7):381–400.
5. Khalili-Mahani N, Rombouts SA, van Osch MJ, Duff EP, Carbonell F, Nickerson LD, et al. Biomarkers, designs, and interpretations of resting-state fMRI in translational pharmacological research: a review of state-of-the-Art, challenges, and opportunities for studying brain chemistry. Hum Brain Mapp. 2017;38(4):2276–325.
6. Smith SM, Dworkin RH, Turk DC, Baron R, Polydefkis M, Tracey I, et al. The potential role of sensory testing, skin biopsy, and functional brain imaging as biomarkers in chronic pain clinical trials: IMMPACT considerations. J Pain. 2017;18(7):757–77.
7. Papanastasiou E, Shergill SS. Why should pharmacological trials in schizophrenia employ functional magnetic resonance imaging (fMRI)? J Psychopharmacol. 2021;35(9):1158–60.
8. De Stefano N, Giorgio A. Advanced MRI measures like DTI or fMRI should be outcome measures in future clinical trials – commentary. Mult Scler. 2017;23(11):1458–60.
9. Diaz-Caneja CM, State MW, Hagerman RJ, Jacquemont S, Marin O, Bagni C, et al. A white paper on a neurodevelopmental framework for drug discovery in autism and other neurodevelopmental disorders. Eur Neuropsychopharmacol. 2021;48:49–88.
10. Schwarz AJ, Becerra L, Upadhyay J, Anderson J, Baumgartner R, Coimbra A, et al. A procedural framework for good imaging practice in pharmacological fMRI studies applied to drug development# 1: processes and requirements. Drug Discov Today. 2011;16(13):583–93.
11. Schwarz AJ, Becerra L, Upadhyay J, Anderson J, Baumgartner R, Coimbra A, et al. A procedural framework for good imaging practice in pharmacological fMRI studies applied

to drug development# 2: protocol optimization and best practices. Drug Discov Today. 2011;16(15):671–82.
12. Carmichael O, Schwarz AJ, Chatham CH, Scott D, Turner JA, Upadhyay J, et al. The role of fMRI in drug development. Drug Discov Today. 2018;23(2):333–48.
13. Logothetis NK, Pfeuffer J. On the nature of the BOLD fMRI contrast mechanism. Magn Reson Imaging. 2004;22(10):1517–31.
14. Logothetis NK, Wandell BA. Interpreting the BOLD signal. Annu Rev Physiol. 2004;66:735–69.
15. Petersen ET, Zimine I, Ho YC, Golay X. Non-invasive measurement of perfusion: a critical review of arterial spin labelling techniques. Br J Radiol. 2006;79(944):688–701.
16. van den Heuvel MP, Hulshoff Pol HE. Exploring the brain network: a review on resting-state fMRI functional connectivity. Eur Neuropsychopharmacol. 2010;20(8):519–34.
17. Wandschneider B, Koepp MJ. Pharmaco fMRI: determining the functional anatomy of the effects of medication. Neuroimage Clin. 2016;12:691–7.
18. Yang J, Gohel S, Vachha B. Current methods and new directions in resting state fMRI. Clin Imaging. 2020;65:47–53.
19. Mathew SJ, Manji HK, Charney DS. Novel drugs and therapeutic targets for severe mood disorders. Neuropsychopharmacology. 2008;33(9):2080–92.
20. Wong EH, Yocca F, Smith MA, Lee C-M. Challenges and opportunities for drug discovery in psychiatric disorders: the drug hunters' perspective. Int J Neuropsychopharmacol. 2010;13(9):1269–84.
21. Perez-Caballero L, Torres-Sanchez S, Bravo L, Mico JA, Berrocoso E. Fluoxetine: a case history of its discovery and preclinical development. Expert Opin Drug Discovery. 2014;9(5):567–78.
22. Goineau S, Lemaire M, Froget G. Overview of safety pharmacology. In: Current protocols in pharmacology. Hoboken: John Wiley & Sons, Inc.; 2001.
23. Gunn RN, Rabiner EA, editors. Imaging in central nervous system drug discovery. In: Seminars in nuclear medicine. Cambridge: Elsevier; 2017.
24. Combes R, Berridge T, Connelly J, Eve M, Garner R, Toon S, et al. Early microdose drug studies in human volunteers can minimise animal testing: proceedings of a workshop organised by Volunteers in Research and Testing. Eur J Pharm Sci. 2003;19(1):1–11.
25. LoRusso PM. Phase 0 clinical trials: an answer to drug development stagnation? J Clin Oncol Off J Am Soc Clin Oncol. 2009;27(16):2586–8.
26. Hansen AR, Graham DM, Pond GR, Siu LL. Phase 1 trial design: is 3 + 3 the best? Cancer Control. 2014;21(3):200–8.
27. Le Tourneau C, Lee JJ, Siu LL. Dose escalation methods in phase I cancer clinical trials. J Natl Cancer Inst. 2009;101(10):708–20.
28. Cook N, Hansen AR, Siu LL, Abdul Razak AR. Early phase clinical trials to identify optimal dosing and safety. Mol Oncol. 2015;9(5):997–1007.
29. Lesko LJ, Rowland M, Peck CC, Blaschke TF. Optimizing the science of drug development: opportunities for better candidate selection and accelerated evaluation in humans. J Clin Pharmacol. 2000;40(8):803–14.
30. Biomarkers Definitions Working Group. Biomarkers and surrogate endpoints: preferred definitions and conceptual framework. Clin Pharmacol Ther. 2001;69(3):89–95.
31. Arneric SP, Kern VD, Stephenson DT. Regulatory-accepted drug development tools are needed to accelerate innovative CNS disease treatments. Biochem Pharmacol. 2018;151:291–306.
32. McGonigle P. Animal models of CNS disorders. Biochem Pharmacol. 2014;87(1):140–9.
33. Enna SJ, Williams M. Challenges in the search for drugs to treat central nervous system disorders. J Pharmacol Exp Ther. 2009;329(2):404–11.
34. Belzung C, Lemoine M. Criteria of validity for animal models of psychiatric disorders: focus on anxiety disorders and depression. Biol Mood Anxiety Disord. 2011;1:9.
35. Markou A, Chiamulera C, Geyer MA, Tricklebank M, Steckler T. Removing obstacles in neuroscience drug discovery: the future path for animal models. Neuropsychopharmacology. 2008;34(1):74–89.

36. Kaffman A, White JD, Wei L, Johnson FK, Krystal JH. Enhancing the utility of preclinical research in neuropsychiatry drug development. Methods Mol Biol. 2011;2019:3–22.
37. Zerbi V, Markicevic M, Gasparini F, Schroeter A, Rudin M, Wenderoth N. Inhibiting mGluR5 activity by AFQ056/Mavoglurant rescues circuit-specific functional connectivity in Fmr1 knockout mice. Neuroimage. 2019;191:392–402.
38. Li J, Yang R, Xia K, Wang T, Nie B, Gao K, et al. Effects of stress on behavior and resting-state fMRI in rats and evaluation of Telmisartan therapy in a stress-induced depression model. BMC Psychiatry. 2018;18(1):337.
39. Paasonen J, Salo RA, Ihalainen J, Leikas JV, Savolainen K, Lehtonen M, et al. Dose-response effect of acute phencyclidine on functional connectivity and dopamine levels, and their association with schizophrenia-like symptom classes in rat. Neuropharmacology. 2017;119:15–25.
40. Paasonen J. Optimization and implementation of preclinical pharmacologic fMRI for drug research and development. Dissertation. Itä-Suomen yliopisto; 2017.
41. English BA, Thomas K, Johnstone J, Bazih A, Gertsik L, Ereshefsky L. Use of translational pharmacodynamic biomarkers in early-phase clinical studies for schizophrenia. Biomark Med. 2014;8(1):29–49.
42. Wilson FJ, Leiser SC, Ivarsson M, Christensen SR, Bastlund JF. Can pharmaco-electroencephalography help improve survival of central nervous system drugs in early clinical development? Drug Discov Today. 2014;19(3):282–8.
43. Borsook D, Becerra L, Fava M. Use of functional imaging across clinical phases in CNS drug development. Transl Psychiatry. 2013;3:e282.
44. Hargreaves RJ, Hoppin J, Sevigny J, Patel S, Chiao P, Klimas M, et al. Optimizing central nervous system drug development using molecular imaging. Clin Pharmacol Ther. 2015;98(1):47–60.
45. Wong DF, Tauscher J, Grunder G. The role of imaging in proof of concept for CNS drug discovery and development. Neuropsychopharmacology. 2009;34(1):187–203.
46. Nathan PJ, Phan KL, Harmer CJ, Mehta MA, Bullmore ET. Increasing pharmacological knowledge about human neurological and psychiatric disorders through functional neuroimaging and its application in drug discovery. Curr Opin Pharmacol. 2014;14:54–61.
47. Paulus MP, Feinstein JS, Castillo G, Simmons AN, Stein MB. Dose-dependent decrease of activation in bilateral amygdala and insula by lorazepam during emotion processing. Arch Gen Psychiatry. 2005;62(3):282–8.
48. Upadhyay J, Anderson J, Baumgartner R, Coimbra A, Schwarz AJ, Pendse G, et al. Modulation of CNS pain circuitry by intravenous and sublingual doses of buprenorphine. Neuroimage. 2012;59(4):3762–73.
49. Miller B, Marks LA, Koller JM, Newman BJ, Bretthorst GL, Black KJ. Prolactin and fMRI response to SKF38393 in the baboon. PeerJ. 2013;1:e195.
50. Borsook D, Upadhyay J, Hargreaves R, Wager T. Enhancing choice and outcomes for therapeutic trials in chronic pain: N-of-1 + Imaging (+ i). Trends Pharmacol Sci. 2020;41(2):85–98.
51. Krystal AD, Pizzagalli DA, Mathew SJ, Sanacora G, Keefe R, Song A, et al. The first implementation of the NIMH FAST-FAIL approach to psychiatric drug development. Nat Rev Drug Discov. 2019;18(1):82–4.
52. Krystal AD, Pizzagalli DA, Smoski M, Mathew SJ, Nurnberger J, Lisanby SH, et al. A randomized proof-of-mechanism trial applying the 'fast-fail' approach to evaluating kappa-opioid antagonism as a treatment for anhedonia. Nat Med. 2020;26(5):760–8.
53. Gandhi KD, Mansukhani MP, Karpyak VM, Schneekloth TD, Wang Z, Kolla BP. The impact of varenicline on alcohol consumption in subjects with alcohol use disorders: systematic review and meta-analyses. J Clin Psychiatry. 2020;81(2):19r12924.
54. Cannella N, Ubaldi M, Masi A, Bramucci M, Roberto M, Bifone A, et al. Building better strategies to develop new medications in Alcohol Use Disorder: learning from past success and failure to shape a brighter future. Neurosci Biobehav Rev. 2019;103:384–98.
55. Cummings J, Zhong K, Cordes D. Drug development in Alzheimer's disease: the role of default mode network assessment in phase II. US Neurol. 2017;13(2):67–9.

56. Orloff J, Douglas F, Pinheiro J, Levinson S, Branson M, Chaturvedi P, et al. The future of drug development: advancing clinical trial design. Nat Rev Drug Discov. 2009;8(12):949–57.
57. Sadraee A, Paulus M, Ekhtiari H. fMRI as an outcome measure in clinical trials: a systematic review in clinicaltrials.gov. Brain Behav. 2021;11(5):e02089.
58. EMA. European Medicines Agency website 2016. Available from: http://www.ema.europa.eu/ema/
59. FDA. Food and Drug Administration website 2016. Available from: https://www.fda.gov
60. FDA. Food and Drug Administration: development & approval process (drugs) 2016. Available from: https://www.fda.gov/Drugs/DevelopmentApprovalProcess/default.htm
61. EMA. European Medicines Agency: scientific guidelines 2016. Available from: http://www.ema.europa.eu/ema/index.jsp?curl=pages/regulation/general/general_content_000043.jsp&mid=WC0b01ac05800240cb
62. EMA. European Medicines Agency: qualification of novel methodologies for medicine development 2016. Available from: http://www.ema.europa.eu/ema/index.jsp?curl=pages/regulation/document_listing/document_listing_000319.jsp
63. FDA. Food and Drug Administration: drug development tools (DDT) qualification programs 2016. Available from: https://www.fda.gov/Drugs/DevelopmentApprovalProcess/DrugDevelopmentToolsQualificationProgram/default.htm
64. FDA. Food and Drug Administration: biomarker qualification program 2016. Available from: https://www.fda.gov/Drugs/DevelopmentApprovalProcess/DrugDevelopmentToolsQualificationProgram/BiomarkerQualificationProgram/ucm20086360.htm
65. FDA. Qualification of biomarker-total kidney volume in studies for treatment of autosomal dominant polycystic kidney disease; guidance for industry; availability. Federal Register: Food and Drug Administration; 2016.
66. Wise RG, Williams P, Tracey I. Using fMRI to quantify the time dependence of remifentanil analgesia in the human brain. Neuropsychopharmacology. 2004;29(3):626.
67. Schweinhardt P, Bountra C, Tracey I. Pharmacological FMRI in the development of new analgesic compounds. NMR Biomed. 2006;19(6):702–11.
68. Oertel B, Preibisch C, Wallenhorst T, Hummel T, Geisslinger G, Lanfermann H, et al. Differential opioid action on sensory and affective cerebral pain processing. Clin Pharmacol Ther. 2008;83(4):577–88.
69. De Simoni S, Schwarz AJ, O'Daly OG, Marquand AF, Brittain C, Gonzales C, et al. Test–retest reliability of the BOLD pharmacological MRI response to ketamine in healthy volunteers. Neuroimage. 2013;64:75–90.
70. Doyle OM, De Simoni S, Schwarz AJ, Brittain C, O'Daly OG, Williams SCR, Mehta MA. Quantifying the attenuation of the ketamine pharmacological magnetic resonance imaging response in humans: a validation using antipsychotic and glutamatergic agents. J Pharmacol Exp Ther. 2013;345(1):151–60.
71. Clarke LP, Sriram RD, Schilling LB. Imaging as a biomarker: standards for change measurements in therapy workshop summary. Acad Radiol. 2008;15(4):501–30.
72. Kessler LG, Barnhart HX, Buckler AJ, Choudhury KR, Kondratovich MV, Toledano A, et al. The emerging science of quantitative imaging biomarkers terminology and definitions for scientific studies and regulatory submissions. Stat Methods Med Res. 2015;24(1):9–26.
73. Friedman L, Glover GH. Report on a multicenter fMRI quality assurance protocol. J Magn Reson Imaging. 2006;23(6):827–39.
74. Mitsis GD, Iannetti GD, Smart TS, Tracey I, Wise RG. Regions of interest analysis in pharmacological fMRI: how do the definition criteria influence the inferred result? Neuroimage. 2008;40(1):121–32.
75. Buck R, Singhal H, Arora J, Schlitt H, Constable RT. Detecting change in BOLD signal between sessions for atlas-based anatomical ROIs. Neuroimage. 2008;40(3):1157–65.
76. Jack CR Jr, Bernstein MA, Fox NC, Thompson P, Alexander G, Harvey D, et al. The Alzheimer's disease neuroimaging initiative (ADNI): MRI methods. J Magn Reson Imaging. 2008;27(4):685–91.

77. Van Essen DC, Ugurbil K, Auerbach E, Barch D, Behrens TE, Bucholz R, et al. The IMAGEN study: reinforcement-related behavior in normal brain function and pathology. Mol Psychiatry. 2010;15(12):1128–39.
78. Van Essen DC, Ugurbil K, Auerbach E, Barch D, Behrens TE, Bucholz R, et al. The Human Connectome Project: a data acquisition perspective. Neuroimage. 2012;62(4):2222–31.
79. Chen JI, Ha B, Bushnell MC, Pike B, Duncan GH. Differentiating noxious- and innocuous-related activation of human somatosensory cortices using temporal analysis of fMRI. J Neurophysiol. 2002;88(1):464–74.
80. Glover GH, Mueller BA, Turner JA, van Erp TG, Liu TT, Greve DN, et al. Function biomedical informatics research network recommendations for prospective multicenter functional MRI studies. J Magn Reson Imaging. 2012;36(1):39–54.
81. Friedman L, Glover GH, Krenz D, Magnotta V, First B. Reducing inter-scanner variability of activation in a multicenter fMRI study: role of smoothness equalization. Neuroimage. 2006;32(4):1656–68.
82. Browne RH. On the use of a pilot sample for sample size determination. Stat Med. 1995;14(17):1933–40.
83. Van Dijk KR, Hedden T, Venkataraman A, Evans KC, Lazar SW, Buckner RL. Intrinsic functional connectivity as a tool for human connectomics: theory, properties, and optimization. J Neurophysiol. 2010;103(1):297–321.
84. Hutton C, Bork A, Josephs O, Deichmann R, Ashburner J, Turner R. Image distortion correction in fMRI: a quantitative evaluation. Neuroimage. 2002;16(1):217–40.
85. De Zwart JA, Ledden PJ, van Gelderen P, Bodurka J, Chu R, Duyn JH. Signal-to-noise ratio and parallel imaging performance of a 16-channel receive-only brain coil array at 3.0 Tesla. Magn Reson Med. 2004;51(1):22–6.
86. Pruessmann KP, Weiger M, Scheidegger MB, Boesiger P. SENSE: sensitivity encoding for fast MRI. Magn Reson Med. 1999;42(5):952–62.
87. Nagy Z, Weiskopf N. Efficient fat suppression by slice-selection gradient reversal in twice-refocused diffusion encoding. Magn Reson Med. 2008;60(5):1256–60.
88. Smith SM, Vidaurre D, Beckmann CF, Glasser MF, Jenkinson M, Miller KL, et al. Functional connectomics from resting-state fMRI. Trends Cogn Sci. 2013;17(12):666–82.
89. Larkman DJ, Hajnal JV, Herlihy AH, Coutts GA, Young IR, Ehnholm G. Use of multicoil arrays for separation of signal from multiple slices simultaneously excited. J Magn Reson Imaging. 2001;13(2):313–7.
90. Stocker T, Schneider F, Klein M, Habel U, Kellermann T, Zilles K, et al. Automated quality assurance routines for fMRI data applied to a multicenter study. Hum Brain Mapp. 2005;25(2):237–46.
91. Power JD, Mitra A, Laumann TO, Snyder AZ, Schlaggar BL, Petersen SE. Methods to detect, characterize, and remove motion artifact in resting state fMRI. Neuroimage. 2014;84:320–41.
92. Grodin EN, Ray LA. The use of functional magnetic resonance imaging to test pharmacotherapies for alcohol use disorder: a systematic review. Alcohol Clin Exp Res. 2019;43(10):2038–56.
93. Mahajan KR, Ontaneda D. The role of advanced magnetic resonance imaging techniques in multiple sclerosis clinical trials. Neurotherapeutics. 2017;14(4):905–23.
94. Sormani MP, Pardini M. Assessing repair in multiple sclerosis: outcomes for phase II clinical trials. Neurotherapeutics. 2017;14(4):924–33.
95. Moningka H, Lichenstein S, Worhunsky PD, DeVito EE, Scheinost D, Yip SW. Can neuroimaging help combat the opioid epidemic? A systematic review of clinical and pharmacological challenge fMRI studies with recommendations for future research. Neuropsychopharmacology. 2019;44(2):259–73.
96. Xiao F, Koepp MJ, Zhou D. Pharmaco-fMRI: a tool to predict the response to antiepileptic drugs in epilepsy. Front Neurol. 2019;10:1203.
97. Duff EP, Vennart W, Wise RG, Howard MA, Harris RE, Lee M, et al. Learning to identify CNS drug action and efficacy using multistudy fMRI data. Sci Transl Med. 2015;7(274):274ra16.

98. Liu CH, Greve DN, Dai G, Marota JJ, Mandeville JB. Remifentanil administration reveals biphasic phMRI temporal responses in rat consistent with dynamic receptor regulation. Neuroimage. 2007;34(3):1042–53.
99. Wager TD, Atlas LY, Lindquist MA, Roy M, Woo CW, Kross E. An fMRI-based neurologic signature of physical pain. N Engl J Med. 2013;368(15):1388–97.
100. Upadhyay J, Anderson J, Schwarz AJ, Coimbra A, Baumgartner R, Pendse G, et al. Imaging drugs with and without clinical analgesic efficacy. Neuropsychopharmacology. 2011;36(13):2659–73.
101. Gear R, Becerra L, Upadhyay J, Bishop J, Wallin D, Pendse G, et al. Pain facilitation brain regions activated by nalbuphine are revealed by pharmacological fMRI. PLoS One. 2013;8(1):e50169.
102. Elman I, Zubieta JK, Borsook D. The missing p in psychiatric training: why it is important to teach pain to psychiatrists. Arch Gen Psychiatry. 2011;68(1):12–20.
103. Baliki MN, Petre B, Torbey S, Herrmann KM, Huang L, Schnitzer TJ, et al. Corticostriatal functional connectivity predicts transition to chronic back pain. Nat Neurosci. 2012;15(8):1117–9.
104. Maleki N, Linnman C, Brawn J, Burstein R, Becerra L, Borsook D. Her versus his migraine: multiple sex differences in brain function and structure. Brain. 2012;135(Pt 8):2546–59.
105. Napadow V, Kim J, Clauw DJ, Harris RE. Decreased intrinsic brain connectivity is associated with reduced clinical pain in fibromyalgia. Arthritis Rheum. 2012;64(7):2398–403.
106. Kulkarni B, Bentley DE, Elliott R, Julyan PJ, Boger E, Watson A, et al. Arthritic pain is processed in brain areas concerned with emotions and fear. Arthritis Rheum. 2007;56(4):1345–54.
107. Xu A, Larsen B, Baller EB, Scott JC, Sharma V, Adebimpe A, et al. Convergent neural representations of experimentally-induced acute pain in healthy volunteers: a large-scale fMRI meta-analysis. Neurosci Biobehav Rev. 2020;112:300–23.
108. Deakin JF, Lees J, McKie S, Hallak JE, Williams SR, Dursun SM. Glutamate and the neural basis of the subjective effects of ketamine: a pharmaco-magnetic resonance imaging study. Arch Gen Psychiatry. 2008;65(2):154–64.
109. Joules R, Doyle OM, Schwarz AJ, O'Daly OG, Brammer M, Williams SC, et al. Ketamine induces a robust whole-brain connectivity pattern that can be differentially modulated by drugs of different mechanism and clinical profile. Psychopharmacology (Berl). 2015;232(21-22):4205–18.
110. Yu H, Li Q, Wang D, Shi L, Lu G, Sun L, et al. Mapping the central effects of chronic ketamine administration in an adolescent primate model by functional magnetic resonance imaging (fMRI). Neurotoxicology. 2012;33(1):70–7.
111. Corlett PR, Honey GD, Krystal JH, Fletcher PC. Glutamatergic model psychoses: prediction error, learning, and inference. Neuropsychopharmacology. 2011;36(1):294–315.
112. Krystal JH, Abi-Saab W, Perry E, D'Souza DC, Liu N, Gueorguieva R, et al. Preliminary evidence of attenuation of the disruptive effects of the NMDA glutamate receptor antagonist, ketamine, on working memory by pretreatment with the group II metabotropic glutamate receptor agonist, LY354740, in healthy human subjects. Psychopharmacology (Berl). 2005;179(1):303–9.
113. Large CH. Do NMDA receptor antagonist models of schizophrenia predict the clinical efficacy of antipsychotic drugs? J Psychopharmacol. 2007;21(3):283–301.
114. Chin CL, Upadhyay J, Marek GJ, Baker SJ, Zhang M, Mezler M, et al. Awake rat pharmacological magnetic resonance imaging as a translational pharmacodynamic biomarker: metabotropic glutamate 2/3 agonist modulation of ketamine-induced blood oxygenation level dependence signals. J Pharmacol Exp Ther. 2011;336(3):709–15.
115. Gozzi A, Large CH, Schwarz A, Bertani S, Crestan V, Bifone A. Differential effects of antipsychotic and glutamatergic agents on the phMRI response to phencyclidine. Neuropsychopharmacology. 2008;33(7):1690–703.
116. Haaf M, Leicht G, Curic S, Mulert C. Glutamatergic deficits in schizophrenia – biomarkers and pharmacological interventions within the ketamine model. Curr Pharm Biotechnol. 2018;19(4):293–307.

117. Delaveau P, Jabourian M, Lemogne C, Guionnet S, Bergouignan L, Fossati P. Brain effects of antidepressants in major depression: a meta-analysis of emotional processing studies. J Affect Disord. 2011;130(1-2):66–74.
118. van Wingen GA, Tendolkar I, Urner M, van Marle HJ, Denys D, Verkes RJ, et al. Short-term antidepressant administration reduces default mode and task-positive network connectivity in healthy individuals during rest. Neuroimage. 2013;88C:47–53.
119. Sarpal DK, Robinson DG, Lencz T, Argyelan M, Ikuta T, Karlsgodt K, et al. Antipsychotic treatment and functional connectivity of the striatum in first-episode schizophrenia. JAMA Psychiat. 2015;72(1):5–13.
120. Risacher SL, Wang Y, Wishart HA, Rabin LA, Flashman LA, McDonald BC, et al. Cholinergic enhancement of brain activation in mild cognitive impairment during episodic memory encoding. Front Psychiatry. 2013;4:105.
121. Breiter HC, Gollub RL, Weisskoff RM, Kennedy DN, Makris N, Berke JD, et al. Acute effects of cocaine on human brain activity and emotion. Neuron. 1997;19(3):591–611.
122. Stein EA, Pankiewicz J, Harsch HH, Cho JK, Fuller SA, Hoffmann RG, et al. Nicotine-induced limbic cortical activation in the human brain: a functional MRI study. Am J Psychiatry. 1998;155(8):1009–15.
123. Hooker BA, Tobon G, Baker SJ, Zhu C, Hesterman J, Schmidt K, et al. Gabapentin-induced pharmacodynamic effects in the spinal nerve ligation model of neuropathic pain. Eur J Pain. 2014;18(2):223–37.
124. Governo RJ, Morris PG, Marsden CA, Chapman V. Gabapentin evoked changes in functional activity in nociceptive regions in the brain of the anaesthetized rat: an fMRI study. Br J Pharmacol. 2008;153(7):1558–67.
125. Upadhyay J, Baker SJ, Rajagovindan R, Hart M, Chandran P, Hooker BA, et al. Pharmacological modulation of brain activity in a preclinical model of osteoarthritis. Neuroimage. 2013;64:341–55.
126. Kocsis P, Gyertyan I, Eles J, Laszy J, Hegedus N, Gajari D, et al. Vascular action as the primary mechanism of cognitive effects of cholinergic, CNS-acting drugs, a rat phMRI BOLD study. J Cereb Blood Flow Metab. 2014;34(6):995–1000.
127. Byun NE, Grannan M, Bubser M, Barry RL, Thompson A, Rosanelli J, et al. Antipsychotic drug-like effects of the selective M4 muscarinic acetylcholine receptor positive allosteric modulator VU0152100. Neuropsychopharmacology. 2014;39(7):1578–93.
128. Baker S, Chin CL, Basso AM, Fox GB, Marek GJ, Day M. Xanomeline modulation of the blood oxygenation level-dependent signal in awake rats: development of pharmacological magnetic resonance imaging as a translatable pharmacodynamic biomarker for central activity and dose selection. J Pharmacol Exp Ther. 2012;341(1):263–73.
129. Rech J, Hess A, Finzel S, Kreitz S, Sergeeva M, Englbrecht M, et al. Association of brain functional magnetic resonance activity with response to tumor necrosis factor inhibition in rheumatoid arthritis. Arthritis Rheum. 2013;65(2):325–33.
130. Kantrowitz JT, Javitt DC, Freedman R, Sehatpour P, Kegeles LS, Carlson M, et al. Double blind, two dose, randomized, placebo-controlled, cross-over clinical trial of the positive allosteric modulator at the alpha7 nicotinic cholinergic receptor AVL-3288 in schizophrenia patients. Neuropsychopharmacology. 2020;45(8):1339–45.
131. Bruns A, Mueggler T, Kunnecke B, Risterucci C, Prinssen EP, Wettstein JG, et al. "Domain gauges": a reference system for multivariate profiling of brain fMRI activation patterns induced by psychoactive drugs in rats. Neuroimage. 2015;112:70–85.
132. Harris RE, Napadow V, Huggins JP, Pauer L, Kim J, Hampson J, et al. Pregabalin rectifies aberrant brain chemistry, connectivity, and functional response in chronic pain patients. Anesthesiology. 2013;119(6):1453–64.
133. Sanders D, Krause K, O'Muircheartaigh J, Thacker MA, Huggins JP, Vennart W, et al. Pharmacologic modulation of hand pain in osteoarthritis: a double-blind placebo-controlled functional magnetic resonance imaging study using naproxen. Arthritis Rheumatol. 2015;67(3):741–51.

134. Becerra L, Schwartzman RJ, Kiefer RT, Rohr P, Moulton EA, Wallin D, et al. CNS measures of pain responses pre- and post-anesthetic ketamine in a patient with complex regional pain syndrome. Pain Med 2015;16(12):2368–85.
135. Wanigasekera V, Mezue M, Andersson J, Kong Y, Tracey I. Disambiguating pharmacodynamic efficacy from behavior with neuroimaging: implications for analgesic drug development. Anesthesiology. 2016;124(1):159–68.
136. Flodin P, Martinsen S, Altawil R, Waldheim E, Lampa J, Kosek E, et al. Intrinsic brain connectivity in chronic pain: a resting-state fMRI study in patients with rheumatoid arthritis. Front Hum Neurosci. 2016;10:107.
137. Petersen CS, Miskowiak KW. Toward a transdiagnostic neurocircuitry-based biomarker model for pro-cognitive effects: challenges, opportunities, and next steps. CNS Spectr. 2021;26(4):333–7.
138. Barron HC, Mars RB, Dupret D, Lerch JP, Sampaio-Baptista C. Cross-species neuroscience: closing the explanatory gap. Philos Trans R Soc Lond B Biol Sci. 2021;376(1815):20190633.
139. Robbins TW. Cross-species studies of cognition relevant to drug discovery: a translational approach. Br J Pharmacol. 2017;174(19):3191–9.
140. Francois J, Grimm O, Schwarz AJ, Schweiger J, Haller L, Risterucci C, et al. Ketamine suppresses the ventral striatal response to reward anticipation: a cross-species translational neuroimaging study. Neuropsychopharmacology. 2016;41(5):1386–94.
141. Becker R. Brain networks in pharmacological fMRI of NMDA antagonists. Dissertation. Heidelberg: Universitätsbibliothek Heidelberg; 2019.
142. Duncan GE, Leipzig JN, Mailman RB, Lieberman JA. Differential effects of clozapine and haloperidol on ketamine-induced brain metabolic activation. Brain Res. 1998;812(1-2):65–75.
143. Becerra L, Upadhyay J, Chang PC, Bishop J, Anderson J, Baumgartner R, et al. Parallel buprenorphine phMRI responses in conscious rodents and healthy human subjects. J Pharmacol Exp Ther. 2013;345(1):41–51.
144. Grimm O, Gass N, Weber-Fahr W, Sartorius A, Schenker E, Spedding M, et al. Acute ketamine challenge increases resting state prefrontal-hippocampal connectivity in both humans and rats. Psychopharmacology (Berl). 2015;232(21-22):4231–41.
145. Masaki Y, Kashiwagi Y, Watabe H, Abe K. (R)- and (S)-ketamine induce differential fMRI responses in conscious rats. Synapse. 2019;73(12):e22126.
146. Masaki Y, Kashiwagi Y, Rokugawa T, Ito M, Iimori H, Abe K. Pharmacological MRI responses of raclopride in rats: The relationship with D2 receptor occupancy and cataleptic behavior. Synapse. 2020;74(12):e22180.
147. Gao L, Huang P, Dong Z, Gao T, Huang S, Zhou C, et al. Modified Xiaoyaosan (MXYS) exerts anti-depressive effects by rectifying the brain blood oxygen level-dependent fMRI signals and improving hippocampal neurogenesis in mice. Front Pharmacol. 2018;9:1098.
148. Erb J, Armendariz M, De Martino F, Goebel R, Vanduffel W, Formisano E. Homology and specificity of natural sound-encoding in human and monkey auditory cortex. Cereb Cortex. 2019;29(9):3636–50.
149. Hori Y, Clery JC, Selvanayagam J, Schaeffer DJ, Johnston KD, Menon RS, et al. Interspecies activation correlations reveal functional correspondences between marmoset and human brain areas. Proc Natl Acad Sci U S A. 2021;118(37):e2110980118.
150. Schaeffer DJ, Hori Y, Gilbert KM, Gati JS, Menon RS, Everling S. Divergence of rodent and primate medial frontal cortex functional connectivity. Proc Natl Acad Sci U S A. 2020;117(35):21681–9.
151. Specht K. Current challenges in translational and clinical fMRI and future directions. Front Psychiatry. 2019;10:924.
152. Gelman A, Loken E. The garden of forking paths: why multiple comparisons can be a problem, even when there is no "fishing expedition" or "p-hacking" and the research hypothesis was posited ahead of time. Department of Statistics, Columbia University; 2013.
153. Kriegeskorte N, Simmons WK, Bellgowan PS, Baker CI. Circular analysis in systems neuroscience: the dangers of double dipping. Nat Neurosci. 2009;12(5):535–40.

154. Bullmore E, Sporns O. Complex brain networks: graph theoretical analysis of structural and functional systems. Nat Rev Neurosci. 2009;10(3):186–98.
155. Calhoun VD, Liu J, Adalı T. A review of group ICA for fMRI data and ICA for joint inference of imaging, genetic, and ERP data. Neuroimage. 2009;45(1):S163–S72.
156. Poldrack RA. Region of interest analysis for fMRI. Soc Cogn Affect Neurosci. 2007;2(1/4):67.
157. Vatansever S, Schlessinger A, Wacker D, Kaniskan HU, Jin J, Zhou MM, et al. Artificial intelligence and machine learning-aided drug discovery in central nervous system diseases: state-of-the-arts and future directions. Med Res Rev. 2021;41(3):1427–73.
158. Friedman L, Stern H, Brown GG, Mathalon DH, Turner J, Glover GH, et al. Test-retest and between-site reliability in a multicenter fMRI study. Hum Brain Mapp. 2008;29(8):958–72.
159. Harvey JL, Demetriou L, McGonigle J, Wall MB. A short, robust brain activation control task optimised for pharmacological fMRI studies. PeerJ. 2018;6:e5540.
160. Upadhyay J, Lemme J, Anderson J, Bleakman D, Large T, Evelhoch JL, et al. Test-retest reliability of evoked heat stimulation BOLD fMRI. J Neurosci Methods. 2015;253:38–46.
161. Plichta MM, Schwarz AJ, Grimm O, Morgen K, Mier D, Haddad L, et al. Test-retest reliability of evoked BOLD signals from a cognitive-emotive fMRI test battery. Neuroimage. 2012;60(3):1746–58.
162. Kampa M, Schick A, Sebastian A, Wessa M, Tuscher O, Kalisch R, et al. Replication of fMRI group activations in the neuroimaging battery for the Mainz Resilience Project (MARP). Neuroimage. 2020;204:116223.
163. Holiga S, Sambataro F, Luzy C, Greig G, Sarkar N, Renken RJ, et al. Test-retest reliability of task-based and resting-state blood oxygen level dependence and cerebral blood flow measures. PLoS One. 2018;13(11):e0206583.
164. Elliott ML, Knodt AR, Ireland D, Morris ML, Poulton R, Ramrakha S, et al. Poor test-retest reliability of task-fMRI: new empirical evidence and a meta-analysis. BioRxiv. 2019:681700.
165. Nord CL, Gray A, Charpentier CJ, Robinson OJ, Roiser JP. Unreliability of putative fMRI biomarkers during emotional face processing. Neuroimage. 2017;156:119–27.
166. Fiecas M, Ombao H, van Lunen D, Baumgartner R, Coimbra A, Feng D. Quantifying temporal correlations: a test-retest evaluation of functional connectivity in resting-state fMRI. Neuroimage. 2013;65:231–41.
167. Shehzad Z, Kelly AM, Reiss PT, Gee DG, Gotimer K, Uddin LQ, et al. The resting brain: unconstrained yet reliable. Cereb Cortex. 2009;19(10):2209–29.
168. Baumgartner R, Cho W, Coimbra A, Chen C, Wang Z, Struyk A, et al. Evaluation of an fMRI USPIO-based assay in healthy human volunteers. J Magn Reson Imaging. 2017;46(1):124–33.
169. Petersson KM, Nichols TE, Poline JB, Holmes AP. Statistical limitations in functional neuroimaging. II. Signal detection and statistical inference. Philos Trans R Soc Lond B Biol Sci. 1999;354(1387):1261–81.
170. Henson R. What can functional neuroimaging tell the experimental psychologist? Q J Exp Psychol A. 2005;58(2):193–233.
171. Vul E, Harris C, Winkielman P, Pashler H. Puzzlingly high correlations in fmri studies of emotion, personality, and social cognition. Perspect Psychol Sci. 2009;4(3):274–90.
172. Yarkoni T. Big correlations in little studies: inflated fMRI correlations reflect low statistical power—commentary on Vul et al. (2009). Perspect Psychol Sci. 2009;4(3):294–8.
173. David SP, Ware JJ, Chu IM, Loftus PD, Fusar-Poli P, Radua J, et al. Potential reporting bias in fMRI studies of the brain. PLoS One. 2013;8(7):e70104.
174. Button KS, Ioannidis JP, Mokrysz C, Nosek BA, Flint J, Robinson ES, et al. Power failure: why small sample size undermines the reliability of neuroscience. Nat Rev Neurosci. 2013;14(5):365–76.
175. Nichols TE, Das S, Eickhoff SB, Evans AC, Glatard T, Hanke M, et al. Best practices in data analysis and sharing in neuroimaging using MRI. bioRxiv. 2016:054262.
176. Poldrack RA, Fletcher PC, Henson RN, Worsley KJ, Brett M, Nichols TE. Guidelines for reporting an fMRI study. Neuroimage. 2008;40(2):409–14.

177. Frohner JH, Teckentrup V, Smolka MN, Kroemer NB. Addressing the reliability fallacy in fMRI: similar group effects may arise from unreliable individual effects. Neuroimage. 2019;195:174–89.
178. Kragel PA, Han X, Kraynak TE, Gianaros PJ, Wager TD. Functional MRI can be highly reliable, but it depends on what you measure: a commentary on Elliott et al. (2020). Psychol Sci. 2021;32(4):622–6.
179. Elliott ML, Knodt AR, Hariri AR. Striving toward translation: strategies for reliable fMRI measurement. Trends Cogn Sci. 2021;25(9):776–87.
180. Sheth SA, Nemoto M, Guiou M, Walker M, Pouratian N, Toga AW. Linear and nonlinear relationships between neuronal activity, oxygen metabolism, and hemodynamic responses. Neuron. 2004;42(2):347–55.
181. Zilles K, Mohlberg H, Amunts K, Palomero-Gallagher N, Bludau S. Cytoarchitecture and maps of the human cerebral cortex. Strukturelle und funktionelle Organisation des Gehirns; 2015. Report No.: 0123970253.
182. Amunts K, Zilles K. Advances in cytoarchitectonic mapping of the human cerebral cortex. Neuroimaging Clin N Am. 2001;11(2):151–69, vii.
183. Wohlschläger AM, Specht K, Lie C, Mohlberg H, Wohlschläger A, Bente K, et al. Linking retinotopic fMRI mapping and anatomical probability maps of human occipital areas V1 and V2. Neuroimage. 2005;26(1):73–82.
184. Toga AW. Brain mapping: an encyclopedic reference. London: Academic Press; 2015.
185. Ogawa S, Menon R, Tank D, Kim S, Merkle H, Ellermann J, et al. Functional brain mapping by blood oxygenation level-dependent contrast magnetic resonance imaging. A comparison of signal characteristics with a biophysical model. Biophys J. 1993;64(3):803.
186. Mathiesen C, Caesar K, Akgören N, Lauritzen M. Modification of activity-dependent increases of cerebral blood flow by excitatory synaptic activity and spikes in rat cerebellar cortex. J Physiol. 1998;512(2):555–66.
187. Shmuel A, Grinvald A. Functional organization for direction of motion and its relationship to orientation maps in cat area 18. J Neurosci. 1996;16(21):6945–64.
188. Smith AJ, Blumenfeld H, Behar KL, Rothman DL, Shulman RG, Hyder F. Cerebral energetics and spiking frequency: the neurophysiological basis of fMRI. Proc Natl Acad Sci. 2002;99(16):10765–70.
189. Shulman RG, Rothman DL, Behar KL, Hyder F. Energetic basis of brain activity: implications for neuroimaging. Trends Neurosci. 2004;27(8):489–95.
190. Ances B, Vaida F, Ellis R, Buxton R. Test–retest stability of calibrated BOLD-fMRI in HIV− and HIV+ subjects. Neuroimage. 2011;54(3):2156–62.
191. Goodwin JA, Vidyasagar R, Balanos GM, Bulte D, Parkes LM. Quantitative fMRI using hyperoxia calibration: reproducibility during a cognitive Stroop task. Neuroimage. 2009;47(2):573–80.
192. Leontiev O, Buxton RB. Reproducibility of BOLD, perfusion, and CMRO 2 measurements with calibrated-BOLD fMRI. Neuroimage. 2007;35(1):175–84.
193. Wey H-Y, Wang DJ, Duong TQ. Baseline CBF, and BOLD, CBF, and CMRO2 fMRI of visual and vibrotactile stimulations in baboons. J Cereb Blood Flow Metab. 2011;31(2):715–24.
194. He X, Yablonskiy DA. Quantitative BOLD: mapping of human cerebral deoxygenated blood volume and oxygen extraction fraction: default state. Magn Reson Med. 2007;57(1):115–26.
195. Lu H, Ge Y. Quantitative evaluation of oxygenation in venous vessels using T2-Relaxation-Under-Spin-Tagging MRI. Magn Reson Med. 2008;60(2):357–63.
196. Bolar DS, Rosen BR, Sorensen A, Adalsteinsson E. QUantitative Imaging of eXtraction of oxygen and TIssue consumption (QUIXOTIC) using venular-targeted velocity-selective spin labeling. Magn Reson Med. 2011;66(6):1550–62.
197. Yablonskiy DA, Sukstanskii AL, He X. Blood oxygenation level-dependent (BOLD)-based techniques for the quantification of brain hemodynamic and metabolic properties–theoretical models and experimental approaches. NMR Biomed. 2013;26(8):963–86.

198. Wehrl HF, Hossain M, Lankes K, Liu C-C, Bezrukov I, Martirosian P, et al. Simultaneous PET-MRI reveals brain function in activated and resting state on metabolic, hemodynamic and multiple temporal scales. Nat Med. 2013;19(9):1184–9.
199. Bruinsma TJ, Sarma VV, Oh Y, Jang DP, Chang SY, Worrell GA, et al. The relationship between dopamine neurotransmitter dynamics and the blood-oxygen-level-dependent (BOLD) signal: a review of pharmacological functional magnetic resonance imaging. Front Neurosci. 2018;12:238.
200. Mitsis GD, Governo RJ, Rogers R, Pattinson KT. The effect of remifentanil on respiratory variability, evaluated with dynamic modeling. J Appl Physiol (Bethesda, Md: 1985). 2009;106(4):1038–49.
201. Khalili-Mahani N, Chang C, van Osch MJ, Veer IM, van Buchem MA, Dahan A, et al. The impact of "physiological correction" on functional connectivity analysis of pharmacological resting state fMRI. Neuroimage. 2013;65:499–510.
202. Ekhtiari H, Kuplicki R, Yeh HW, Paulus MP. Physical characteristics not psychological state or trait characteristics predict motion during resting state fMRI. Sci Rep. 2019;9(1):419.
203. Olszowy W. Validation of statistical methods used in task fMRI studies. Dissertation. Cambridge: University of Cambridge; 2019.
204. Sirmpilatze N, Baudewig J, Boretius S. Temporal stability of fMRI in medetomidine-anesthetized rats. Sci Rep. 2019;9(1):16673.
205. Steiner AR, Rousseau-Blass F, Schroeter A, Hartnack S, Bettschart-Wolfensberger R. Systematic review: anaesthetic protocols and management as confounders in rodent blood oxygen level dependent functional magnetic resonance imaging (BOLD fMRI)-Part A: effects of changes in physiological parameters. Front Neurosci. 2020;14:577119.
206. Steiner AR, Rousseau-Blass F, Schroeter A, Hartnack S, Bettschart-Wolfensberger R. Systematic review: anesthetic protocols and management as confounders in rodent blood oxygen level dependent functional magnetic resonance imaging (BOLD fMRI)-Part B: effects of anesthetic agents, doses and timing. Animals (Basel). 2021;11(1):199.
207. Sorger B, Scharnowski F, Linden DEJ, Hampson M, Young KD. Control freaks: towards optimal selection of control conditions for fMRI neurofeedback studies. Neuroimage. 2019;186:256–65.
208. Iannetti G, Wise RG. BOLD functional MRI in disease and pharmacological studies: room for improvement? Magn Reson Imaging. 2007;25(6):978–88.
209. Walter SA, Forsgren M, Lundengård K, Simon R, Nilsson MT, Söderfeldt B, et al. Positive allosteric modulator of GABA lowers BOLD responses in the cingulate cortex. PLoS One. 2016;11(3):e0148737.
210. Black KJ, Koller JM, Miller BD. Rapid quantitative pharmacodynamic imaging by a novel method: theory, simulation testing and proof of principle. PeerJ. 2013;1:e117.
211. Koller JM, Vachon MJ, Bretthorst GL, Black KJ. Rapid quantitative pharmacodynamic imaging with Bayesian estimation. Front Neurosci. 2016;10:144.
212. Mandeville JB, Sander CY, Jenkins BG, Hooker JM, Catana C, Vanduffel W, et al. A receptor-based model for dopamine-induced fMRI signal. Neuroimage. 2013;75:46–57.
213. Marota JJ, Mandeville JB, Weisskoff RM, Moskowitz MA, Rosen BR, Kosofsky BE. Cocaine activation discriminates dopaminergic projections by temporal response: an fMRI study in rat. Neuroimage. 2000;11(1):13–23.
214. Sidén P, Eklund A, Bolin D, Villani M. Fast Bayesian whole-brain fMRI analysis with spatial 3D priors. Neuroimage. 2017;146:211–25.
215. Borsook D, Upadhyay J, Klimas M, Schwarz AJ, Coimbra A, Baumgartner R, et al. Decision-making using fMRI in clinical drug development: revisiting NK-1 receptor antagonists for pain. Drug Discov Today. 2012;17(17-18):964–73.
216. Rabiner EA, Beaver J, Makwana A, Searle G, Long C, Nathan P, et al. Pharmacological differentiation of opioid receptor antagonists by molecular and functional imaging of target occupancy and food reward-related brain activation in humans. Mol Psychiatry. 2011;16(8):826.

217. Nathan PJ, O'neill BV, Bush MA, Koch A, Tao WX, Maltby K, et al. Opioid receptor modulation of hedonic taste preference and food intake: a single-dose safety, pharmacokinetic, and pharmacodynamic investigation with GSK1521498, a novel μ-opioid receptor inverse agonist. J Clin Pharmacol. 2012;52(4):464–74.
218. Yeomans MR, Gray RW. Opioid peptides and the control of human ingestive behaviour. Neurosci Biobehav Rev. 2002;26(6):713–28.
219. Yeomans MR, Wright P. Lower pleasantness of palatable foods in nalmefene-treated human volunteers. Appetite. 1991;16(3):249–59.
220. Hirvonen J, Aalto S, Hagelberg N, Maksimow A, Ingman K, Oikonen V, et al. Measurement of central mu-opioid receptor binding in vivo with PET and [11C]carfentanil: a test-retest study in healthy subjects. Eur J Nucl Med Mol Imaging. 2009;36(2):275–86.
221. Leppa M, Korvenoja A, Carlson S, Timonen P, Martinkauppi S, Ahonen J, et al. Acute opioid effects on human brain as revealed by functional magnetic resonance imaging. Neuroimage. 2006;31(2):661–9.
222. Becerra L, Harter K, Gonzalez RG, Borsook D. Functional magnetic resonance imaging measures of the effects of morphine on central nervous system circuitry in opioid-naive healthy volunteers. Anesth Analg. 2006;103(1):208–16, table of contents.
223. Murdaugh DL, Cox JE, Cook EW, Weller RE. fMRI reactivity to high-calorie food pictures predicts short-and long-term outcome in a weight-loss program. Neuroimage. 2012;59(3):2709–21.
224. Etkin A, Maron-Katz A, Wu W, Fonzo GA, Huemer J, Vertes PE, et al. Using fMRI connectivity to define a treatment-resistant form of post-traumatic stress disorder. Sci Transl Med. 2019;11(486):eaal3236.
225. Szaflarski JP, Gloss D, Binder JR, Gaillard WD, Golby AJ, Holland SK, et al. Practice guideline summary: use of fMRI in the presurgical evaluation of patients with epilepsy: report of the Guideline Development, Dissemination, and Implementation Subcommittee of the American Academy of Neurology. Neurology. 2017;88(4):395–402.
226. Tu Y, Zeng F, Lan L, Li Z, Maleki N, Liu B, et al. An fMRI-based neural marker for migraine without aura. Neurology. 2020;94(7):e741–e51.
227. Loth E, Charman T, Mason L, Tillmann J, Jones EJH, Wooldridge C, et al. The EU-AIMS Longitudinal European Autism Project (LEAP): design and methodologies to identify and validate stratification biomarkers for autism spectrum disorders. Mol Autism. 2017;8:24.
228. Webb SJ, Shic F, Murias M, Sugar CA, Naples AJ, Barney E, et al. biomarker acquisition and quality control for multi-site studies: the Autism Biomarkers Consortium for clinical trials. Front Integr Neurosci. 2019;13:71.
229. Ness S, Pandina G, Jagannatha S, Wathen K, Bangerter A, Manyakov NV, et al. ASPI: a public-private partnership to develop treatments for autism. Nat Rev Drug Discov. 2020;19(4):219–20.
230. Koroshetz W, Gordon J, Adams A, Beckel-Mitchener A, Churchill J, Farber G, et al. The state of the NIH BRAIN initiative. J Neurosci. 2018;38(29):6427–38.
231. Parekh A, Buckman-Garner S, McCune S, ONeill R, Geanacopoulos M, Amur S, et al. Catalyzing the critical path initiative: FDA's progress in drug development activities. Clin Pharmacol Ther. 2015;97(3):221–33.
232. The National Academies of Sciences, Engineering, and Medicine. Advancing therapeutic development for pain and opioid use disorders through public-private partnerships: proceedings of a workshop. 2018.
233. Brady LS, Potter WZ, Gordon JA. Redirecting the revolution: new developments in drug development for psychiatry. Expert Opin Drug Discov. 2019;14(12):1213–9.
234. Bach P, Weil G, Pompili E, Hoffmann S, Hermann D, Vollstadt-Klein S, et al. FMRI-based prediction of naltrexone response in alcohol use disorder: a replication study. Eur Arch Psychiatry Clin Neurosci. 2021;271(5):915–27.
235. Neville J, Kopko S, Romero K, Corrigan B, Stafford B, LeRoy E, et al. Accelerating drug development for Alzheimer's disease through the use of data standards. Alzheimers Dement (N Y). 2017;3(2):273–83.

236. Spellman BA. Introduction to the special section data, data, everywhere... especially in my file drawer. Perspect Psychol Sci. 2012;7(1):58–9.
237. Eickhoff S, Nichols TE, Van Horn JD, Turner JA. Sharing the wealth: neuroimaging data repositories. Neuroimage. 2016;124(Pt B):1065–8.
238. The Research Domain Criteria Database (RDoCdb) [Internet]. Available from: http://rdocdb.nimh.nih.gov/about/
239. Ekhtiari H, Zare-Bidoky M, Sangchooli A, Janes AC, Kaufman MJ, Oliver JA, et al. A methodological checklist for fMRI drug cue reactivity studies: development and expert consensus. Nat Protoc. 2022;17(3):567–95.
240. Klapwijk ET, van den Bos W, Tamnes CK, Raschle NM, Mills KL. Opportunities for increased reproducibility and replicability of developmental neuroimaging. Dev Cogn Neurosci. 2021;47:100902.
241. Heinrichs HS, Beyer F, Medawar E, Prehn K, Ordemann J, Floel A, et al. Effects of bariatric surgery on functional connectivity of the reward and default mode network: a pre-registered analysis. Hum Brain Mapp. 2021;42(16):5357–73.
242. Weiss F, Zhang J, Aslan A, Kirsch P, Gerchen MF. Feasibility of training the dorsolateral prefrontal-striatal network by real-time fMRI neurofeedback. Sci Rep. 2022;12(1):1669.
243. Park L, Furey M, Nugent AC, Farmer C, Ellis J, Szczepanik J, et al. Neurophysiological changes associated with antidepressant response to ketamine not observed in a negative trial of scopolamine in major depressive disorder. Int J Neuropsychopharmacol. 2019;22(1):10–8.

Chapter 14
Monoamine Oxidase B (MAO-B): A Target for Rational Drug Development in Schizophrenia Using PET Imaging as an Example

Kankana Nisha Aji, Jeffrey H. Meyer, Pablo M. Rusjan, and Romina Mizrahi

Abstract Monoamine oxidase B (MAO-B) is an important high-density enzyme involved in the generation of oxidative stress and central in the catabolism of dopamine, particularly in brain subcortical regions with putative implications in the pathophysiology of schizophrenia. In this chapter, we review postmortem studies, preclinical models, and peripheral and genetic studies implicating MAO-B in psychosis. A literature search in PubMed was conducted and 64 studies were found to be eligible for systematic review. We found that MAO-B could be identified as a potential target in schizophrenia. Evidence comes mostly from studies of peripheral markers, showing reduced platelet MAO-B activity in schizophrenia, together with preclinical results from MAO-B knock-out mice resulting in a hyperdopaminergic state and behavioral disinhibition. However, whether brain MAO-B is altered in vivo in patients with schizophrenia remains unknown. We therefore review methodological studies involving MAO-B positron emission tomography (PET) radioligands used to quantify MAO-B in vivo in the human brain. Given the limitations of

K. Nisha Aji
Douglas Research Centre, Clinical and Translational Sciences Lab, Montreal, QC, Canada

Department of Pharmacology & Toxicology, University of Toronto, Toronto, ON, Canada
e-mail: kankana.aji@mail.utoronto.ca

J. H. Meyer
Department of Pharmacology & Toxicology, University of Toronto, Toronto, ON, Canada

Department of Psychiatry, University of Toronto, Toronto, ON, Canada

Campbell Family Mental Health Research Institute, Centre for Addiction and Mental Health, Toronto, ON, Canada
e-mail: jeff.meyer@camhpet.ca

P. M. Rusjan · R. Mizrahi (✉)
Douglas Research Centre, Clinical and Translational Sciences Lab, Montreal, QC, Canada

Department of Psychiatry, McGill University, Montreal, QC, Canada
e-mail: pablo.rusjan@mcgill.ca; romina.mizrahi@mcgill.ca

M. Macaluso et al. (eds.), *Drug Development in Psychiatry*, Advances in Neurobiology 30,
https://doi.org/10.1007/978-3-031-21054-9_14

currently available treatments for schizophrenia, elucidating whether MAO-B could be used as a target for risk stratification or clinical staging in schizophrenia could allow for a rational search for newer antipsychotics and the development of new treatments.

Keywords Psychosis · MAO-B · Dopamine · Striatum · PET · Antipsychotics · Astroglial dysfunction

Abbreviations

[11C]	Carbon-11
[11C]DMPEA	[11C]N,N-dimethylphenylethylamine
2-TCM	2-Tissue compartmental model (2TCM)
5-HT	5-Hydroxytryptamine
CNS	Central nervous system
CSF	Cerebrospinal fluid
D1	Dopamine (D1)
D2/3	Dopamine (D2/3)
DNA	Deoxyribonucleic acid
GABRB3	Gamma-aminobutyric acid receptor subunit beta-3
GFAP	Glial fibrillary acidic protein
GWAS	Genome-wide association studies
KO	Knock-out
LOD	Logarithm of the odds
MAO	Monoamine oxidase
MAO-A	Monoamine oxidase A
MAO-B	Monoamine oxidase B
MHPG	3-methoxy-4-hydroxyphenylglycol
NDP	Norrie-disease gene
PAH	Phenylalanine hydroxylase
PEA	2-Phenylethylamine
PET	Positron emission tomography
r2	Coefficient of determination
SCZ	Schizophrenia
SLC6A4	Solute carrier family 6 member 4
SNP	Single nucleotide polymorphism
VT	Total distribution volume

14.1 General

Schizophrenia (SCZ) is a debilitating mental disorder affecting about 1% of the world population. It is characterized by a complex heterogeneous set of thought, perceptual, and cognitive deficits. Significant evidence from neuroimaging and postmortem studies suggests increased dopamine synthesis and/or stress-induced increases in striatal dopamine release, particularly the dorsal caudate or associative striatum [1]. The treatment of schizophrenia involves antipsychotic medications targeting dopamine (D2/3) receptors. Unfortunately, these current treatments for schizophrenia are not effective in about 20–35% of patients [2]. The persistence of positive symptoms (e.g., hallucinations and delusions) despite ≥2 trials of adequate treatment is a serious clinical problem resulting in significant disability. The psychopathology of schizophrenia is multifactorial, and a rational search for alternative treatment targets is urgently needed.

Monoamine oxidase (MAO) isoenzymes A and B are proteins mainly found on the outer mitochondrial membranes, which catalyze the oxidative deamination of monoamine neurotransmitters, including dopamine (Fig. 14.1). In rodent brains, monoamine oxidase A (MAO-A) has high affinity for serotonin and to a lesser degree norepinephrine [3], which is the predominant active MAO enzyme [4], while MAO-B primarily serves the catabolism of 2-phenylethylamine (PEA) and benzylamine [5]. Dopamine [6], norepinephrine, epinephrine, and other trace amines are oxidized by both forms of the enzyme in most species in the presynaptic terminal [7–9]. In human brains, dopamine is a substrate for both isoenzymes [10]. In human postmortem brain samples, MAO-B concentration is generally highest in the substantia nigra, followed by the basal ganglia, thalamus, and cerebral cortex, with substantially lower levels in the cerebellum and white matter [11].

Neuroimaging via positron emission tomography (PET) can quantify specific radioligand binding, which in this case represents an index of regional MAO density in the living human brain. For example, over the past three decades since the first radiotracers were developed and the PET images of MAOs were carried out, radioligands for in vivo quantification and localization of MAO-A [12–15] and MAO-B [11, 16–19] in the living human brain have been identified. However, no study to date has investigated MAO-B density in vivo in patients with schizophrenia.

In this chapter, we (I) provide a concise synthesis of findings from multiple lines of evidence including postmortem studies, preclinical research on dopamine, and peripheral and genetic data implicating MAO-B in schizophrenia and related disorders and (II) summarize the novel PET radiotracers targeting MAO-B in vivo in the human brain, which could serve as a potential stratification tool to quantify MAO-B in patients with schizophrenia. Given the limitations of current medical treatment for schizophrenia, alternative targets such as MAO-B may provide the fundamental knowledge, which will permit a rational search for and discovery of newer antipsychotics or other psychotherapeutic approaches for the treatment of schizophrenia patients with complex symptomatology. Of note, MAO-B inhibitors such as selegiline and rasagiline have been shown to improve the negative symptoms of schizophrenia such as alogia and avolition [20–23].

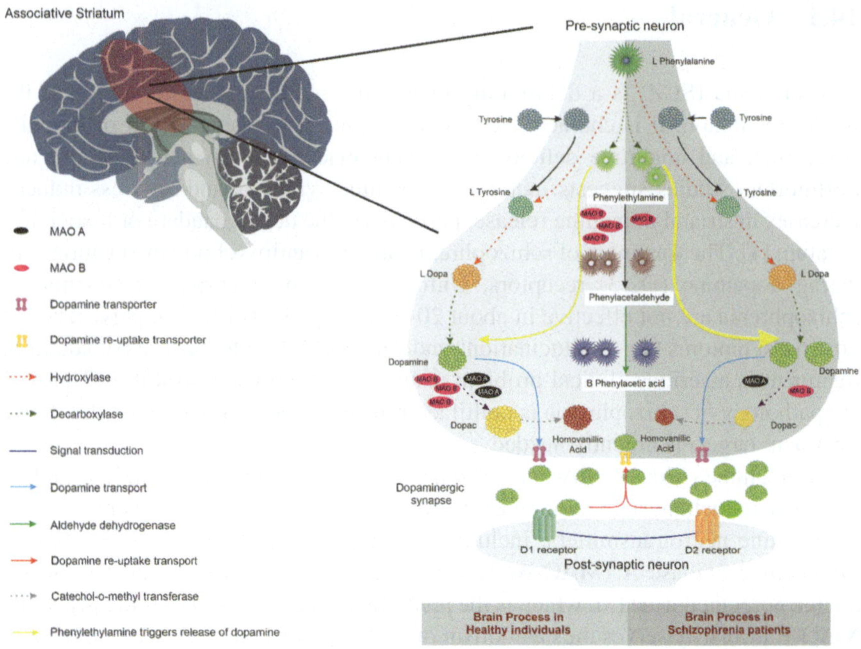

Fig. 14.1 Schematic illustration of MAO-B hypothesis in psychosis. The schematic diagram illustrates the dopamine metabolism via MAO-B in healthy and schizophrenia human brain. Given that dopaminergic dysfunction is greatest in the associative striatum (AST), AST is the primary region of interest in psychosis represented in the diagram. In a healthy brain, dopamine synthesis begins with the amino acid L-phenylalanine and proceeds sequentially through L-tyrosine, L-DOPA (dihydroxyphenylalanine), and then dopamine. Tyrosine hydroxylase is the rate-limiting enzyme, which converts the amino acid L-tyrosine to L-dopa. L-dopa is then metabolized to dopamine by aromatic amino acid dopa decarboxylase. Dopamine catabolism occurs via monoamine oxidase enzymes, MAO-A, and MAO-B. MAO-B predominantly metabolizes dopamine to its primary metabolite DOPAC (3,4-dihydroxyphenylacetic acid) and further to homovanillic acid by catechol-o-methyl transferase (COMT). Dopamine released into the dopaminergic synapse via dopamine transporters (DAT) is cleared from synapse via the following: (1) dopamine reuptake via DAT provides the primary mechanism through which dopamine is cleared from synapse. In the cytosol, other transporters sequester the dopamine into vesicles for storage and later release. (2) Dopamine binds to the dopamine receptors (D1 and D2 primarily) on the postsynaptic neurons and cause dopamine signalling. Further, dopamine comes off the receptor and is taken back into the terminals via dopamine reuptake transporters. Alternatively, amino acid L-phenylalanine decarboxylates to β-phenylethylamine, a substrate of the dopamine transporter (DAT) via aromatic amino acid decarboxylase. β-Phenylacetic acid is the primary urinary metabolite of β-phenethylamine and is produced via MAO-B metabolism and subsequent aldehyde dehydrogenase metabolism. β-phenylethylamine (PEA) acts both to inhibit dopamine reuptake and to cause its release from storage granules

14.2 Experimental Materials and Methods

14.2.1 Eligibility Criteria

Studies were included if they presented original data published before February 29, 2020, which examined the role of altered MAO-B activity and its implication in psychosis. Studies assessing MAO-B and/or altered dopaminergic activity in alternate clinical populations such as Parkinson's, Alzheimer's, Huntington's disease, anxiety, and depression were excluded. Additionally, the PET literature in schizophrenia was briefly reviewed, with a focus on MAO-B as the molecular target. While we acknowledge that other molecular targets may be important in the pathophysiology of schizophrenia, we decided to focus on MAO-B, which has been largely overlooked historically.

14.2.2 Literature Search

A computerized literature search was conducted in PubMed on March 2, 2020, for articles using the following search query in schizophrenia: "MAO-B and dopamine metabolism," "MAO-B and dopamine release," "amphetamine-induced dopamine release," "MAO-B activity in postmortem brains," "MAO-B inhibition and behavioral disinhibition in mice/rats/guinea-pigs/primates," "MAO-B gene knock-out and behavioral studies," "MAO-B inhibition in preclinical studies," "MAO-B and striatum," "MAO-B and caudate nucleus," "MAO-B activity and cerebrospinal fluid/CSF," "platelet MAO-B activity," "MAO-B and cigarette smoking," "MAO-B gene polymorphism," "astroglia dysfunction and schizophrenia," "astrocyte markers and schizophrenia," "MAO-B and astrogliosis," "MAO-B and glial fibrillary acidic protein (GFAP)," "schizophrenia and associative striatum," and "MAO-B inhibitors and schizophrenia."

In parallel, a similar search strategy was conducted in PubMed using the following search query for novel MAO radioligands: "Imaging of MAO-B and postmortem brain," "MAO-B and imaging of cigarette smoking," MAO-B and positron emission tomography/PET," "MAO-B density and PET radioligands/radiotracers," "MAO-A density and PET radioligands/radiotracers," "MAO-B density and PET human in vivo studies," and "MAO-A density and PET human in vivo studies." Additionally, references cited in the identified papers were also reviewed to find additional relevant studies.

14.2.3 Study Selection

After removing 650 duplicate articles or nonspecific articles, we screened the titles and abstracts from the remaining 598 articles and selected 156 potentially eligible studies for full text review. A list of eligible 82 full text articles was developed by consensus review of the authors, applying the eligibility criteria outlined above. Full text articles were then downloaded, and relevant data was extracted and rigorously analyzed following which we ended up with 64 articles for inclusion in the review after excluding 48 articles that were not relevant to our study.

14.2.4 Data Extraction

We extracted the following data from each study: name of first author, year of publication, number of participants per diagnostic category, illness severity, medication exposure, data acquisition or assay parameters, PET parameters, and brain regions of interest assessed in each study and main study outcomes.

14.2.5 Study Identification

The search yielded 82 potentially relevant articles that met the inclusion criteria. This included 6 postmortem studies that determine MAO-B activity via assay of brain specimens in individuals with psychosis, 10 studies using either preclinical models related to MAO-B and dopamine release via pharmacological inhibition or behavioral studies using MAO-B gene knock-out, 39 peripheral MAO-B studies that assessed platelet MAO-B activity in schizophrenia via MAO assays that detect MAO-B activity by employing specific substrates and inhibitors, 14 genetic association studies assessing the potential role of altered MAO-B genes in schizophrenia; following the second search pertaining to MAO radioligands, 13PET in vivo studies were included in the discussion, which sheds light on the potential usage of specific MAO-B radiotracers. Among these 82 articles, 64 articles were found relevant to MAO-B dysregulation that precipitated schizophrenia symptomatology (Fig. 14.2). The search for in vivo quantification of MAO-B in psychosis via PET yielded no results.

14.3 Review of Studies

14.3.1 Postmortem Findings (Table 14.1)

Several postmortem studies have examined MAO-B activity in patients with schizophrenia via MAO assays of postmortem brain specimens. A study involving 19 patients with either chronically treated schizophrenia or psychosis but without a schizophrenia diagnosis and 24 controls revealed increased activity of MAO-B in the pons in both chronic schizophrenia and non-schizophrenia-related psychosis as

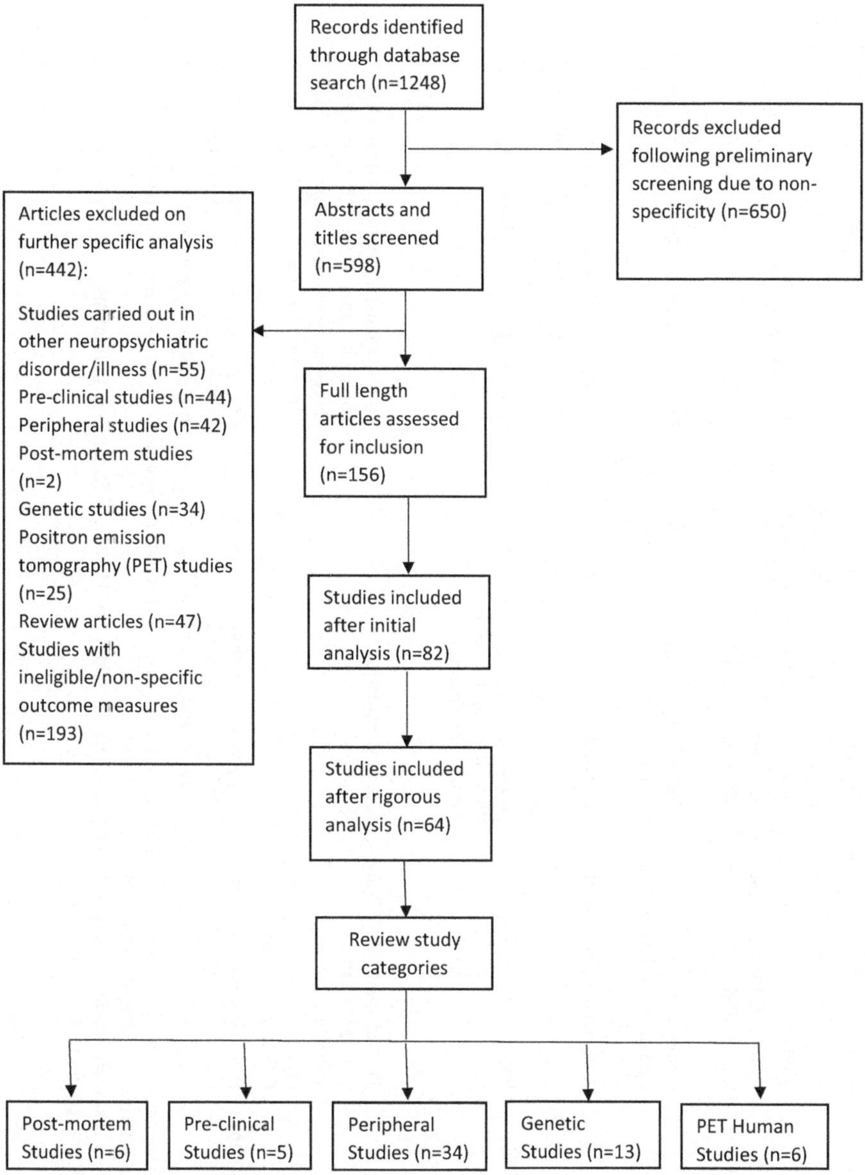

Fig. 14.2 Schematic review flowchart as per study selection criteria

compared with age-matched controls [24]. Further, deoxyribonucleic acid (DNA) methylation of MAO-A and MAO-B genes revealed that MAO-B genes were highly methylated in postmortem brains of treated female patients with schizophrenia in both the nucleus accumbens and prefrontal cortex [25]. In contrast, a study involving treated schizophrenia patients and 44 controls showed a significant decrease in

Table 14.1 Postmortem findings: determination of MAO-B activity in different brain regions in psychosis

Reference	Subject (SCZ/HC)	Medication	Method (MAO-B assay - substrate)	MAO-B activity	Regions of interest
Fowler et al. 1981	19 chronic SCZ/ non-SCZ psychoses (13 M, 6 F)/24 (11 M, 13 F)	Antipsychotic treatment—long periods	Radiochemical method (14C Benzylamine)	↑	Pons
Qiaohui Yang et al. 2012	6/7	Chlorpromazine equivalents—114/190/225/ 325/550 mg/d	Bisulfite sequencing method	↑DNA methylation of MAO-B gene	PFC, nucleus accumbens
Owen et al. 1987	39/44	Antipsychotic treatment	Benzylamine	↓	↓Frontal cortex, ↓temporal cortex, ↓amygdala, caudate, putamen, hippocampus, hypothalamus
Michael et al. 1981	Caudate nucleus: 10 SCZ, 10 SCZ-like psychosis/20 Frontal cortex: 7 SCZ, 6 SCZ-like psychosis/11	NR	Dopamine, phenylethylamine, and 5-hydroxytryptamine	(−)	Caudate nucleus, frontal cortex
Schwartz, et al. 1974	9/9	Thioridazine 200/400 mg/d, Chlorpromazine 45/333/600 mg/d, procyclidine 20 mg/d, fluphenazine 6 mg/d	Liquid scintillation technique	(−)	Cortex, caudate, putamen, GP, medial thalamus, hypothalamus, hippocampus, amygdala, septal region, periaqueductal pons–mesencephalon, substantia nigra
Crow et al. 1979	Nine (4M, 5F)/10 (5M, 5F)	Antipsychotic treatment	Radiometric technique (serotonin, benzylamine, tyramine, and dopamine)	(−)	Cortex, caudate, putamen, nucleus accumbens, thalamus hypothalamus, hippocampus, amygdala, substantia nigra, cerebellum

Method, MAO-B assay of brain specimens and the substrates employed; ↓MAO-B activity in patients than controls; ↑ MAO-B activity in patients than controls; (−) MAO-B activity, no change/difference in MAO-B activity between groups

Abbreviations: *SCZ* schizophrenia cases; *HC* healthy controls; *MAO-B* monoamine oxidase B; *M* male; *F* female; *mg* milligrams; *d* day; *DNA* deoxyribonucleic acid; *PFC* prefrontal cortex; *NR* not reported; *GP* globus pallidus

the activity of MAO-B in the frontal cortex, temporal cortex, and amygdala, which was not accounted for by antipsychotic medication, age, gender, or postmortem variables. In a subsample of 22 patients assessed prior to death, there was a significant correlation of reduction in MAO-B activity with negative symptoms such as flattening of affect and paucity of speech [26].

However, there is wide variability across studies as others found no significant difference in MAO-B activity between schizophrenia patients and controls [27–29]. This is most likely due to the challenges inherent in postmortem studies: (a) small samples from deceased individuals with schizophrenia, (b) inclusion of schizophrenia subjects with medical comorbidities, (c) chronic antipsychotic medication, (d) confounding effects of suicide, (e) confounding effects of cigarette smoking [30], (f) additional effects of age and chronic treatment in elderly schizophrenia subjects, (g) limited retrospective clinical information, and (h) some variability in postmortem interval and sample preparation. Based on postmortem samples, it is plausible that dysregulation of MAO-B activity and/or level may be involved in schizophrenia, but this should be evaluated in larger samples of medication and substance free (no cigarette smokers) subjects in the earlier stages of schizophrenia. Notably, to date, no single study has investigated MAO-B in the living brain of patients with schizophrenia.

14.3.2 Preclinical Findings in MAO-B Knock-Out Mice (Table 14.2)

Several preclinical models have examined the neurochemical divergence (increase/decrease in dopamine/PEA levels) and/or behavioral abnormalities of MAO-B deficient rodents. Previous studies reveal that MAO-A has higher affinity for the substrate serotonin (5-hydroxytryptamine (5-HT)) and is preferentially inhibited by clorgyline, whereas MAO-B has higher affinity for phenylethylamine (PEA) and benzylamine and is preferentially inhibited by L-deprenyl [5, 7, 31]. The MAO-B substrate, PEA, sometimes regarded as an endogenous amphetamine [32], with regard to its similar chemical structure and effects in vivo was implicated in schizophrenia in very early studies [33].

Mice carrying genetic knock-out (KO) of MAO-B had significantly higher levels of PEA in the brain [34], which is related to dopamine function [35], particularly in the striatum and prefrontal cortex [36]. MAO-B KO mice exhibit behavioral disinhibition such as novelty seeking behavior and reduced anxiety-like behaviors but had comparatively less aggressive behavior compared with MAO-A KO mice in several behavioral paradigms targeting emotional reactivity [37–41]. These behavioral observations coincide with numerous findings of low MAO-B platelet activity and novelty-seeking personality in humans [42]. Of interest, extracellular dopamine levels remain unaltered in MAO-B knock-out mice in a preclinical study [43] and was presumed to be due to the significant adaptive upregulation of the D2-like dopamine receptors and hypersensitivity of dopamine (D1) receptors [44–46].

Table 14.2 Preclinical findings in MAO-B knock-out mice: Behavioral effects of MAO-B inhibition on brain dopamine and phenylethylamine levels

Reference	Sex (M/F)/strain/species	Animal model	Methods of quantification	Target	Behavioral	Regions of interest
Grimsby et al. 1997	M/WT and MAO-B KO mice (6–20 weeks old)	Genetic model; behavioral schizophrenia model	W and ICC; behavioral assay; MAO-B assay - PEA (substrate)	↑ PEA, ↓MAO-B	↑ stress	Cortex, hippocampus, raphe nucleus, substantia nigra, thalamus
Bortolato et al. 2009	M/experimentally naive (adult; $n = 166$); WT and MAO-B KO mice	Genetic model; behavioral schizophrenia model	Behavioral assays (anxiety-related manifestations); brain-regional levels of PEA measurement - GC	↑ PEA, ↓MAO-B	↓anxiety, behavioral disinhibition	↑cortex, ↑striatum, hippocampus, thalamus, cerebellum
Bortolato et al. (Review article) 2011	WT and MAO-B KO mice	Genetic model	NR	↑ PEA, ↓MAO-B	N/A	Striatum
Holschneider et al. 2001	WT and MAO-B KO mice (Adult)	Genetic model; behavioral schizophrenia model	W and ICC; behavioral assays; MAO-B assays - PEA (substrate)	↑ PEA, ↓MAO-B	↓anxiety, behavioral disinhibition	↑ cortex, ↑striatum, substantia nigra
Chen, He et al. 1999	M/WT and MAO-B KO mice (6–8 weeks old; $n = 3$)	Genetic model	In vivo microdialysis; MAO-B assay- HPLC-ED; autoradiography	(−) DA, ↓MAO-B	N/A	Striatum

Genetic model – MAO-B knock-out; ↑ DA/PEA due to reduced MAO-B activity or release of amphetamine-like substances; (−) DA, no effect or no change; ↓MAO-B activity is lower due to knockout/inactivation of MAO-B gene; (−) DA - (upregulation of D2R)

Abbreviations: *M* male; *F* female; *WT* wild type; *MAO-B* monoamine oxidase B; *KO* knock-out; *W & ICC* Western and immunocytochemical analysis; *PEA* phenylethylamine; *GC* gas chromatography; *NR* not reported; *N/A* not applicable; *HPLC* high performance liquid chromatography; *ED* electrochemical detection; *DA* dopamine; *D2R* dopamine 2 receptors

Overall, while some studies have shown that reduction of MAO-B is associated with an increase in dopamine levels [38, 41, 47], other studies have shown no effect [43, 44]. This discrepancy in findings could be explained by the differences in dopamine catabolism between rodents and primates [44, 48, 49], pointing to a need for human studies.

14.3.3 Peripheral Findings (Table 14.3)

Several studies have reported an abnormal or significant reduction in platelet MAO-B activity in chronic schizophrenia cases with paranoid [50–54], residual, and auditory hallucinations [55] when compared with controls [56–63]. Further, an old review and meta-analysis of studies until 1988 [42] suggests that the greatest reduction in MAO-B activity was observed in paranoid schizophrenia cases (30%) followed by non-paranoid schizophrenia cases (24%) and lastly by schizophrenia cases with predominantly auditory hallucinations (24%). A more recent meta-analysis [64] of medication-free schizophrenia cases showed increased platelet MAO-B activity [65] or similar activity to controls with only a minority of studies reporting decreased platelet MAO-B levels. None of the above findings could be readily attributed to diagnostic, demographic, or methodological factors, nor to the effects of alcohol or antipsychotics. However, other studies have found no significant difference in enzymatic activity between schizophrenia patients and controls [66–73], further suggesting that the reduced MAO-B activity may be secondary to antipsychotic treatment and dose [74–76]. Nevertheless, a study in acute schizophrenia patients revealed that the reduction in platelet MAO-B activity was rather slow suggesting that the reduction may not be due to a direct inhibitory effect of antipsychotic drugs (flupenthixol/chlorpromazine) on platelet MAO activity, [77] but instead may be consistent with a change in platelet physiology or MAO-B synthesis [78, 79].

Importantly, several lines of evidence demonstrated a link between cigarette smoking and MAO-B inhibition, a significant confounding variable, which can provide an alternative explanation for the reduced MAO-B activity observed in patients with schizophrenia. However, previous investigations into MAO-B activity have not reported smoking rates of participants. Recent peripheral findings have demonstrated that regular cigarette smokers have reduced brain levels of MAO-B [80, 81], in line with the evidence from postmortem [30] and PET findings [82] showing low MAO-B platelet activity in heavy smokers. This is of significance in schizophrenia as up to 80% of individuals with chronic schizophrenia smoke tobacco cigarettes [83].

Further, monozygotic twins discordant for schizophrenia had significantly reduced platelet MAO-B activity compared with controls [84] but not as compared with their psychiatrically well, antipsychotic-free co-twins. This suggests that lower platelet MAO-B activity in schizophrenia may be genetically modulated [85, 86].

Overall, these data suggest a potential involvement of MAO-B alterations in schizophrenia patients.

Table 14.3 Peripheral findings: determination of platelet MAO-B activity using blood sample in patients with schizophrenia and healthy volunteers

Reference	*Subject (SCZ/HC)*	*Antipsychotic medication*	*Substrate*	*MAO-B activity*	*Effect of confounders*
Antipsychotic treated					
Mészáros, et al. 1998	50/30	MNR	–	↓ (paranoid and residual SCZ)	(−) TE
Berger et al. 1978	Forty M chronic SCZ (undifferentiated and paranoid)/55 M HC, and 16 M hospitalized controls	MNR	[14C] benzylamine	↓ (chronic SCZ)	(−) TE
Sullivan et al. 1978	Eleven M hospitalized first-episode acute SCZ, 19 chronic SCZ M/20 M HC	MNR	[14C] tryptamine (platelets and lymphocytes)	↓ (chronic SCZ), (−) MAO-B activity (acute SCZ and HC)	(−) TE/NSE
Tachiki et al. 1984	Seventeen paranoid SCZ; residual (8)/undifferentiated group (5)/ (11 M, 2F)	Fluphenazine decanoate 20–100 mg/week IM, fluphenazine 40 mg/day, molindone 150–300 mg/day, chlorpromazine 2000 mg/day, haloperidol 20–30 mg/day, thioridazine 400 mg/day, thiothixene 45 mg/day, loxapine 50–600 mg/day	Whole blood assay - [14C] benzylamine and added pargyline	↓ (chronic paranoid SCZ), (−) MAO-B activity (other SCZ subgroups or HC)	(−) TE/NSE
Gruen et al. 1982	Thirty-one M chronic SCZ	Chlorpromazine equivalents	[14C] benzylamine	↓ (poor prognosis)	(−) TE
Bierer et al.1984	Twenty-one M chronic SCZ/ 16 M controls	Chlorpromazine equivalents (125–1103 mg/day)	Tyramine	↓ (poor prognosis)	(−) TE
Berrettini et.al 1978	Sixteen chronic SCZ (12 M, 4 F)/19 (14 M,5 F)	Chlorpromazine, fluphenazine hydrochloride, haloperidol	Dopamine, serotonin, DMT, 5-MT	↓	(−) TE
Domino et al. 1976	Thirteen M unmedicated chronic SCZ/13 M	MNR	–	↓	(−) TE

Murphy et al. 1974	Thirty-three chronic SCZ, five first0episodic psychoses with schizophreniform features/22	Chlorpromazine, perphenazine, trifluoperazine and/or thioridazine	Tryptamine-2 [14C] (HCl)	↓	(−) TE/NSE
Murphy et al. 1976	Twenty-two chronic SCZ (17 M, 5 F)/(48 M, 16 F)	Chlorpromazine, fluphenazine hydrochloride, haloperidol, benztropine mesylate	Tyramine, benzylamine, and tryptamine	(−)	(−) NSE
Murphy et al. 1972	Thirty-three chronic SCZ (18 M, 4 F)/22 nonhospitalized (29 M, 4 F)	Phenothiazine (perphenazine, trifluoperazine, chlorpromazine)	Tryptamine-2 [14C] (HCl)	↓	(−) TE/SE
Becker et al. 1977	Twenty-nine SCZ inpatients and 26 SCZ outpatients/ 20 drug-free controls	Chlorpromazine	Tryptamine	↓	(−) TE
Zureick et al. 1988 (Review)	Paranoid subtype: 165 paranoid and 152 non-paranoid SCZ/985 HC Hallucinating subtype: 130 hallucinating, 81 Non-hallucinating SCZ/186 HC	MNR	Benzylamine, tryptamine, tyramine, PEA, kynuramine	↓ (percentage decrease in MA0-B activity) -Paranoid SCZ (30%), non-paranoid SCZ (24%), hallucinating SCZ (24%)	(−) TE/NSE
Marco Antonio et al. (review article). 1992	Antipsychotic-treated/drug-free SCZ (all subgroups)	MNR	Benzylamine, Tryptamine, Tyramine, PEA	↓ (only minority of studies)	(+) TE
Lewine et al. 1984	Fifty-nine (42 M, I7 F)/HC	MNR	[14C] benzylamine	↑ (M) correlated with negative symptoms)	(−) TE or outlier effects
Frank Owen et al. 1976	Sixty chronic SCZ (antipsychotic-free)/70 HC	Depot injections – phenothiazine	Tryptamine and tyramine	(−)	NE

(continued)

Table 14.3 (continued)

Reference	*Subject (SCZ/HC)*	*Antipsychotic medication*	*Substrate*	*MAO-B activity*	*Effect of confounders*
White et al. 1976	Twelve chronic SCZ (6 M, 6F)/12 (6 M, 6F)	Thiothixene HCI (45), thioridazine (600), mesoridazine (800), chlorpromazine (200–1600), fluphenazine (25), trifluoperazine HCI (30) mg/day	[14C] tryptamine	(−)	(−) TE
Mann et al. 1979	One hundred and fifteen (56 M, 59 F); 46 acute SCZ; institutionalized SCZ: 18 (medicated), 18 (unmedicated) and 37 noninstitutionalized cases (medicated) and 41 (unmedicated); hallucinating cases: 31 (medicated) and 31 (unmedicated) and non-hallucinating cases: two (medicated) and four (unmedicated)/ (58 M, 57 F)/28 brain-damaged controls	Phenothiazine (chlorpromazine, fluphenazine)	[14C] benzylamine and tyramine	(−) (groups and subgroups)	(−) SE
Spivak et al. 1994	Ten chronic SCZ (6 F, 4M)/17 (8 F, 9 M)	Chronic (10-week) perphenazine (24–64 mg/day	[14C] benzylamine	(−)	(−) TE
Del Vecchio et al. 1983	Seventy-five unmedicated schizophrenia cases (35 M, 40 F); second patient group - 55 (24 M and 31 F)/67 (31 M and 36 F)	Haloperidol (6–25 mg/day)	[14C] tryptamine and [14C] phenylethylamine	↓ (↑haloperidol-treated group, unmedicated patient group)	(+) TE (chronic haloperidol treatment)

Maj et al. 1984	First group: 100 (45 M, 55 F)/75 (35 M, 40 F) Second group: 65 (27 M, 38 F); 25 (10 M, 15 F)	Second group: haloperidol (6–25 mg/day) -2 months	[14C] tryptamine and [14C] phenylethylamine	↓ (acute and chronic haloperidol-treated group); (−) MAO-B activity (drug-free SCZ and HC or other subgroups)	(+) TE
Takahashi et al. 1975	76/50 Thirty-three acute schizophrenia cases (13 M, 20 F - previously unmedicated); 43 chronic cases (21 M, 22 F)	Thirty-three acute cases: chlorpromazine 75–150 mg per day; 43 chronic cases: phenothiazines, chlorpromazine, levomepromazine, perphenazine, perazine, propericiazine, and thioridazine	Serotonin	↓ (phenothiazine-treated chronic SCZ); (−) MAO-B activity (acute cases and controls)	(+) TE (phenothiazine)
Frank Owen et al. 1981	Thirty-five acute SCZ	Flupenthixol, chlorpromazine	[14C] tyramine	↓ (slow reduction consistent with change in platelet or enzyme synthesis)	NE
Lynn Eleanor DeLisi et al. 1981	Sixteen chronic SCZ (14 M, 2 F)	Haloperidol/fluphenazine/ trifluoperazine	*[14C] phenylethylamine	Change in MAO-B activity	(+) TE and dose of medication
Orologas et al. 1986	Thirteen chronic paranoid SCZ/8 controls	MNR	[14C] benzylamine	↓	NE
Wyatt. et al. 1973	Thirteen monozygotic twin pairs SCZ/23	Phenothiazine	[14C] tryptamine	↓	(−) TE
Reveley et al. 1983	Five pairs (Monozygotic SCZ and non-SCZ co-twins), 10 pairs (monozygotic, dizygotic control twins, unrelated individuals)	MNR	[14C] tyramine	↓ (↓non-SCZ co-twins, monozygotic SCZ)	(−) TE or illness

(continued)

Table 14.3 (continued)

Reference	Subject (SCZ/HC)	Antipsychotic medication	Substrate	MAO-B activity	Effect of confounders
Antipsychotic-free					
Reference	*Subject (SCZ/HC)*	*Substrate*	*MAO-B activity*	*Effect of confounders*	
Meltzer et al. 1987	One hundred one (37 F, 64 M)/(71 F, 65 M)	[14C] benzylamine	↓ (paranoid and auditory hallucinations subtype, black M cases), paranoid subtype (white M cases)	(+) race and gender	
Baron et.al 1980	Twenty-four M chronic paranoid SCZ/ 22 HC	[14C] benzylamine	↓	(−) TE/NSE	
Mann et al. 1981	Fifteen chronic paranoid SCZ/HC	Assay and enzyme kinetics	(−)	(−) TE	
Carpenter et al. 1975	Forty-four (25 F, 19 M)/131 HC	[14C] tryptamine	(−)	(−) TE/NSE	
Duncavage et al. 1982	Fifty SCZ subgrouped family history of SCZ/56 age-matched HC	[14C] benzylamine	(−) (both groups and subgroups)	NE	
Bridge et al. 1984	Thirty-six (22 F, 14 M)	Tryptamine	↓ (SCZ symptomatology)	(+) TE/NSE	
Baron et.al 1980	One hundred fifteen nonhospitalized, previously undiagnosed volunteers (72 M, 43 F) grouped into i) low ii) High platelet MAO-B proband	–	↓ Predisposed to borderline SCZ (+ family history)	NE	

SCZ schizophrenia cases; *HC* healthy controls; method, platelets harvested for MAO assay using specific substrates; ↓MAO-B activity in patients than controls; (−) MAO-B activity, no difference in MAO-B activity between groups; (−) TE, not accounted for by various treatment effects (TE); (+) TE, accounted for by various treatment effects (TE); (−)NSE, not accounted for by nonspecific effects; (−)SE, not accounted for by specific effects (severity or duration of illness, duration of medication, the presence of symptoms or institutionalization); NE, no effect noted; *in vitro, haloperidol and chlorpromazine preincubated with platelet pellets harvested for MAO assay in HC and in vivo MAO assay

Abbreviations: *MAO-B* Monoamine oxidase B; *14[C]* 14 Carbon; *HCl* hydrogen chloride; *M* male; *F* female; *mg* milligram; *MNR* medication not reported; *DMT* N,N-dimethyltryptamine; *5-MT* 5-methoxytryptamine; *PEA* phenylethylamine

14.3.4 Genetic Findings

MAO-B genes were not analyzed in most of the genome-wide association studies (GWAS) on psychiatric disorders since it is located on chromosome Xp11. A linkage study of schizophrenia patients to identify markers within Xp11 near the MAO-B gene utilized 92 families with a maternal pattern of inheritance and 34 families unselected for parental mode of transmission measuring logarithm of the odds (LOD) score and investigating the likely proximity of a gene and disease gene with its potential for inheritance. The association study revealed positive LOD scores for MAO-B with significant allele sharing mapped within a small region of Xp11 [87].

While anecdotal, gene polymorphism studies have also reported significant associations [88]. For example, a study assessing psychosis-proneness ("schizotypy") and altered gene regulation for dopaminergic neurotransmission revealed a negative correlation between the MAO-B gene (among other genes) and positive schizotypy scale (O-Life) [89]. Furthermore, an association study involving 110 schizophrenia patients and 87 controls found an association between allele 1 of the MAO-B gene and paranoid schizophrenia [90]. Another study involving 532 schizophrenia cases and 597 controls revealed that a haplotype of MAO-B in concordance with the ancestral haplotypes were significantly overrepresented in schizophrenia but was restricted to males [91]. Another allelic association study between dinucleotide repeats at the MAO loci and schizophrenia revealed significant differences in frequency distribution between transmitted (higher allelic frequencies) and non-transmitted repeats in the families of male schizophrenia patients [92]. Further, a study conducted in a Spanish population investigating the association of A/G polymorphism in intron 13 identified the G allele as a risk factor for schizophrenia in women [93], in line with the results observed in the Han-Chinese population and associated with the MAO-B polymorphism (rs1799836) [94]. Findings from another study suggests altered monoamine turnover rates in the central nervous system (CNS) reflecting associations between the MAO-B rs5905512 single nucleotide polymorphism (SNP) and 3-methoxy-4-hydroxyphenylglycol (MHPG) concentrations in schizophrenia [95]. Another study investigated the association of four gene regions: (1) gamma-aminobutyric acid receptor subunit beta-3 (GABRB3), (2) MAO-B, (3) phenylalanine hydroxylase (PAH), and (4) solute carrier family 6 member 4 (SLC6A4) with five symptoms in schizophrenia and revealed that the MAO-B/Norrie disease (NDP) gene region was significantly associated with delusions [96].

However, other studies failed to establish a significant association between MAO-B gene variant and schizophrenia susceptibility. For example, 100 African Americans male patients with schizophrenia screened for the MAO-B gene failed to identify functionally significant sequence changes [97], which is consistent with a similar study of schizophrenia subjects with severe aggression [98]. Similarly, a study in a Turkish population investigated the effect of the MAO-B A644G variant and found no significant effect [99].

Overall, while anecdotal, these early studies suggest a potential involvement of MAO-B in schizophrenia.

14.3.5 PET Findings: Review of Human MAO-B Studies (Table 14.4)

In vivo quantification of brain proteins (e.g., receptors, transporters, and enzymes) is possible using positron emission tomography (PET) [100]. Several PET radioligands [16–18, 101–108] have been developed to quantify MAO-B density in vivo in the human brain, but so far only [11C] labeled compounds such as [11C]N,N-dimethyl phenylethylamine ([11C]DMPEA) [16], [11C]L-deprenyl [17], [11C]L-deprenyl-D2 [18], and [11C]SL25.1188 [109] have been translated to humans.

Among the first generation radioligands for MAO-B, [11C]DMPEA [110] and [11C]L-deprenyl [17] were promising radioligands for MAO-B; however, there was significant trapping of the parent compound and its metabolites in the human brain. To improve the quantification, [11C]L-deprenyl-D2 [18], a second-generation MAO-B radioligand, was developed which had a reduced rate of trapping and displayed improved PET tracer characteristics including a more reversible time activity curve and good reproducibility [111, 112]. However, the lack of full reversibility and presence of radioactive brain-penetrant metabolites remained as potential limitations [18, 111]. A third generation radioligand, [11C]SL25.1188, presented improved PET tracer characteristics [113] including excellent reversibility in humans and TACs fitting remarkably well with a 2-tissue compartmental model (2TCM) with total distribution volume (VT) as an outcome with good test-retest values [109]. Importantly, regional MAO-B VT as measured with [11C]SL25.1188 PET in the human brain correlates highly with postmortem MAO-B protein concentration [(coefficient of determination) $r2 > 0.9$)] [11].

14.4 Conclusion and Clinical Translation

The studies reviewed here pertain only to the potential relevance of MAO-B in psychosis and schizophrenia. Major limitations include that most studies are not necessarily matched for sex, age, smoking, or body mass index. MAO-B is sensitive to the effects of age, sex, and cigarette smoking [114], as demonstrated in postmortem [25], genetic [91, 92], and peripheral studies [64]. Further limitations among clinical studies are the use of different combinations of antipsychotic treatments and smoking status among patient groups that may affect the results. Also, chronic treatment with antipsychotics may compromise psychosis severity quantification, especially in postmortem samples. Other limitations inherent to postmortem studies may contribute to the variable results.

With respect to peripheral findings, while MAO-B activity may be related to alterations in platelets rather than central MAO-B, there are other considerations [42]. When results are reported in equivalent units, values of MAO-B activity for similar diagnostic groups may vary considerably from one study to another,

Table 14.4 Clinical PET studies quantifying MAO-B density in different human brain regions

Reference	Subjects (HC)	Radiotracer	MAO-B activity	Regions of interest (ROIs)	Pros	Cons
[1]Shinotoh et al. 1987	Four M (age range, 48–70 years)	$[^{14}C]$ DMPEA	↑ (all ROIs except brain stem)	Frontal cortex, temporal cortex, occipital cortex, basal ganglia (caudate, putamen, and pallidum), thalamus, cerebellum, brain stem	↑ selectivity, brain uptake, ↑ specificity (L-deprenyl (10 mg/kg) reduced radioactivity)	Irreversible trapping (nonspecific binding), too quickly metabolized
[2]Fowler et al. 1987	Five (age range, 26–86 years)	^{11}C-labeled suicide enzyme inactivators, clorgyline, L-deprenyl (1st generation)	↑ (ROIs)	corpus striatum, thalamus, cortex, and brainstem	↑ selectivity, ↑ specificity (phenelzine reduced brain uptake (one subject); prior L-deprenyl treatment blocked brain retention)	Irreversible trapping, nonpolar metabolites, sensitive to blood flow
Fowler et al. 1995	Five (age range, 43–64 years)	$[^{11}C]$ L-deprenyl & $[^{11}C]$ L-deprenyl-D2 (2nd generation)	↑ $[^{11}C]$ L-deprenyl- D2 radioactivity	Occipital cortex, cingulate cortex, thalamus, basal ganglia, cerebellum, whole brain (global)	Reduced rate of trapping, ↑ MAO-B sensitivity & selectivity	Brain-penetrant metabolites
Logan et al. 2000	Five (4 F, 1 M) (age range, 23–73 years)	$[^{11}C]$ L-deprenyl-D2	↑ $[^{11}C]$ L-deprenyl- D2 radioactivity	Frontal cortex, occipital cortex, temporal cortex, parietal cortex, basal ganglia, thalamus, cerebellum, global	Reduced rate of trapping, ↑ MAO-B sensitivity, ↑ selectivity, good reproducibility (test-retest performed (7–27 days apart) showed acceptable results[¥])	Brain-penetrant metabolites: R(—) methamphetamine and R(—) amphetamine)

(continued)

Table 14.4 (continued)

Reference	Subjects (HC)	Radiotracer	MAO-B activity	Regions of interest (ROIs)	Pros	Cons
Rusjan, et al. 2014	Seven (4 M, 3 F; age range, 19–49 years)	[^{11}C]SL25.1188	Measurement of Vt showed ↑: caudate>thalamus>putamen>anterior cingulate cortex>frontal cortex>temporal cortex>Pons>cerebellar cortex	Cerebellar cortex, caudate, putamen, frontal cortex, temporal cortex, anterior cingulate cortex, thalamus, and pons	good identifiability, polar metabolites, ↑ MAO-B selectivity, ↑ sensitivity, good reproducibility (test-retest performed (5 ± 1 weeks apart showed acceptable variation[£])	No specificity quantification
Arakawa et al. 2017	Six (4 M, 2 F) (age range, 21–62 years)	[^{11}C] L-deprenyl-D2	↑ [^{11}C] L-deprenyl- D2 radioactivity	Cerebellum, hippocampus, lateral frontal cortex, lateral occipital cortex, lateral parietal cortex, lateral temporal cortex, medial frontal cortex, putamen, thalamus	Reduced rate of trapping, ↑ MAO-B sensitivity, ↑ selectivity, good reproducibility (test-retest performed (same day - morning and afternoon) showed acceptable results[€])	Brain-penetrant metabolites

↑ MAO-B activity in the brain regions of interest; [1], Parallel experiment conducted for MAO-B inhibition: C3H adult mice (M): pretreatment with doses (0, 0.01, 0.1, 10 mg/kg, i.v.) of L-deprenyl & clorgyline - 10 mg/kg); [2], MAO-B inhibition performed using Phenelzine (one subject) [L-deprenyl (15 mg)]; λk3*, model term proportional to MAO-B (validated outcome measure for [11C] L-deprenyl-D2); K1, the plasma to brain transfer constant that is related to blood flow; k3, rate constant (MAO-B binding quantified by two tissue compartment model (2TCM) with three rate constants (K1, k2, and k3) using metabolite-corrected plasma radioactivity) and λk3** = (K1/k2) × k3; ¥1) global value (mean difference – 2.1 ± 4.7%) 2) mean % change in eight brain regions (−2.84 ± 7.07%) 3) test-retest variability of λk3* values (mean difference – 2.84 ± 7.07%), high correlation (r2 = 0.98; $p < 0.0001$) 4) K1 values, significant correlation (r2 = 0.61; $p < 0.0001$) though scan two values < scan 1 5), regional brain uptake values differ; (1) TRV, 12%; (2) between-subject COV, 20%; (3) within-subject variability, around 10%; (1) k3** and λk3 values in ROIs (mean absolute percent difference, 5.3% and 7.0%, respectively); (2) All ROIs showed <10% variabilities except hippocampus in k3 (10.3%); (3)test-retest variability of λk3 (% difference: −1.1 ± 8.3%); (4) Mean ICC for k3 and λk3 are 0.94 and 0.89, respectively

Abbreviations: *MAO-B* monoamine oxidase B; *ROIs* regions of interest; *HC* healthy control; *M* male; *F* female; *[^{11}C]DMPEA* carbon-11-labeled N,N-dimethylphenylethylamine; *[^{11}C]* Carbon-11-labeled; *mg/kg* milligram/kilogram; *D-2* dopamine 2 receptor; *Vt* total distribution volume; *R* R nomenclature of chiral center (Clockwise); *%* percent; r^2 correlation coefficient; *p* p-value (significance constant); *TRV* time-retest variability; *COV* coefficient of variation; *ICC* intraclass correlation coefficient presumably because of variations in platelet preparation, assay procedures, and the apparently large normal range of human platelet MAO-B activity. Large variations from study to study are also due to differences in laboratory equipment, assay procedures, and statistical analyses. Furthermore, among genetic studies, power concerns are important caveats.

In preclinical studies, MAO-B knock-out mice may not display altered extracellular levels of dopamine owing to the significant adaptive upregulation of the D2-like dopamine receptors and hypersensitivity of D1 receptors [44]. The degradation of dopamine is mediated by both MAOs, but the relative contribution of each isoenzyme differs in relation to the species and the tissue under consideration [49, 115].

Finally, MAO-B is predominantly found in astrocytes (and serotonin releasing neurons), and its overexpression in activated astrocytes has led to the proposition that MAO-B could be a reliable biomarker for astrocytosis in disease states [116–118]. Greater MAO-B levels occur in neurodegenerative diseases with astrogliosis. MAO-B was also significantly associated with other astroglial markers such as glial fibrillary acidic protein (GFAP) [119]. For example, a postmortem study showed increased levels of MAO-B in the plaque-associated astrocytes in Alzheimer's disease in temporal, parietal and frontal cortices [120].

The findings summarized in this chapter indicate that although the role of MAO-B in schizophrenia is still inconclusive, preclinical, peripheral, postmortem, and genetic studies suggest it is possible that there is altered MAO-B activity in schizophrenia. The most supported finding is the reduced MAO-B activity in the peripheral tissues of patients, particularly in chronic paranoid schizophrenia; however, previous studies are confounded by cigarette smoking. Additionally, some MAO-B knock-out mice studies resulted in hyperdopaminergic states and behavioral disinhibition, which is broadly in line with the hyperdopaminergic hypothesis of schizophrenia. Moreover, preclinical studies utilizing MAO-B inhibitors (selegiline/L-deprenyl) suggest that dopamine metabolism is altered with higher concentrations (10 mg/kg) [39] and chronic administration (21 days) [121]. Additionally, several human studies have evaluated the selective MAO-B inhibitor, selegiline, in the treatment of negative symptoms [20, 21]. Bodkin et al. [22] found that selegiline was significantly more effective than placebo for the treatment of predominant negative symptoms [122]. Further, a selective MAO-B inhibitor, rasagiline which is up to 15 times more potent than selegiline, may be of clinical benefit for negative symptoms [23]. Together, this indicates the potential involvement of MAO-B in the pathophysiology of schizophrenia and warrants the need for in vivo human studies, potentially usable as a biomarker or stratification tool via the use of highly selective PET radioligands, which target MAO-B.

Acknowledgments This review chapter was done as part of the research undertaken, thanks in part to funding from the Canada First Research Excellence Fund, awarded to the Healthy Brains for Healthy Lives initiative at McGill University. I would like to acknowledge and thank my research colleagues for their continued support and guidance throughout the preparation of this review chapter.

Declaration of Interest The authors declare that they have no known competing financial interests or personal relationships that could have appeared to influence the work reported in this chapter.

Credit Authorship Contribution Statement **Kankana Nisha Aji:** conceptualization, methodology, and writing – original draft. **Jeff Meyer:** validation and writing – review and editing. **Pablo M. Rusjan:** validation and writing – review and editing for the PET section. **Romina Mizrahi:** supervision, validation, and writing – review and editing.

Source of Funding The preparation of this review chapter received no specific grant from any funding agency in the public, commercial, or not-for-profit sectors.

References

1. Weinstein JJ, van de Giessen E, Rosengard RJ, Xu X, Ojeil N, Brucato G, et al. PET imaging of dopamine-D2 receptor internalization in schizophrenia. Mol Psychiatry. 2017. https://doi.org/10.1038/mp.2017.157
2. Gillespie AL, Samanaite R, Mill J, et al. Is treatment-resistant schizophrenia categorically distinct from treatment-responsive schizophrenia? A systematic review. BMC Psychiatry. 2017;17(1):12. https://doi.org/10.1186/s12888-016-1177-y
3. Cases O, Seif I, Grimsby J, Gaspar P, Chen K, Pournin S, et al. Aggressive behavior and altered amounts of brain serotonin and norepinephrine in mice lacking MAOA. Science. 1995;268(80):1763–66. https://doi.org/10.1126/science.7792602
4. Goridis C, Neff NH. Evidence for a specific monoamine oxidase associated with sympathetic nerves. Neuropharmacology. 1971;10(5):557–64.
5. Owen F, Cross AJ, Lofthouse R, Glover V. Distribution and inhibition characteristics of human brain monoamine oxidase. Biochem Pharmacol. 1979;28(7):1077–80. https://doi.org/10.1016/0006-2952(79)90307-1
6. Glover V, Sandler M, Owen F, Riley GJ. Dopamine is a monoamine oxidase B substrate in man [33]. Nature. 1977;265(5589):80–1.
7. Chen K, Shih JC. Monoamine oxidase A and B: structure, function, and behavior. Adv Pharmacol. 1998;42:292–6.
8. Youdim MBH, Edmondson D, Tipton KF. The therapeutic potential of monoamine oxidase inhibitors. Nat Rev Neurosci. 2006;7(4):295–309.
9. Ramsay RR, Majekova M, Medina MVM. Key targets for multi-target ligands designed to combat neurodegeneration. Front Neurosci. 2016;10:375.
10. Saura Marti J, Kettler R, Da Prada M, Richards JG. Molecular neuroanatomy of MAO-A and MAO-B. J Neural Transm Suppl. 1990;32:49–53.
11. Tong J, Meyer JH, Furukawa Y, Boileau I, Chang LJ, Wilson AA, et al. Distribution of monoamine oxidase proteins in human brain: implications for brain imaging studies. J Cereb Blood Flow Metab. 2013;33(6):863–71.
12. Meyer JH, Ginovart N, Boovariwala A, Sagrati S, Hussey D, Garcia A, et al. Elevated monoamine oxidase A levels in the brain: an explanation for the monoamine imbalance of major depression. Arch Gen Psychiatry. 2006;63(11):1209–16.
13. Fowler JS, Logan J, Ding YS, Franceschi D, Wang GJ, Volkow ND, et al. Non-MAO A binding of clorgyline in white matter in human brain. J Neurochem. 2001;79(5):1039–46.
14. Ginovart N, Wilson AA, Meyer JH, Hussey D, Houle S. Positron emission tomography quantification of [11C]-DASB binding to the human serotonin transporter: modeling strategies. J Cereb Blood Flow Metab. 2001;27(4):857–71.

15. Zanotti-Fregonara P, Leroy C, Rizzo G, Roumenov D, Trichard C, Martinot JL, et al. Imaging of monoamine oxidase-A in the human brain with [11C]befloxatone: quantification strategies and correlation with mRNA transcription maps. Nucl Med Commun. 2014;35:1254–61.
16. Shinotoh H, Inoue O, Hirayama K, Aotsuka A, Asahina M, Suhara T, et al. Dopamine D 1 receptors in Parkinson's disease and striatonigral degeneration: a positron emission tomography study. J Neurol Neurosurg Psychiatry. 1993;56(5):467–72.
17. Fowler JS, MacGregor RR, Wolf AP, Arnett CD, Dewey SL, Schlyer D, et al. Mapping human brain monoamine oxidase a and B with 11C-labeled suicide inactivators and PET. Science. 1987;235(4787):481–5.
18. Fowler JS, Wang GJ, Logan J, Xie S, Volkow ND, MacGregor RR, et al. Selective reduction of radiotracer trapping by deuterium substitution: comparison of carbon-11-L-deprenyl and carbon-11-deprenyl-D2 for MAO B mapping. J Nucl Med. 1995;36(7):1255–62.
19. Moriguchi S, Wilson AA, Miler L, Rusjan PM, Vasdev N, Kish SJ, et al. Monoamine oxidase b total distribution volume in the prefrontal cortex of major depressive disorder: an 11csl25.1188 positron emission tomography study. JAMA Psychiat. 2019;76(6):634–41.
20. Fohey KD, Hieber R, Nelson LA. The role of selegiline in the treatment of negative symptoms associated with schizophrenia. Ann Pharmacother. 2007;41(5):851–6.
21. Amiri A, Noorbala AA, Nejatisafa AA, Ghoreishi A, Derakhshan MK, Khodaie-Ardakani MR, Hajiazim M, Raznahan M, Akhondzadeh S. Efficacy of selegiline add on therapy to risperidone in the treatment of the negative symptoms of schizophrenia: a double-blind randomized placebo-controlled study. Hum Psychopharmacol. 2008;23(2):79–86.
22. Bodkin JA, Siris SG, Bermanzohn PC, Hennen J, Cole JO. Double-blind, placebo-controlled, multicenter trial of selegiline augmentation of antipsychotic medication to treat negative symptoms in outpatients with schizophrenia. Am J Psychiatry. 2005;162(2):288–90.
23. Buchanan RW, Weiner E, Kelly DL, Gold JM, Keller WR, Waltz JA, McMahon RP, DAG. Rasagiline in the treatment of the persistent negative symptoms of Schizophrenia. Schizophr Bull. 2015;41(4):900–8.
24. Fowler CJ, Carlsson A, Winblad B. Monoamine oxidase-A and -B activities in the brain stem of schizophrenics and non-schizophrenic psychotics. J Neural Transm. 1981;52:23–32.
25. Yang Q, Ikemoto K, Nishino S, Yamaki J, Kunii Y, Wada A, et al. DNA methylation of the Monoamine Oxidases A and B genes in postmortem brains of subjects with schizophrenia. Open. J Psychiatry. 2012;2:374–83.
26. Owen F, Crow TJ, Frith CD, Johnson JA, Johnstone EC, Lofthouse R, et al. Selective decreases in MAO-B activity in post-mortem brains from schizophrenic patients with type II syndrome. Br J Psychiatry. 1987;151:514–9.
27. Reveley MA, Glover V, Sandler M, Spokes EG. Brain monoamine oxidase activity in Schizophrenics and controls. Arch Gen Psychiatry. 1981;38(6):663–5.
28. Schwartz MA, Wyatt RJ, Yang HYT, Neff NH. Multiple forms of brain monoamine oxidase in Schizophrenic and normal individuals. Arch Gen Psychiatry. 1974;31(4):557–60.
29. Crow TJ, Baker HF, Cross AJ, Joseph MH, Lofthouse R, Longden A, et al. Monoamine mechanisms in chronic schizophrenia: post-mortem neurochemical findings. Br J Psychiatry. 1979;134:249–56.
30. Karolewicz B, Klimek V, Zhu H, Szebeni K, Nail E, Stockmeier CA, et al. Effects of depression, cigarette smoking, and age on monoamine oxidase B in amygdaloid nuclei. Brain Res. 2005;1043(1–2):57–64.
31. Fowler CJ, Oreland L, Marcusson JWB. Titration of human brain monoamine oxidase -A and -B by clorgyline and L-deprenil. Naunyn Schmiedeberg's Arch Pharmacol. 1980;311(3):263–72.
32. Baud P, Arbilla S, Cantrill RC, Scatton B, Langer SZ. Trace amines inhibit the electrically evoked release of [3H]acetylcholine from slices of rat striatum in the presence of pargyline: similarities between β-phenylethylamine and amphetamine. J Pharmacol Exp Ther. 1985;235(1):220–9.

33. O'Reilly R, Davis BA, Durden DA, Thorpe L, Machnee H, Boulton AA. Plasma phenylethylamine in schizophrenic patients. Biol Psychiatry. 1991;30(2):145–50.
34. Grimsby J, Toth M, Chen K, Kumazawa T, Klaidman L, Adams JD, et al. Increased stress response and p-phenylethylamine in maob-def icient mice. Nat Genet. 1997;17(2):206–10.
35. Kuroki T, Tsutsumi T, Hirano M, et al. Behavioral sensitization to beta-phenylethylamine (PEA): enduring modifications of specific dopaminergic neuron systems in the rat. Psychopharmacol. 1990;102(1):5–10.
36. Van Gaalen MM, Brueggeman RJ, Bronius PFC, Schoffelmeer ANM, Vanderschuren LJMJ. Behavioral disinhibition requires dopamine receptor activation. Psychopharmacology. 2006;187(1):73–85.
37. Sabelli HC, Javaid JI. Phenylethylamine modulation of affect: therapeutic and diagnostic implications. J Neuropsychiatry Clin Neurosci. 1995;7(1):6–14.
38. Bortolato M, Godar SC, Davarian S, Chen K, Shih JC. Behavioral disinhibition and reduced anxiety-like behaviors in monoamine oxidase b-deficient mice. Neuropsychopharmacology. 2009;34(13):2746–57.
39. Butcher SP, Fairbrother IS, Kelly JS, Arbuthnott GW. Effects of selective monoamine oxidase inhibitors on the in vivo release and metabolism of dopamine in the rat striatum. J Neurochem. 1990;55(3):981–8.
40. Bortolato M, Shih JC. Behavioral outcomes of monoamine oxidase deficiency: preclinical and clinical evidence. Int Rev Neurobiol. 2011;100:13–42.
41. Holschneider DP, Chen K, Seif I, Shih JC. Biochemical, behavioral, physiologic, and neurodevelopmental changes in mice deficient in monoamine oxidase A or B. Brain Res Bull. 2001;56(5):453–62.
42. Zureick JL, Meltzer HY. Platelet MAO activity in hallucinating and paranoid schizophrenics: a review and meta-analysis. Biol Psychiatry. 1988;24(1):63–78.
43. Fornai F, Chen K, Giorgi FS, Gesi M, Alessandri MG, Shih JC. Striatal dopamine metabolism in monoamine oxidase B-deficient mice: a brain dialysis study. J Neurochem. 1999;73(6):2434–40.
44. Chen L, He M, Sibille E, Thompson A, Sarnyai Z, Baker H, et al. Adaptive changes in postsynaptic dopamine receptors despite unaltered dopamine dynamics in mice lacking monoamine oxidase B. J Neurochem. 1999;22:197–217.
45. Peters JR, Vallie B, Difronzo M, Donaldson ST. Role of dopamine D1 receptors in novelty seeking in adult female Long-Evans rats. Brain Res Bull. 2007;74(4):232–236.
46. Olsen CM, Winder DG. Operant sensation seeking engages similar neural substrates to operant drug seeking in C57 mice. Neuropsychopharmacology. 2009;34(7):1685–94.
47. Okuda C, Segal DS, Kuczenski R. Deprenyl alters behavior and caudate dopamine through an amphetamine-like action. Pharmacol Biochem Behav. 1992;43(4):1075–80.
48. O'Carroll AM, Fowler CJ, Phillips JP, Tobbia I, Tipton KF. The deamination of dopamine by human brain monoamine oxidase – specificity for the two enzyme forms in seven brain regions. Naunyn Schmiedeberg's Arch Pharmacol. 1983;322(3):198–202.
49. Garrick NA, Murphy DL. Species differences in the deamination of dopamine and other substrates for monoamine oxidase in brain. Psychopharmacology. 1980;72:27–33.
50. Mészáros Z, Borcsiczky D, Máté M, Tarcali J, Tekes K, Magyar K. MAO inhibitory side effects of neuroleptics and platelet serotonin content in schizophrenic patients. J Neural Transm Suppl. 1998;52:79–85.
51. Berger PA, Ginsburg RA, Barchas JD, Murphy DLWR. Platelet monoamine oxidase in chronic schizophrenic patients. Am J Psychiatry. 1978;135(1):95–9.
52. Sullivan JL, Cavenar JO, Stanfield CN, Hammett EB. Reduced MAO activity in platelets and lymphocytes of chronic schizophrenics. Am J Psychiatry. 1978;135(5):597–8.
53. Baron M, Perlman R, Levitt M. Paranoid schizophrenia and platelet MAO activity. Am J Psychiatry. 1980;137(11):1465–6.
54. Tachiki KH, Kurtz N, Kling AS, Hullett FJ. Blood monoamine oxidases and CT scans in subgroups of chronic schizophrenics. J Psychiatr Res. 1984;18(3):233–43.

55. Meltzer HY, Zureick JL. Relationship of auditory hallucinations and paranoia to platelet MAO activity in schizophrenics: sex and race interactions. Psychiatry Res. 1987;22:99–109.
56. Gruen R, Baron M, Levitt M, Asnis L. Platelet MAO activity and schizophrenic prognosis. Am J Psychiatry. 1982;139(2):240–1.
57. Bierer L, Docherty JP, Young JG, Giller E, Cohen D. Poor outcome and low platelet MAO activity in chronic schizophrenia. Am J Psychiatry. 1984;141(2):57–260.
58. Berrettini WH, Prozialeck W, Vogel WH. Decreased platelet monoamine oxidase activity in chronic Schizophrenia, shown with novel substrates. Arch Gen Psychiatry.1978;35(5):600–5.
59. Domino EF, Sampath KS. Decreased blood platelet MAO activity in unmedicated chronic schizophrenic patients. Am J Psychiatry. 1976;133(3):323–6.
60. Murphy DL, Belmaker R, Wyatt RJ. Monoamine oxidase in schizophrenia and other behavioral disorders. J Psychiatr Res. 1974;11:221–47.
61. Murphy DL, Donnelly CH, Miller L, Wyatt RJ. Platelet Monamine oxidase in chronic Schizophrenia: some enzyme characteristics relevant to reduced activity. Arch Gen Psychiatry. 1976;33(11):1377–81.
62. Murphy DL, Wyatt RJ. Reduced monoamine oxidase activity in blood platelets from schizophrenic patients. Nature. 1972;238(5361):225–6.
63. Becker RE, Shaskan EG. Platelet monoamine oxidase activity in schizophrenic patients. Am J Psychiatry. 1976;133(10):1191–3.
64. Marcolin MA, Davis JM. Platelet monoamine oxidase in schizophrenia: a meta-analysis. Schizophr Res. 1992;7(3):249–67.
65. Lewine RJ, Meltzer HY. Negative symptoms and platelet monoamine oxidase activity in male schizophrenic patients. Psychiatry Res. 1984;12(2):99–109.
66. Mann JJ, Kaplan RD, Georgotas A, Friedman E, Branchey M, Gershon S. Monoamine oxidase activity and enzyme kinetics in three subpopulations of density-fractionated platelets in chronic paranoid schizophrenics. Psychopharmacology. 1981;74(4):344–8.
67. Owen F, Bourne R, Crow TJ, Johnstone EC, Bailey AR, Hershon HI. Platelet Monamine oxidase in Schizophrenia: an investigation in drug-free chronic hospitalized patients. Arch Gen Psychiatry. 1976;33(11):1370–3.
68. Carpenter WT, Murphy DLWR. Platelet monoamine oxidase activity in acute schizophrenia. Am J Psychiatry. 1975;132(4):438–41.
69. White HL, McLeod MN, Davidson JRT. Platelet monoamine oxidase activity in schizophrenia. Am J Psychiatry. 1976;133(10):1191–3.
70. Mann J, Thomas KM. Platelet monoamine oxidase activity in schizophrenia. Relationship to disease, treatment, institutionalization and outcome. Br J Psychiatry. 1979;134:366–71.
71. Duncavage M, Luchins DJ, Meltzer HY. Platelet MAO activity and family history of schizophrenia. Psychiatry Res. 1982;7(1):47–51.
72. Spivak B, Kosower N, Zipser Y, Shreiber-Schul N, Apter A, Tyano S, et al. Platelet monoamine oxidase activity in neuroleptic-naive schizophrenic patients: lack of influence of chronic perphenazine treatment. Clin Neuropharmacol. 1994;17(1):83–8.
73. Bridge TP, Jeste DV, Wise CD, Potkin SG, Phelps BH, Wyatt RJ. Schizophrenic outcome in late life: symptom state and platelet monoamine oxidase activity. Psychiatry Res. 1984;11(2):91–7.
74. Del Vecchio M, Maj M, D'Ambrosio A, Kemali D. Low platelet MAO activity in chronic schizophrenics: a long-term effect of neuroleptic treatment? Psychopharmacology. 1983;79(2–3):177–9.
75. Maj M, Ariano MG, Pirozzi R, Salvati A, Kemali D. Platelet monoamine oxidase activity in schizophrenia: relationship to family history of the illness and neuroleptic treatment. J Psychiatr Res. 1984;18(2):131–7.
76. Takahashi S, Yamane H, Tani N. Reduction of blood platelet monoamine oxidase activity in Schizophrenic patients on Phenothiazines. Psychiatry Clin Neurosci. 1975;29(3):207–14.

77. Owen F, Bourne RC, Crow TJ, Fadhli AA, Johnstone EC. Platelet monoamine oxidase activity in acute schizophrenia: relationship to symptomatology and neuroleptic medication. Br J Psychiatry. 1981;139:16–22.
78. DeLisi LE, Wise CD, Bridge TP, Rosenblatt JE, Wagner RL, Morihisa J, et al. A probable neuroleptic effect on platelet monoamine oxidase in chronic schizophrenic patients. Psychiatry Res. 1981;4(1):95–107.
79. Orologas AG, Buckman TD, Eiduson S. A comparison of platelet monoamine oxidase activity and phosphatidylserine content between chronic paranoid schizophrenics and normal controls. Neurosci Lett. 1986;68:293–8.
80. Fowler JS, Volkow ND, Wang GJ, Pappas N, Logan J, MacGregor R, et al. Inhibition of monoamine oxidase B in the brains of smokers. Nature. 1996;93(24):14065–9.
81. Berlin I, Said S, Spreux-Varoquaux O, Olivares R, Launay JM, Puech AJ. Monoamine oxidase A and B activities in heavy smokers. Biol Psychiatry. 1995;38(11):56–761.
82. Fowler JS, Wang GJ, Volkow ND, Franceschi D, Logan J, Pappas N, et al. Maintenance of brain monoamine oxidase B inhibition in smokers after overnight cigarette abstinence. Am J Psychiatry. 2000;157(11):1864–6.
83. De Leon J, Dadvand M, Canuso C, White AO, Stanilla JKSG. Schizophrenia and smoking: an epidemiologic survey in a state hospital. Am J Psychiatry. 1995;152:453–5.
84. Wyatt RJ, Murphy DL, Belmaker R, Cohen S, Donnelly CH, Pollin W. Reduced monoamine oxidase activity in platelets: a possible genetic marker for vulnerability to schizophrenia. Science.1973;179:916–8.
85. Reveley MA, Reveley AM, Clifford CA, Murray RM. Genetics of platelet MAO activity in discordant schizophrenic and normal twins. Br J Psychiatry. 1983;142:560–5.
86. Baron M, Levitt MPR. Low platelet monoamine oxidase activity: a possible biochemical correlate of borderline schizophrenia. Psychiatry Res. 1980;3(3):329–35.
87. Dann J, DeLisi LE, Devoto M, Laval S, Nancarrow DJ, Shields G, et al. A linkage study of schizophrenia to markers within Xp11 near the MAOB gene. Psychiatry Res. 1997;70(3):131–43.
88. Wei J, Hemmings GP. A study of linkage disequilibrium between polymorphic loci for monamine oxidases A and B in schizophrenia. Psychiatr Genet. 1999;9(4):177–81.
89. Grant P, Gabriel F, Kuepper Y, Wielpuetz C, Hennig J. Psychosis-proneness correlates with expression levels of dopaminergic genes. Eur Psychiatry. 2014;29:304–6.
90. Carrera N, Sanjuán J, Moltó MD, Carracedo Á, Costas J. Recent adaptive selection at MAOB and ancestral susceptibility to schizophrenia. Am J Med Genet Part B Neuropsychiatr Genet. 2009;150B(3):369–74
91. Coron B, Campion D, Thibaut F, Dollfus S, Preterre P, Langlois S, et al. Association study between schizophrenia and monoamine oxidase A and B DNA polymorphisms. Psychiatry Res. 1996;62(3):221–6.
92. Wei J, Hammings GP. Allelic association between dinucleotide repeats at the monoamine oxidase loci and schizophrenia. Eur Psychiatry. 1998;13(8):407–10.
93. Gassó P, Bernardo M, Mas S, Crescenti A, Garcia C, Parellada ELA. Association of A/G polymorphism in intron 13 of the monoamine oxidase B gene with schizophrenia in a Spanish population. Neuropsychobiology. 2008;58(2):65–70.
94. Wei YL, Li CX, Li SB, Liu YHL. Association study of monoamine oxidase A/B genes and schizophrenia in Han Chinese. Behav Brain Funct. 2011;7(1):42.
95. Andreou D, Söderman E, Axelsson T, Sedvall GC, Terenius L, Agartz I, et al. Polymorphisms in genes implicated in dopamine, serotonin and noradrenalin metabolism suggest association with cerebrospinal fluid monoamine metabolite concentrations in psychosis. Behav Brain Funct. 2014;10:26.
96. Sun J, Jayathilake K, Zhao Z, Meltzer HY. Investigating association of four gene regions (GABRB3, MAOB, PAH, and SLC6A4) with five symptoms in schizophrenia. Psychiatry Res. 2012;198:202–6.

97. Sobell JL. Screening the monoamine oxidase B gene in 100 male patients with schizophrenia: a cluster of polymorphisms in African-Americans but lack of functionally significant sequence changes. Am J Med Genet - Neuropsychiatr Genet. 1997;74(1):44–9.
98. Perkovic MN, Strac DS, Erjavec GN, Uzun S, Podobnik J, Kozumplik O, Vlatkovic SPN. Monoamine oxidase and agitation in psychiatric patients. Prog Neuro-Psychopharmacol Biol Psychiatry. 2016;69:131–46.
99. Pehlivan S, Aydin PC, Uysal MA, Ciftci HS, Sever U, Yavuz FK, Aydin NNA. Effect of monoamine oxidase B A644G variant on nicotine dependence and/or schizophrenia risk. Arch Clin Psychiatry. 2019;46(1):21–4.
100. Jones T, Rabiner EA. The development, past achievements, and future directions of brain PET. J Cereb Blood Flow Metab. 2012;32(7):1426–54.
101. Ishiwata K, Ido T, Yanai K, Kawashima K, Miura Y, Monma M, et al. Biodistribution of a positron-emitting suicide inactivator of monoamine oxidase, carbon-11 pargyline, in mice and a rabbit. J Nucl Med. 1985;26(6):630–6.
102. Plenevaux A, Dewey SL, Fowler JS, Guillaume MWA. Synthesis of (R)-(-)-and (S)-(+)-4-fluorodeprenyl,(R)-(-)-and (S)-(+)-[N-11C-methyl]-4-fluorodeprenyl and PET studies in baboon brain. J Med Chem. 33(7):2015–9.
103. Bläuenstein P, Rémy N, Buck A, Ametamey S, Häberli MSP. In vivo properties of N-(2-aminoethyl)-5-halogeno-2-pyridinecarboxamide 18F-and 123I-labelled reversible inhibitors of monoamine oxidase B. Nucl Med Biol. 1998;25(1):47–52.
104. Mukherjee J, Yang ZY, Lew R. N-(6-18F-fluorohexyl)-N-methylpropargylamine: a fluorine-18-labeled monoamine oxiduse B inhibitor for potential use in PET studies. Nucl Med Biol. 1999;26(1):111–6.
105. Hirata M, Kagawa S, Yoshimoto M, Ohmomo Y. Synthesis and characterization of radioiodinated MD-230254: a new ligand for potential imaging of monoamine oxidase B activity by single photon emission computed tomography. Chem Pharm Bull. 2002;50(5):609–14.
106. Saba W, Valette H, Peyronneau MA, Bramoullé Y, Coulon C, Curet O, George P, Dollé FBM. [11C] SL25. 1188, a new reversible radioligand to study the monoamine oxidase type B with PET: preclinical characterisation in nonhuman primate. Synapse. 2010;64(1):61–9.
107. Nag S, Lehmann L, Kettschau G, Heinrich T, Thiele A, Varrone A, Gulyas BHC. Synthesis and evaluation of [18F] fluororasagiline, a novel positron emission tomography (PET) radioligand for monoamine oxidase B (MAO-B). Bioorg Med Chem. 2012;20(9):3065–71.
108. Hicks JW, Sadovski O, Parkes J, Houle S, Hay BA, Carter RL, Wilson AAVN. Radiosynthesis and ex vivo evaluation of [18F]-(S)-3-(6-(3-fluoropropoxy) benzo [d] isoxazol-3-yl)-5-(methoxymethyl) oxazolidin-2-one for imaging MAO-B with PET. Bioorg Med Chem Lett. 2015;25(2):288–91.
109. Rusjan PM, Wilson AA, Miler L, Fan I, Mizrahi R, Houle S, Vasdev NMJ. Kinetic modeling of the monoamine oxidase B radioligand [11C] SL25. 1188 in human brain with high-resolution positron emission tomography. J Cereb Blood Flow Metab. 2014;34(5):883–9.
110. Inoue O, Tominaga T, Yamasaki T, Kinemuchi H, Radioactive N. N-Dimethylphenylethylamine: a selective radiotracer for in vivo measurement of monoamine oxidase-B activity in the brain. J Neurochem. 1985; 44(1):210–6.
111. Logan J, Fowler JS, Volkow ND, Wang GJ, MacGregor RRSC. Reproducibility of repeated measures of deuterium substituted [11C] L-deprenyl ([11C] L-deprenyl-D2) binding in the human brain. Nucl Med Biol. 2000;27(1):43–9.
112. Arakawa R, Stenkrona P, Takano A, Nag S, Maior RSHC. Test-retest reproducibility of [11 C]-l-deprenyl-D 2 binding to MAO-B in the human brain. EJNMMI Res. 2017;7(1):54.
113. Curet O, Sontag N, Aubin N, Marc C, Jegham S, George PB, J SB. Biochemical profile of SL25.1188, a new reversible MAO-B inhibitor. In: Proceedings of the 29th annual meeting of society neuroscience. 1999;848–914.
114. Fowler JS, Volkow ND, Wang GJ, Pappas N, Logan J, Macgregor R, et al. Neuropharmacological actions of cigarette smoke: brain monoamine oxidase B (MAO B) inhibition. J Addict Dis. 1998;17(1):23–34.

115. Tipton KF, O'Carroll AMMJ. The catalytic behavior of monoamine oxidase. J Neural Transm. 1987;23:25–35.
116. Levitt P, Pintar JE, Breakefield XO. Immunocytochemical demonstration of monoamine oxidase B in brain astrocytes and serotonergic neurons. Proc Natl Acad Sci U S A. 1982;79(20): 6385–9.
117. Nakamura S, Kawamata T, Akiguchi I, Kameyama M, Nakamura N, Kimura H. Expression of monoamine oxidase B activity in astrocytes of senile plaques. Acta Neuropathol. 1990; 80(4):419–25.
118. Ekblom J, Jossan SS, Oreland L, Walum E, Aquilonius SM. Reactive gliosis and monoamine oxidase B. J Neural Transm. 1994;41:253–8.
119. Tong J, Rathitharan G, Meyer JH, Furukawa Y, Ang LC, Boileau I, et al. Brain monoamine oxidase B and A in human parkinsonian dopamine deficiency disorders. Brain. 2017; 33(6):863–71.
120. Saura J, Luque JM, Cesura AM, Da PM, Chan-Palay V, Huber G, et al. Increased monoamine oxidase b activity in plaque-associated astrocytes of Alzheimer brains revealed by quantitative enzyme radioautography. Neuroscience. 1994;62(1):15–30.
121. Lamensdorf I, Youdim MBFJ. Effect of long-term treatment with selective monoamine oxidase A and B inhibitors on dopamine release from rat striatum in vivo. J Neurochem. 1996;67(4):1532–9.
122. Jungerman T, Rabinowitz D, Klein E. Deprenyl augmentation for treating negative symptoms of schizophrenia: a double-blind, controlled study. J Clin Psychopharmacol. 1999;19(6):522–5.

Chapter 15
Genomics in Treatment Development

Yogesh Dwivedi and Richard C. Shelton

Abstract The Human Genome Project mapped the 3 billion base pairs in the human genome, which ushered in a new generation of genomically focused treatment development. While this has been very successful in other areas, neuroscience has been largely devoid of such developments. This is in large part because there are very few neurological or mental health conditions that are related to single-gene variants. While developments in pharmacogenomics have been somewhat successful, the use of genetic information in practice has to do with drug metabolism and adverse reactions. Studies of drug metabolism related to genetic variations are an important part of drug development. However, outside of cancer biology, the actual translation of genomic information into novel therapies has been limited. Epigenetics, which relates in part to the effects of the environment on DNA, is a promising newer area of relevance to CNS disorders. The environment can induce chemical modifications of DNA (e.g., cytosine methylation), which can be induced by the environment and may represent either shorter- or longer-term changes. Given the importance of environmental influences on CNS disorders, epigenetics may identify important treatment targets in the future.

Keywords Cytochrome enzymes · Drug–drug interactions · Drug metabolism · Enrichment strategies · Epigenetics · Genetics · Genomics · Genome-wide association studies · Pharmacogenomics · Uridine 5′-diphospho-glucuronosyltransferases

Y. Dwivedi · R. C. Shelton (✉)
Department of Psychiatry and Behavioral Neurobiology, Heersink School of Medicine, University of Alabama at Birmingham, Birmingham, AL, USA
e-mail: rcshelton@uabmc.edu

M. Macaluso et al. (eds.), *Drug Development in Psychiatry*, Advances in Neurobiology 30,
https://doi.org/10.1007/978-3-031-21054-9_15

The Human Genome Project was an international effort to determine the base pair sequence of the human genome and mapping of individual genes. The project began with great promise in 1990, and the initial publication of most of the sequence and a tentative set of genes occurred in 2003 [1]. The complete mapping of the 3 billion base pairs in the human genome was an enormous task that required the efforts of many labs distributed across the globe. The promises of the Project were ambitious and included the ability to understand diseases at the most fundamental level. The Project was expected to discover mutations linked to diseases, allowing scientists to develop new medicines and other treatments, to predict response to treatment, and to both predict and prevent disease [2–5].

The results of the Project were no less than revolutionary, and it is hard to image a science without the sequence of human DNA. There certainly have been many advances in the understanding of human diseases and the development of treatments as a result. This has been particularly true in cancer genomics, in which mutations have been linked to specific cancers, which has led to an improved understanding of the underlying physiology of disease, leading to new treatments [6]. Other areas significantly impacted by genomics have included anticoagulant therapy [7, 8], infectious disease surveillance and treatment [9], Alzheimer's disease [10], and many others. A very large number of drugs now include pharmacogenomic recommendations in FDA-approved drug labeling [118]. However, much of the use of genomic information in practice now has to do with drug metabolism and adverse reactions. Outside of cancer biology, the actual translation of genomic information into novel therapies has been limited.

The Human Genome Project produced some significant surprises. A key unexpected finding is the number genes. Prior to the Project, estimates of the number of genes in the human genome ranged from 100,000 to as high as 150,000, which roughly corresponds to the number of proteins in the body [12]. However, during various stages of the Project, this number declined progressively to the current estimate of about 22,300 [13]. The large number of proteins can be explained by the fact that single genes can code for several proteins through alternative splicing in which exon-coded RNAs can be cleaved into multiple variations representing different types of proteins. Also surprising is that only about 1.5% of the human genome is in the form of genes [14], leaving massive amounts of genetic material uncharacterized – as much as 2.95 billion bases. This intergenic (initially called "junk") DNA was a mystery that has gradually emerged as being important in human diseases. In addition, environmental influences on DNA referred to as epigenetics have also become very important in our understanding of human disease and potential treatment targets. These will be discussed more below.

15.1 Pharmacogenomics and Drug Development

Drug metabolism is the process by which the body chemically modifies medications to facilitate elimination. This process is divided into two phases: Phase I reactions inactivate drugs through direct chemical modifications; this process can also change drugs into active metabolites or convert prodrugs to their active forms. These reactions typically involve oxidation, reduction, or hydrolysis of the parent drug. Phase II reactions convert drugs or metabolites into polar forms that can be eliminated [15]. This can involve the addition of many different molecules to the basic structure of the drug. Examples of these processes include methylation, sulfation, acetylation, glucuronidation, glutathione conjugation, and glycine conjugation. There can also be phase III reactions involving further chemical modifications. For most drugs, this involves conversion into water soluble (i.e., hydrophilic) forms that can be eliminated. The most prominent of the phase I enzymes are members of the cytochrome P450 (CYP) superfamily of proteins. These are monooxygenases that serve a wide range of biological processes in addition to their roles in drug metabolism [16]. Most psychotropics (in fact most medications) are metabolized via these enzymes. These comprise a large number of members; the human genome codes for 116 genes and pseudogenes across 18 families of cytochrome P450 genes and 43 subtypes [17]. Each family involves several related subtypes; for example, the CYP2 family includes 16 subtypes. For psychotropics, the most significant of those include CYP2B6, CYP2C9, CYP2C19, and CYP2D6. Other important psychotropic metabolizing enzymes include CYP1A2 and CYP3A4. Medications can also be metabolized by other enzymes (e.g., monoamine oxidase) or directly conjugated by several enzymes including uridine 5′-diphospho-glucuronosyltransferases (UGT) that are responsible for glucuronidation, the addition of glucuronic acid to a drug. Together, these and other enzymatic processes are responsible for drug elimination.

The human genome contains a very large number of mutations, gene duplications, or deletions involving drug metabolizing enzymes. This leads to wide variation in the metabolism of drugs. Metabolic activity is divided into several categories, including poor (i.e., little to no metabolic activity), intermediate (i.e., low), extensive (normal), rapid (high), or ultrarapid (very high) metabolism. Not surprisingly, understanding the metabolic pathways for drug elimination is an important part of drug development.

Cytochrome P450 enzymes were first discovered in rat liver in 1955 [18] and therefore were known long before gene sequencing existed. The existence of subfamilies of cytochrome enzymes and that there were large variations in their activity has likewise been known for decades. The assessment of drug metabolism has traditionally relied on several components. The most basic step is preclinical, which will be discussed in greater detail below. The first step in humans, which is completed in phase I of drug development, involves a full characterization of drug pharmacokinetics. This earliest studies in humans usually involve single ascending dose studies in which groups of volunteers are given various doses and medication levels are measured in the blood. In this step, variations in blood levels are determined, and side effects are identified. Typically, the occurrence of side effects identifies a maximum tolerated dose (MTD) in humans. These are typically followed by multiple ascending dose studies, which more fully describe the kinetics of the expected dose range. In addition, other aspects of pharmacokinetics can be determined, including the effects of age, sex, race, and the presence or absence of food.

A classic approach to determine both important enzymes that metabolize medications and potential drug–drug interactions are drug coadministration studies. These involve studies in which drugs that are known to be metabolized by a particular enzyme are administered along with the medication under development. A classic example of this is the concomitant administration of a new medication with debrisoquine, a medication known to be metabolized by CYP2D6 [19]. These drugs can be coadministered, and the impact of the new drug on debrisoquine blood levels can determine if there are significant drug–drug interactions; CYP2D6 inhibitors will increase debrisoquine levels. In addition, drugs that are known to induce (i.e., increase the activity of) specific enzymes can be coadministered along with the new medication. For example, ketoconazole is a potent inhibitor and rifampicin is an inducer of CYP3A4, and coadministration with a drug in development can indicate whether this enzyme is involved in the metabolism of the new medication [20]. There are a number of examples of medications that can be coadministered to determine CYP activity. This process can also identify which metabolic enzymes have the greatest impact on a drug, designated the primary pathway, and which are less important but still involved in metabolism, referred to as secondary pathways. This process is important not only in predicting potential drug–drug interactions but also in determining whether specific genetic variants of drug metabolizing genes will affect drug blood levels.

Some enzymes are subject to induction, which is an increase in metabolic activity caused by a drug or other substance. A classic example is the induction of CYP3A4. The increase in enzymatic action of CYP3A4 is the product of the binding of the drug or other compound to the pregnane X receptor, which forms a heterodimer with the retinoid X receptor. This complex then binds to the XREM portion of the gene for CYP3A4, which increases gene transcription. This results in a larger than normal amount of CYP3A4 production, increasing activity of the enzyme. Well-known inducers include carbamazepine, oxcarbazepine, topiramate, rifampicin, efazirenz, nevirapine, and modafinil, although the potency of induction can vary considerably.

In recent years, preclinical models have been used to characterize drug absorption, distribution, metabolism, and excretion. With regard to drug metabolism, these have included in vitro, in vivo, or in silico methods [15]. In vivo methods involve the administration of drugs to animals such as rodents or zebra fish. In vitro approaches can be done at large scale to test for drug metabolism, interactions, toxicity, or other properties [15]. With regard to drug metabolism, model systems have been developed to characterize enzyme activity. These include hepatic cell cultures and the extraction of microsomes, which are subcellular fractions of endoplasmic reticulum that contain CYP and UGT enzymes. These screening methods allow for a full characterization of metabolism prior to administration to humans. This, in turn, simplifies early stage drug testing, which can focus on known metabolic pathways.

A final approach involves in silico (meaning, in a computer) methods. These involve computational tools that can match drug structure to enzyme, which have been determined by X-ray crystallography and nuclear magnetic resonance methods [15]. Drug models can then be fitted into enzyme structures to determine likely metabolic pathways. This can be used with other proteins like receptors or transporters. While these and related computational models may not identify all such interactions, they can also reduce the time to develop new molecules. (For a more complete discussion of in vitro and in silico models, see Issa TA et al. [15]).

Once drug metabolic pathways are identified, then variations in drug level can be predicted. Variations in drug metabolism can involve either loss or gain of single nucleotide polymorphisms (SNPs), meaning a single base substitution in the gene or other related genetic changes that can alter the metabolic activity of a protein such as a SNP enzyme. Genes for specific enzymes can also be deleted or duplicated. In the case of deletion, part or all of a gene can be removed from the genome, rendering the body incapable of making the protein. Gene duplication occurs when a person's DNA contains multiple copies of the same gene. All genetic variations can occur in either heterozygous or homozygous forms; in the former, a person can have one of the two copies of a gene that is affected. In the latter, it is two copies. In the case of loss of function mutations, the heterozygous state usually involves a reduction, but not total loss of enzymatic activity. When homozygous, this can involve a severe reduction or even complete loss of activity of that enzyme. For example, in the case of a homozygous deletion of the CYP2D6 enzyme, there is no metabolic activity, meaning that people with the double deletion can run extremely high blood levels of drug. The gain of function alleles or gene duplications has the opposite effect. The heterozygous state will produce a reduction in expected blood levels while the homozygous state can cause dramatic reductions in levels. Therefore, determining metabolic pathways is a critical element in the drug development process.

The effects of gain or loss of function of a metabolizing enzyme is made more complicated by two factors. The first are heterologous gene combinations, which involve varying combinations of normal, gain, or loss of function alleles. The combination of an extensive metabolizing allele with either a loss or gain is the typical heterozygous state described above. However, other heterozygous combinations can create more complexity. An example would be the combination of gain and loss of function alleles. The most extreme would be a combination of a gene deletion with a gene duplication. These kinds of combinations result in a variety of metabolism that do not fit into the individual poor to ultrarapid metabolizer states noted earlier. This is why blood levels tend to be continuously distributed and not into five neat metabolism groups. The second factor is that while medications usually have a primary metabolic pathway involving a single enzyme, there are also secondary pathways that can compensate for a reduction in metabolic activity. For example, while the primary metabolic pathway for the tricyclic antidepressant amitriptyline is CYP2D6, it is also metabolized by CYP1A2, CYP2C9, CYP2C19, CYP3A4, and UGT1A4. Therefore, a complete loss of function of CYP2D6 does not necessarily result in inevitable toxicity (although the levels would be higher). By contrast, gain of function alleles or gene duplication of the primary metabolizing pathway invariably result in significantly lower blood levels, although this can vary considerably depending on whether it is heterozygous or homozygous and whether it is a SNP or gene duplication. Even in situations in which the primary pathway predicts extensive (normal) metabolism, a rapid or ultrarapid metabolizer status of a secondary pathway may result in lower than expected blood levels. For this reason, in the case of an unexpected negative outcome such as poor response or toxicity, a blood level is warranted (if available) even if the predicted metabolizer status is extensive (i.e., normal). Taken together, all these factors indicate that a thorough understanding of the enzymatic pathways involved in drug metabolism is a critical aspect of drug development.

15.2 Genomics and Drug Development

The decoding of the human genome came with tremendous hope that it would lead to novel treatments for a full range of medical disorders. The expectation was that the whole drug discovery process would change. Genomic information could lead to the discovery of disease pathways, molecular diagnostics, novel drug target prediction, drug response markers, methods for optimizing drug choice (i.e., personalized medicine), improved safety and efficacy, and other effects [3]. In some ways, that promise has been realized. A variety of mutations of specific genes has been shown to be associated with increased breast cancer risk, including mutations of *BRCA1* and *BRCA2*, *p53*, *PTEN*, *STK11*, *CHEK2*, *ATM*, *BRIP1*, and *PALB2* [21], and genetic testing has become a routine medical practice. *BRCA1/2* mutations are also associated with risk for other malignancies, including ovarian cancer. Olaparib, an inhibitor of poly (ADP-ribose) polymerase (PARP), originally approved for advanced ovarian cancer in patients with certain BRCA1/2 mutations [22]. This is an example of genomic personalized (or precision) medicine – i.e., of matching a treatment to cancers associated with specific genetic mutations.

Enrichment Strategies Conducting clinical trials that use a specific genomic or other biomarker that is associated with better response to a particular medication is an example of an *enrichment* strategy. According to the US FDA, "enrichment is the prospective use of any patient characteristic to select a study population in which detection of a drug effect (if one is in fact present) is more likely than it would be in an unselected population." [23] Trials of this type can determine if specific genetic mutations, or any other characteristic, are associated with preferential response to certain medications. The biomarker can then be used to select participants for subsequent trials [18]. If trials are positive, the US FDA and other regulatory agencies then require the product labeling, that is, the language used to describe the agency approved indication, to indicate that the enrichment marker, genomic, other biomarker, or other patient characteristic should be present for the medication to be used. Such enrichment strategies not only provide information for clinical practice but also (ideally) increase the likelihood of trial success (i.e., increase study power). (For more information on enrichment, see the relevant FDA guidance document [24]).

There are at least three possible enrichment approaches: (1) strategies to decrease variability, (2) prognostic enrichment strategies, and (3) predictive enrichment [18]. Strategies to decrease variability narrow the participant pool using a biomarker or other characteristic, thereby excluding participants who are unlikely to contribute to a positive study endpoint [24]. Examples would be excluding people whose condition is likely to improve spontaneously, those who are likely to show enhanced placebo effect, or participants who are likely to have highly variable outcomes. Most clinical trials employ inclusion and exclusion criteria to reduce variability and could be considered forms of enrichment. For example, trials in patients with treatment resistant major depressive disorder typically exclude people with certain comorbid disorders (e.g., obsessive compulsive disorder or post-traumatic stress disorder) and those with greater than five prior adequate antidepressant trials, since they are less likely to respond to either the active treatment or placebo. Patients with borderline personality disorder are usually excluded because of the expectation that they will have highly variable outcomes not related to treatment. The use of exclusion criteria is an attempt to reduce variability by eliminating characteristics that may be associated with reduced drug-placebo differences.

Strategies to reduce variability are, in essence, exclusion strategies. That is, they eliminate participants that are unlikely to contribute to drug-placebo differences. This would also apply to other treatments as well, such as device, psychotherapy, or other trials. Other approaches to reduce variability include detailed entry criteria to ensure that enrolled participants have the disease under study, selecting participants who are likely to adhere to the study protocol, enrolling only people with consistent baseline values (e.g., blood pressures or depression severity on repeated testing), or placebo lead-in periods to eliminate placebo responders [24].

It is important to consider inclusion and exclusion criteria when interpreting the results of trials. While the treatment may be effective for people who were excluded from studies, the treatment has not been adequately tested for people with those characteristics and, therefore, may not be effective. If people with certain characteristics are systematically excluded from all trials supporting a treatment, then the treatment cannot be considered evidence based for people with those features [24].

Prognostic enrichment strategies involve the selection of participants based on the likelihood of an event occurring. This is typically some type of negative outcome, for example, recurrence of cancer, relapse of depression or psychosis, or the occurrence of an adverse effect. Depression relapse prevention trials often exclude people who are in their first episode of illness, since a relapse in the study is more likely in recurrent depression. This does not change the relative effectiveness of an active treatment and placebo; it simply makes it more likely to detect meaningful differences [24].

Predictive enrichment strategies are the most relevant to biomarker research, including genomics. These are selection approaches in which certain patient characteristics like genotype or other biomarkers are used to select participants who are likely to respond to a treatment. Rather than determining who is *unlikely* to respond, as is often the case with strategies to reduce variability, this method identifies those *likely* to respond to the treatment. More specially, this approach determines who is likely to have an increased active versus inactive treatment difference. Such strategies can also be used to identify people who are likely to experience adverse outcomes and who should not receive a treatment. Biomarkers like genotype or other characteristics can help divide people who have a particular condition into those who are more versus less likely to respond to the active treatment [24]. The example of olaparib given above is an example of a prognostic strategy. In that case, BRCA1/2 genotype is used to select the treatment for people with advanced ovarian cancer. Olaparib does not directly target the protein coded by the BRCA genes, breast cancer type 1 or 2 susceptibility proteins, which are tumor suppressor genes. Rather, the medication inhibits the enzyme poly ADP ribose polymerase (PARP), which is a DNA repair protein that tumor cells require to continue replicating. Olaparib is effective in several BRCA1/2 positive cancers, and it is more effective in BRCA1/2 positive than negative cancers, but it has some efficacy even in certain BRCA1/2 negative cancers [25]. Therefore, in the case of olaparib, the identification of the association of the BRCA1/2 mutation to certain cancers did not lead directly to a treatment that targeted the mutated proteins directly. Rather, the treatment targets a protein involved in tumor cell replication that is downstream from the effects of BRCA1/2. People with those mutations appear to be more responsive to olaparib than people without, and therefore, BRCA1/2 mutations can be used as a selection criterion for treatment. However, there is not a direct homology between mutated gene and protein and treatment target.

Gene Targeting Therapies There have been very few examples of genetic mutations leading directly to an underlying pathophysiology that can be targeted by new therapies. Prime examples are therapies directed to receptor tyrosine-protein kinase erbB-2, also known as human epidermal growth factor receptor 2 or HER2 coded by the *ERBB2* gene. Overexpression of this oncogene plays a role in the progression of certain breast cancers and probably other types of cancers. Cancers that show overexpression of HER2 (called HER2 positive cancers) can be targeted by trastuzumab, a monoclonal antibody. Other medications that have been approved or are in clinical trials include pertuzumab, margetuximab, and the immunotherapy NeuVax, a peptide vaccine that directs killer T cells to HER2+ cancers.

Why then is it difficult to go from a gene mutation to a treatment directly – that is, a treatment that targets the mutated protein? An ideal scenario is one in which a mutation results in a gain of function – that is, the protein product has a greater effect than the non-mutated wild type [26]. A mutation could also result in an increase in the expression of mRNA and protein. In those instances, a medication could be developed that inhibits the mutated protein, treating the underlying disease. However, most mutations involve loss of function, in which the mutated protein either no longer functions or does so at reduced efficiency. It is possible to conceive of a treatment targeting a loss-of-function mutation. This would be the case, for example, for loss-of-function mutations in tumor suppressor genes [27]. This kind of development program could target the gene or protein directly or the downstream targets of the protein action. Olaparib, discussed earlier, is one example, with the drug targeting a downstream effect of the BRCA1/2 mutation. However, this has seldom been the case. The discovery of specific genes and proteins involved in the pathophysiology of a disorder does not necessarily identify a target for a drug, biologic, or other therapy. Although there have many mutations of oncogenes or tumor suppressor genes identified, most have not been developed as direct targets.

One promising area is Fragile X syndrome, which is an X-linked dominant disorder that is caused by an expansion of a CGC triplet in the fragile X mental retardation 1 (*FMR1*) gene on the X chromosome [28]. The protein coded by *FMR1* is an RNA binding protein involved in the maturation and pruning of synapses [29, 30]. The *FMR1* triplet results in gene silencing, which affects neuroplasticity. The downstream effects are complex (for a review see Maurin et al. [29] and Mila et al. [30]). The downstream effects of *FMR1* silencing are possible targets for treatment development.

Single-Gene Therapies Single-gene diseases have been the focus of intense research. Gene replacement therapies have been approved for a range of disorders, including AAV2-hRPE65v2 (also called voretigene neparvovec) that treats a specific type of Leber's congenital amaurosis type 2, which is related to a mutation in the retinal pigment epithelium-specific 65 kDa protein (*RPE65*) gene, [31] onasemnogene abeparvovec that treats spinal muscular atrophy related to a mutation in the *SMN1* gene, [32] and others [33]. However, relatively few single-gene therapies have been developed.

Multifactorial (complex) diseases: Another limitation of genomic approaches, particularly gene therapies, to treatment development is that most illnesses are multifactorial in origin, often called complex diseases. Multifactorial diseases can result from effects of multiple genes and environmental effects. These include not only the most common illnesses like diabetes, hypertension, or cardiovascular disease but also conditions like depression, autism, or schizophrenia. One approach to dealing with these issues is to take a "genotype first" approach, that is, to identify all genomic variants associated with a disease prior to identifying gene by environment effects. The most common genotype first approaches are genome-wide association studies (GWAS). GWA studies evaluate the whole genome to discover associations between gene variants and specific traits, including complex diseases. GWA studies have been applied to many diseases including schizophrenia, bipolar disorder, depression, and other disorders [33–39]. GWAS can identify multiple genes with very small effects on risk, or rare variants with large effects. It would be difficult to develop treatments from polygenic associations unless a coherent pathophysiological model could be constructed, which is often unlikely. Rare genes may be more promising targets, but by their rarity, it might be difficult to identify a large enough population for clinical trials. Converting genomic associations with particular disease states poses a major challenge.

One recent development has been to take the results of GWAS analysis to identify predictors of response to individual medications. This approach is neutral with regard to the physiological link between the actual GWAS SNP(s) associated with treatment response and either the underlying disease process or the mechanism of action of a given medication. Rather, it capitalizes on a statistical association between one or more SNPs and better or worse response to a medication, making it a predictive enrichment approach as described above. While this has not yielded any approved medications, it is now being used in clinical trials to identify participants who are more likely to respond to a treatment and give a better drug-placebo difference. It is also being used to "resurrect" medications that failed in previous development programs.

Traditional genomics, specifically the focus on the sequence of individual genes, may not succeed in developing treatments for most complex diseases. As noted earlier, however, only a tiny fraction of DNA is in the form of genes, and the genome is subject to the epigenetic effects of the environment. The next section will deal with epigenetic targets of treatment development.

15.3 Epigenetic Targets of Drug Development

Several lines of evidence show the role of gene x environment interaction in various mental disorders and associated epigenetic interference in the functioning of neuronal circuits [41]. The term, epigenetics, is referred to as long-standing changes in gene expression that are regulated via transcriptional, posttranscriptional, translational, and/or post-translational mechanisms, which does not entail any change in DNA sequence. Epigenetic processes, therefore, are nongenetic and can be impacted by both internal and external stimuli such as hormones, neurotransmitters, drugs, toxins, maternal care, and trauma. In psychiatric illness, most studies have concentrated on epigenetic changes influenced by trauma, stress, and maternal care. Since the changes associated with epigenetic interferences are dynamic, a correlation of the episodic modulations in mental disorders can be correlated with underlying epigenetic changes [42, 43].

15.3.1 Epigenetic Modifications: General Aspects

Several types of epigenetic modifications have been proposed. Classic epigenetics involves DNA methylation and histone modification. One of the most common epigenetic modifications is DNA methylation, which involves methylation of the cytosine residue at the 5-position (C5). This occurs by transfer of a methyl group from S-adenosyl methionine to cytosine residues in the dinucleotide sequence of CpG initiated by DNA methyltransferase (DNMT). In general, methylation of cytosine at the 5-position is very stable, which causes reversible changes in gene expression at the transcriptional level [44]. Thus, DNA methylation is inversely correlated with gene expression changes. DNMTs are called "writers" and exist in various isoforms: Dnmt1, 3a, 3b, 2, and 3L. Interestingly, each DNMT has its own regulatory function, which is primarily ascribed to the lack of sequence homology at the N-terminal regulatory domains [45]. Once the cytosine sites are methylated by DNMTs, methylated CPGs are targets for DNA-binding domain proteins, also known as "readers." These include methyl-CPG-binding domain (MBD) proteins and MeCP2, which bind to methylated DNA to induce transcriptional repression [46]. Both of these proteins cause transcriptional repression by recruiting histone deacetylase (HDAC) machinery that further remodel chromatin in such a state that it facilitates a repressed state [47, 48]. DNA methylation can occur in promoter regions and in the gene body; however, more often, CPG methylation in the promoter region represses gene expression [49]. More recently, it has been shown that DNA methylation can also directly silence genes that have non-CPG island (CGI) promoters. In fact, in certain disease conditions, differentially methylated regions occur more frequently within CGI shores or shelves representing relatively low CpG density that flank traditional CPG islands compared with within CPG islands themselves. MeCP2 has the capability to bind to non-CpG methylation sites [49–51] and can assist in gene repression. In addition to its role in gene regulation, DNA methylation also maintains genomic stability by controlling the expression of highly repetitive regions in the genome such as retrotransposons and satellite DNA [53].

Besides traditional CPG methylation, DNAs are hydroxymethylated at the C5 position of a cytosine base, that is, the addition of a CH_2OH group at the C5 position. Hydroxymethylation is highly enriched in promoter regions and in intragenic regions; however, it is largely absent from non-gene regions [54]. The mechanism of cytosine hydroxymethylation and its impact on gene expression is not fully known, but a dynamic balance between cytosine methylation and hydroxymethylation exists [55]. Hydroxymethylation is implicated in demethylation and is considered to be a necessary intermediary for methylation by allowing the promoter sites to be prepared for activation [56]. Hydroxymethylation is also assumed to play a role in compensating for the repressing effect of hypermethylation by increasing gene transcription.

Histone modifications are a type of epigenetic alteration that involves reversible chromatin rearrangements, which can have a dramatic effect on transcription without affecting the DNA sequence. As is well known, histones are the structural framework for eukaryotic chromosomes and provide three-dimensional architecture to the genome. Histone proteins are basic in nature and have four isoforms: H2A, H2B, H3, and H4. The N-terminal tails of histones are susceptible to various reversible covalent modifications including acetylation, methylation, phosphorylation, ubiquitination, and sumoylation [56–59]. Distinct histone modifications, such as histone 3 lysine 4 (H3K4) dimethylation and trimethylation and histone 3 lysine 27 (H3K27) acetylation at promoter regions and H3K4 monomethylation in enhancer regions of genes, are associated with active gene transcription. On the other hand, H3K9 and H3K27 dimethylation and trimethylation are involved in repressing promoter activity. Various histone tail modifications and associated enzymatic modifiers, such as histone methyltransferase (HMT) and histone deacetylase (HDAC), participate in epigenetic mechanisms that transduce active changes in gene transcription [61, 62]. The function of HMTs is to add methyl groups to lysine residues; on the other hand, histone demethylases (HMDs) remove methyl groups. There are separate HMTs and HDMs for various lysine residues, each with specific abilities to catalyze mono-, di-, or trimethylated states. Histone acetylation occurs most frequently on the lysine residues at H3 and H4 of the NH2-terminal. This is dynamically regulated by specific classes of enzymes known as histone acetyl transferases (HATs), which catalyze the addition of acetyl groups. HDACs catalyze the removal of acetyl groups from lysine residues in the NH2 terminal tails of histones. Primarily, increased histone acetylation causes the decondensation of chromatin and subsequent increase in the expression of genes, whereas lower acetylation leads to repression of chromatin and lower gene expression [63].

15.3.2 Epigenetic Modifications: Role in Psychiatric Disorders

At the functional level, epigenetic processes are involved in both embryonic and postnatal neural development. Most importantly, they participate in neurogenesis [64], neuronal differentiation, cell survival [65], synaptic, and structural plasticity [65–67]. Earlier, it was believed that epigenetic marks obtained in utero remain identical throughout the adult life; however, it is now clear that epigenetic mechanisms are dynamically regulated and epigenetic remodeling takes place throughout the adult life. Because of the dynamic nature, epigenetic modifications are critically involved in susceptibility or resiliency to both internal and external cues. A large body of evidence shows that epigenetic modifications of genes are significantly involved in various psychiatric disorders, including major depressive disorder (MDD), schizophrenia, bipolar disorder (BD), and suicidal behavior [68–72]. DNA methylation and subsequent alteration in the expression of specific genes associated with GABAergic (GABAA α1), polyamine (SMS and SMOX and SAT1), neurotrophin (BDNF, TrkB, and TrkBT1), and stress (NR3C1, SKA2, and FKBP5) signaling have been shown in various stress-related disorders such as post-traumatic stress disorder, major depressive disorders, and suicidal behavior. In BD and schizophrenia, several studies showed downregulation of RELN and GAD67 genes, which were associated with hypermethylation of their respective promoter CPG islands (CGIs) [74]. Interestingly, hypermethylation of these genes were correlated with increased expression of DNA methyltransferase 1 (DNMT1) in cortical GABAergic interneurons [74–76]. A whole-genome DNA methylation study showed that epigenetic modifications can influence neurocognitive functions associated with suicidal behavior [78]. In this study, NR2E1, GRM7, CHRNB2, and DBH genes coding for membrane receptors and membrane-associated enzymes were correlated with hyperresponsive behavioral phenotypes and were considered risk factors for suicidal behavior. Astrocyte-specific genome-wide methylation study showed differentially methylated regions (DMRs) in the prefrontal cortex of patients who had died by suicide [79]. Interestingly, 90% of DMRs were associated with non-promoter regions and localized in the vicinity of gene body. Early-life adversity profoundly impacts gene transcription through epigenetic modifications [80, 81]. One study showed that epigenetic and transcriptomic alterations significantly affected oligodendrocytes in the gray matter of cingulate cortex of subjects with early life trauma and identified oligodendrocyte-specific epigenetic reprograming of POU3F1 and LINGO3 genes (Lutz, 2017). A twin study identified promoter hypermethylation of serotonin transporter gene SLC6A4 in bipolar patients [82]. In postmortem brain samples from BD patients, it has been shown that S/S genotype of HTTLPR was associated with promoter hypermethylation of SLC6A4, which led to the downregulation of its mRNA level. Methylation status of 5-HT3AR (5-hydroxytryptamine 3A) was also reported to mediate the effect of childhood trauma and its impact on adult psychopathology such as BD, borderline personality disorder, and attention deficit disorder [83]. Among various CpG sites, the methylation status of CpG2 III was found to be associated with the number of previous mood episodes, previous suicide attempts, and the polymorphism in single-nucleotide polymorphism rs1062613, regardless of underlying diagnosis. Recently, voltage-gated channel gene KCNQ3 gene, which has been the focus of genetic linkage studies [84, 85], showed significantly lower methylation level and correspondingly higher mRNA level in BD patients [86].

As far as histone modifications are concerned, HDAC4 mRNA showed increased expression in a depressed state of BD patients, whereas expression levels of HDAC6 and HDAC8 were decreased [87]. It has been reported that histone acetylation of CREB protein increases its transcription, which in turn, is involved in MDD and BD [88]. Another family of deacetylases, sirtuins, also target histone marks. The gene expression of sirtuin 1–7 [89] has been investigated in mood disorder patients. One study found state-dependent alterations in sirtuin 1, 2, and 6 in peripheral blood cells of BD and MDD patients [90]. The level of acetylated histone 3 (H3K9/K14ac) was investigated between a mixed patient sample from BD and schizophrenia, targeting psychosis candidate gene promoters. Acetylation levels of the mixed patients sample differed significantly to the controls [91]. Another postmortem study showed increased global histone H3 acetylation in BD subjects compared with controls [92]. A significant increase in type 3 histone (H3) lysine (K) methylation in the core octamer of nucleosomes close to the TRKB.T1 promoter was found, which was responsible for the downregulation of TRKB and TRKB.T1 in the brain of suicide subjects [93]. Both TRKB and TRKB.T1 play critical roles in neurotrophin signaling. An association of increased H3K4 trimethylation and higher risk of suicide has been reported [94]. Altogether, these studies suggest that the epigenetic modifications of genes can have far reaching impact on behavior associated with various psychiatric illnesses.

15.3.3 Epigenetic Pharmacotherapy

Since epigenetic modifications can have significant behavioral consequences relevant to psychiatric illnesses, there has been an enormous interest in developing drugs that can specifically target individual molecules involved in epigenetic modifications. In addition, mechanisms of action of existing psychiatric drugs have been extensively examined for their association with epigenetic modifications; such examination not only provides their mechanisms of action but also offers novel targets for future drug development.

Psychoactive Drugs and Their Epigenetic Effects Valproic acid (VPA), a highly effective drug in BD, is one of most studied drug for its epigenetic targets [95]. The major function of VPA is to increase GABAergic activity by inhibiting GABA transaminase and blocking voltage-gated sodium channels. At the epigenetic level, VPA inhibits histone acetylation via interacting with HDACs, particularly HDAC class I and II, and increases the levels of acetylated histone H3 and H4, thereby stimulating gene expression [95–97]. VPA also interacts with HDAC2 and HDAC9; however, these actions are primarily effective in pain modulation [99] and preventing ischemic stroke [100], respectively. VPA can increase hippocampal neurogenesis, which is attributed to its increased histone acetylation activity [100–102]. VPA treatment also increases acetylation of H3 and H4 [74, 104] and decreases the expression of Dnmt1 and 3A and B, thereby increasing the expression of reelin (RELN) and glutamate decarboxylase (GAD), the two genes involved in BD and schizophrenia [105]. In cortical astrocytes and hippocampus, VPA increased the acetylation of histones H3 and H4 and decreased the levels of inhibitory H3K9 dimethylation. The increased acetylation and reduced DNA methylation were associated with increased expression of glutamate transporter-1 [106], a gene associated BD.

Haloperidol, a widely used antipsychotic drug and D2 receptor antagonist, causes phosphorylation of histone H3 at serine 10, acetylation of H3K14, and phosphoacetylation [107]. Histone phosphoacetylation was also found to be associated with raclopride, another D2 receptor blocker [108]. Striatal H3 phosphorylation, in response to haloperidol, has also been reported. Interestingly, a benzamine derivate, MS-275 ((pyridin-3-ylmethyl N-[[4 [(2aminophenyl)carbamoyl]phenyl]methyl]carbamate)), acts as HDAC inhibitor and is highly effective in increasing the acetylation of H3 associated with Reelin and GAD67 gene promoters [109]. Interestingly, MS-275 was much more potent than VPA in increasing acetylhistone 3 (Ac-H3), suggesting that this benzamine derivative may have greater efficacy when used adjunctive to antipsychotics [109]. Another benzamine derivative, sulpiride, increased H3K9 and H3K14 acetylation in the promoter region of reelin gene [109]. Risperidone, an antipsychotic, can induce global phosphoacetylation of H3 in the striatum [107]. Fluoxetine, an antidepressant, decreases histone H3K9 trimethylation induced by chronic restraint stress [110]. In serotonin projection areas, fluoxetine induced expression of MBD1 and Mecp2 transcripts [111]. Induction of the MBD proteins was accompanied with enhanced HDAC2 labeling intensity and mRNA synthesis [111]. Antidepressants can also reverse repressive histone modification patterns to elevate *Bdnf* expression, a gene involved in stress, mood disorders, and the mechanisms of action of antidepressants [112]. For example, escitalopram reduces HDAC expression, thereby increasing the acetylation of histones in mice which were previously exposed to maternal stress. Maternal stress reduces H3 and H4 acetylation at BDNF promoter IV, whereas escitalopram reduces depressive behavior by increasing acetylation and causing subsequent increase in

BDNF exon IV expression. Antidepressant treatment can also modify patterns of histone modifications to elevate BDNF expression [113]. Imipramine treatment reduces HDAC5 expression and elevates H3 and H4 acetylation, which leads to the alleviation of anxiety-like behavior [114]. Amitriptyline, another antidepressant drug, induces cytosine demethylation, along with a reduction in the enzymatic activity of DNMT.

Drugs Targeting Specific Epigenetic Pathways and Their Potential in the Treatment of Psychiatric Disorders Several drugs that are involved in epigenetic processes are being developed for various disorders including psychiatric disorders. Some of these drugs have been tested in an animal model, and some are still in the conceptual phase; however, the development of these drugs is exciting and may have the potential to provide personalized treatment for psychiatric illnesses. Sodium butyrate is a widely used HDAC inhibitor that exerts antidepressant-like effects [115]. It has been shown that sodium butyrate can upregulate both *BDNF* and *GDNF* genes in astrocytes via histone H3 acetylation in the promoter regions of these genes [116]. In mice, it has shown high efficiency in enhancing long-term memory and memory formation, which was primarily driven by elevation in trimethylation and simultaneous diminution of dimethylation of H3K9 [117]. Also, in the genetically depressed mice, which show lower levels of 10–11 translocation methyl cytosine dioxygenase 1 (TET1), sodium butyrate not only exhibited antidepressant activity but also increased levels of TET1. TET1 upregulation was accompanied by decrease in methylation and increase in hydroxymethylation of *Bdnf* gene [118]. TET1 catalyzes the conversion of DNA methylation to hydroxymethylation. Interestingly, combined treatment with sodium butyrate and fluoxetine was superior to fluoxetine alone. Thus, the combination of a HDAC inhibitor together with an SSRI might be a promising novel antidepressant treatment strategy. Trichostatin A (TSA; 7-[4-(dimethylamino) phenyl]-N-hydroxy-4,6-dimethyl-7-oxohepta-2,4-dienamide) is an HDAC inhibitor and targets classes I and II HDACs [119]. TSA show differential effects on two activation dependent regions of the *Bdnf* gene physically linked to transcription sites for exons I and IV. TSA treatment of cultures of hippocampal neurons produced a stronger response at promoter 1, which was correlated with increased occupancy of the promoter by acetylated histones (H3AcK9/K14). TSA treatment also produced a time-dependent increase in the level of H3AcK9 and H3AcK14 protein and HDAC1 mRNA levels and HDAC1 protein levels. These results suggest that the inhibition of HDAC activity by TSA activates BDNF transcription and a compensatory change in HDAC1 expression in neurons. TSA also increased the expression of GDNF along with GDNF promoter activity and promoter-associated H3 acetylation [116]. Interestingly, TSA treatment reversed the adverse early life experience induced by poor maternal care [120]. TSA was also effective in regulating GAD67, RELN, and GLET-1 genes through demethylation and acetylation [74, 104, 121, 122]. HDAC inhibitors such as sirtinol (2-[(2-hydroxynaphthalen-1-ylmethylene)amino]-N-(1-phenethyl)benzamide) and MS-275 have also been investigated as potential antidepressants in a rodent model [123]. To date, no clinical studies have been conducted to evaluate the effect of newly designed HDAC inhibitors in psychiatric patients. L-Methylfolate as methyl donor has been tested as an adjunctive therapy in several clinical trials [124] and has been found to be safe and effective in MDD patients [125]. Another methyl donor, S-adenosyl methionine ((2S)-2-amino-4-[[(2S,3S,4R,5R)-5-(6-aminopurin-9-yl)-3,4-dihydroxy oxolan-2-yl]methyl-methyl sulfonio]butanoate), has been shown to restore normal gene expression in neuroblastoma cells [126]. It has also been demonstrated that methionine administration increases the methylation levels of GAD67 and RELN with a consequent downregulation of their corresponding mRNAs. In clinical trials, SAMe was not different from placebo and established antidepressants; the exception was that compared to imipramine, fewer participants experienced adverse effects when treated with parenteral SAMe [127].

One of the shortcomings of global epigenetic modifiers is their broad effect on the epigenome. Also, epigenetic changes are tissue and cell-type specific. Recent developments in finding effective epigenetic targets revolve around epigenome manipulation at specified loci. The purpose of this approach is to adjust only the specific pathogenic marks rather than altering the entire epigenome. This new technique is based on generating targeted EpiEffectors, which are engineered transcription factors such as transcription activator-like effectors or zinc-finger-proteins, which have been designed to bind at specific loci in the genome [128]. Using this approach, studies have shown locus-specific epigenetic remodeling and its impact on correcting addiction and depression-related behaviors [129]. In addition, engineered zinc finger protein activator of endogenous glial cell line-derived neurotrophic factor gene provided functional neuroprotection in a rat model of Parkinson's disease [130]. CRISPR-dCas9 is another highly promising strategy, which facilitates the design of DNA recognition domains. Recent studies suggest the usefulness of this strategy in both in vitro and in animals where it induced long-lasting changes in DNA methylation [130–132] or histone modifications [134, 135]. Through the use of another innovative approach, a recent study demonstrated that fusion of Tet1 or Dnmt3a with a catalytically inactive Cas9 enabled targeted DNA methylation editing of *Bdnf* gene in a long-lasting manner [136]. These techniques can be highly effective in delivering enduring epigenetic marks that will be crucial in the patient population.

15.4 Conclusion

The decoding of the human genome has been truly revolutionary, ushering in a new era of drug development. The association of specific gene variants with human diseases has allowed the identification of disease risk and provided some targets for new therapies. Specific examples include *BRCA1/2* and *ERBB2* (HER2) variants in breast cancer. However, most human illnesses are not single-gene diseases; most are *complex disease traits* that not only can involve contributions from many genes but also have significant environmental components. Examples of diseases that have some underlying biological (presumably genetic) predisposing factors and also strong environmental influences include such common conditions as not only hypertension and type 2 diabetes but also depression, post-traumatic stress disorder, and schizophrenia. These complex diseases are not amenable to treatment development that target abnormalities in single-gene sequences (i.e., the gene product itself or downstream effects). Epigenetics has emerged as an important potential target of treatment development given the important role of the environment in disease pathogenesis. It remains to be seen if epigenetic changes will emerge as bona fide treatment targets. However, the development of molecular approaches that target specific epigenetic marks may ultimately treat specific environmentally induced disease. However, this approach may also be able to improve a broad set of disorders that are affected by similar environmental antecedents. As an example, early life trauma induces DNA methylation [137], and it predisposes to the subsequent development of separate disorders such as depression [138], PTSD [139], and suicide [140]. The epigenetic changes may serve as common and possibly modifiable vulnerability factors for trauma-related disorders more broadly. Targeting these changes might lead to novel treatments and, more importantly, ways of reversing the effects of early life stress, thereby reducing risk for a range of mental disorders. This could bring a new generation of risk modifying approaches that may provide ways of preventing and not just treating diseases.

References

1. Collins FS, Morgan M, Patrinos A. The Human Genome Project: lessons from large-scale biology. Science. 2003;300(5617):286–90.
2. Collins FS, Fink L. The Human Genome Project. Alcohol Health Res World. 1995;19(3):190–5.
3. Emilien G, Ponchon M, Caldas C, Isacson O, Maloteaux JM. Impact of genomics on drug discovery and clinical medicine. QJM. 2000;93(7):391–423.
4. Gottesman MM, Collins FS. The role of the human genome project in disease prevention. Prev Med. 1994;23(5):591–4.
5. Keleher C. Translating the genetic library: the goals, methods, and applications of the Human Genome Project. Bull Med Libr Assoc. 1993;81(3):274–7.
6. Petersen I. Classification and treatment of diseases in the age of genome medicine based on pathway pathology. Int J Mol Sci. 2021;22(17):9418.
7. Iqbal O. Pharmacogenomics in anticoagulant drug development. Pharmacogenomics. 2002;3(6):823–8.
8. Wang L, McLeod HL, Weinshilboum RM. Genomics and drug response. N Engl J Med. 2011;364(12):1144–53.
9. Bah SY, Moranga CM, Kengne-Ouafo JA, Amenga-Etego L, Awandare GA. Highlights on the application of genomics and bioinformatics in the fight against infectious diseases: challenges and opportunities in Africa. Front Genet. 2018;9:575.
10. Mahley RW, Weisgraber KH, Huang Y. Apolipoprotein E: structure determines function, from atherosclerosis to Alzheimer's disease to AIDS. J Lipid Res. 2009;50 Suppl(Suppl):S183–188.
11. Salzberg SL. Open questions: how many genes do we have? BMC Biol. 2018;16(1):94.
12. Pertea M, Salzberg SL. Between a chicken and a grape: estimating the number of human genes. Genome Biol. 2010;11(5):206.

13. Lander ES, Linton LM, Birren B, Nusbaum C, Zody MC, Baldwin J, et al. Initial sequencing and analysis of the human genome. Nature. 2001;409(6822):860–921.
14. Issa NT, Wathieu H, Ojo A, Byers SW, Dakshanamurthy S. Drug metabolism in preclinical drug development: a survey of the discovery process, toxicology, and computational tools. Curr Drug Metab. 2017;18(6):556–65.
15. Danielson PB. The cytochrome P450 superfamily: biochemistry, evolution and drug metabolism in humans. Curr Drug Metab. 2002;3(6):561–97.
16. Cytochrome P450. https://drnelson.uthsc.edu/, 2021, Accessed Date Accessed 2021 Accessed.
17. Green JP, Sondergaard E, Dam H. Liver respiration, succinoxidase and DPN-cytochrome c reductase activity in vitamin K-deficiency and after treatment with long-acting anticoagulants. Acta Pharmacol Toxicol (Copenh). 1955;11(1):79–89.
18. Llerena A, Dorado P, Peñas-Lledó EM. Pharmacogenetics of debrisoquine and its use as a marker for CYP2D6 hydroxylation capacity. Pharmacogenomics. 2009;10(1):17–28.
19. Liu Y, Zhou S, Wan Y, Wu A, Palmisano M. The impact of co-administration of ketoconazole and rifampicin on the pharmacokinetics of apremilast in healthy volunteers. Br J Clin Pharmacol. 2014;78(5):1050–7.
20. Gage M, Wattendorf D, Henry LR. Translational advances regarding hereditary breast cancer syndromes. J Surg Oncol. 2012;105(5):444–51.
21. Wiggans AJ, Cass GK, Bryant A, Lawrie TA, Morrison J. Poly(ADP-ribose) polymerase (PARP) inhibitors for the treatment of ovarian cancer. Cochrane Database Syst Rev (Online). 2015;2015(5):Cd007929.
22. FDA U. Enrichment strategiex for clinical trials to support determination of effectiveness of human drugs and biological products. Guidance for Industry. US Deparrtment of Health and Human Services, Food and Drug Administration, 2019.
23. FDAU. Enrichment strategiexsx for clinical trials to support determination of effectiveness of human drugs and biological products. Guidance for Industry. US Deparrtment of Health and Human Services, Food and Drug Administration 2019.
24. Ledermann J, Harter P, Gourley C, Friedlander M, Vergote I, Rustin G, et al. Olaparib maintenance therapy in patients with platinum-sensitive relapsed serous ovarian cancer: a preplanned retrospective analysis of outcomes by BRCA status in a randomised phase 2 trial. Lancet Oncol. 2014;15(8):852–61.
25. Li Y, Zhang Y, Li X, Yi S, Xu J. Gain-of-function mutations: an emerging advantage for cancer biology. Trends Biochem Sci. 2019;44(8):659–74.
26. Wang H, Han H, Mousses S, Von Hoff DD. Targeting loss-of-function mutations in tumor-suppressor genes as a strategy for development of cancer therapeutic agents. Semin Oncol. 2006;33(4):513–20.
27. Hagerman RJ, Berry-Kravis E, Hazlett HC, Bailey DB Jr, Moine H, Kooy RF, et al. Fragile X syndrome. Nat Rev Dis Primers. 2017;3:17065.
28. Maurin T, Zongaro S, Bardoni B. Fragile X syndrome: from molecular pathology to therapy. Neurosci Biobehav Rev. 2014;46(Pt 2):242–55.
29. Mila M, Alvarez-Mora MI, Madrigal I, Rodriguez-Revenga L. Fragile X syndrome: an overview and update of the FMR1 gene. Clin Genet. 2018;93(2):197–205.
30. Russell S, Bennett J, Wellman JA, Chung DC, Yu ZF, Tillman A, et al. Efficacy and safety of voretigene neparvovec (AAV2-hRPE65v2) in patients with RPE65-mediated inherited retinal dystrophy: a randomised, controlled, open-label, phase 3 trial. Lancet. 2017;390(10097):849–60.
31. Stevens D, Claborn MK, Gildon BL, Kessler TL, Walker C. Onasemnogene Abeparvovec-xioi: gene therapy for spinal muscular atrophy. Ann Pharmacother. 2020;54(10):1001–9.
32. Sinclair A, Islam S, Jones S. Gene therapy: an overview of approved and pipeline technologies. CADTH Issues in Emerging Health Technologies. Canadian Agency for Drugs and Technologies in Health Copyright © CADTH 2018. You are permitted to reproduce this document for non-commercial purposes, provided it is not modified when reproduced and appropriate credit is given to CADTH.: Ottawa, 2016, pp. 1–23.

33. Bulik CM, Blake L, Austin J. Genetics of eating disorders: what the clinician needs to know. Psychiatr Clin North Am. 2019;42(1):59–73.
34. Gordovez FJA, McMahon FJ. The genetics of bipolar disorder. Mol Psychiatry. 2020;25(3):544–59.
35. Gottschalk MG, Domschke K. Genetics of generalized anxiety disorder and related traits. Dialogues Clin Neurosci. 2017;19(2):159–68.
36. Grimm O, Kranz TM, Reif A. Genetics of ADHD: what should the clinician know? Curr Psychiatry Rep. 2020;22(4):18.
37. Horwitz T, Lam K, Chen Y, Xia Y, Liu C. A decade in psychiatric GWAS research. Mol Psychiatry. 2019;24(3):378–89.
38. Mullins N, Lewis CM. Genetics of depression: progress at last. Curr Psychiatry Rep. 2017;19(8):43.
39. Reitz C. Genetic diagnosis and prognosis of Alzheimer's disease: challenges and opportunities. Expert Rev Mol Diagn. 2015;15(3):339–48.
40. Klengel T, Binder EB. Gene-environment interactions in major depressive disorder. Can J Psychiatr. 2013;58(2):76–83.
41. Autry AE, Monteggia LM. Epigenetics in suicide and depression. Biol Psychiatry. 2009;66(9):812–3.
42. Turecki G. Epigenetics and suicidal behavior research pathways. Am J Prev Med. 2014;47(3 Suppl 2):S144–51.
43. Moore LD, Le T, Fan G. DNA methylation and its basic function. Neuropsychopharmacology. 2013;38(1):23–38.
44. Bestor TH, Ingram VM. Two DNA methyltransferases from murine erythroleukemia cells: purification, sequence specificity, and mode of interaction with DNA. Proc Natl Acad Sci U S A. 1983;80(18):5559–63.
45. Hendrich B, Bird A. Identification and characterization of a family of mammalian methyl-CpG binding proteins. Mol Cell Biol. 1998;18(11):6538–47.
46. Fuks F, Hurd PJ, Wolf D, Nan X, Bird AP, Kouzarides T. The methyl-CpG-binding protein MeCP2 links DNA methylation to histone methylation. J Biol Chem. 2003;278(6):4035–40.
47. Jones PL, Veenstra GJ, Wade PA, Vermaak D, Kass SU, Landsberger N, et al. Methylated DNA and MeCP2 recruit histone deacetylase to repress transcription. Nat Genet. 1998;19(2):187–91.
48. Jones PA. Functions of DNA methylation: islands, start sites, gene bodies and beyond. Nat Rev Genet. 2012;13(7):484–92.
49. Gabel HW, Kinde B, Stroud H, Gilbert CS, Harmin DA, Kastan NR, et al. Disruption of DNA-methylation-dependent long gene repression in Rett syndrome. Nature. 2015;522(7554):89–93.
50. Guo JU, Su Y, Shin JH, Shin J, Li H, Xie B, et al. Distribution, recognition and regulation of non-CpG methylation in the adult mammalian brain. Nat Neurosci. 2014;17(2):215–22.
51. Mellén M, Ayata P, Dewell S, Kriaucionis S, Heintz N. MeCP2 binds to 5hmC enriched within active genes and accessible chromatin in the nervous system. Cell. 2012;151(7):1417–30.
52. Woodcock DM, Lawler CB, Linsenmeyer ME, Doherty JP, Warren WD. Asymmetric methylation in the hypermethylated CpG promoter region of the human L1 retrotransposon. J Biol Chem. 1997;272(12):7810–6.
53. Jin SG, Wu X, Li AX, Pfeifer GP. Genomic mapping of 5-hydroxymethylcytosine in the human brain. Nucleic Acids Res. 2011;39(12):5015–24.
54. Grayson DR, Guidotti A. The dynamics of DNA methylation in schizophrenia and related psychiatric disorders. Neuropsychopharmacology. 2013;38(1):138–66.
55. Pastor WA, Pape UJ, Huang Y, Henderson HR, Lister R, Ko M, et al. Genome-wide mapping of 5-hydroxymethylcytosine in embryonic stem cells. Nature. 2011;473(7347):394–7.
56. Cao J, Yan Q. Histone ubiquitination and deubiquitination in transcription, DNA damage response, and cancer. Front Oncol. 2012;2:26.
57. Lachner M, Jenuwein T. The many faces of histone lysine methylation. Curr Opin Cell Biol. 2002;14(3):286–98.

58. Machado-Vieira R, Frey BN, Andreazza AC, Quevedo J. Translational research in bipolar disorders. Neural Plast. 2015;2015:576978.
59. Seeler JS, Dejean A. Nuclear and unclear functions of SUMO. Nat Rev Mol Cell Biol. 2003;4(9):690–9.
60. Kouzarides T. Chromatin modifications and their function. Cell. 2007;128(4):693–705.
61. Mosammaparast N, Shi Y. Reversal of histone methylation: biochemical and molecular mechanisms of histone demethylases. Annu Rev Biochem. 2010;79:155–79.
62. Roth SY, Denu JM, Allis CD. Histone acetyltransferases. Annu Rev Biochem. 2001;70:81–120.
63. Hahn MA, Qiu R, Wu X, Li AX, Zhang H, Wang J, et al. Dynamics of 5-hydroxymethylcytosine and chromatin marks in Mammalian neurogenesis. Cell Rep. 2013;3(2):291–300.
64. Fan G, Beard C, Chen RZ, Csankovszki G, Sun Y, Siniaia M, et al. DNA hypomethylation perturbs the function and survival of CNS neurons in postnatal animals. J Neurosci. 2001;21(3):788–97.
65. Bredy TW, Wu H, Crego C, Zellhoefer J, Sun YE, Barad M. Histone modifications around individual BDNF gene promoters in prefrontal cortex are associated with extinction of conditioned fear. Learn Mem. 2007;14(4):268–76.
66. Korzus E, Rosenfeld MG, Mayford M. CBP histone acetyltransferase activity is a critical component of memory consolidation. Neuron. 2004;42(6):961–72.
67. Vecsey CG, Hawk JD, Lattal KM, Stein JM, Fabian SA, Attner MA, et al. Histone deacetylase inhibitors enhance memory and synaptic plasticity via CREB:CBP-dependent transcriptional activation. J Neurosci. 2007;27(23):6128–40.
68. Kuehner JN, Bruggeman EC, Wen Z, Yao B. Epigenetic regulations in neuropsychiatric disorders. Front Genet. 2019;10:268.
69. Kular L, Kular S. Epigenetics applied to psychiatry: clinical opportunities and future challenges. Psychiatry Clin Neurosci. 2018;72(4):195–211.
70. Ludwig B, Dwivedi Y. Dissecting bipolar disorder complexity through epigenomic approach. Mol Psychiatry. 2016;21(11):1490–8.
71. Roy B, Dwivedi Y. Understanding epigenetic architecture of suicide neurobiology: a critical perspective. Neurosci Biobehav Rev. 2017;72:10–27.
72. Smigielski L, Jagannath V, Rössler W, Walitza S, Grünblatt E. Epigenetic mechanisms in schizophrenia and other psychotic disorders: a systematic review of empirical human findings. Mol Psychiatry. 2020;25(8):1718–48.
73. Chen Y, Sharma RP, Costa RH, Costa E, Grayson DR. On the epigenetic regulation of the human reelin promoter. Nucleic Acids Res. 2002;30(13):2930–9.
74. Dong E, Ruzicka WB, Grayson DR, Guidotti A. DNA-methyltransferase1 (DNMT1) binding to CpG rich GABAergic and BDNF promoters is increased in the brain of schizophrenia and bipolar disorder patients. Schizophr Res. 2015;167(1–3):35–41.
75. Veldic M, Caruncho HJ, Liu WS, Davis J, Satta R, Grayson DR, et al. DNA-methyltransferase 1 mRNA is selectively overexpressed in telencephalic GABAergic interneurons of schizophrenia brains. Proc Natl Acad Sci U S A. 2004;101(1):348–53.
76. Veldic M, Kadriu B, Maloku E, Agis-Balboa RC, Guidotti A, Davis JM, et al. Epigenetic mechanisms expressed in basal ganglia GABAergic neurons differentiate schizophrenia from bipolar disorder. Schizophr Res. 2007;91(1–3):51–61.
77. Labonté B, Suderman M, Maussion G, Lopez JP, Navarro-Sánchez L, Yerko V, et al. Genome-wide methylation changes in the brains of suicide completers. Am J Psychiatry. 2013;170(5):511–20.
78. Nagy C, Suderman M, Yang J, Szyf M, Mechawar N, Ernst C, et al. Astrocytic abnormalities and global DNA methylation patterns in depression and suicide. Mol Psychiatry. 2015;20(3):320–8.
79. Li M, Fu X, Xie W, Guo W, Li B, Cui R, et al. Effect of early life stress on the epigenetic profiles in depression. Front Cell Dev Biol. 2020;8:867.
80. Lux V. Epigenetic programming effects of early life stress: a dual-activation hypothesis. Curr Genomics. 2018;19(8):638–52.

81. Sugawara H, Iwamoto K, Bundo M, Ueda J, Miyauchi T, Komori A, et al. Hypermethylation of serotonin transporter gene in bipolar disorder detected by epigenome analysis of discordant monozygotic twins. Transl Psychiatry. 2011;1(7):e24.
82. Perroud N, Zewdie S, Stenz L, Adouan W, Bavamian S, Prada P, et al. Methylation of serotonin receptor 3A in ADHD, borderline personality, and bipolar disorders: link with severity of the disorders and childhood maltreatment. Depress Anxiety. 2016;33(1):45–55.
83. Avramopoulos D, Willour VL, Zandi PP, Huo Y, MacKinnon DF, Potash JB, et al. Linkage of bipolar affective disorder on chromosome 8q24: follow-up and parametric analysis. Mol Psychiatry. 2004;9(2):191–6.
84. Zhang P, Xiang N, Chen Y, Sliwerska E, McInnis MG, Burmeister M, et al. Family-based association analysis to finemap bipolar linkage peak on chromosome 8q24 using 2,500 genotyped SNPs and 15,000 imputed SNPs. Bipolar Disord. 2010;12(8):786–92.
85. Kaminsky Z, Jones I, Verma R, Saleh L, Trivedi H, Guintivano J, et al. DNA methylation and expression of KCNQ3 in bipolar disorder. Bipolar Disord. 2015;17(2):150–9.
86. Hobara T, Uchida S, Otsuki K, Matsubara T, Funato H, Matsuo K, et al. Altered gene expression of histone deacetylases in mood disorder patients. J Psychiatr Res. 2010;44(5):263–70.
87. Gaspar L, van de Werken M, Johansson AS, Moriggi E, Owe-Larsson B, Kocks JW, et al. Human cellular differences in cAMP--CREB signaling correlate with light-dependent melatonin suppression and bipolar disorder. Eur J Neurosci. 2014;40(1):2206–15.
88. Bosch-Presegué L, Vaquero A. Sirtuin-dependent epigenetic regulation in the maintenance of genome integrity. FEBS J. 2015;282(9):1745–67.
89. Abe N, Uchida S, Otsuki K, Hobara T, Yamagata H, Higuchi F, et al. Altered sirtuin deacetylase gene expression in patients with a mood disorder. J Psychiatr Res. 2011;45(8):1106–12.
90. Tang B, Dean B, Thomas EA. Disease- and age-related changes in histone acetylation at gene promoters in psychiatric disorders. Transl Psychiatry. 2011;1(12):e64.
91. Rao JS, Keleshian VL, Klein S, Rapoport SI. Epigenetic modifications in frontal cortex from Alzheimer's disease and bipolar disorder patients. Transl Psychiatry. 2012;2(7):e132.
92. Ernst C, Deleva V, Deng X, Sequeira A, Pomarenski A, Klempan T, et al. Alternative splicing, methylation state, and expression profile of tropomyosin-related kinase B in the frontal cortex of suicide completers. Arch Gen Psychiatry. 2009;66(1):22–32.
93. Fiori LM, Gross JA, Turecki G. Effects of histone modifications on increased expression of polyamine biosynthetic genes in suicide. Int J Neuropsychopharmacol. 2012;15(8):1161–6.
94. Mello MLS. Sodium valproate-induced chromatin remodeling. Front Cell Dev Biol. 2021;9:645518.
95. Göttlicher M, Minucci S, Zhu P, Krämer OH, Schimpf A, Giavara S, et al. Valproic acid defines a novel class of HDAC inhibitors inducing differentiation of transformed cells. EMBO J. 2001;20(24):6969–78.
96. Phiel CJ, Zhang F, Huang EY, Guenther MG, Lazar MA, Klein PS. Histone deacetylase is a direct target of valproic acid, a potent anticonvulsant, mood stabilizer, and teratogen. J Biol Chem. 2001;276(39):36734–41.
97. Eyal S, Yagen B, Sobol E, Altschuler Y, Shmuel M, Bialer M. The activity of antiepileptic drugs as histone deacetylase inhibitors. Epilepsia. 2004;45(7):737–44.
98. Krämer OH, Zhu P, Ostendorff HP, Golebiewski M, Tiefenbach J, Peters MA, et al. The histone deacetylase inhibitor valproic acid selectively induces proteasomal degradation of HDAC2. EMBO J. 2003;22(13):3411–20.
99. Brookes RL, Crichton S, Wolfe CDA, Yi Q, Li L, Hankey GJ, et al. Sodium valproate, a histone deacetylase inhibitor, is associated with reduced stroke risk after previous ischemic stroke or transient ischemic attack. Stroke. 2018;49(1):54–61.
100. Ookubo M, Kanai H, Aoki H, Yamada N. Antidepressants and mood stabilizers effects on histone deacetylase expression in C57BL/6 mice: brain region specific changes. J Psychiatr Res. 2013;47(9):1204–14.

101. Park SW, Lee JG, Seo MK, Cho HY, Lee CH, Lee JH, et al. Effects of mood-stabilizing drugs on dendritic outgrowth and synaptic protein levels in primary hippocampal neurons. Bipolar Disord. 2015;17(3):278–90.
102. Yu IT, Park JY, Kim SH, Lee JS, Kim YS, Son H. Valproic acid promotes neuronal differentiation by induction of proneural factors in association with H4 acetylation. Neuropharmacology. 2009;56(2):473–80.
103. Mitchell CP, Chen Y, Kundakovic M, Costa E, Grayson DR. Histone deacetylase inhibitors decrease reelin promoter methylation in vitro. J Neurochem. 2005;93(2):483–92.
104. Dong E, Guidotti A, Grayson DR, Costa E. Histone hyperacetylation induces demethylation of reelin and 67-kDa glutamic acid decarboxylase promoters. Proc Natl Acad Sci U S A. 2007;104(11):4676–81.
105. Perisic T, Zimmermann N, Kirmeier T, Asmus M, Tuorto F, Uhr M, et al. Valproate and amitriptyline exert common and divergent influences on global and gene promoter-specific chromatin modifications in rat primary astrocytes. Neuropsychopharmacology. 2010;35(3):792–805.
106. Li J, Guo Y, Schroeder FA, Youngs RM, Schmidt TW, Ferris C, et al. Dopamine D2-like antagonists induce chromatin remodeling in striatal neurons through cyclic AMP-protein kinase A and NMDA receptor signaling. J Neurochem. 2004;90(5):1117–31.
107. Bertran-Gonzalez J, Bosch C, Maroteaux M, Matamales M, Hervé D, Valjent E, et al. Opposing patterns of signaling activation in dopamine D1 and D2 receptor-expressing striatal neurons in response to cocaine and haloperidol. J Neurosci. 2008;28(22):5671–85.
108. Simonini MV, Camargo LM, Dong E, Maloku E, Veldic M, Costa E, et al. The benzamide MS-275 is a potent, long-lasting brain region-selective inhibitor of histone deacetylases. Proc Natl Acad Sci U S A. 2006;103(5):1587–92.
109. Hunter RG, McCarthy KJ, Milne TA, Pfaff DW, McEwen BS. Regulation of hippocampal H3 histone methylation by acute and chronic stress. Proc Natl Acad Sci U S A. 2009;106(49):20912–7.
110. Cassel S, Carouge D, Gensburger C, Anglard P, Burgun C, Dietrich JB, et al. Fluoxetine and cocaine induce the epigenetic factors MeCP2 and MBD1 in adult rat brain. Mol Pharmacol. 2006;70(2):487–92.
111. Dwivedi Y. Brain-derived neurotrophic factor: role in depression and suicide. Neuropsychiatr Dis Treat. 2009;5:433–49.
112. Seo MK, Ly NN, Lee CH, Cho HY, Choi CM, Nhu LH, et al. Early life stress increases stress vulnerability through BDNF gene epigenetic changes in the rat hippocampus. Neuropharmacology. 2016;105:388–97.
113. Tsankova NM, Berton O, Renthal W, Kumar A, Neve RL, Nestler EJ. Sustained hippocampal chromatin regulation in a mouse model of depression and antidepressant action. Nat Neurosci. 2006;9(4):519–25.
114. Schroeder FA, Lin CL, Crusio WE, Akbarian S. Antidepressant-like effects of the histone deacetylase inhibitor, sodium butyrate, in the mouse. Biol Psychiatry. 2007;62(1):55–64.
115. Wu X, Chen PS, Dallas S, Wilson B, Block ML, Wang CC, et al. Histone deacetylase inhibitors up-regulate astrocyte GDNF and BDNF gene transcription and protect dopaminergic neurons. Int J Neuropsychopharmacol. 2008;11(8):1123–34.
116. Gupta S, Kim SY, Artis S, Molfese DL, Schumacher A, Sweatt JD, et al. Histone methylation regulates memory formation. J Neurosci. 2010;30(10):3589–99.
117. Wei Y, Melas PA, Wegener G, Mathé AA, Lavebratt C. Antidepressant-like effect of sodium butyrate is associated with an increase in TET1 and in 5-hydroxymethylation levels in the Bdnf gene. Int J Neuropsychopharmacol. 2014;18(2):pyu032.
118. Tian F, Marini AM, Lipsky RH. Effects of histone deacetylase inhibitor Trichostatin A on epigenetic changes and transcriptional activation of Bdnf promoter 1 by rat hippocampal neurons. Ann N Y Acad Sci. 2010;1199:186–93.
119. Weaver IC, Meaney MJ, Szyf M. Maternal care effects on the hippocampal transcriptome and anxiety-mediated behaviors in the offspring that are reversible in adulthood. Proc Natl Acad Sci U S A. 2006;103(9):3480–5.

120. Kundakovic M, Chen Y, Guidotti A, Grayson DR. The reelin and GAD67 promoters are activated by epigenetic drugs that facilitate the disruption of local repressor complexes. Mol Pharmacol. 2009;75(2):342–54.
121. Gavin DP, Kartan S, Chase K, Jayaraman S, Sharma RP. Histone deacetylase inhibitors and candidate gene expression: an in vivo and in vitro approach to studying chromatin remodeling in a clinical population. J Psychiatr Res. 2009;43(9):870–6.
122. Covington HE 3rd, Maze I, LaPlant QC, Vialou VF, Ohnishi YN, Berton O, et al. Antidepressant actions of histone deacetylase inhibitors. J Neurosci. 2009;29(37):11451–60.
123. Fava M, Shelton RC, Zajecka JM. Evidence for the use of l-methylfolate combined with antidepressants in MDD. J Clin Psychiatry. 2011;72(8):e25.
124. Shelton RC, Sloan Manning J, Barrentine LW, Tipa EV. Assessing effects of l-methylfolate in depression management: results of a real-world patient experience trial. Prim Care Companion CNS Disord. 2013:**15**(4).
125. Fuso A, Seminara L, Cavallaro RA, D'Anselmi F, Scarpa S. S-adenosylmethionine/homocysteine cycle alterations modify DNA methylation status with consequent deregulation of PS1 and BACE and beta-amyloid production. Mol Cell Neurosci. 2005;28(1):195–204.
126. Galizia I, Oldani L, Macritchie K, Amari E, Dougall D, Jones TN, et al. S-adenosyl methionine (SAMe) for depression in adults. Cochrane Database Syst Rev. 2016;10(10):Cd011286.
127. Peedicayil J. Epigenetic approaches for bipolar disorder drug discovery. Expert Opin Drug Discov. 2014;9(8):917–30.
128. Heller EA, Cates HM, Peña CJ, Sun H, Shao N, Feng J, et al. Locus-specific epigenetic remodeling controls addiction- and depression-related behaviors. Nat Neurosci. 2014;17(12):1720–7.
129. Laganiere J, Kells AP, Lai JT, Guschin D, Paschon DE, Meng X, et al. An engineered zinc finger protein activator of the endogenous glial cell line-derived neurotrophic factor gene provides functional neuroprotection in a rat model of Parkinson's disease. J Neurosci. 2010;30(49):16469–74.
130. Choudhury SR, Cui Y, Lubecka K, Stefanska B, Irudayaraj J. CRISPR-dCas9 mediated TET1 targeting for selective DNA demethylation at BRCA1 promoter. Oncotarget. 2016;7(29):46545–56.
131. Vojta A, Dobrinić P, Tadić V, Bočkor L, Korać P, Julg B, et al. Repurposing the CRISPR-Cas9 system for targeted DNA methylation. Nucleic Acids Res. 2016;44(12):5615–28.
132. Xu X, Tao Y, Gao X, Zhang L, Li X, Zou W, et al. A CRISPR-based approach for targeted DNA demethylation. Cell Discov. 2016;2:16009.
133. Hilton IB, D'Ippolito AM, Vockley CM, Thakore PI, Crawford GE, Reddy TE, et al. Epigenome editing by a CRISPR-Cas9-based acetyltransferase activates genes from promoters and enhancers. Nat Biotechnol. 2015;33(5):510–7.
134. Thakore PI, D'Ippolito AM, Song L, Safi A, Shivakumar NK, Kabadi AM, et al. Highly specific epigenome editing by CRISPR-Cas9 repressors for silencing of distal regulatory elements. Nat Methods. 2015;12(12):1143–9.
135. Liu XS, Wu H, Ji X, Stelzer Y, Wu X, Czauderna S, et al. Editing DNA methylation in the mammalian genome. Cell. 2016;167(1):233–247.e217.
136. Reynolds GP. Early life trauma, DNA methylation and mental illness. Epigenomics. 2021;13(11):825–7.
137. Mandelli L, Petrelli C, Serretti A. The role of specific early trauma in adult depression: a meta-analysis of published literature. Childhood trauma and adult depression. Eur Psychiatry. 2015;30(6):665–80.
138. Copeland WE, Shanahan L, Hinesley J, Chan RF, Aberg KA, Fairbank JA, et al. Association of childhood trauma exposure with adult psychiatric disorders and functional outcomes. JAMA Netw Open. 2018;1(7):e184493.
139. Zatti C, Rosa V, Barros A, Valdivia L, Calegaro VC, Freitas LH, et al. Childhood trauma and suicide attempt: a meta-analysis of longitudinal studies from the last decade. Psychiatry Res. 2017;256:353–8.

Chapter 16
Increased Inflammation and Treatment of Depression: From Resistance to Reuse, Repurposing, and Redesign

Jennifer C. Felger

Abstract Based on mounting clinical and translational evidence demonstrating the impact of exogenously administered inflammatory stimuli on the brain and behavior, increased endogenous inflammation has received attention as one pathophysiologic process contributing to psychiatric illnesses and particularly depression. Increased endogenous inflammation is observed in a significant proportion of depressed patients and has been associated with reduced responsiveness to standard antidepressant therapies. This chapter presents recent evidence that inflammation affects neurotransmitters and neurocircuits to contribute to specific depressive symptoms including anhedonia, motor slowing, and anxiety, which may preferentially improve after anti-cytokine therapies in patients with evidence of increased inflammation. Existing and novel pharmacological strategies that target inflammation or its downstream effects on the brain and behavior will be discussed in the context of a need for intelligent trial design in order to meaningfully translate these concepts and develop more precise therapies for depressed patients with increased inflammation.

Keywords Inflammation · Cytokines · Depression · Anhedonia · Dopamine · Glutamate · Antidepressants

16.1 Introduction

Based on mounting clinical and translational evidence demonstrating the impact of exogenously administered inflammatory stimuli on the brain and behavior, increased endogenous inflammation has received attention as a pathophysiologic process that may contribute to psychiatric illnesses and particularly depression. A rich literature describes increased inflammation in patients with depression and other psychiatric

J. C. Felger (✉)
Emory University School of Medicine, Atlanta, GA, USA
e-mail: jfelger@emory.edu

M. Macaluso et al. (eds.), *Drug Development in Psychiatry*, Advances in Neurobiology 30,
https://doi.org/10.1007/978-3-031-21054-9_16

disorders, as evidenced by elevated peripheral and central inflammatory cytokines and acute phase proteins. This endogenous inflammation may arise from numerous sources including risk factors for psychiatric illness (e.g., stress, obesity or metabolic dysfunction, genetics, and lifestyle factors) and has been associated with reduced responsiveness to standard antidepressant therapies. As both increased inflammation and treatment resistance occur in a significant proportion of patients with major depressive disorder (MDD), new conceptual frameworks are needed to identify relevant targets and develop novel therapies.

This chapter will present recent evidence that inflammation affects neurotransmitters and neurocircuits to contribute to specific depressive symptoms including anhedonia and motor retardation. Converging findings from neuroimaging studies involving the administration of exogenous inflammatory stimuli or characterization of depressed patients with increased endogenous inflammation will be discussed in relation to growing evidence that inhibition of inflammation with anti-cytokine therapies in patient with mood disorder including depression specifically reduces anhedonia in patients with evidence of increased inflammation. This chapter will then focus on potential pharmacological strategies based on the neurobiological mechanisms by which inflammation affects neurotransmitters, circuits, and symptoms. Such strategies include the use of existing compounds, either by employing biomarkers to guide selection of antidepressants for patients with high inflammation or by repurposing of existing compounds indicated for other conditions as novel therapies to target inflammation or its downstream effects on the brain and behavior. The need for novel immune-modulatory or redesigned, next-generation anti-cytokine therapies will also be discussed in light of design considerations for clinical trials that are required to meaningfully translate these concepts and develop more precise therapies for patients with increased inflammation.

16.2 Increased Inflammation in Depression: Sources, Symptoms, and Role in Treatment Resistance

16.2.1 Inflammation in Depression: Causes and Consequences

Numerous studies including meta-analyses have reported increased peripheral and central inflammatory markers like the acute phase reactant, C-reactive protein (CRP) and the inflammatory cytokines interleukin (IL)-1, IL-6, and tumor necrosis factor (TNF), in MDD [1–4]. Longitudinal studies have also found that increased inflammatory markers predicted subsequent depression symptoms [5–7], even above and beyond prior depression severity [8]. Of note, similar increases in inflammatory markers have been described in other psychiatric disorders that share commons symptom domains like anhedonia, motor slowing, and anxiety, including bipolar disorder, schizophrenia, anxiety disorders, and post-traumatic stress disorder (PTSD) [9–12]. In addition inflammation-related genetic risk [13], gene expression in peripheral blood immune cells of MDD patients have also revealed activation of oxidative stress pathways and inflammatory cytokines as well as canonical inflammatory signaling pathways including toll-like receptors, nuclear factor kappa B, and the NLRP3 inflammasome complex [14–18]. Activation of these pathways reflects innate immune responses to both pathogen-associated molecular patterns and danger-associated molecular patterns that are generated by host cells under stress, including psychological stress [19]. Indeed, in MDD patients who are otherwise medically healthy, genetic predisposition may interact with not only stress and trauma but also a range of environmental and lifestyle factors to activate the innate immune system and contribute to low-grade systemic "sterile" inflammation in the absence of pathogens including disturbed sleep, physical inactivity, obesity and metabolic disturbances, Western diet, aging, and smoking [20]. Many of these causes of inflammation are risk also factors for both psychiatric and major medical illnesses, suggesting shared pathophysiologic processes that may explain notable comorbidity between psychiatric disorders and cardiovascular disease, diabetes, and cancer [21].

While most studies report relationships between increased peripheral inflammatory markers and depression and many sources of inflammation per above are from the body, inflammatory cytokines and activated immune cells can access the CNS to directly influence neurotransmitters and circuits and to activate local inflammatory processes. Increased inflammatory markers have been described in the cerebrospinal fluid (CSF) in MDD [22, 23]. Postmortem studies have identified evidence of increased inflammatory signaling in brain parenchyma, including increased TLR expression, expression of inflammatory cytokines, and evidence of both peripheral immune cell trafficking to the brain and activation of microglia [24–27]. Positron emission tomography (PET) imaging of increased translocator protein (TSPO), which is thought to reflect activated microglia and macrophage, has also been reported throughout the brain in MDD but did not relate to peripheral inflammatory markers [28]. While TSPO binds activated microglia in response to acute inflammation [29], it is not clear whether this is specific to inflammatory microglia versus those that activated to perform physiologic roles like synaptic pruning, how much of the signal is due to binding to other cells including astrocytes and neurons [30], or whether signal is confounded by uptake in the periphery of patients with increased inflammation [31] in MDD. It should be noted however that recent data show that inflammatory pathways can disrupt the blood brain barrier (BBB) in discrete subcortical brain regions, particularly those that regulate motivation and reward [32] and correspond to the impact of inflammation on specific circuits and symptoms, as detailed below.

16.2.2 Increased Inflammation and Antidepressant Treatment Response

Whereas not every patient with MDD has increased inflammation, higher concentrations of inflammatory markers have been reliably observed in patients with reduced responsiveness to conventional antidepressants [33]. Indeed, approximately 25–40% of MDD patients depending on the sample exhibit CRP >3 mg/L, considered high risk for developing cardiovascular disease per American Heart Association guidelines [3, 4, 34, 35], with even more falling in the moderate risk range of CRP 1–3 mg/L, while <1 mg/L is considered normal. Retrospective analyses of longitudinal studies have shown that patients with CRP >1 mg/L prior to therapy are less responsive to antidepressants, especially not only selective serotonin reuptake inhibitors (SSRIs) but also serotonin norepinephrine reuptake inhibitors (SNRIs), over the course of an adequate trial [35–38]. Measurement of inflammatory cytokine mRNA expression in peripheral immune cells was even more predictive of this effect than CRP [37, 39]. Similarly, in MDD patients with a history of antidepressant nonresponse, higher levels of inflammatory markers including IL-6, TNF, and its soluble receptor 2 and CRP were associated with increasing number of prior failed trials [40, 41]. Moreover, some studies have found that patients with CRP >1 mg/L are more responsive to drugs that affect dopaminergic and noradrenergic pathways including bupropion and nortriptyline [35, 42]. Additionally, MDD patients with higher levels of inflammatory markers have shown better response to adjuvant or therapies that boost monoamine availability or target glutamate [42–44] and also electroconvulsive therapy [45]. Together, these data suggest that (1) inflammatory markers may help guide antidepressant treatment selection and (2) better understanding the mechanisms by which inflammation affects the brain may lead to development of novel therapies targeted to the many MDD patients with higher levels of inflammation. Therefore, it is it is important to prospectively consider the role of inflammation in future antidepressant trials and to design studies examining novel treatment avenues using appropriately selected biomarkers and outcome measures relevant to specific symptoms associated with high inflammation in MDD.

16.2.3 Relationships Between Inflammation and Symptom Domains

Consistent with impact of inflammatory cytokines on specific circuits and symptoms as described below, evidence of increased inflammation in MDD has been associated with specific symptoms that are common to other disorders including anhedonia, motor slowing, and anxiety [46–49]. For example, our group recently identified clusters of cytokines and their soluble receptors in CSF that were associated with higher levels of plasma CRP in otherwise medically stable patients with MDD [4]. These CSF markers, in turn, associated with symptom severity with the strongest relationships between CSF TNF and reduced motivation per a subscale from the Multidimensional Fatigue Inventory and CSF IL-6 soluble receptor and anhedonia per a subscale from the Inventory of Depressive Symptomatology Self-Report (IDS-SR) that correlates with the both the self and clinician-administered Snaith-Hamilton pleasure scale (SHAPS) [50, 51]. These results were confirmed and extended by a study demonstrating that both T- and non-T-cell cytokines were associated with anhedonia severity per the IDS-SR subscale [52]. Furthermore, longitudinal associations between cytokines and anhedonia have been reported in MDD where higher baseline plasma TNF predicted greater severity of anhedonia both at baseline and at a four-month follow-up [53]. Similar relationships between psychomotor slowing and inflammatory markers have also been observed in MMD [54], and many studies have found associations between increased inflammatory markers in schizophrenia and negative symptoms, which include motivational deficits, blunted affect, and social withdrawal among others [55]. In regard to anxiety, a growing literature reports correlations between increased CRP and inflammatory cytokines and symptoms of anxiety [46, 56], including in a longitudinal study [57] and in patients with MDD [58]. Together, these studies provide a clinical framework for the potential role of inflammation in symptoms of anhedonia, motor slowing and anxiety in MDD, and other psychiatric disorders and support the need for mechanistic studies to better understand the impact of inflammation on the brain.

16.2.4 Inhibition of Inflammation in Depression and Symptom Specificity

Numerous studies treating psychiatric patients with rather nonspecific anti-inflammatory agents having multiple off-target effects, e.g., nonsteroidal anti-inflammatory drugs and minocycline [59–61], were not targeted to patients with increased inflammation and yield mixed results. Although having limited viability as antidepressants for a myriad of reasons [62, 63], more specific anti-cytokine therapies have shown efficacy for reducing specific depressive symptoms in depressed or medically ill patients with high inflammation. For example, treatment of patients with autoimmune or inflammatory disorders with anti-cytokine therapies reduces depression symptom severity [64]. The TNF antagonist infliximab reduced depression severity with respect to placebo in treatment-resistant MDD patients with higher concentrations of plasma CRP, and anhedonia (*work and activities*) was the most improved symptom followed by motor slowing (*retardation*) and anxiety (*psychic anxiety*) [3]. Moreover, similar results have been seen in two recent studies reporting that anti-TNF or IL-6 therapies in unipolar or bipolar depressed patients with evidence of increased inflammation were primarily effective in reducing anhedonia assessed by SHAPS [3, 65, 66]. These data reinforce specificity for the effects of inflammation on neurobiological pathways that contribute to anhedonia, as well as motor slowing and anxiety, as discussed below.

16.3 Inflammation Effects on the Brain and Behavior

Neuroimaging studies have consistently found that administration of a variety of peripheral inflammatory stimuli, including cytokines and cytokine inducers (e.g., vaccination and subfebrile doses of endotoxin), impact corticostriatal reward and motor circuits to drive reduced motivation and motor slowing as well as anxiety-related brain regions including amygdala, insula, and anterior cingulate cortex (ACC), which may result from cytokine effects on monoamines and glutamate (Fig. 16.1) [67]. Causal evidence for the effects of inflammation on neural circuits and neurotransmitters were initially revealed in patients administered chronic inflammatory cytokines, such as the antiviral and antiproliferative cytokine interferon (IFN)-α, which caused clinical depression in up to half and depressive symptoms in nearly all patients over weeks to months of treatment for infectious diseases or cancer [68, 69]. Like IFN-α, endotoxin and vaccination induce release of classic inflammatory cytokines IL-6, IL-1, and TNF in association with transient increases

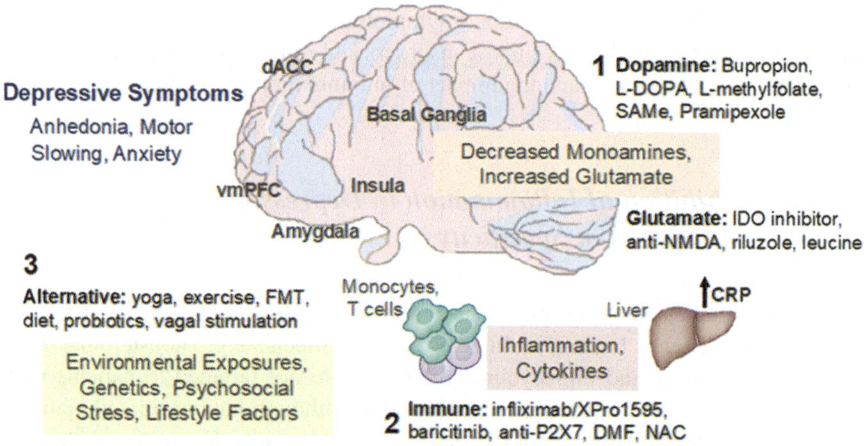

Fig. 16.1 Mechanisms of inflammation effects on the brain and behavior and targets for intervention in depression. Inflammation is increased in otherwise medically stable patients with major depressive disorder (MDD) due to environmental exposures, genetics, psychosocial stressors, diet, and other lifestyle factors. Innate immune cell activation and the release of inflammatory cytokines cause both increased CRP production from the liver and effects on brain neurotransmitters and circuits to drive relevant behavioral changes. Evidence indicates that inflammation and cytokines may preferentially affects dopamine and glutamate systems to disrupt circuits involved in reward and motor activity, as well as those involved anxiety and emotional regulation. In terms of potential novel therapies that may target inflammation or its effects on the brain, there is current interest in (1) compounds that increase dopamine or decrease glutamate signaling, (2) therapies that directly target the immune system to decrease inflammation, and (3) alternative strategies via lifestyle changes or treatments that modify the sources of inflammation. CRP C-reactive protein, dACC dorsal anterior cingulate cortex, DMF dimethyl fumarate, FMT fecal microbiota transplant, IDO indoleamine 2,3 dioxygenase, L-DOPA levodopa, NAC N-acetyl cysteine, NMDA n-methyl-d-aspartate, P2X7 purinergic ATP receptor 7, SAMe S-adenosylmethionine, vmPFC ventromedial prefrontal cortex

in depressive symptoms and are commonly used in lab settings to understand their acute effects on the brain [70], as reviewed below. Recent work has also translated findings from these studies investigating causal effects of exogenously administered inflammatory stimuli to study relationships between endogenous inflammation, neurotransmitters, and circuits in MDD patients.

16.3.1 Impact of Inflammation on Reward and Motor Regions and Circuits

Early positron emission tomography (PET) studies investigating broad effects of chronic inflammatory cytokines on the brain found that resting glucose metabolism was increased in basal ganglia and decreased in frontal cortex [71, 72], whereby increased metabolism in the left putamen and left nucleus accumbens correlated with IFN-α-induced anergia and fatigue [71]. This pattern of increased glucose metabolism in basal ganglia nuclei is similar to that seen in patients with Parkinson's disease (PD) [73], which thought to indicate increased oscillatory burst activity secondary to loss of inhibitory dopamine input [74]. Complementary PET using radiolabeled dopamine precursor, [^{18}F]fluorodopa, in IFN-α-treated patients also showed both increased uptake and decreased turnover of FDOPA, reflecting decreased availability of dopamine/precursor and impaired packaging or release of newly synthesized dopamine, in the caudate, putamen, and VS [68]. Magnetic resonance spectroscopy (MRS) further showed increased glutamate in left basal ganglia in patients treated with IFN-α that correlated with reduced motivation [75]. Complementary to these chronic studies, acute challenge with IFN-α caused rapid (4 hours) changes in striatal microstructure that predicted subsequent development of fatigue [76, 77].

Functional impact of the effects of peripheral inflammation on brain regions relevant to reduced motivation and motor activity and involving dopamine and glutamate have also been revealed by functional MRI (fMRI). Indeed, decreased ventral striatal (VS) neural activation to win versus loss was seen in a gambling task after chronic IFN-α, which correlated with self-reported reduced motivation [68]. Studies in healthy controls using vaccination and subfebrile doses of endotoxin have also assessed acute effects of inflammation on reward processing. Reduced activation of VS to reward-predicting cues during a monetary incentive delay task (MIDT) were associated with increased self-reported depressed mood [78] and with cytokine responses in women but not men hours after endotoxin [79]. In a probabilistic instrumental learning task combined with fMRI, typhoid vaccine compared with saline control reduced behavioral attractiveness of rewards while making punishments more aversive, in association with opposing change in VS responses that were decreased to positive feedback but increased to negative feedback [80]. This corresponds with a study showing that greater inflammatory responses to laboratory stress correlated with decreased VS sensitivity to positive feedback [81]. Additionally, typhoid vaccination affected task-based activity in the substantia nigra that correlated with both psychomotor slowing and increased peripheral blood concentrations of IL-6 [82, 83]. Finally, acute administration of IFN-α or typhoid vaccination has been shown to acutely decrease functional connectivity (FC) within motivation-relevant brain regions including VS and the ventromedial prefrontal cortex (vmPFC) [84, 85].

16.3.2 Impact of Inflammation on Regions and Circuits for Fear, Anxiety, and Emotional Processing

Similar to reports of increased reactivity in MDD as well as anxiety disorders and PTSD, exogenous administration of peripheral inflammatory stimuli has been shown to increase neural activity in amygdala, dorsal ACC, and insula [86]. For example, acute IFN-α (4 hour) administration enhanced right amygdala responses to sad versus neutral faces, which correlated with subsequent IFN-α-induced depression severity [87]. Increased IL-6 and TNF after administration of endotoxin to healthy subjects was also shown to increase amygdala activity in response to socially threatening images, which correlated with feelings of social disconnection [88]. Greater dorsal ACC activation was also seen in IFN-α -treated patients that highly correlated with task-related errors [89], and this may be due to increased glutamate in dorsal ACC as measured by MRS in patients administered IFN-a that correlated with depressive symptom severity [90]. In participants administered endotoxin prior to a neuroimaging session in which they were socially excluded during a virtual ball-tossing game, increases in IL-6 were associated with increases in social pain-related neural activity in both dorsal ACC and anterior insula in females but not males [91]. Another study administering typhoid vaccination also reported increased activation of amygdala and dorsal ACC as well as insula, during presentation of congruent and incongruent stimuli [92]. Given the role of the insula in interoception, it is not surprising that this brain region also had increased resting glucose metabolism as measured by PET after endotoxin [93]. These findings suggest that increased inflammatory cytokines in the periphery may contribute to altered neural activity in circuits involving amygdala, dorsal ACC, and insula to disrupt emotional processing in MDD and anxiety-related disorders.

16.3.3 Endogenous Inflammation and Circuit Dysfunction in Patients with Depression

In light of converging evidence of the impact of exogenously induced inflammation on circuits and symptoms relevant to reduced motivation, motor slowing, and anxiety (as described above), recent studies have examined a potential role for increased endogenous inflammation in relevant circuit deficits that are frequently observed in patients with MDD and other psychiatric disorders [94–96]. For example, in medically stable and unmedicated MDD patients, endogenous inflammation as measured by plasma CRP and inflammatory cytokines was associated with lower left VS to vmPFC FC, which in turn positively correlated with anhedonia per the IDS-SR subscale [51]. These targeted findings were corroborated by parcellation-based network analysis in MDD revealing vmPFC and VS (a region parcellated as anterior ventral caudate) as the two most significant hubs, respectively, in a widely distributed network of low FC within 63 features in relation to CRP, subsets of which were highly predictive of anhedonia as measured by the IDS-SR subscale and SHAPS [97]. Of note, relationships between increased inflammation and low FC among several regions, dorsal striatal regions, the vmPFC and pre-supplementary motor area, and key components of corticostriatal circuitry involved in linking motivation to motor output [98, 99], were correlated with objective measures of psychomotor slowing in these studies [51, 97]. Furthermore, FC was shown to mediate relationships between CRP and anhedonia and psychomotor symptom severity [51].

Similar relationships between increased inflammation and low FC in primarily left VS to vmPFC reward circuitry were also observed in treatment-resistant MDD patients [100] and in trauma-exposed women in relation to an anhedonia subscale via Beck Depression Inventory [101]. Further evidence of associations between increased endogenous inflammation and functional changes in reward circuits include reduced striatal activation during reward anticipation in MDD patients with higher CRP and inflammatory cytokines [102, 103]. While the above findings generally indicate a role for reduced dopamine signaling, high inflammation in MDD was associated with increased glutamate concentrations in left basal ganglia that correlated with anhedonia [104]. Patients with combined elevations in CRP and glutamate displayed both high anhedonia and low regional homogeneity in left basal ganglia, indicating disrupted local coherence of activity that may be driven by increased glutamate [105].

Regarding threat and anxiety-related circuitry, inflammatory markers have been associated with similar deficits in amygdala-vmPFC circuitry as those reported to characterize individuals with high anxiety, MDD, and/or PTSD [106–108]. For example, we previously found that higher concentrations of plasma CRP and inflammatory cytokines correlated with lower right amygdala-vmPFC FC in patients with a primary diagnosis of MDD in association with anxiety symptoms, particularly in patients with comorbid anxiety-related disorders including PTSD [109]. Recent studies in adolescents have also found relationships between endogenous inflammation and altered FC in circuits relevant to threat, anxiety, and emotional processing [110, 111]. Interestingly, acute blockade of TNF in inflammatory arthritis patients with infliximab decreased right amygdala reactivity to emotional (sad, happy, and neutral) faces in association with reduced depressive symptoms at 24 hours [87], suggesting emotional reactivity as a potential target for efficacy of anti-inflammatory therapies.

16.4 Treatment Targets for Depressed Patients with Increased Inflammation

Given the mounting evidence of differential response to antidepressants in MDD patients with higher versus lower levels of inflammation and the reproducible effects of inflammation on behavior, there is a need to consider inflammation in studies examining antidepressant outcomes and for development of novel therapies that block inflammation or its consequences on the brain (Fig. 16.1).

Numerous clinical trials have addressed this concern by the use of agents with putative anti-inflammatory activity as adjuvant or therapy in patients with psychiatric disorders, primarily in depression or schizophrenia [112–114]. Meta-analyses of the use of such drugs including cyclooxygenase (COX)-2 inhibitors, anti-cytokine therapies, minocycline, statins, pioglitazone, glucocorticoids, and omega-3 fatty acids suggest modest efficacy [60, 113–117] despite small sample sizes, heterogeneity across studies, and numerous design issues. Most studies used therapies as adjuvant to conventional antidepressants and did compare with placebo, patients were rarely selected to have high inflammation, only a few studies measured inflammatory markers to stratify patients or establish anti-inflammatory effect, and importantly, many of the chosen therapies convey only mild anti-inflammatory activity in the context of numerous "off target" effects that can confound data interpretation [63, 118]. In the largest randomized controlled trial to date using the COX-2 inhibitor celecoxib and minocycline (an antibiotic thought to stabilize microglia but also disrupt microbiota [119]), both failed to separate from placebo in reducing depressive symptoms in depressed bipolar patients [61]. Two studies did, however, consider inflammation levels in treatment-resistant or bipolar depression and found that higher CRP (>3 mg/L) or IL-6 concentrations prior to treatment were predictive of response to minocycline [120, 121], with reduced serum IL-6 after treatment seen only in bipolar depressed responders [121]. While existing cytokine antagonists may not be viable antidepressants [62, 63], they have demonstrated efficacy in depressed patients with high inflammation with specificity for symptoms consistent with the known effects of inflammation on the brain [3, 65, 66], thus providing a foundation for enrolment and outcomes strategies for testing novel therapies.

Given the above-described inconsistencies and challenges in studying anti-inflammatory therapies for depression, existing or novel compounds that target the neurotransmitters impacted by inflammation, like dopamine and glutamate, may serve as more proximal approaches for translating these concepts into patients (Fig. 16.1(1)). Discussed below are the multiple pharmacological interventions that can be used to block inflammation or its downstream effects on the brain, starting with the potential for informed selection of available antidepressants and reuse or repurposing of existing compounds that affect neurotransmitters. Because the above-described use of COX-2 inhibitors, anti-cytokine therapies, minocycline, fatty acids, and the like have been reviewed extensively elsewhere, discussion on immune targets will focus on novel agents or redesign of existing therapies (Fig. 16.1(2)), along with mention of alternative treatments that may exert efficacy via effects on neural and physiologic processes that modulate inflammation (Fig. 16.1(3)). Biomarker and study design considerations for patient selection and target engagement of the brain and behavior will also be discussed.

16.4.1 Compounds That Increase Dopamine Synthesis, Synaptic Availability, and Receptor Signaling

Dopamine reuptake Decreased response to conventional antidepressant therapies like SSRIs in patients with high inflammation may be due to decreased monoamine synthesis and availability, or to facilitatory effects of cytokines on serotonin transporters, which may circumvent or interfere with their action [122, 123]. Alternatively, reduced response may be due to a preferential access and effects of peripheral inflammation on reward and motor-related brain regions that receive primarily dopamine input [67, 124], consistent with evidence of increased responsiveness in these patients to antidepressants with catecholamine activity and particularly bupropion [42]. Bupropion, an FDA approved and effective medication for MDD [125] that functions primarily by inhibiting dopamine and norepinephrine reuptake, has been shown to increase high effort activity in rats [126]. Some trials also suggest preferential response of anhedonia to bupropion [127, 128]. Together, these findings warranted further investigation of potential efficacy in MDD with high inflammation, such as a recent trial prospectively examining the ability of bupropion versus escitalopram to increase FC in reward circuitry and improve motivation in patients with high CRP (NCT04352101). It should be noted that although classical psychostimulant medications with potent effects on dopamine reuptake or release increase motivation in rodent models and acutely in healthy humans [126, 129, 130], they have demonstrated only limited efficacy in chronically treating fatigue and other dopamine-related symptoms in trials for patients with cancer and other medical illnesses that are associated with inflammation [131–140], or as augmentation therapy for depression [141–145].

Dopamine synthesis While compounds that inhibit dopamine reuptake may exert efficacy in high inflammation patients, the primary mechanisms of inflammation effects are likely through the inhibition of key components of dopamine synthesis like tetrahydrobiopterin (BH4) [146, 147], a pivotal cofactor for the enzymes that synthesize dopamine and other monoamines. Indeed, inflammation reduces BH4 availability through oxidation and excessive conversion to BH2 during generation of nitric oxide by nitric oxide synthase [148]. Therefore, depressed patients with high inflammation may benefit from therapies that increase BH4 stability or activity including sapropterin and folic acid, L-methylfolate, and S-adenosylmethionine (SAMe). Low serum folate has been associated with increased risk of depression and non-response to antidepressant treatment and an increased likelihood of depression relapse [149], yet clinical trials using L-methylfolate (marketed as Deplin and Zervalx) and SAMe have shown mixed results [150, 151]. Post hoc analysis of two parallel-sequential adjuvant trials of L-methylfolate in patients with MDD [150] did however reveal that a combination of increased concentrations of leptin, CRP, and inflammatory cytokines or high BMI was associated with greater symptom improvement [44], supporting potential value of targeting such therapies to MDD patients with high inflammation.

Another strategy to address impaired dopamine synthesis is the administration of its precursor, levodopa (L-DOPA). Indeed, in monkeys experiencing similar behavioral responses including reduced effort-based sucrose consumption after chronic IFN-α exposure [152], decreases in striatal dopamine release were reversed by L-DOPA administered via reverse in vivo microdialysis [153]. Replacement of dopamine with L-DOPA improves motor function and was also shown to increase motivation in patients with PD [154]. Whether L-DOPA (in combination with carbidopa) versus placebo improves FC in reward circuitry in association with improved motivation and anhedonia in MDD patients with higher levels of CRP is currently being studied (NCT04723147). Open-label L-DOPA-carbidopa administration to aged depressed patients with motor slowing, a group likely to exhibit increased inflammation, also showed a positive antidepressant response [155], and a similar ongoing study in this population aims to better understand mechanisms of these findings including the potential role of inflammation (NCT04469959).

Dopamine agonists Dopamine receptor agonists have received attention as efficacious augmentation strategies for depression. For example, antiparkinsonian agents like pramipexole have demonstrated efficacy to reduce depressive symptoms in patients with treatment-resistant depression [156–159]. Although it is unknown whether this effect is specific to high inflammation in depressed patients [160], it has been shown to block endotoxin-induced degeneration of nigrostriatal dopamine cells in rodents [161].

A growing body of evidence also supports the use of atypical antipsychotics as an augmentation strategy in MDD, including meta-analyses suggesting these agents are more effective than placebo for both response and remission [162, 163]. In addition to more serotonergic activity, newer generation antipsychotics like aripiprazole and amisulpride appear to act as D2/3 partial agonists by facilitate dopamine signaling at lower doses or in states of low endogenous ligand [164], such as with increased inflammation, while preventing overstimulation when endogenous dopamine levels are high.

16.4.2 Therapies That Target Glutamate Transmission

Modulation of the kynurenine pathway Immune-mediated activation of indoleamine 2,3 dioxygenase (IDO) catabolizes tryptophan (TRP), the primary amino-acid precursor of serotonin, to kynurenine (KYN), downstream metabolites of which affect glutamate transmission in the brain [165, 166]. Peripheral blood KYN/

TRP ratio in combination with TNF defines a population of MDD patients with increased anhedonia and treatment resistance [167]. With regard to preventing activation of the KYN pathway, the IDO antagonist, 1-methyl tryptophan (1-MT) has been shown to abrogate depressive-like behavior in animal models of inflammatory challenge or infection [168, 169]. Given the importance of serotonin in T-cell activation, there has been interest in developing IDO inhibitors as a pharmacologic strategy to enhance T-cell function against cancer [170], but compounds like 1-MT have not yet been translated outside of oncology. As leucine competes with KYN for the large amino acid transporter, it can inhibit transport of KYN into the brain and reduce production of neuroactive metabolites like quinolinic acid (QUIN) from KYN in microglia [171]. Thus, high dose leucine (8 mg/d for 2 weeks) is currently being tested in MMD patients, although inflammatory markers that have been associated with KP metabolites in the periphery and CNS will only be examined in post hoc analyses (NCT03079297).

Glutamate receptor modulators Inflammation can promote excitotoxic glutamate transmission through several mechanisms including decreased buffering by astrocytic expression of excitatory amino acid transporters (EEATs), increase release of glutamate from astrocytes and activated microglia [165, 172–174], and increased QUIN, as described above, which directly activates the n-methyl-d-aspartate (NMDA) receptor [175, 176]. Therefore, glutamate receptor antagonists may be useful in preventing excitotoxicity and oxidative stress and may reverse or prevent inflammation-related behavioral change. Indeed, in rodents, the NMDA antagonist ketamine reversed endotoxin-induced depressive-like behavior including anhedonic behavior, while having no effect on inflammation or activation of IDO in the brain [177]. Moreover, blockade of AMPA receptors was able to reverse ketamine's effects on endotoxin-induced depressive-like behavior, indicating that the effects of ketamine were specific to its impact on glutamate signaling. Moreover, in an animal model of treatment-resistant depression, ketamine responsiveness was predicted by baseline peripheral blood levels of CRP and TNF [178].

In humans, one study in treatment-resistant depression found that patients who were most responsive to ketamine were those with the highest concentrations of serum IL-6 [43]. However, another study found that although treatment-resistant depressed patients exhibited increased IL-6 compared with controls, IL-6 and other inflammatory cytokines were not associated with response to ketamine [179]. Given the restrictions to ketamine and esketamine use (i.e., administration route, post-dose monitoring), alternative agents with equal efficacy and favorable tolerability and safety profile are being actively investigated. One example is AXS-05, a combination oral pill containing the NMDA receptor antagonist/sigma-1 receptor agonist dextromethorphan given with bupropion (to boost dextromethorphan blood levels through CYP2D6 inhibition). As phase 2 results appeared promising, with significant improvements in response and remission at 6 weeks compared with bupropion alone [180], this therapy might be particularly well-suited for treatment of depressed patients with high inflammation.

Glutamate stabilizers As downstream effects of inflammation both stimulate glutamate receptors and disrupt balance of intracellular and extracellular glutamate, strategies targeting reuptake mechanisms via EAATs may be beneficial in patients with high inflammation. Riluzole is one agent that may support glutamate by facilitating EAAT activity to protect against excitotoxicity [181] and has some evidence of benefit in MDD. One small open-label study found benefit in treatment-resistant depressed patients over the course of 6 weeks with riluzole monotherapy [182], while another small trial showed benefit of riluzole augmentation [183], thus providing support for future studies in MDD with high inflammation.

16.4.3 Therapies That Affect the Immune System

Anti-inflammatory drugs Results from many trials in psychiatry using anti-inflammatory therapies are mixed at best [59], and only a handful of studies have enriched for patients with evidence of increased inflammation and/or used drugs with known anti-inflammatory activity and little off-target effects [184]. Studies using cytokine antagonists that have potent anti-inflammatory effects have reported encouraging results for improved symptoms of anhedonia in depressed patients with increased inflammation [3, 65, 66]. Translation of these therapies is unfortunately limited, for example, by increased risk for infection, and blockade of potentially beneficial effects of innate immune signaling on other neurobiological pathways such a myelination [62, 63]. Fortunately, immunotherapies are evolving with even more specificity for inflammatory signaling pathways [185]. For example, XPro1595, a novel, first in class selective “dominant-negative” mutant variant of the human TNF protein [186], rapidly binds to and inhibits “inflammatory” signaling driven by the soluble form of TNF through soluble TNF receptors, while having no effect on the immunologic and neuro“protective” signaling driven by the transmembrane form of TNF [186–188]. XPro1595 has demonstrated preclinical efficacy in multiple laboratory animal models of depression [189–191], including a treatment-resistance model where XPro1595 neutralized TNF of the rat and decreased peripheral blood CRP concentrations [192]. Of note, laboratory animal studies have exemplified that crossing the BBB is not required for the antidepressant effects of traditional cytokine antagonists [193], which are relatively large molecules, as reducing peripheral inflammation in MDD is the primary target [3]. However, the novel TNF antagonist XPro1595 has significant brain penetrance [194], suggesting potential human benefit for diseases like depression above and beyond the reduced risk of infection or demyelination as compared with traditional anti-cytokine therapies.

Drugs targeting inflammatory signaling pathways, such as baricitinib, an oral Janus Kinase (JAK 1/2) inhibitor FDA-approved for the treatment of rheumatoid arthritis with recent additional FDA-approval for emergency use authorization for the treatment of patients hospitalized with COVID-19, are being studied [195–199]. Baricitinib significantly reduces plasma IL-6, TNF, and CRP within days after administration in humans across disease states [196–202] and might exert benefit in MDD with high inflammation. Other similar drugs in development include those that inhibit other intracellular immune pathways, toll-like receptor signaling, cell adhesion molecules, or chemokine receptors. Additionally, inhibitors of inflammasome activation via the purinergic P2X7 receptor are also being explored, including current testing in depressed patients with incomplete response to monoaminergic antidepressants and CRP ≥1 mg/L (NCT04116606).

Despite considerable interest in the role of the immune system in MDD and other psychiatric disorders and its therapeutic implications, little information exists regarding the specific immunologic mechanisms required to design therapies engaging specific immune cell types. While a monocyte phenotype has traditionally been thought to represent high inflammation in MDD, a recent study clustering patients into inflammatory subgroups suggested distinct populations of high inflammation patients represented either by myeloid cells in one case or lymphoid populations in the other [203]. Accordingly, vast array of other anti-cytokine therapies that selectively target T cell cytokines include anti-IL-17 and anti-IL-12/23, which are FDA-approved; maraviroc which inhibits CCR5 for prevention of HIV; and plerixafor, a CXCR4 antagonist that mobilizes stem cells [204].

Immunometabolic modulation Evidence of the metabolic and energetic reprograming used by immune cells to sustain inflammatory activation has been observed in medically stable MDD patients who had both high CRP and significant anhedonia [205]. Furthermore, immunometabolic pathways in specific cell types are being targeted for new therapies in autoimmune and inflammatory disorders [206] and align with recent data that rapamycin, an inhibitor of mTORC1 signaling involved in such processes, prolonged the antidepressant benefit of ketamine therapy [207]. Additionally, first-line treatment with dimethyl fumarate (DMF) in patients with multiple sclerosis has recently been shown to inhibit Warburg-like metabolism in immune cells via inhibiting the glyceraldehyde 3-phosphate dehydrogenase, a key enzyme of the glycolytic pathway [208]. Consistent with the role of aerobic glycolysis in activation of specific immune cell subsets, DMF inhibited inflammatory cytokines and lactate production from activated macrophages, Th1 and Th17 cells, while sparing the function of resting macrophages and regulatory T cells. Additionally, N-acetyl cysteine (NAC), a precursor to the antioxidant glutathione that reduces oxidative stress/reactive oxygen species bioproducts of intracellular immunometabolic shifts and inflammation, may both improve mitochondrial function and reduce inflammation [209–211]. NAC may also exert antidepressant activity via effects on glutamate signaling through AMPA receptors and the astroglial glutamate exchanger xCT [212, 213]. Given evidence of antidepressant efficacy of NAC, there are ongoing clinical trials including one with inclusion criterion of a CRP >0.85 mg/L (NCT02972398).

Alternative therapies: lifestyle factors, stress reduction, neuroimmunomodulation, and microbiome The efficacy of modifying environmental exposures and lifestyle factors that contribute to inflammation in depression have also been investigated including exercise, weight reduction, yoga, massage, tai chi, and cognitive behavioral therapy and meditation. Many of these interventions have been shown to induce a variety of immune changes including reduced inflammation [214]. For example, mindfulness meditation over 4 months increased FC between the posterior cingulate cortex and the left dorsolateral PFC, which in turn was associated with decreases in IL-6 [215]. In addition, both cognitive behavioral therapy and tai chi were associated with reduced CRP, monocyte production of inflammatory cytokines, and inflammatory gene expression in elderly patients with insomnia [216]. Both diet and exercise programs have been shown to have antidepressant and anti-anxiety effects [217, 218] and to reduce a variety of inflammatory markers in longitudinal studies [219, 220]. A 3-month hatha yoga program also reduced endotoxin-induced peripheral blood mononuclear cell production of TNF, IL-6, and IL-1beta as well as fatigue in breast cancer survivors [221]. These studies have, however, failed to determine whether changes in inflammation or immune function are required for the efficacy of these interventions [63], or if they are generalizable to MDD patients with high inflammation and anhedonia.

An elegant body of work has described a direct mechanism for neural regulation of the immune response mechanisms by stimulation of efferent vagal fibers to provide acetylcholine-mediated inhibition of the release of TNF and other cytokines from immune cells such as macrophages [222]. This anti-inflammatory cholinergic reflex can be activated by electrically stimulating the vagus and is now being capitalized on via a novel bioelectric platform that recently received designation as an FDA breakthrough device for the treatment of rheumatoid arthritis in patients intolerant of or exhibiting incomplete response to biologic drugs [223]. There is also growing interest in the role of the microbiome in a variety of neuropsychiatric disorders including MDD and its potential as a therapeutic target [224]. While the precise mechanisms by which the microbiome influences behavior are unknown, evidence suggests effects on inflammation as a plausible pathophysiologic pathway [225, 226]. These relationships have been illuminated by translational work showing induction of depressive-like behavior in mice after fecal microbiota transplant (FMT) from donors with major depression [227] and improvements in depressive and anxiety symptoms after FMT in patients with inflammatory or functional bowel disease [228, 229]. Trials examining probiotic supplementation show small but significant improvement of depressive and anxiety symptoms [230] and research on more directed therapies like FMT is warranted.

16.5 Summary and Translational Conclusions

In this chapter, extensive clinical and translational evidence is presented supporting increased inflammation and its effects on the brain as one pathophysiologic pathway to symptoms of anhedonia, reduced motivation and motor slowing as well as anxiety in MDD and other psychiatric disorders. Decreased dopamine availability and excessive glutamate may serve as mechanisms of inflammation's impact on the brain and potential therapeutic targets for symptom reduction in psychiatric patients with elevated biomarkers of inflammation using existing or novel antidepressant strategies. Moreover, studies employing anti-cytokine therapies in depression have consistently found anhedonia to be the symptom most improved, with some support for motor retardation and anxiety. While translation potential of previously tested anti-inflammatory agents may be limited, there are a myriad of new immune-targeted therapies with promise for therapeutic benefit in depression including redesigned, next-generation cytokine antagonists with improved specificity for inflammatory signaling, intracellular signal transduction inhibitors FDA approved for inflammatory illness, and alternative therapies that modulate immune function via manipulation of the vagal nerve or microbiome to name a few.

Despite consistent causal findings of associations between inflammation and alterations in neurotransmitters and neurocircuits relevant to MDD and other psychiatric illnesses, several challenges and considerations for translation of these concepts exist. Crucial in this regard is the need for informed trial design. For one, given strong evidence of relationships between inflammation and treatment resistance, inflammatory markers should be considered in studies examining predictors of response or selection of patients for existing antidepressant therapies in a prospective rather than post hoc fashion. For treatment development, biomarker-driven approaches should target specific therapies to patients with evidence of high inflammation (i.e., using CRP) and/or relevant symptoms like anhedonia, motor slowing, and/or anxiety and assess not only response and remission but also target engagement of relevant circuits and symptoms. Given that FC in reward and motor circuits has been identified to mediate relationships between endogenous inflammatory markers like CRP and anhedonia and psychomotor speed, functional neuroimaging biomarkers that associate with symptoms of anhedonia, motor slowing, or anxiety can serve as relevant brain biomarkers. Finally, while substantial work has established such relationships in depression, future work is needed to better understand the role of inflammation in the brain and specific behaviors and how it relates to treatment response in other disorders like schizophrenia.

Another challenge for translation is that the FDA and other regulatory bodies do not currently recognize individual symptom domains as appropriate criteria for drug development, despite recent appreciation in the field of symptoms subdomains that cut across disorders and have a similar, well-defined pathophysiological basis [231]. Future efforts will clearly require advocacy in this area, particularly as it relates to treatment resistance and residual symptoms, and considering that the FDA has recognized drugs for cancer based on surrogate markers of specific genes and proteins involved in the growth and survival of cancer cells irrespective of clinical outcome [232]. A similar agnostic approach to treatment based on an emerging understanding of biomarkers such as genes, proteins, neurotransmitters, and circuits that underlie the biological bases of behaviors may be needed to facilitate development and approval of new drugs for treatment of psychiatric disorders.

In sum, an emerging understanding of the mechanisms by which peripheral inflammation can affect neurotransmitters and relevant circuits to impact behavior and contribute to symptoms of MDD and other psychiatric disorders has provided a framework for development of novel therapies. Further identification of a platform of neuroimaging, behavioral, and peripheral biomarkers that can be used to test these therapies lends potential for future personalization of treatments targeted to biologically based subgroups of patients with transdiagnostic presentation of symptoms like anhedonia, motor slowing, and anxiety.

Financial Disclosure Dr. Felger previously consulted for Otsuka on a topic unrelated to this work.

References

1. Dowlati Y, Herrmann N, Swardfager W, Liu H, Sham L, Reim EK, et al. A meta-analysis of cytokines in major depression. Biol Psychiatry. 2010;67(5):446–57.
2. Howren MB, Lamkin DM, Suls J. Associations of depression with C-reactive protein, IL-1, and IL-6: a meta-analysis. Psychosom Med. 2009;71(2):171–86.
3. Raison CL, Rutherford RE, Woolwine BJ, Shuo C, Schettler P, Drake DF, et al. A randomized controlled trial of the tumor necrosis factor antagonist infliximab for treatment-resistant depression: the role of baseline inflammatory biomarkers. JAMA Psychiat. 2013;70(1):31–41.
4. Felger JC, Haroon E, Patel TA, Goldsmith DR, Wommack EC, Woolwine BJ, et al. What does plasma CRP tell us about peripheral and central inflammation in depression? Mol Psychiatry. 2020;25(6):1301–11.
5. Wium-Andersen MK, Orsted DD, Nielsen SF, Nordestgaard BG. Elevated C-reactive protein levels, psychological distress, and depression in 73, 131 individuals. JAMA Psychiat. 2013;70(2):176–84.
6. Gimeno D, Kivimaki M, Brunner EJ, Elovainio M, De Vogli R, Steptoe A, et al. Associations of C-reactive protein and interleukin-6 with cognitive symptoms of depression: 12-year follow-up of the Whitehall II study. Psychol Med. 2009;39(3):413–23.
7. Au B, Smith KJ, Gariepy G, Schmitz N. The longitudinal associations between C-reactive protein and depressive symptoms: evidence from the English Longitudinal Study of Ageing (ELSA). Int J Geriatr Psychiatry. 2015;30(9):976–84.
8. Bondy E, Norton SA, Voss MD, Marks RB, Bourdreaux MJ, Treadway MT, et al. Inflammation is associated with future depressive symptoms among older adults. Brain Behav Immun Health. 2021;13:100226.
9. Costello H, Gould RL, Abrol E, Howard R. Systematic review and meta-analysis of the association between peripheral inflammatory cytokines and generalised anxiety disorder. BMJ Open. 2019;9(7):e027925.
10. Passos IC, Vasconcelos-Moreno MP, Costa LG, Kunz M, Brietzke E, Quevedo J, et al. Inflammatory markers in post-traumatic stress disorder: a systematic review, meta-analysis, and meta-regression. Lancet Psychiatry. 2015;2(11):1002–12.

11. Goldsmith DR, Rapaport MH, Miller BJ. A meta-analysis of blood cytokine network alterations in psychiatric patients: comparisons between schizophrenia, bipolar disorder and depression. Mol Psychiatry. 2016;21(12):1696–709.
12. Munkholm K, Vinberg M, Vedel KL. Cytokines in bipolar disorder: a systematic review and meta-analysis. J Affect Disord. 2013;144(1–2):16–27.
13. Barnes J, Mondelli V, Pariante CM. Genetic contributions of inflammation to depression. Neuropsychopharmacology. 2017;42(1):81–98.
14. Hung YY, Kang HY, Huang KW, Huang TL. Association between toll-like receptors expression and major depressive disorder. Psychiatry Res. 2014;220(1–2):283–6.
15. Hung YY, Lin CC, Kang HY, Huang TL. TNFAIP3, a negative regulator of the TLR signaling pathway, is a potential predictive biomarker of response to antidepressant treatment in major depressive disorder. Brain Behav Immun. 2017;59:265–72.
16. Chen RA, Huang TL, Huang KW, Hung YY. TNFAIP3 mRNA level is associated with psychological anxiety in major depressive disorder. Neuroimmunomodulation. 2017;24(4–5):271–5.
17. Hajebrahimi B, Bagheri M, Hassanshahi G, Nazari M, Bidaki R, Khodadadi H, et al. The adapter proteins of TLRs, TRIF and MYD88, are upregulated in depressed individuals. Int J Psychiatry Clin Pract. 2014;18(1):41–4.
18. Keri S, Szabo C, Kelemen O. Expression of Toll-Like Receptors in peripheral blood mononuclear cells and response to cognitive-behavioral therapy in major depressive disorder. Brain Behav Immun. 2014;40:235–43.
19. Fleshner M, Crane CR. Exosomes, DAMPs and miRNA: features of stress physiology and immune homeostasis. Trends Immunol. 2017;38(10):768–76.
20. Berk M, Williams LJ, Jacka FN, O'Neil A, Pasco JA, Moylan S, et al. So depression is an inflammatory disease, but where does the inflammation come from? BMC Med. 2013;11:200.
21. Furman D, Campisi J, Verdin E, Carrera-Bastos P, Targ S, Franceschi C, et al. Chronic inflammation in the etiology of disease across the life span. Nat Med. 2019;25(12):1822–32.
22. Franzen AD, Lam TT, Williams KR, Nairn AC, Duman RS, Sathyanesan M, et al. Cerebrospinal fluid proteome evaluation in major depressive disorder by mass spectrometry. BMC Psychiatry. 2020;20(1):481.
23. Felger JC, Haroon E, Patel TA, Goldsmith DR, Wommack EC, Woolwine BJ, et al. What does plasma CRP tell us about peripheral and central inflammation in depression? Mol Psychiatry. 2018;25:1301–11.
24. Torres-Platas SG, Cruceanu C, Chen GG, Turecki G, Mechawar N. Evidence for increased microglial priming and macrophage recruitment in the dorsal anterior cingulate white matter of depressed suicides. Brain Behav Immun. 2014;42:50–9.
25. Torres-Platas SG, Comeau S, Rachalski A, Bo GD, Cruceanu C, Turecki G, et al. Morphometric characterization of microglial phenotypes in human cerebral cortex. J Neuroinflammation. 2014;11:12.
26. Pandey GN, Rizavi HS, Ren X, Bhaumik R, Dwivedi Y. Toll-like receptors in the depressed and suicide brain. J Psychiatr Res. 2014;53:62–8.
27. Pandey GN. Inflammatory and innate immune markers of neuroprogression in depressed and teenage suicide brain. Mod Trends Pharmacopsychiatry. 2017;31:79–95.
28. Setiawan E, Wilson AA, Mizrahi R, Rusjan PM, Miler L, Rajkowska G, et al. Role of translocator protein density, a marker of neuroinflammation, in the brain during major depressive episodes. JAMA Psychiat. 2015;72(3):268–75.
29. Hannestad J, Gallezot JD, Schafbauer T, Lim K, Kloczynski T, Morris ED, et al. Endotoxin-induced systemic inflammation activates microglia: [(1)(1)C]PBR28 positron emission tomography in nonhuman primates. NeuroImage. 2012;63(1):232–9.
30. Notter T, Schalbetter SM, Clifton NE, Mattei D, Richetto J, Thomas K, et al. Neuronal activity increases translocator protein (TSPO) levels. Mol Psychiatry. 2021;26(6):2025–37.
31. Nettis MA, Veronese M, Nikkheslat N, Mariani N, Lombardo G, Sforzini L, et al. PET imaging shows no changes in TSPO brain density after IFN-alpha immune challenge in healthy human volunteers. Transl Psychiatry. 2020;10(1):89.

32. Dudek KA, Dion-Albert L, Lebel M, LeClair K, Labrecque S, Tuck E, et al. Molecular adaptations of the blood-brain barrier promote stress resilience vs. depression. Proc Natl Acad Sci U S A. 2020;117(6):3326–36.
33. Arteaga-Henríquez G, Simon MS, Burger B, Weidinger E, Wijkhuijs A, Arolt V, et al. Low-grade inflammation as a predictor of antidepressant and anti-inflammatory therapy response in MDD patients: a systematic review of the literature in combination with an analysis of experimental data collected in the EU-MOODINFLAME Consortium. Front Psych. 2019;10:458.
34. Rapaport MH, Nierenberg AA, Schettler PJ, Kinkead B, Cardoos A, Walker R, et al. Inflammation as a predictive biomarker for response to omega-3 fatty acids in major depressive disorder: a proof-of-concept study. Mol Psychiatry. 2016;21(1):71–9.
35. Uher R, Tansey KE, Dew T, Maier W, Mors O, Hauser J, et al. An inflammatory biomarker as a differential predictor of outcome of depression treatment with escitalopram and nortriptyline. Am J Psychiatry. 2014;171(12):1278–86.
36. Jha MK, Minhajuddin A, Chin-Fatt C, Greer TL, Carmody TJ, Trivedi MH. Sex differences in the association of baseline c-reactive protein (CRP) and acute-phase treatment outcomes in major depressive disorder: findings from the EMBARC study. J Psychiatr Res. 2019;113:165–71.
37. Cattaneo A, Gennarelli M, Uher R, Breen G, Farmer A, Aitchison KJ, et al. Candidate genes expression profile associated with antidepressants response in the GENDEP study: differentiating between baseline 'predictors' and longitudinal 'targets'. Neuropsychopharmacology. 2013;38(3):377–85.
38. Zhang J, Yue Y, Thapa A, Fang J, Zhao S, Shi W, et al. Baseline serum C-reactive protein levels may predict antidepressant treatment responses in patients with major depressive disorder. J Affect Disord. 2019;250:432–8.
39. Cattaneo A, Ferrari C, Uher R, Bocchio-Chiavetto L, Riva MA, Consortium MRCI, et al. Absolute measurements of macrophage migration inhibitory factor and interleukin-1-beta mRNA levels accurately predict treatment response in depressed patients. Int J Neuropsychopharmacol. 2016;19(10):pyw045.
40. Haroon E, Daguanno AW, Woolwine BJ, Goldsmith DR, Baer WM, Wommack EC, et al. Antidepressant treatment resistance is associated with increased inflammatory markers in patients with major depressive disorder. Psychoneuroendocrinology. 2018;95:43–9.
41. Chamberlain SR, Cavanagh J, de Boer P, Mondelli V, Jones DNC, Drevets WC, et al. Treatment-resistant depression and peripheral C-reactive protein. Br J Psychiatry. 2018;214:11–9.
42. Jha MK, Minhajuddin A, Gadad BS, Greer T, Grannemann B, Soyombo A, et al. Can C-reactive protein inform antidepressant medication selection in depressed outpatients? Findings from the CO-MED trial. Psychoneuroendocrinology. 2017;78:105–13.
43. Yang JJ, Wang N, Yang C, Shi JY, Yu HY, Hashimoto K. Serum interleukin-6 is a predictive biomarker for ketamine's antidepressant effect in treatment-resistant patients with major depression. Biol Psychiatry. 2015;77(3):e19–20.
44. Shelton RC, Pencina MJ, Barrentine LW, Ruiz JA, Fava M, Zajecka JM, et al. Association of obesity and inflammatory marker levels on treatment outcome: results from a double-blind, randomized study of adjunctive L-methylfolate calcium in patients with MDD who are inadequate responders to SSRIs. J Clin Psychiatry. 2015;76(12):1635–41.
45. Kruse JL, Congdon E, Olmstead R, Njau S, Breen EC, Narr KL, et al. Inflammation and improvement of depression following electroconvulsive therapy in treatment-resistant depression. J Clin Psychiatry. 2018;79(2):17m11597.
46. Milaneschi Y, Kappelmann N, Ye Z, Lamers F, Moser S, Jones PB, et al. Association of inflammation with depression and anxiety: evidence for symptom-specificity and potential causality from UK Biobank and NESDA cohorts. Mol Psychiatry. 2021;26(12):7393–402.
47. Swardfager W, Rosenblat JD, Benlamri M, McIntyre RS. Mapping inflammation onto mood: inflammatory mediators of anhedonia. Neurosci Biobehav Rev. 2016;64:148–66.

48. Lucido MJ, Bekhbat M, Goldsmith DR, Treadway MT, Haroon E, Felger JC, et al. Aiding and abetting anhedonia: impact of inflammation on the brain and pharmacological implications. Pharmacol Rev. 2021;73(3):1084–117.
49. Michopoulos V, Powers A, Gillespie CF, Ressler KJ, Jovanovic T. Inflammation in fear- and anxiety-based disorders: PTSD, GAD, and beyond. Neuropsychopharmacology. 2017;42(1):254–70.
50. Ameli R, Luckenbaugh DA, Gould NF, Holmes MK, Lally N, Ballard ED, et al. SHAPS-C: the Snaith-Hamilton pleasure scale modified for clinician administration. PeerJ. 2014;2:e429.
51. Felger JC, Li Z, Haroon E, Woolwine BJ, Jung MY, Hu X, et al. Inflammation is associated with decreased functional connectivity within corticostriatal reward circuitry in depression. Mol Psychiatry. 2016;21(10):1358–65.
52. Jha MK, Miller AH, Minhajuddin A, Trivedi MH. Association of T and non-T cell cytokines with anhedonia: role of gender differences. Psychoneuroendocrinology. 2018;95:1–7.
53. Rengasamy M, Marsland A, McClain L, Kovats T, Walko T, Pan L, et al. Longitudinal relationships of cytokines, depression and anhedonia in depressed adolescents. Brain Behav Immun. 2021;91:74–80.
54. Bekhbat M, Goldsmith DR, Woolwine BJ, Haroon E, Miller AH, Felger JC. Transcriptomic signatures of psychomotor slowing in peripheral blood of depressed patients: evidence for immunometabolic reprogramming. Mol Psychiatry. 2021;26(12):7384–92.
55. Goldsmith DR, Rapaport MH. Inflammation and negative symptoms of schizophrenia: implications for reward processing and motivational deficits. Front Psych. 2020;11:46.
56. Ye Z, Kappelmann N, Moser S, Davey Smith G, Burgess S, Jones PB, et al. Role of inflammation in depression and anxiety: tests for disorder specificity, linearity and potential causality of association in the UK Biobank. EClinicalMedicine. 2021;38:100992.
57. van Eeden WA, El Filali E, van Hemert AM, Carlier IVE, Penninx B, Lamers F, et al. Basal and LPS-stimulated inflammatory markers and the course of anxiety symptoms. Brain Behav Immun. 2021;98:378–87.
58. Gaspersz R, Lamers F, Wittenberg G, Beekman ATF, van Hemert AM, Schoevers RA, et al. The role of anxious distress in immune dysregulation in patients with major depressive disorder. Transl Psychiatry. 2017;7(12):1268.
59. Eyre HA, Air T, Proctor S, Rositano S, Baune BT. A critical review of the efficacy of non-steroidal anti-inflammatory drugs in depression. Prog Neuro-Psychopharmacol Biol Psychiatry. 2015;57:11–6.
60. Rosenblat JD, McIntyre RS. Efficacy and tolerability of minocycline for depression: a systematic review and meta-analysis of clinical trials. J Affect Disord. 2018;227:219–25.
61. Husain MI, Chaudhry IB, Khoso AB, Husain MO, Hodsoll J, Ansari MA, et al. Minocycline and celecoxib as adjunctive treatments for bipolar depression: a multicentre, factorial design randomised controlled trial. Lancet Psychiatry. 2020;7(6):515–27.
62. Dreyer L, Magyari M, Laursen B, Cordtz R, Sellebjerg F, Locht H. Risk of multiple sclerosis during tumour necrosis factor inhibitor treatment for arthritis: a population-based study from DANBIO and the Danish Multiple Sclerosis Registry. Ann Rheum Dis. 2016;75(4):785–6.
63. Miller AH, Haroon E, Felger JC. Therapeutic implications of brain-immune interactions: treatment in translation. Neuropsychopharmacology. 2017;42(1):334–59.
64. Kappelmann N, Lewis G, Dantzer R, Jones PB, Khandaker GM. Antidepressant activity of anti-cytokine treatment: a systematic review and meta-analysis of clinical trials of chronic inflammatory conditions. Mol Psychiatry. 2018;23(2):335–43.
65. Salvadore G, Nash A, Bleys C, Hsu B, Saad Z, Gause A, et al. A double-blind, placebo-controlled, multicenter study of Sirukumab as adjunctive treatment to a monoaminergic antidepressant in adults with major depressive disorder, in ACNP 57th annual meeting: poster session II, Hollywood, FL. Neuropsychopharmacology: official publication of the American College of Neuropsychopharmacology. 2018;43(1):228–382.

66. Lee Y, Mansur RB, Brietzke E, Carmona NE, Subramaniapillai M, Pan Z, et al. Efficacy of adjunctive infliximab vs. placebo in the treatment of anhedonia in bipolar I/II depression. Brain Behav Immun. 2020;88:631–9.
67. Felger JC, Treadway MT. Inflammation effects on motivation and motor activity: role of dopamine. Neuropsychopharmacology. 2017;42(1):216–41.
68. Capuron L, Pagnoni G, Drake DF, Woolwine BJ, Spivey JR, Crowe RJ, et al. Dopaminergic mechanisms of reduced basal ganglia responses to hedonic reward during interferon alfa administration. Arch Gen Psychiatry. 2012;69(10):1044–53.
69. Capuron L, Gumnick JF, Musselman DL, Lawson DH, Reemsnyder A, Nemeroff CB, et al. Neurobehavioral effects of interferon-alpha in cancer patients: phenomenology and paroxetine responsiveness of symptom dimensions. Neuropsychopharmacology. 2002;26(5):643–52.
70. Lasselin J, Lekander M, Benson S, Schedlowski M, Engler H. Sick for science: experimental endotoxemia as a translational tool to develop and test new therapies for inflammation-associated depression. Mol Psychiatry. 2021;26(8):3672–83.
71. Capuron L, Pagnoni G, Demetrashvili MF, Lawson DH, Fornwalt FB, Woolwine B, et al. Basal ganglia hypermetabolism and symptoms of fatigue during interferon-alpha therapy. Neuropsychopharmacology. 2007;32(11):2384–92.
72. Juengling FD, Ebert D, Gut O, Engelbrecht MA, Rasenack J, Nitzsche EU, et al. Prefrontal cortical hypometabolism during low-dose interferon alpha treatment. Psychopharmacology. 2000;152(4):383–9.
73. Mentis MJ, McIntosh AR, Perrine K, Dhawan V, Berlin B, Feigin A, et al. Relationships among the metabolic patterns that correlate with mnemonic, visuospatial, and mood symptoms in Parkinson's disease. Am J Psychiatry. 2002;159(5):746–54.
74. Wichmann T, DeLong MR. Functional neuroanatomy of the basal ganglia in Parkinson's disease. Adv Neurol. 2003;91:9–18.
75. Haroon E, Felger JC, Woolwine BJ, Chen X, Parekh S, Spivey JR, et al. Age-related increases in basal ganglia glutamate are associated with TNF, reduced motivation and decreased psychomotor speed during IFN-alpha treatment: preliminary findings. Brain Behav Immun. 2015;46:17–22.
76. Dowell NG, Bouyagoub S, Tibble J, Voon V, Cercignani M, Harrison NA. Interferon-alpha-induced changes in NODDI predispose to the development of fatigue. Neuroscience. 2019;403:111–7.
77. Dowell NG, Cooper EA, Tibble J, Voon V, Critchley HD, Cercignani M, et al. Acute changes in striatal microstructure predict the development of interferon-alpha induced fatigue. Biol Psychiatry. 2016;79(4):320–8.
78. Eisenberger NI, Berkman ET, Inagaki TK, Rameson LT, Mashal NM, Irwin MR. Inflammation-induced anhedonia: endotoxin reduces ventral striatum responses to reward. Biol Psychiatry. 2010;68(8):748–54.
79. Moieni M, Irwin MR, Jevtic I, Olmstead R, Breen EC, Eisenberger NI. Sex differences in depressive and socioemotional responses to an inflammatory challenge: implications for sex differences in depression. Neuropsychopharmacology. 2015;40(7):1709–16.
80. Harrison NA, Voon V, Cercignani M, Cooper EA, Pessiglione M, Critchley HD. A neurocomputational account of how inflammation enhances sensitivity to punishments versus rewards. Biol Psychiatry. 2016;80(1):73–81.
81. Treadway MT, Admon R, Arulpragasam AR, Mehta M, Douglas S, Vitaliano G, et al. Association between interleukin-6 and striatal prediction-error signals following acute stress in healthy female participants. Biol Psychiatry. 2017;82(8):570–7.
82. Brydon L, Harrison NA, Walker C, Steptoe A, Critchley HD. Peripheral inflammation is associated with altered substantia nigra activity and psychomotor slowing in humans. Biol Psychiatry. 2008;63(11):1022–9.
83. Harrison NA, Cercignani M, Voon V, Critchley HD. Effects of inflammation on hippocampus and substantia nigra responses to novelty in healthy human participants. Neuropsychopharmacology. 2015;40(4):831–8.

84. Dipasquale O, Cooper EA, Tibble J, Voon V, Baglio F, Baselli G, et al. Interferon-alpha acutely impairs whole-brain functional connectivity network architecture - a preliminary study. Brain Behav Immun. 2016;58:31–9.
85. Harrison NA, Brydon L, Walker C, Gray MA, Steptoe A, Critchley HD. Inflammation causes mood changes through alterations in subgenual cingulate activity and mesolimbic connectivity. Biol Psychiatry. 2009;66(5):407–14.
86. Felger JC. Imaging the role of inflammation in mood and anxiety-related disorders. Curr Neuropharmacol. 2018;16(5):533–58.
87. Davies KA, Cooper E, Voon V, Tibble J, Cercignani M, Harrison NA. Interferon and anti-TNF therapies differentially modulate amygdala reactivity which predicts associated bidirectional changes in depressive symptoms. Mol Psychiatry. 2021;26(9):5150–60.
88. Inagaki TK, Muscatell KA, Irwin MR, Cole SW, Eisenberger NI. Inflammation selectively enhances amygdala activity to socially threatening images. NeuroImage. 2012;59(4):3222–6.
89. Capuron L, Pagnoni G, Demetrashvili M, Woolwine BJ, Nemeroff CB, Berns GS, et al. Anterior cingulate activation and error processing during interferon-alpha treatment. Biol Psychiatry. 2005;58(3):190–6.
90. Haroon E, Woolwine B, Chen X, Pace T, Parekh S, Spivey J, et al. IFN-alpha-induced cortical and subcortical glutamate changes assessed by magnetic resonance spectroscopy. Neuropsychopharmacology. 2014;39(7):1777–85.
91. Eisenberger NI, Lieberman MD, Satpute AB. Personality from a controlled processing perspective: an fMRI study of neuroticism, extraversion, and self-consciousness. Cogn Affect Behav Neurosci. 2005;5(2):169–81.
92. Harrison NA, Brydon L, Walker C, Gray MA, Steptoe A, Dolan RJ, et al. Neural origins of human sickness in interoceptive responses to inflammation. Biol Psychiatry. 2009;66(5):415–22.
93. Hannestad J, Subramanyam K, Dellagioia N, Planeta-Wilson B, Weinzimmer D, Pittman B, et al. Glucose metabolism in the insula and cingulate is affected by systemic inflammation in humans. J Nucl Med. 2012;53(4):601–7.
94. Whitton AE, Treadway MT, Pizzagalli DA. Reward processing dysfunction in major depression, bipolar disorder and schizophrenia. Curr Opin Psychiatry. 2015;28(1):7–12.
95. Kaiser RH, Andrews-Hanna JR, Wager TD, Pizzagalli DA. Large-scale network dysfunction in major depressive disorder: a meta-analysis of resting-state functional connectivity. JAMA Psychiat. 2015;72(6):603–11.
96. Cullen KR, Westlund MK, Klimes-Dougan B, Mueller BA, Houri A, Eberly LE, et al. Abnormal amygdala resting-state functional connectivity in adolescent depression. JAMA Psychiat. 2014;71(10):1138–47.
97. Yin L, Xu X, Chen G, Mehta ND, Haroon E, Miller AH, et al. Inflammation and decreased functional connectivity in a widely-distributed network in depression: centralized effects in the ventral medial prefrontal cortex. Brain Behav Immun. 2019;80:657–66.
98. Haber SN, Knutson B. The reward circuit: linking primate anatomy and human imaging. Neuropsychopharmacology. 2010;35(1):4–26.
99. Samanez-Larkin GR, Knutson B. Decision making in the ageing brain: changes in affective and motivational circuits. Nat Rev Neurosci. 2015;16(5):278–89.
100. Rengasamy M, Brundin L, Griffo A, Panny B, Capan C, Forton C, et al. Cytokine and reward circuitry relationships in treatment-resistant depression. Biol Psych Global Open Sci. 2022;2(1):45–53.
101. Mehta ND, Stevens JS, Li Z, Gillespie CF, Fani N, Michopoulos V, et al. Inflammation, reward circuitry and symptoms of anhedonia and PTSD in trauma-exposed women. Soc Cogn Affect Neurosci. 2020;15(10):1046–55.
102. Burrows K, Stewart JL, Kuplicki R, Figueroa-Hall L, Spechler PA, Zheng H, et al. Elevated peripheral inflammation is associated with attenuated striatal reward anticipation in major depressive disorder. Brain Behav Immun. 2021;93:214–25.

103. Costi S, Morris LS, Collins A, Fernandez NF, Patel M, Xie H, et al. Peripheral immune cell reactivity and neural response to reward in patients with depression and anhedonia. Transl Psychiatry. 2021;11(1):565.
104. Haroon E, Fleischer CC, Felger JC, Chen X, Woolwine BJ, Patel T, et al. Conceptual convergence: increased inflammation is associated with increased basal ganglia glutamate in patients with major depression. Mol Psychiatry. 2016;21(10):1351–7.
105. Haroon E, Chen X, Li Z, Patel T, Woolwine BJ, Hu XP, et al. Increased inflammation and brain glutamate define a subtype of depression with decreased regional homogeneity, impaired network integrity, and anhedonia. Transl Psychiatry. 2018;8(1):189.
106. Tang Y, Kong L, Wu F, Womer F, Jiang W, Cao Y, et al. Decreased functional connectivity between the amygdala and the left ventral prefrontal cortex in treatment-naive patients with major depressive disorder: a resting-state functional magnetic resonance imaging study. Psychol Med. 2013;43(9):1921–7.
107. Stevens JS, Jovanovic T, Fani N, Ely TD, Glover EM, Bradley B, et al. Disrupted amygdala-prefrontal functional connectivity in civilian women with posttraumatic stress disorder. J Psychiatr Res. 2013;47(10):1469–78.
108. Kim MJ, Gee DG, Loucks RA, Davis FC, Whalen PJ. Anxiety dissociates dorsal and ventral medial prefrontal cortex functional connectivity with the amygdala at rest. Cereb Cortex. 2011;21(7):1667–73.
109. Mehta ND, Haroon E, Xu X, Woolwine BJ, Li Z, Felger JC. Inflammation negatively correlates with amygdala-ventromedial prefrontal functional connectivity in association with anxiety in patients with depression: preliminary results. Brain Behav Immun. 2018;73:725–30.
110. Swartz JR, Carranza AF, Tully LM, Knodt AR, Jiang J, Irwin MR, et al. Associations between peripheral inflammation and resting state functional connectivity in adolescents. Brain Behav Immun. 2021;95:96–105.
111. Nusslock R, Brody GH, Armstrong CC, Carroll AL, Sweet LH, Yu T, et al. Higher peripheral inflammatory Signaling associated with lower resting-state functional brain connectivity in emotion regulation and central executive networks. Biol Psychiatry. 2019;86(2):153–62.
112. Deakin B, Suckling J, Barnes TRE, Byrne K, Chaudhry IB, Dazzan P, et al. The benefit of minocycline on negative symptoms of schizophrenia in patients with recent-onset psychosis (BeneMin): a randomised, double-blind, placebo-controlled trial. Lancet Psychiatry. 2018;5(11):885–94.
113. Zheng W, Cai DB, Yang XH, Ungvari GS, Ng CH, Muller N, et al. Adjunctive celecoxib for schizophrenia: a meta-analysis of randomized, double-blind, placebo-controlled trials. J Psychiatr Res. 2017;92:139–46.
114. Xiang YQ, Zheng W, Wang SB, Yang XH, Cai DB, Ng CH, et al. Adjunctive minocycline for schizophrenia: a meta-analysis of randomized controlled trials. Eur Neuropsychopharmacol. 2017;27(1):8–18.
115. Cai DB, Zheng W, Zhang QE, Ng CH, Ungvari GS, Huang X, et al. Minocycline for depressive symptoms: a meta-analysis of randomized, double-blinded, placebo-controlled trials. Psychiatry Q. 2020;91(2):451–61.
116. Liao Y, Xie B, Zhang H, He Q, Guo L, Subramaniapillai M, et al. Efficacy of omega-3 PUFAs in depression: a meta-analysis. Transl Psychiatry. 2019;9(1):190.
117. Kohler-Forsberg O, Nicolaisen Lydholm C, Hjorthoj C, Nordentoft M, Mors O, Benros ME. Efficacy of anti-inflammatory treatment on major depressive disorder or depressive symptoms: meta-analysis of clinical trials. Acta Psychiatr Scand. 2019;139(5):404–19.
118. Miller AH, Raison CL. Are anti-inflammatory therapies viable treatments for psychiatric disorders? Where the rubber meets the road. JAMA Psychiatry. 2015;72(6):527–8.
119. Thompson KG, Rainer BM, Antonescu C, Florea L, Mongodin EF, Kang S, et al. Minocycline and its impact on microbial dysbiosis in the skin and gastrointestinal tract of acne patients. Ann Dermatol. 2020;32(1):21–30.
120. Nettis MA, Lombardo G, Hastings C, Zajkowska Z, Mariani N, Nikkheslat N, et al. Augmentation therapy with minocycline in treatment-resistant depression patients with low-grade peripheral inflammation: results from a double-blind randomised clinical trial. Neuropsychopharmacology. 2021;46(5):939–48.

121. Savitz JB, Teague TK, Misaki M, Macaluso M, Wurfel BE, Meyer M, et al. Treatment of bipolar depression with minocycline and/or aspirin: an adaptive, 2x2 double-blind, randomized, placebo-controlled, phase IIA clinical trial. Transl Psychiatry. 2018;8(1):27.
122. Kitagami T, Yamada K, Miura H, Hashimoto R, Nabeshima T, Ohta T. Mechanism of systemically injected interferon-alpha impeding monoamine biosynthesis in rats: role of nitric oxide as a signal crossing the blood-brain barrier. Brain Res. 2003;978(1–2):104–14.
123. Zhu CB, Lindler KM, Owens AW, Daws LC, Blakely RD, Hewlett WA. Interleukin-1 receptor activation by systemic lipopolysaccharide induces behavioral despair linked to MAPK regulation of CNS serotonin transporters. Neuropsychopharmacology. 2010;35(13):2510–20.
124. Menard C, Pfau ML, Hodes GE, Kana V, Wang VX, Bouchard S, et al. Social stress induces neurovascular pathology promoting depression. Nat Neurosci. 2017;20(12):1752–60.
125. Clayton AH, Croft HA, Horrigan JP, Wightman DS, Krishen A, Richard NE, et al. Bupropion extended release compared with escitalopram: effects on sexual functioning and antidepressant efficacy in 2 randomized, double-blind, placebo-controlled studies. J Clin Psychiatry. 2006;67(5):736–46.
126. Randall PA, Lee CA, Podurgiel SJ, Hart E, Yohn SE, Jones M, et al. Bupropion increases selection of high effort activity in rats tested on a progressive ratio/chow feeding choice procedure: implications for treatment of effort-related motivational symptoms. Int J Neuropsychopharmacol. 2015;18(2):pyu017.
127. Tomarken AJ, Dichter GS, Freid C, Addington S, Shelton RC. Assessing the effects of bupropion SR on mood dimensions of depression. J Affect Disord. 2004;78(3):235–41.
128. Thase ME, Clayton AH, Haight BR, Thompson AH, Modell JG, Johnston JA. A double-blind comparison between bupropion XL and venlafaxine XR: sexual functioning, antidepressant efficacy, and tolerability. J Clin Psychopharmacol. 2006;26(5):482–8.
129. Yohn SE, Thompson C, Randall PA, Lee CA, Muller CE, Baqi Y, et al. The VMAT-2 inhibitor tetrabenazine alters effort-related decision making as measured by the T-maze barrier choice task: reversal with the adenosine A2A antagonist MSX-3 and the catecholamine uptake blocker bupropion. Psychopharmacology. 2015;232(7):1313–23.
130. Soder HE, Cooper JA, Lopez-Gamundi P, Hoots JK, Nunez C, Lawlor VM, et al. Dose-response effects of d-amphetamine on effort-based decision-making and reinforcement learning. Neuropsychopharmacology. 2021;46(6):1078–85.
131. Mar Fan HG, Clemons M, Xu W, Chemerynsky I, Breunis H, Braganza S, et al. A randomised, placebo-controlled, double-blind trial of the effects of d-methylphenidate on fatigue and cognitive dysfunction in women undergoing adjuvant chemotherapy for breast cancer. Support Care Cancer. 2008;16(6):577–83.
132. Moraska AR, Sood A, Dakhil SR, Sloan JA, Barton D, Atherton PJ, et al. Phase III, randomized, double-blind, placebo-controlled study of long-acting methylphenidate for cancer-related fatigue: North Central Cancer Treatment Group NCCTG-N05C7 trial. J Clin Oncol. 2010;28(23):3673–9.
133. Butler JM Jr, Case LD, Atkins J, Frizzell B, Sanders G, Griffin P, et al. A phase III, double-blind, placebo-controlled prospective randomized clinical trial of d-threo-methylphenidate HCl in brain tumor patients receiving radiation therapy. Int J Radiat Oncol Biol Phys. 2007;69(5):1496–501.
134. Sugawara Y, Akechi T, Shima Y, Okuyama T, Akizuki N, Nakano T, et al. Efficacy of methylphenidate for fatigue in advanced cancer patients: a preliminary study. Palliat Med. 2002;16(3):261–3.
135. Pucci E, Branas P, D'Amico R, Giuliani G, Solari A, Taus C. Amantadine for fatigue in multiple sclerosis. Cochrane Database Syst Rev (Online). 2007;(1):CD002818.
136. Stankoff B, Waubant E, Confavreux C, Edan G, Debouverie M, Rumbach L, et al. Modafinil for fatigue in MS: a randomized placebo-controlled double-blind study. Neurology. 2005;64(7):1139–43.

137. Bruera E, Yennurajalingam S, Palmer JL, Perez-Cruz PE, Frisbee-Hume S, Allo JA, et al. Methylphenidate and/or a nursing telephone intervention for fatigue in patients with advanced cancer: a randomized, placebo-controlled, phase II trial. J Clin Oncol Off J Am Soc Clin Oncol. 2013;31(19):2421–7.
138. Ruddy KJ, Barton D, Loprinzi CL. Laying to rest psychostimulants for cancer-related fatigue? J Clin Oncol Off J Am Soc Clin Oncol. 2014;32(18):1865–7.
139. Escalante CP, Meyers C, Reuben JM, Wang X, Qiao W, Manzullo E, et al. A randomized, double-blind, 2-period, placebo-controlled crossover trial of a sustained-release methylphenidate in the treatment of fatigue in cancer patients. Cancer J. 2014;20(1):8–14.
140. Gong S, Sheng P, Jin H, He H, Qi E, Chen W, et al. Effect of methylphenidate in patients with cancer-related fatigue: a systematic review and meta-analysis. PLoS One. 2014;9(1):e84391.
141. Patkar AA, Masand PS, Pae C-U, Peindl K, Hooper-Wood C, Mannelli P, et al. A randomized, double-blind, placebo-controlled trial of augmentation with an extended release formulation of methylphenidate in outpatients with treatment-resistant depression. J Clin Psychopharmacol. 2006;26(6):653–6.
142. Ravindran AV, Kennedy SH, O'Donovan MC, Fallu A, Camacho F, Binder CE. Osmotic-release oral system methylphenidate augmentation of antidepressant monotherapy in major depressive disorder: results of a double-blind, randomized, placebo-controlled trial. J Clin Psychiatry. 2008;69(1):87.
143. Michelson D, Adler LA, Amsterdam JD, Dunner DL, Nierenberg AA, Reimherr FW, et al. Addition of atomoxetine for depression incompletely responsive to sertraline: a randomized, double-blind, placebo-controlled study. J Clin Psychiatry. 2007;68(4):582–7.
144. Pary R, Scarff JR, Jijakli A, Tobias C, Lippmann S. A review of psychostimulants for adults with depression. Fed Pract. 2015;32(Suppl 3):30S–7S.
145. Dunlop BW, Crits-Christoph P, Evans DL, Hirschowitz J, Solvason HB, Rickels K, et al. Coadministration of modafinil and a selective serotonin reuptake inhibitor from the initiation of treatment of major depressive disorder with fatigue and sleepiness: a double-blind, placebo-controlled study. J Clin Psychopharmacol. 2007;27(6):614–9.
146. Felger JC, Li L, Marvar PJ, Woolwine BJ, Harrison DG, Raison CL, et al. Tyrosine metabolism during interferon-alpha administration: association with fatigue and CSF dopamine concentrations. Brain Behav Immun. 2013;31:153–60.
147. Zoller H, Schloegl A, Schroecksnadel S, Vogel W, Fuchs D. Interferon-alpha therapy in patients with hepatitis C virus infection increases plasma phenylalanine and the phenylalanine to tyrosine ratio. J Interferon Cytokine Res. 2012;32(5):216–20.
148. Cunnington C, Channon KM. Tetrahydrobiopterin: pleiotropic roles in cardiovascular pathophysiology. Heart. 2010;96(23):1872–7.
149. Papakostas GI, Petersen T, Mischoulon D, Ryan JL, Nierenberg AA, Bottiglieri T, et al. Serum folate, vitamin B12, and homocysteine in major depressive disorder, part 1: predictors of clinical response in fluoxetine-resistant depression. J Clin Psychiatry. 2004;65(8):1090–5.
150. Papakostas GI, Shelton RC, Zajecka JM, Etemad B, Rickels K, Clain A, et al. L-methylfolate as adjunctive therapy for SSRI-resistant major depression: results of two randomized, double-blind, parallel-sequential trials. Am J Psychiatry. 2012;169(12):1267–74.
151. Sarris J, Price LH, Carpenter LL, Tyrka AR, Ng CH, Papakostas GI, et al. Is S-adenosyl methionine (SAMe) for depression only effective in males? A re-analysis of data from a randomized clinical trial. Pharmacopsychiatry. 2015;48(4–5):141–4.
152. Felger JC, Mun J, Kimmel HL, Nye JA, Drake DF, Hernandez CR, et al. Chronic interferon-alpha decreases dopamine 2 receptor binding and striatal dopamine release in association with anhedonia-like behavior in nonhuman primates. Neuropsychopharmacology: official publication of the American college of. Neuropsychopharmacology. 2013;38(11):2179–87.
153. Felger JC, Hernandez CR, Miller AH. Levodopa reverses cytokine-induced reductions in striatal dopamine release. Int J Neuropsychopharmacol. 2015;18(4):1–5.
154. Czernecki V, Pillon B, Houeto JL, Pochon JB, Levy R, Dubois B. Motivation, reward, and Parkinson's disease: influence of dopatherapy. Neuropsychologia. 2002;40(13):2257–67.

155. Rutherford BR, Slifstein M, Chen C, Abi-Dargham A, Brown PJ, Wall MW, et al. Effects of L-DOPA monotherapy on psychomotor speed and [(11)C]Raclopride binding in high-risk older adults with depression. Biol Psychiatry. 2019;86(3):221–9.
156. Franco-Chaves JA, Mateus CF, Luckenbaugh DA, Martinez PE, Mallinger AG, Zarate CA Jr. Combining a dopamine agonist and selective serotonin reuptake inhibitor for the treatment of depression: a double-blind, randomized pilot study. J Affect Disord. 2013;149(1–3):319–25.
157. Cusin C, Iovieno N, Iosifescu DV, Nierenberg AA, Fava M, Rush AJ, et al. A randomized, double-blind, placebo-controlled trial of pramipexole augmentation in treatment-resistant major depressive disorder. J Clin Psychiatry. 2013;74(7):636–41.
158. Corrigan MH, Denahan AQ, Wright CE, Ragual RJ, Evans DL. Comparison of pramipexole, fluoxetine, and placebo in patients with major depression. Depress Anxiety. 2000;11(2):58–65.
159. Cassano P, Lattanzi L, Fava M, Navari S, Battistini G, Abelli M, et al. Ropinirole in treatment-resistant depression: a 16-week pilot study. Can J Psychiatr. 2005;50(6):357–60.
160. Escalona R, Fawcett J. Pramipexole in treatment resistant-depression, possible role of inflammatory cytokines. Neuropsychopharmacology. 2017;42(1):363.
161. Iravani MM, Sadeghian M, Leung CC, Tel BC, Rose S, Schapira AH, et al. Continuous subcutaneous infusion of pramipexole protects against lipopolysaccharide-induced dopaminergic cell death without affecting the inflammatory response. Exp Neurol. 2008;212(2):522–31.
162. Nelson JC, Papakostas GI. Atypical antipsychotic augmentation in major depressive disorder: a meta-analysis of placebo-controlled randomized trials. Am J Psychiatr. 2009;166(9):980–91.
163. Zhou X, Keitner GI, Qin B, Ravindran AV, Bauer M, Del Giovane C, et al. Atypical antipsychotic augmentation for treatment-resistant depression: a systematic review and network meta-analysis. Int J Neuropsychopharmacol. 2015;18(11):pyv060.
164. Admon R, Kaiser RH, Dillon DG, Beltzer M, Goer F, Olson DP, et al. Dopaminergic enhancement of striatal response to reward in major depression. Am J Psychiatry. 2017;174(4):378–86.
165. Dantzer R, Walker AK. Is there a role for glutamate-mediated excitotoxicity in inflammation-induced depression? J Neural Transm. 2014;
166. Dantzer R, O'Connor JC, Lawson MA, Kelley KW. Inflammation-associated depression: from serotonin to kynurenine. Psychoneuroendocrinology. 2011;36(3):426–36.
167. Haroon E, Welle JR, Woolwine BJ, Goldsmith DR, Baer W, Patel T, et al. Associations among peripheral and central kynurenine pathway metabolites and inflammation in depression. Neuropsychopharmacology. 2020;45(6):998–1007.
168. O'Connor JC, Lawson MA, Andre C, Briley EM, Szegedi SS, Lestage J, et al. Induction of IDO by bacille Calmette-Guerin is responsible for development of murine depressive-like behavior. J Immunol. 2009;182(5):3202–12.
169. O'Connor JC, Lawson MA, Andre C, Moreau M, Lestage J, Castanon N, et al. Lipopolysaccharide-induced depressive-like behavior is mediated by indoleamine 2,3-dioxygenase activation in mice. Mol Psychiatry. 2008;
170. Katz JB, Muller AJ, Prendergast GC. Indoleamine 2,3-dioxygenase in T-cell tolerance and tumoral immune escape. Immunol Rev. 2008;222:206–21.
171. Walker AK, Wing EE, Banks WA, Dantzer R. Leucine competes with kynurenine for blood-to-brain transport and prevents lipopolysaccharide-induced depression-like behavior in mice. Mol Psychiatry. 2019;24(10):1523–32.
172. Tilleux S, Hermans E. Neuroinflammation and regulation of glial glutamate uptake in neurological disorders. J Neurosci Res. 2007;85(10):2059–70.
173. Ida T, Hara M, Nakamura Y, Kozaki S, Tsunoda S, Ihara H. Cytokine-induced enhancement of calcium-dependent glutamate release from astrocytes mediated by nitric oxide. Neurosci Lett. 2008;432(3):232–6.
174. Takaki J, Fujimori K, Miura M, Suzuki T, Sekino Y, Sato K. L-glutamate released from activated microglia downregulates astrocytic L-glutamate transporter expression in neuroinflammation: the 'collusion' hypothesis for increased extracellular L-glutamate concentration in neuroinflammation. J Neuroinflammation. 2012;9:275.

175. Schwarcz R, Pellicciari R. Manipulation of brain kynurenines: glial targets, neuronal effects, and clinical opportunities. J Pharmacol Exp Ther. 2002;303(1):1–10.
176. Tavares RG, Schmidt AP, Abud J, Tasca CI, Souza DO. In vivo quinolinic acid increases synaptosomal glutamate release in rats: reversal by guanosine. Neurochem Res. 2005;30(4):439–44.
177. Walker AK, Budac DP, Bisulco S, Lee AW, Smith RA, Beenders B, et al. NMDA receptor blockade by ketamine abrogates lipopolysaccharide-induced depressive-like behavior in C57BL/6J mice. Neuropsychopharmacology. 2013;38(9):1609–16.
178. Walker AJ, Foley BM, Sutor SL, McGillivray JA, Frye MA, Tye SJ. Peripheral proinflammatory markers associated with ketamine response in a preclinical model of antidepressant-resistance. Behav Brain Res. 2015;293:198–202.
179. Kiraly DD, Horn SR, Van Dam NT, Costi S, Schwartz J, Kim-Schulze S, et al. Altered peripheral immune profiles in treatment-resistant depression: response to ketamine and prediction of treatment outcome. Transl Psychiatry. 2017;7(3):e1065.
180. Anderson A, Iosifescu DV, Jacobson M, Jones A, Kennon K, O'Gorman C, et al. in ASCP Annual Meeting 28–31 (2019).
181. Fumagalli E, Funicello M, Rauen T, Gobbi M, Mennini T. Riluzole enhances the activity of glutamate transporters GLAST, GLT1 and EAAC1. Eur J Pharmacol. 2008;578(2–3):171–6.
182. Zarate CA Jr, Payne JL, Quiroz JA, Sporn J, Denicoff KK, Luckenbaugh DA, et al. An open-label trial of riluzole in patients with treatment-resistant major depression. Am J Psychiatr. 2004;161(1):171–4.
183. Sanacora G, Kendell SF, Levin Y, Simen AA, Fenton LR, Coric V, et al. Preliminary evidence of riluzole efficacy in antidepressant-treated patients with residual depressive symptoms. Biol Psychiatry. 2007;61(6):822–5.
184. Manabe T, Togashi H, Uchida N, Suzuki SC, Hayakawa Y, Yamamoto M, et al. Loss of cadherin-11 adhesion receptor enhances plastic changes in hippocampal synapses and modifies behavioral responses. Mol Cell Neurosci. 2000;15(6):534–46.
185. Iwakura Y, Ishigame H. The IL-23/IL-17 axis in inflammation. J Clin Invest. 2006;116(5):1218–22.
186. Steed PM, Tansey MG, Zalevsky J, Zhukovsky EA, Desjarlais JR, Szymkowski DE, et al. Inactivation of TNF signaling by rationally designed dominant-negative TNF variants. Science. 2003;301(5641):1895–8.
187. Zalevsky J, Secher T, Ezhevsky SA, Janot L, Steed PM, O'Brien C, et al. Dominant-negative inhibitors of soluble TNF attenuate experimental arthritis without suppressing innate immunity to infection. J Immunol. 2007;179(3):1872–83.
188. Alexopoulou L, Kranidioti K, Xanthoulea S, Denis M, Kotanidou A, Douni E, et al. Transmembrane TNF protects mutant mice against intracellular bacterial infections, chronic inflammation and autoimmunity. Eur J Immunol. 2006;36(10):2768–80.
189. Eidson LN, deSousa Rodrigues ME, Johnson MA, Barnum CJ, Duke BJ, Yang Y, et al. Chronic psychological stress during adolescence induces sex-dependent adulthood inflammation, increased adiposity, and abnormal behaviors that are ameliorated by selective inhibition of soluble tumor necrosis factor with XPro1595. Brain Behav Immun. 2019;81:305–16.
190. Bedrosian TA, Weil ZM, Nelson RJ. Chronic dim light at night provokes reversible depression-like phenotype: possible role for TNF. Mol Psychiatry. 2013;18(8):930–6.
191. Aguilar-Valles A, Haji N, De Gregorio D, Matta-Camacho E, Eslamizade MJ, Popic J, et al. Translational control of depression-like behavior via phosphorylation of eukaryotic translation initiation factor 4E. Nat Commun. 2018;9(1):2459.
192. Price J, Maltais D, Butters K, McGee S, Tye SJ. Antidepressant and anxiolytic effects of tumor necrosis factor inhibition with XPro1595 in a rodent model of antidepressant-resistance. Soc Neurosci. 2018;15:650833.
193. Karson A, Demirtas T, Bayramgurler D, Balci F, Utkan T. Chronic administration of infliximab (TNF-alpha inhibitor) decreases depression and anxiety-like behaviour in rat model of chronic mild stress. Basic Clin Pharmacol Toxicol. 2013;112(5):335–40.

194. Karamita M, Barnum C, Möbius W, Tansey MG, Szymkowski DE, Lassmann H, et al. Therapeutic inhibition of soluble brain TNF promotes remyelination by increasing myelin phagocytosis by microglia. JCI Insight. 2017;2(8):e87455.
195. Gavegnano C, Haile WB, Hurwitz S, Tao S, Jiang Y, Schinazi RF, et al. Baricitinib reverses HIV-associated neurocognitive disorders in a SCID mouse model and reservoir seeding in vitro. J Neuroinflammation. 2019;16(1):182.
196. Hoang TN, Pino M, Boddapati AK, Viox EG, Starke CE, Upadhyay AA, et al. Baricitinib treatment resolves lower-airway macrophage inflammation and neutrophil recruitment in SARS-CoV-2-infected rhesus macaques. Cell. 2021;184(2):460–75 e21.
197. Kalil AC, Patterson TF, Mehta AK, Tomashek KM, Wolfe CR, Ghazaryan V, et al. Baricitinib plus remdesivir for hospitalized adults with Covid-19. N Engl J Med. 2020;384:795–807.
198. Marconi VC, Moser C, Gavegnano C, Deeks SG, Lederman MM, Overton ET, et al. Randomized trial of ruxolitinib in antiretroviral-treated adults with HIV. Clin Infect Dis. 2022;74(1):95–104.
199. Titanji BK, Farley MM, Mehta A, Connor-Schuler R, Moanna A, Cribbs SK, et al. Use of baricitinib in patients with moderate to severe coronavirus disease 2019. Clin Infect Dis. 2021;72(7):1247–50.
200. Urits I, Israel J, Hakobyan H, Yusin G, Lassiter G, Fackler N, et al. Baricitinib for the treatment of rheumatoid arthritis. Reumatologia. 2020;58(6):407–15.
201. Gavegnano C, Detorio M, Montero C, Bosque A, Planelles V, Schinazi RF. Ruxolitinib and tofacitinib are potent and selective inhibitors of HIV-1 replication and virus reactivation in vitro. Antimicrob Agents Chemother. 2014;58(4):1977–86.
202. Haile WB, Gavegnano C, Tao S, Jiang Y, Schinazi RF, Tyor WR. The Janus kinase inhibitor ruxolitinib reduces HIV replication in human macrophages and ameliorates HIV encephalitis in a murine model. Neurobiol Dis. 2016;92(Pt B):137–43.
203. Lynall ME, Turner L, Bhatti J, Cavanagh J, de Boer P, Mondelli V, et al. Peripheral blood cell-stratified subgroups of inflamed depression. Biol Psychiatry. 2020;88(2):185–96.
204. Szollosi DE, Manzoor MK, Aquilato A, Jackson P, Ghoneim OM, Edafiogho IO. Current and novel anti-inflammatory drug targets for inhibition of cytokines and leucocyte recruitment in rheumatic diseases. J Pharm Pharmacol. 2018;70:18–26.
205. Bekhbat M, Treadway MT, Goldsmith DR, Woolwine BJ, Haroon E, Miller AH, et al. Gene signatures in peripheral blood immune cells related to insulin resistance and low tyrosine metabolism define a sub-type of depression with high CRP and anhedonia. Brain Behav Immun. 2020;88:161–5.
206. O'Neill LA, Kishton RJ, Rathmell J. A guide to immunometabolism for immunologists. Nat Rev Immunol. 2016;16(9):553–65.
207. Abdallah CG, Averill LA, Gueorguieva R, Goktas S, Purohit P, Ranganathan M, et al. Modulation of the antidepressant effects of ketamine by the mTORC1 inhibitor rapamycin. Neuropsychopharmacology. 2020;45(6):990–7.
208. Kornberg MD, Bhargava P, Kim PM, Putluri V, Snowman AM, Putluri N, et al. Dimethyl fumarate targets GAPDH and aerobic glycolysis to modulate immunity. Science. 2018;360(6387):449–53.
209. Mocelin R, Marcon M, D'Ambros S, Mattos J, Sachett A, Siebel AM, et al. N-acetylcysteine reverses anxiety and oxidative damage induced by unpredictable chronic stress in zebrafish. Mol Neurobiol. 2019;56(2):1188–95.
210. Wright DJ, Renoir T, Smith ZM, Frazier AE, Francis PS, Thorburn DR, et al. N-acetylcysteine improves mitochondrial function and ameliorates behavioral deficits in the R6/1 mouse model of Huntington's disease. Transl Psychiatry. 2015;5:e492.
211. Zhang Q, Ju Y, Ma Y, Wang T. N-acetylcysteine improves oxidative stress and inflammatory response in patients with community acquired pneumonia: a randomized controlled trial. Medicine. 2018;97(45).
212. Nasca C, Bigio B, Zelli D, de Angelis P, Lau T, Okamoto M, et al. Role of the astroglial glutamate exchanger xCT in ventral hippocampus in resilience to stress. Neuron. 2017;96(2):402–13 e5.

213. Linck VM, Costa-Campos L, Pilz LK, Garcia CR, Elisabetsky E. AMPA glutamate receptors mediate the antidepressant-like effects of N-acetylcysteine in the mouse tail suspension test. Behav Pharmacol. 2012;23(2):171–7.
214. Bower JE, Irwin MR. Mind-body therapies and control of inflammatory biology: a descriptive review. Brain Behav Immun. 2016;51:1–11.
215. Creswell JD, Taren AA, Lindsay EK, Greco CM, Gianaros PJ, Fairgrieve A, et al. Alterations in resting-state functional connectivity link mindfulness meditation with reduced Interleukin-6: a randomized controlled trial. Biol Psychiatry. 2016;80(1):53–61.
216. Irwin MR, Olmstead R, Breen EC, Witarama T, Carrillo C, Sadeghi N, et al. Cognitive behavioral therapy and tai chi reverse cellular and genomic markers of inflammation in late-life insomnia: a randomized controlled trial. Biol Psychiatry. 2015;78(10):721–9.
217. Schuch FB, Vancampfort D, Richards J, Rosenbaum S, Ward PB, Stubbs B. Exercise as a treatment for depression: a meta-analysis adjusting for publication bias. J Psychiatr Res. 2016;77:42–51.
218. Fabricatore AN, Wadden TA, Higginbotham AJ, Faulconbridge LF, Nguyen AM, Heymsfield SB, et al. Intentional weight loss and changes in symptoms of depression: a systematic review and meta-analysis. Int J Obes (Lond). 2011;35(11):1363–76.
219. Forsythe LK, Wallace JM, Livingstone MB. Obesity and inflammation: the effects of weight loss. Nutr Res Rev. 2008;21(2):117–33.
220. Woods JA, Vieira VJ, Keylock KT. Exercise, inflammation, and innate immunity. Immunol Allergy Clin N Am. 2009;29(2):381–93.
221. Kiecolt-Glaser JK, Bennett JM, Andridge R, Peng J, Shapiro CL, Malarkey WB, et al. Yoga's impact on inflammation, mood, and fatigue in breast cancer survivors: a randomized controlled trial. J Clin Oncol Off J Am Soc Clin Oncol. 2014;32(10):1040–9.
222. Tracey KJ. The inflammatory reflex. Nature. 2002;420(6917):853–9.
223. Pavlov VA, Chavan SS, Tracey KJ. Bioelectronic medicine: from preclinical studies on the inflammatory reflex to new approaches in disease diagnosis and treatment. Cold Spring Harb Perspect Med. 2020;10(3):a034140.
224. Bastiaanssen TFS, Cussotto S, Claesson MJ, Clarke G, Dinan TG, Cryan JF. Gutted! Unraveling the role of the microbiome in major depressive disorder. Harv Rev Psychiatry. 2020;28(1):26–39.
225. Rooks MG, Garrett WS. Gut microbiota, metabolites and host immunity. Nat Rev Immunol. 2016;16(6):341–52.
226. Foster JA, Baker GB, Dursun SM. The relationship between the gut microbiome-immune system-brain axis and major depressive disorder. Front Neurol. 2021;12(1660):1–9.
227. Zheng P, Zeng B, Zhou C, Liu M, Fang Z, Xu X, et al. Gut microbiome remodeling induces depressive-like behaviors through a pathway mediated by the host's metabolism. Mol Psychiatry. 2016;21(6):786–96.
228. Kilinçarslan S, Evrensel A. The effect of fecal microbiota transplantation on psychiatric symptoms among patients with inflammatory bowel disease: an experimental study. Actas Esp Psiquiatr. 2020;48(1):1–7.
229. Kurokawa S, Kishimoto T, Mizuno S, Masaoka T, Naganuma M, Liang KC, et al. The effect of fecal microbiota transplantation on psychiatric symptoms among patients with irritable bowel syndrome, functional diarrhea and functional constipation: an open-label observational study. J Affect Disord. 2018;235:506–12.
230. Liu RT, Walsh RFL, Sheehan AE. Prebiotics and probiotics for depression and anxiety: a systematic review and meta-analysis of controlled clinical trials. Neurosci Biobehav Rev. 2019;102:13–23.
231. Cuthbert BN. Research domain criteria: toward future psychiatric nosologies. Dialogues Clin Neurosci. 2015;17(1):89–97.
232. Baudino TA. Targeted cancer therapy: the next generation of cancer treatment. Curr Drug Discov Technol. 2015;12(1):3–20.

Chapter 17
Experimental Medicine Approaches in Early-Phase CNS Drug Development

Brett A. English and Larry Ereshefsky

Abstract Traditionally, Phase 1 clinical trials were largely conducted in healthy normal volunteers and focused on collection of safety, tolerability, and pharmacokinetic data. However, in the CNS therapeutic area, with more drugs failing in later phase development, Phase 1 trials have undergone an evolution that includes incorporation of novel approaches involving novel study designs, inclusion of biomarkers, and early inclusion of patients to improve the pharmacologic understanding of novel CNS-active compounds early in clinical development with the hope of improving success in later phase pivotal trials. In this chapter, the authors will discuss the changing landscape of Phase 1 clinical trials in CNS, including novel trial methodology, inclusion of pharmacodynamic biomarkers, and experimental medicine approaches to inform early decision-making in clinical development.

Keywords Biomarkers · Experimental medicine · Functional magnetic resonance imaging · Electrophysiology · Early phase

17.1 The Evolving Landscape in Early-Phase Clinical Trials in CNS

Drug development for CNS indications continues to represent a significant unmet medical need, with psychiatric and addictive disorders representing 7% of all global burden of disease [1, 2]. In fact, the development of pharmacotherapeutics for CNS diseases has lagged other therapeutic areas with CNS drugs taking up to 20 months longer than drugs in other therapeutic areas to get toward commercial launch, with as many as 50% failing in clinical development prior to Phase 3 [3, 4]. In fact, a subsequent study by the Tufts Center for Drug Development found that success

B. A. English (✉)
Clinical Sciences at CenExel, Brentwood, TN, USA

L. Ereshefsky
Clinical Sciences at CenExel, Marina Del Rey, CA, USA

M. Macaluso et al. (eds.), *Drug Development in Psychiatry*, Advances in Neurobiology 30,
https://doi.org/10.1007/978-3-031-21054-9_17

rates for CNS drugs to achieve final market approval by a regulatory agency was less than half of the approval rates for non-CNS drugs for the period 1995–2007 [3]. This has led to several of the large pharmaceutical companies to either reduce or eliminate discovery and development efforts in the CNS therapeutic area [5, 6].

The challenges to the success rates of CNS drug development are likely multifactorial to include) (1) the poor predictive validity of nonclinical rodent and primate models, (2) poor understanding of the neurobiological substrates underlying psychiatric disorders, (3) heterogeneity in the psychiatric patient population, (4) high placebo response rates, and (5) gaps in the methodologies that permit characterization of pharmacodynamic (PD) effects and their prediction of clinical end points [1, 5, 7]. These challenges have led some pharmaceutical companies to reevaluate their CNS pipeline and to retrospectively review the type and extent of data necessary to improve success through Phase 2 [6]. For example, Pfizer conducted a retrospective review of compounds in early development and found that their Phase 2 success rate was 50% less than the median success rates of other companies [8, 9]. Key findings from their review of 44 programs at Pfizer were that the majority of failures were due to lack of efficacy; however, in 43% of cases, it was not possible to conclude that the lack of clinical success was directly related to the pharmacology [8]. In fact, for those programs that achieved clinical success in Phase 2, those programs had extensive data related to (1) exposure at site of action, (2) target engagement, and (3) expression of functional pharmacologic activity, later termed the "3-pillars of survival" [8]. By focusing on programs that achieved data associated with the "3-pillars" during early development, Pfizer went from 2% success in 2010, to an impressive 21% by 2020 [9, 10].

A similar approach was taken by AstraZeneca after watching their R&D productivity drop below industry averages [11]. The results of their comprehensive review similarly not only identified obtaining data regarding the "right" exposure at the site of action and "right" target engagement improved success in Phase 3 but also identified other successful key determinants, which included selection of the "right" patient, the "right safety profile, and the "right" commercial potential in early development contributed to success [11]. These key technical determinants of success leading to improved Phase 3 completion were labeled the "five-dimensional framework" (5R framework) [11]. Adoption of the 5R-framework model, where obtaining key data related to pharmacology early in development, similarly resulted in improved candidate molecule selection and Phase 3 success, improving from 4% in 2005–10 to 19% in 2012–19 [12].

17.1.1 Challenges with Traditional Approach to Phase 1 Trials

After completion of the initial nonclinical studies and submission of the data to regulatory authorities, the initiation of the Phase 1 (Ph1) clinical study represents an important milestone in the development of a new chemical entity (NCE). Phase 1 clinical studies, also known as "first-in-human" (FiH) studies, have traditionally consisted of single ascending dose (SAD) and multiple ascending dose (MAD) studies, conducted primarily in healthy normal volunteers (HNVs) [13]. The primary objective of the Ph1 SAD/MAD study has primarily focused on characterization of the safety and tolerability profile of the NCE through collection of adverse events, clinical laboratory and vital signs. Secondary objectives of Ph1 studies include characterization of the plasma and urine pharmacokinetic (PK) parameters (e.g., absorption, distribution, metabolism, and excretion; ADME) [14, 15].

The Ph1 SAD and MAD studies consist of small cohorts (e.g. $n = 8$; 6 active: 2 placebo) of HNV subjects that are dosed in a sequential manner, followed by a review of the safety data and subsequent escalation of dose in a new cohort of subjects. Typically, the goal of the Ph1 SAD and MAD study has been to dose escalate up to the maximum tolerated dose (MTD), generally defined as the highest dose that achieves the intended pharmacologic effect without unacceptable adverse events (AEs), although the need to dose the NCE up to MTD in Ph1 studies conducted in healthy populations has been questioned [16–19].

The classical view of Ph1 studies was that of an initial hurdle in the sequential clinical development process prior to getting the drug into the patient population, where the latter would focus on early clinical endpoints in Ph2 proof-of-concept (PoC) studies [20]. These traditional Ph1 studies focus on the characterization of MTD for identifying the Ph2 recommended dose (P2RD), often only in HNVs, of which the MTD dose may not be the optimal dose, particularly in the target patient population [21, 22]. For drugs that are being developed for CNS indications, there is disagreement in the field on the need to determine MTD in Ph1 studies, with some points of view proposing that exploring the MTD permits a wider range of doses to be explored in subsequent development [4, 23, 24]. Lastly, some regulatory agencies such as the European Medicines Agency (EMA) have issued recent guidance regarding FiH studies that have questioned the ethics of pushing doses up to MTD in HNVs [25].

Outside collection of safety, tolerability, PK data, and a handful of rating scales, few early Ph1 study protocols included biomarker (BM) end points that may further characterize the understanding of the pharmacology of the novel drug in terms of target engagement and expression of pharmacology, or adverse event profile [26]. This was largely due to the lack of selective, reliable, or predictive BMs that could be utilized during Ph1 development to inform Ph2 PoC study considerations such as dose, patient subpopulation, etc. Thus, the traditional approach to early-phase development while obtaining initial safety and PK data left many unanswered questions regarding a novel drug's pharmacology, which can lead to costly failures in subsequent Ph2 studies. However, advances in various fluid biomarker platforms (e.g. genomics, proteomics, and metabolomics), electrophysiology, and imaging involving academia, pharmaceutical industry, and various public-private initiatives have expanded the list of BM options that can be explored during Ph1 studies and potentially serve to validate those from the nonclinical data [27–30]. The following sections outline evolving trends that include novel study designs, inclusion of patient population in Ph1 trials, and inclusion of BMs during the SAD and MAD study.

17.1.2 Evolution of Phase 1 Study Designs and Concepts

As previously mentioned, Phase 1 study designs generally are simple, single, or multiple, sequential-dose designs where subsequent escalations to higher doses are achieved based upon the planned dose escalation scheme after review of safety data [31–33]. However, other Ph1 study designs have included grouped cross-over dose escalation, alternating cross-over, algorithm-based "3 + 3" design, and the model-based continuous reassessment method (CRM), with the latter using a Bayesian statistical approach to model dose escalations [14, 34, 35]. While most of the novelty in study design methodology in Ph1 studies has been applied to oncology, the statistical methods and recommendations for implementation in early-phase protocols have been published [36, 37].

The US Food and Drug Administration (FDA) and the European Medicines Agency have issued regulatory guidance on the use of adaptive design methods in clinical trials. Unlike traditional parallel-dose studies where studies are conducted to completion, adaptive design studies allow numerous approaches to be modified during study conduct that may include (a) changes in the subject allocation ratio (e.g., active vs placebo), (b) total sample size, (c) modification of eligibility criteria that can be either clinical entry criteria or BM criteria, and (d) treatment arms (e.g., dose groups) that may be dropped or added [38].

While more commonly seen in oncology early-phase studies, some Ph1 studies in CNS have incorporated "adaptive" design considerations for dose escalations based upon collection of ongoing BM data and/or safety data. In the Ph1 single ascending dose study of risdiplam (RG7616), an orally administered survival of motor neuron 2 (SMN2) mRNA slice modifier under development for the treatment of spinal muscular atrophy (SMA), a Bayesian adaptive design approach was taken to guide dose escalations based upon risdiplam's effect on *SMN2* mRNA levels [39]. Similarly, RG7342, a metabotropic glutamate receptor 5 (mGlu5) PAM for the treatment of schizophrenia, a modified CRM using Bayesian approach was taken to avoid exposing subjects to doses above the MTD and guide subsequent dose escalations [40].

In addition to adaptive study designs, some Ph1 programs have included a novel Ph1 type translational (TxM) or experimental (ExM) medicine study concept referred to as "proof-of-mechanism" (PoM) study that is conducted outside of the initial Ph1 SAD/MAD. These Ph1 TxM/ExM PoM studies can be conducted in HNVs, HNVs that are "enriched" for a specific trait (ie. trait anxiety), or in the target patient population. Ph1 PoM studies are primarily focused on characterizing the pharmacodynamic effects of a novel drug using a particular biomarker that has proximal (or distal) effects related to the pharmacology of the study drug, or in a particular "model" of the targeted disease. Unlike Ph1 SAD/MAD studies where the BM or disease-like model is an exploratory objective, in PoM studies, the effect of the novel drug on a particular biomarker or model is the primary objective and serves to compliment the Ph1 SAD/MAD data and address key questions related to target pharmacology and "de-risk" go/no-go decisions supporting P2RD in Ph2. While most of the Ph1 PoM studies have evaluated the PD effects on a specific biomarker that is related to the pharmacology of the study drug, some have evaluated the effect of the study drug on a disease-like model such as scopolamine or ketamine-reversal of cognitive deficits [41–44].

One example of a Ph1 PoM study using electrophysiology as a biomarker was the α7 nicotinic acetylcholine receptor (nAChR) partial agonist, EVP-6124 being developed for cognitive impairment associated with schizophrenia (CIAS). This was a single-center, randomized, parallel-group, double-blind, placebo-controlled study in medically stable patients with schizophrenia. Study assessments included traditional cognitive tests and event-related potential (ERP) measures, specifically mismatch negativity (MMN) and P300, both shown to be disrupted in patients with schizophrenia and correlated with cognitive deficits [45–47]. In addition, the selection of the ERP end points was based upon the pharmacologic principle that reduction of the α7 nAChR was linked to sensory deficits, including P50 [48]. Data demonstrated that both high and low doses of EVP-6124 produced statistically significant improvements in MMN and P300 relative to placebo [49]. The positive data from this trial was used to support a "go" decision for the initiation of a larger Ph2 study in CIAS [50].

17.1.3 Early Inclusion of Patients into the Phase 1 Study

Increasingly, more Ph1 studies are also including small cohorts of patient populations with the targeted indication [17]. These "bridging studies," also referred to as Ph1b studies, can be included as part of the initial Ph1 SAD/MAD, or as a separate study [27]. Particularly for NCEs being developed for psychiatric or neurologic indications, inclusion of a few cohorts of patients is encouraged, as differences in adverse events and tolerability may impact subsequent development of the NCE, by inadvertently selecting doses too low (or too high) for the target population in later trials or early termination of the development program should the AE profile be unacceptable at a specific dose be below that which is anticipated to produce clinical benefit [17, 51, 52]. The scientific rationale for these disease-dependent differences in tolerability are largely unknown; however, differences in the underlying neurobiology of the disease state and exposure to prior medications have been proposed [53]. Lastly, by characterizing the MTD in the intended population during early development, additional PK/PD modeling can be performed to potentially improve dose selections for the Ph2 proof-of-concept (POC) efficacy studies.

Depending on the objectives and timelines of the clinical development plan (CDP) for the specific NCE, the inclusion of patients into the Ph1 studies can either be part of the initial SAD/MAD study or conducted under a separate protocol in a manner that is generally parallel to the MAD in HNVs. For example, in the development of Astella's ASP4345, a selective dopamine receptor-1 (D1) positive allosteric modulator (PAM), two separate Ph1 studies were run in HNVs and patients with schizophrenia, respectively [54]. These studies consisted of a single ascending dose study in HNVs, while the multiple ascending dose study was conducted in patients diagnosed with schizophrenia or schizoaffective disorder [54]. A similar approach was taken during the development of Pfizer's PF-06412562, a selective D1 and D5 partial agonist. In these studies, two separate BM-focused trials were conducted in HNVs enriched for low working memory deficits and in patients with schizophrenia, of which both studies included an extensive battery of neurocognitive testing; the use of EEG/ERPs and fMRI paradigms focused on reward and working memory [55]. Conversely, in the development of the selective phosphodiesterase 10A (PDE10A) inhibitor, TAK-063, a multiple-dose study included both healthy young Japanese subjects and in subjects with schizophrenia within a single protocol [56].

One downside to the inclusion of patients into the Ph1 study is that depending on the inclusion and exclusion criteria, patient populations generally enroll at a slower rate compared with HNVs. As an alternative to the inclusion of patients with the targeted disease into the Ph1 study, some studies have included subgroup of HNVs that serve as a "surrogate" population because they exhibit a characteristic feature of the main disease, but do not have a diagnosed condition [57]. The central tenant of this concept is that psychopathology represents an extreme variation or disruption of normal cognition and behavioral processes. This concept was highlighted by the US National Institute of Mental Health (NIMH) in 2009, upon the rollout of the Research Domain Criteria (RDoC) project (described in detail below), where mental disorders were studied transdiagnostically by breaking down these extremes in cognition and behavior into various domains and constructs that would be evaluated at the circuitry level [58, 59].

This RDoC-inspired approach was first attempted in the Ph1b experimental medicine study of PF-0641256, a D1/D5 partial agonist under development for the treatment of CIAS [55]. Previous studies had demonstrated that working memory and motivational (reward) deficits involved dopaminergic circuits that contained the D1 receptor [60]; thus, a multimodal approach focused on working memory and reward paradigms that included numerous cognitive assessments, ERP (e.g., contralateral delay activity task), and functional magnetic resonance imaging (fMRI) (e.g., N-back task) in HNVs [55]. The HNV population in this study was unique as study participants had to demonstrate low working memory capacity as determined by performance on the operational span task (O-span). The rationale for this subtype-specific HNV population was that (1) many cognition studies using HNVs fail to demonstrate pharmacologic enhancement and (2) low working memory deficits have demonstrated change with pharmacologic challenge [61–63]. While PF-06412562 did not improve cognitive function across the battery of tasks, further research is needed to define the proper "surrogate" population that may inform go/no-go decisions in later patient intervention trials.

17.1.4 Incorporating Biomarkers into the Phase 1 Clinical Development Plan

As an attempt to increase the pharmacologic knowledge of novel drugs during Ph1 (or Ph1b) development to improve success in later phase clinical development, many pharmaceutical companies are incorporating a myriad of biomarkers into these early clinical studies. The National Institutes of Health (NIH), Biomarkers Definitions Working Group, in 1998, defined a biomarker as "*a characteristic that is objectively measured and evaluated as an indicator of normal biological process, pathogenic process, or pharmacologic response to a therapeutic intervention*" [64]. The NIH working group further outlined how BMs could be utilized as (1) a diagnostic tool, (2) staging of disease, (3) indicator of disease prognosis, and (4) prediction or monitoring of clinical response to an intervention [64]. Table 17.1 describes biomarker nomenclature and some of the methods used. Additionally, the NIH Biomarker and Surrogate Endpoint Working Group has identified three classes of biomarkers: Type 0, biomarkers that track the natural course of the disease; Type 1, biomarkers that examine the effects of an intervention of a known mechanism without strict relationship to clinical outcome; and Type 2, biomarkers considered "surrogate end points" and optimally predictive of clinical outcome [74].

The decision to include BMs into the Ph1 studies should be considered as early in the development program of an NCE as possible to ensure proper integration of the BM plan into the CDP, as the selection of a specific BM may inform crucial questions that impact study designs and the role that the BM will play within the Ph1 development plan. During the development of the early CDP, key questions that a BM strategy is intended to address should be outlined as to support go/no-go decision criteria for Ph2 and then operationalized within the Ph1 study protocols. Key questions include, but are not limited to, (1) what role will the biomarker serve within the Ph1 studies (e.g., exploratory or confirmatory), (2) type of biomarker (e.g., pharmacodynamic and safety), (3) translatability of the biomarker from the nonclinical to clinical, (4) the use of multimodal biomarkers in the Ph1 study (e.g., plasma-based and electrophysiology), and (5) feasibility of incorporating into the Ph1 study (e.g., operational logistics and subject burden) [27, 75, 76]. Additionally, if the BM is a pharmacodynamic BM, additional considerations may be included if the measure(s) will demonstrate proof-of-pharmacology, proof-of-mechanism, correlation with a clinical endpoint, etc. [27, 77]. Lastly, based upon the objectives of the Ph1 development plan, the BM strategy may either be included in the initial Ph1 SAD or MAD study, or conducted separately, in an experimental medicine study that can run in a nested fashion, concurrent with the Ph1 SAD/MAD (Fig. 17.1).

17.2 Leveraging Experimental Medicine to Support Early Decision-Making in Early-Phase Trials

As previously mentioned, despite advances in basic and systems neuroscience, including advances in technologies to better understand psychiatric and neurologic disorders, the discovery of novel drugs and their successful clinical development has been challenging. To address these challenges, a new model of early-phase clinical development was introduced: the experimental medicine study that aimed to merge known mechanism-of-action of the NCE and knowledge of the neurobiology

Table 17.1 Biomarker nomenclature [24, 27, 65–73]

Biomarker class	Use type	Definition	Biomarker methods
Susceptibility	*Risk*	Biomarker that indicates the likelihood of developing a disease	Blood/plasma/CSF (e.g., APOE ε4), qEEG, ERP, PSG, sMRI, rsMRI, fMRI, PET
Mechanistic	*Prognostic*	Biomarker that identifies the likelihood of a clinical event and disease recurrence or progression; can be used to stratify patients in clinical trials	Blood/plasma/CSF (e.g., β-amyloid), qEEG, ERP, PSG, sMRI, rsMRI, fMRI, PET
	Diagnostic	Biomarker used to confirm diagnosis of disease	Blood/plasma/CSF (e.g., β-amyloid), qEEG, sMRI, PET
	Monitoring	Biomarker measured longitudinally to assess the status of disease after an intervention	Blood/plasma/CSF (e.g., β-amyloid), sMRI, PET
Intervention	*Predictive*	Biomarker that predicts that a particular patient will respond to the intervention	Blood/plasma/CSF, qEEG, ERP, fMRI, rsMRI
	Enriched	Selection of a specific population predicted to exhibit a response to an intervention	Blood/plasma/CSF (e.g., β-amyloid), qEEG, ERP, PSG, sMRI, rsMRI (e.g., functional connectivity), fMRI (e.g., task based), PET
	Pharmacodynamic	Biomarker that demonstrates a direct or indirect effect on a biological response to an intervention	Blood/plasma/CSF (e.g., β-amyloid), qEEG, ERP, PSG, sMRI, rsMRI (e.g., functional connectivity), fMRI (e.g., task based), PET
	Safety	Biomarker before or after an intervention indicative of toxicity such as an adverse event (e.g., ARIA)	Blood/plasma/CSF (e.g., GFAP), sMRI (e.g., ARIA)
	Surrogate endpoint	Biomarker that serves an indirect measure and predictive of benefit on a clinical end point	Blood/plasma/CSF (e.g., β-amyloid reduction)

Def: *APOE4* apolipoprotein-4, *ARIA* amyloid related imaging abnormality, *CSF* cerebrospinal fluid, *ERP* evoked related potential, *fMRI* functional magnetic resonance imaging, *GFAP* glial fast acidic protein, *PET* positron emission tomography, *PSG* polysomnography, *qEEG* quantitative electroencephalography, *rsMRI* resting state magnetic resonance imaging, *sMRI* structural magnetic resonance imaging

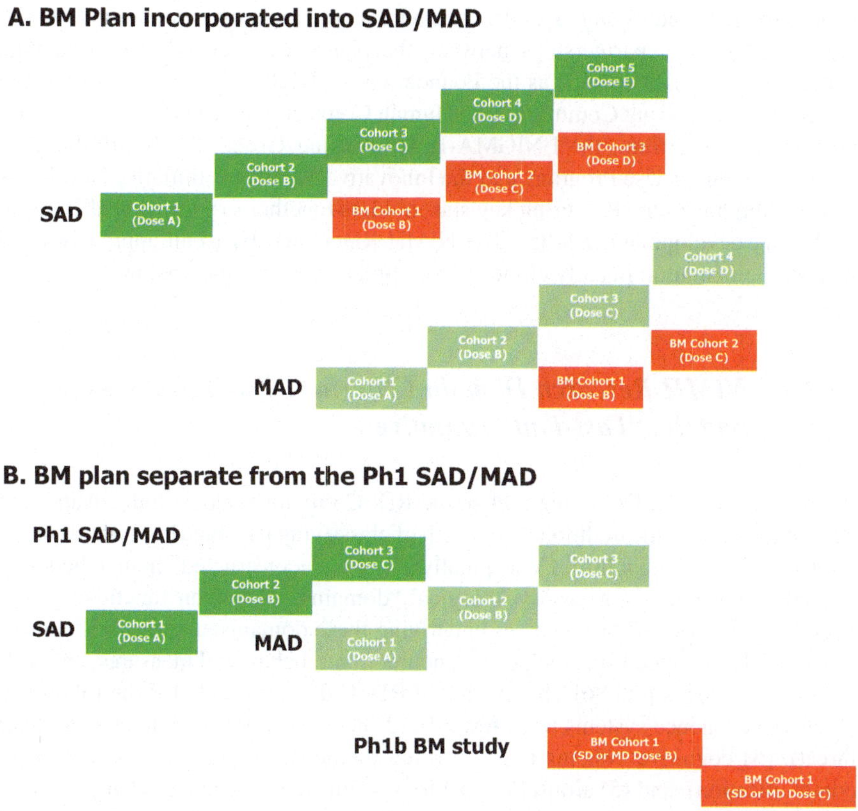

Fig. 17.1 An illustrative approach to the incorporation of a biomarker strategy into either Ph1 SAD/MAD studies (**a**) or as a separate experimental medicine (EM) study (**b**). In the SAD and MAD studies, multimodal biomarkers can be applied without running separate cohorts; however, the investigator is cautioned to overburden subjects in the Ph1 SAD/MAD where the primary objective if safety tolerability. In (**a**), the Ph1 SAD/MAD biomarker cohort could initiate upon safety/tolerability clearance of the previous dose level and run in parallel with safety cohorts. In (**b**), the biomarker cohorts are conducted outside the Ph1 SAD/MAD program, while doses selected for these cohorts are based upon safety/tolerability from Ph1 SAD MAD. Def: BM biomarker, EM experimental medicine, MAD multiple ascending dose, SAD single ascending dose. (Adapted from: English et al. [27])

of the targeted disease state. To support this approach, two major initiatives were introduced by the NIH. The first initiative was the National Institute of Mental Health (NIMH) Research Domain Criteria (RDoC) launched in 2009, which was a novel approach to address shortfalls in the diagnostic nosology of psychiatric disorders and to leverage decades of research into the neurobiological circuits and their dysfunction in psychiatric disorders [58, 59, 78]. The second initiative launched in 2012 was the NIMH "Fast-Fail initiative," aimed at characterizing the pharmacodynamic effects of novel NCEs during early clinical development by incorporating pharmacodynamic biomarkers that could serve as intermediate end points to neurocircuitry underlying specific behavioral or cognitive domains [79–81]. Additionally, other partnerships between the pharmaceutical industry, academia, and private foundations such as the Foundation for NIH's Biomarkers Consortium Neuroscience Steering Committee, the Human Connectome Project, the Psychiatric Genomics Consortium, the ENIGMA-EEG Working Group, the NIMH-Industry New Therapeutics Use Program, and the Innovative Medicines Initiative New Meds Consortium have served to bring key stakeholders together to address challenges in CNS drug development [82–85]. The NIMH RDoC and Fast-Fail approaches and their implementation in early clinical development are described below.

17.2.1 NIMH Research Domain Criteria (RDoC) Framework and the "Fast-Fail" Initiative

As part of the NIMH's Strategic Plan, the RDoC initiative was to link advances in neurobiology with the traditional approach of classifying psychiatric disorders based upon clinical nosology [86]. Conceptually, RDoC "deconstructed" human behavior and cognition into six neuropsychological "domains" of human functioning, followed by "constructs" or processes underlying these domains, and finally, "units of analysis" that included specific genetic, neurocircuit, behavioral measures, and self-report assessments [78, 86] The five initial RDoC domains included the following: (1) positive valence systems (e.g., reward), (2) negative valence systems (e.g., fear/threat), (3) cognitive systems (e.g., working memory), (4) social processes (e.g., communication), and (5) arousal/regulatory systems (e.g., sleep circadian) [86, 87]. A sixth domain, sensorimotor systems was added in 2019 [88]. By "deconstructing" complex psychiatric syndromes into various domains, constructs, and units, researchers could investigate these behaviors transdiagnostically to identify underlying molecular or neural mechanisms that were dysfunctional. While a significant amount of research has been published across psychiatric and neurodevelopmental disorders, some have criticized the limitations of RDoCs and its ability to characterize complex psychiatric disorders vs extremes of normal human behavior [89, 90].

The NIMH "Fast-Fail" Program, initially termed by Paul et al., was an initiative aimed at incorporating novel proof-of-mechanism (PoM)-specific biomarkers in an early phase (Ph1b or Ph2a), experimental medicine focused study with the key objective of characterizing the effects of a novel compound on a PoM-specific BM in order to confirm pharmacology and target engagement and increase confidence in go/no-go decisions [81, 91]. The Fast-Fail PoM approach was to sequentially collect data related to target engagement (e.g., positron emission tomography; PET), followed by effect of the molecule on an established biomarker that measured the underlying neurocircuitry associated with a clinical construct (e.g., anhedonia). Thus, molecules that demonstrated sufficient target occupancy, followed by modulating the neurocircuitry associated with a clinically measurable end point, would then be advanced forward into later phase trials. For those molecules that failed to engage the target or modulate the expected biology, these molecules would be considered as having "failed" in early in clinical development an thus potentially avoid costly later-phase studies [79, 81, 91].

In 2012, the NIMH launched three funded programs under the Fast-Fail program that focused on different psychiatric populations: psychotic (FAST-PS), mood and anxiety (FAST-MAS), and autism spectrum disorder (FAST-AS) [80]. The FAST-PS study evaluated a functional MRI (fMRI)-based pharmacodynamic biomarker using ketamine as a drug-induced surrogate of psychosis that could be utilized in other trials. The FAST-AS study evaluated resting state EEG as a pharmacodynamic BM to evaluate novel therapies in AS, while the FAST-MAS evaluated a novel kappa-opioid receptor (KOR) antagonist on a task-based fMRI task of reward salience [80, 92].

17.2.2 Incorporating RDoC and Fast-Fail Concepts: A Proof-of-Mechanism Study

One of the first studies to implement the NIMH's Fast-Fail approach, incorporating an RDoC-inspired biomarker endpoint, tested a repositioned KOR antagonist, JNJ-67953964, to potentially treat anhedonia in major depressive disorder (MDD) by measuring behavioral performance using the monetary incentive delayed (MID) task in the fMRI as a measure of ventral striatal activation [79]. The Fast-MAS PoM study was a multisite, Ph2a, double-blind, placebo-controlled, fixed-dose, parallel-group study of JNJ-67953964 vs placebo for 8 weeks in patients meeting DSM-5 mood or anxiety disorder criteria who demonstrated significant anhedonia as assessed by the Snaith Hamilton Pleasure Scale score ≥20 [79]. Anhedonia, a construct within the negative valence domain, has been demonstrated to be modulated by reward-related neurocircuitry in the ventral striatum [93]. Additionally, nonclinical data demonstrated that activation of the KOR blunted dopamine release in the striatum and induced negative mood, while inhibition of KOR increased dopamine release and blunted negative behavior [94–96].

To evaluate PoM with the KOR antagonist, the primary outcome measure was the activation of the ventral striatum during anticipation of monetary gain using the MID task in the fMRI [79]. The MID task was selected based upon previous work, including patients with MDD receiving open-label citalopram [97–99]. Results demonstrated that JNJ-67953964 significantly increased ventral striatum activation during reward anticipation on the MID task as measured by fMRI compared with placebo (baseline-adjusted mean: JNJ-67953964, (0.72 (s.d. = 0.67); placebo, 0.33 (s.d. = 0.68); $F\,(1, 86) = 5.58$, $P < 0.01$; effect size = 0.58 (95% CI, 0.13–0.99)) [79]. JNJ-67953964 also improved other measures of anhedonia while not improving specific measures of depression or anxiety (e.g., HAM-D and HAM-A, respectively) [79]. The authors concluded that the positive findings from this study on a PoM study demonstrating target engagement supported further clinical development in a later phase efficacy study.

17.3 Biomarker Technologies in Ph1 Studies to Support PoM

17.3.1 Electrophysiologic Biomarkers in Early-Phase CNS Drug Development

Methods to accurately gauge the clinical benefits of pharmacological interventions for brain disorders are important to efficiently develop new therapies. Yet due to the complexity of these diseases, finding a reliable readout to determine whether (or not) a drug demonstrates target engagement or early clinical benefit in these patient populations has been challenging [100]. Current biomarker approaches (Table 17.2) can be invasive, such as the collection of blood or cerebrospinal fluid (CSF); cost-prohibitive, such as brain imaging techniques; or subjective and lacking in sensitivity, such as cognitive and behavioral assessments. Event-related potentials (ERPs) and quantitative EEG (QEEG) are among the most important translational biomarkers in CNS drug development given their lower cost, low invasiveness, and quantitative end points. Testing and validating reliable and scalable (e.g., multisite) ERP and QEEG approaches will enable wider use of these measures in drug discovery and development [100]. EEG measures the electrical activity of the brain, or brain waves, which is sensitive to a variety of conditions, including the environment, pharmacologic effects, and measures disease-related impact on neuronal function and network connectivity, thereby providing a tool to evaluate within an "RDoC framework" specific diseases (Alzheimer's Disease, schizophrenia and major depression) or transdiagnostically those diseases which share common pathways for impairment (cognitive impairment in schizophrenia, depression, and in neurocognitive disorders – dementia) [45, 86]. The EEG waveform is divided into bands from slowest to highest frequency (delta through gamma). Additionally, the number

Table 17.2 Comparison of current CNS biomarker techniques in Phase 1 studies [45, 65, 77]

Details	qEEG	ERP	PSG	PET (RO)	PET (FDG)	MRI
Technical						
Spatial resolution	+	+	n/a	+++	+++	+++
Temporal resolution	+++	+++	++	+	+	++
Exposure at target site of action?	n/a	n/a	n/a	+++	n/a	++
Binding at target?	n/a	n/a	n/a	+++	n/a	n/a
Expression of pharmacology?	++	++	++	n/a	++	+++
Operational						
Cost	$	$	$	$$	$$	$$$
Integration into Ph1 studies	+++	+++	++	++	++	++
Multisite use	+++	+++	++	+	+	++

Def: *ERP* evoked related potential, *fMRI* functional magnetic resonance imaging, *PET (FDG)* positron emission tomography-fluorodeoxyglucose, *PET (RO)* positron emission tomography-receptor occupancy, *PSG* polysomnography, *qEEG* quantitative electroencephalography, *MRI* magnetic resonance imaging (includes resting state, structural and functional)
Key: +/$ = low; ++/$$ = medium; +++/$$$ = high

of electrodes can vary widely, typically fit for purpose, with the most common recording system including 23 electrodes (international 10–20 system) covering all brain regions accessible via scalp recordings, while other systems use fewer electrodes, and academic investigations can incorporate 64 or more electrodes in an EEG study [101]. Figure 17.2 illustrates the standard five frequency bands measured in QEEG; these data, measuring brain *activity*, are typically obtained in a resting state (with eyes open and then eyes closed). To understand the differences between QEEG and ERP, the graphic below illustrates both QEEG, and an ERP P300 test uses a series of standard tones interspersed with an "oddball" tone. The brain "reacts" to the oddball, with increasing electrical activation; this includes "remembering the oddball is different than the standardized tone and depending on the specific paradigm is indicative of the use of working memory and executive function.

An industry-led ERP Biomarker Qualification Consortium (https://erpbiomarkers.org) was constituted with the objective of bringing together industry, academic, and regulatory stakeholders in a spirit of precompetitive cooperation to demonstrate that robust and reliable ERP and QEEG biomarkers can be collected in target clinical populations, such as patients with schizophrenia, thus ensuring scalability and consistency across studies. (This data has been presented at CNS Summit, Boston 2021 and is available by request (Marco Cecchi – NCT04025502.) There is a regulatory path to qualify biomarkers to support an NDA, and it is hoped that formally qualified and selected ERP and QEEG biomarkers for use in drug trials under the

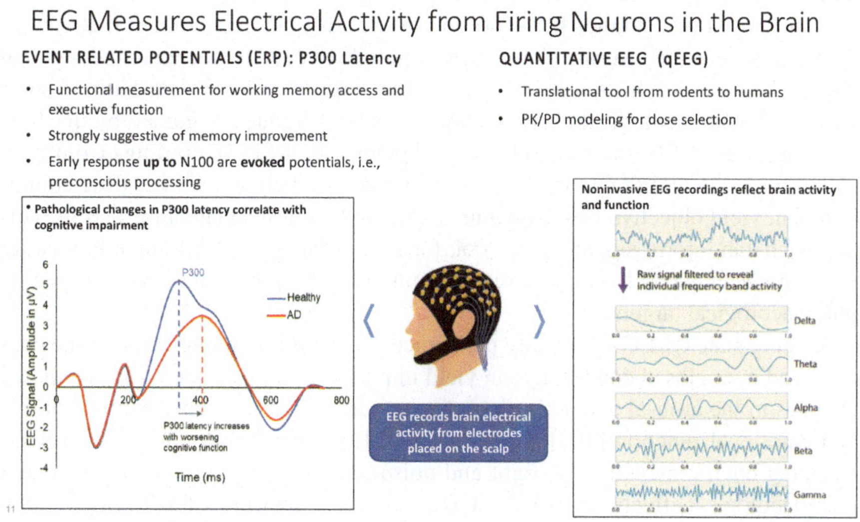

Fig. 17.2 Illustration of electrophysiologic measurements such as QEEG and ERP that have been utilized to characterize electrophysiologic differences between healthy subjects compared with patient populations and have been used to characterize drug effects. (Adapted from: Virtual KOL Event: Reviewing the Predictive Nature of P300 in Determining the Clinical Benefit of Alzheimer's Disease Treatments. October 28, 2020)

FDA Drug Development Tools Qualification Program will become a reality over the next few years.

When properly implemented, ERPs and QEEG can detect target engagement and response to therapeutic intervention [102–105]. ERPs might have the potential to predict response to registration end points for cognitive symptoms, negative symptoms, and global function and to possibly enable stratification of subjects into subpopulations with differential responses to therapy, i.e., with schizophrenia by "biotype" [106].

Quantitative Electroencephalography (QEEG)

QEEG approaches are increasingly incorporated into early-phase trials as noninvasive, cost-effective, and robust strategies to analyze human brain activity in the context of the pharmacological treatment of a CNS disease. EEGs are being used to diagnose patients and evaluate the safety of drugs with possible effects on seizure threshold (beyond the scope of this chapter) and pharmaco-EEGs (phEEG) [77, 107].

QEEG is a readily translatable biomarker from preclinical studies, and with an increasing number of electrodes, now possible on a freely moving animals, changes in electrophysiology can be determined for many brain regions of interest (both QEEG and ERP) and compared between rodent, subhuman primate, and research subjects. In preclinical studies, a typical approach utilizing QEEG spectral analysis determines the changes in specific frequency bands while off and on drug (or transgenic vs wild-type rodent). These finding can guide the design of the early-phase human translational studies by informing exposures (and time course) that engage the CNS target by measuring changes in amplitude of EEG frequencies and regional effects. In more sophisticated analyses using large normative data bases from a variety of pharmacological classes of drugs studied (frequently guided by machine learning), the QEEG effects and local field potentials (LFPs) recordings from typically four to six brain regions, in unanesthetized mice before and after drug administration, yield objective EEG signatures specific to pharmacodynamic action. This approach can inform signatures to look for in translational CNS human biomarker trials and be employed to rapidly screen compounds for potential activity at specific pharmacological targets.

Resting state QEEGs (typically performed for at least 5 minutes each in the eyes open and eyes closed condition) can yield important information on drug-mediated changes in the arousal and activation of neural networks associated with cognitive and emotional function [108]. Changes in QEEG can reflect a variety of stimuli, spanning environmental (e.g., light and noise), arousal state, emotional state, and pharmacological effects. To achieve maximum signal detection of a drug vs placebo effect, these "extraneous" variables that influence electrophysiology need to be controlled, which can be challenging on a busy Ph1 unit. Nonetheless, careful flowcharting of the day's study activities, from the study protocol's schedule of events, can assist the clinical staff in organizing how best to accommodate multiple evaluations while maximizing the environment during the EEG acquisition.

Event-Related Potentials (ERP)

Event-related potentials (ERPs) are task-based EEG measures, where the "brain responds" to sensory stimuli (e.g., visual or audio) or to an activity that requires "processing of information," providing a useful metric of brain function, with specific paradigms reflecting functional measures of working memory, processing speed, emotional response, and executive function [45]. Many of the neurophysiologic ERP measures have undergone extensive validation in various patient populations and can be conducted reliably across multisite trials [109–111]. Common ERP measures performed in early-phase studies include (1) gating measures (e.g., prepulse inhibition (PPI), P50), (2) information-processing measures (e.g., P300 and MMN), and (3) neural-synchrony measures (e.g., γ-auditory steady-state response (ASSR) [45].

In the ERP recording during an auditory oddball paradigm, subjects wear a head cap with attached EEG electrodes connected to a computer that amplifies, digitizes, and filters signal from the electrodes, time-locked to a stimulus generator. Auditory stimuli are presented as a series of tones, the majority of which are low pitched with randomly presented high-pitched oddball target tones. Multiple recordings of the response elicited by the oddball tone are averaged to produce a smooth average waveform. The P300 peak represents the response to the target oddball tone (Fig. 17.3). In Fig. 17.4, blue traces represent individual subject waveforms, where the red trace represents the grand average of the individual waveforms from lead Fz, Cz, and Pz. One can also see how different the standard stimuli tone (on the left panels) is from the oddball stimuli (on the right panels). The brain activates with novel stimuli. So, this stimulus triggers a brain response that is automatic and reflects the functional capacity of the individual to process stimuli. Importantly, the brain processing of the deviant signal requires engagement of memory functions, since the oddball signal relies on the novelty triggering a series of cognitive processes.

P300 latency variability measures have been reported in several studies in AD patients [112–114]. A report from Katada et al., 2003, using a small sample size ($N = 13$), reported standard deviations of 40–50 ms in P300 values [112]. Studies with larger sample sizes ($N = 100$) demonstrate smaller standard deviations of P300 values (approximately <30 ms) [115]. However, a recent study presented by Hua et al. (2019) in patients with Alzheimer's disease demonstrated large effect sizes (Cohen's $d > 1$) from baseline in P300 latency following 8 days of treatment with a novel HGF-MET centrally active agonist, despite a small sample size of 11 subjects, suggesting that exploratory studies in early phase may guide drug development decisions (Ref: Xue Hua, Kevin Church, William Walker, Leen Kawas, Larry Ereshefsky, Philippe L'Hostis, Philippe Danjou, Geoffrey Viardot, Hans Moebius. HGF/MET Receptor Agonist NDX-1017: Translational Phase 1 a and b Results. Presented at 12th Conference Clinical Trials Alzheimer's Disease, December 4–7, 2019, San Diego, USA).

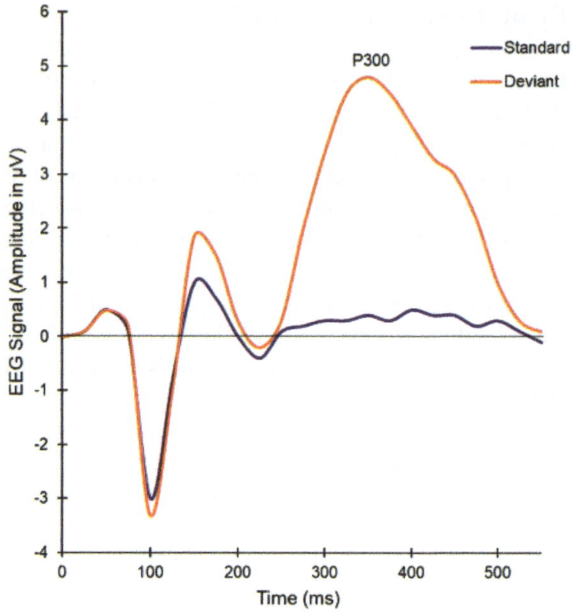

Fig. 17.3 In response to the deviant (oddball) tone during the auditory ERP paradigm, a large positive wave is produced over the 250–450 ms range with a peak around 300–350 ms range (P300). This P300 response is not observed following the standard tone, with little or no deflection observed after 200 ms. (Data on file: Larry Ereshefsky and Jack Johnstone, graphical representation of P300, 2001)

Polysomnography (PSG)

In addition to EEG, polysomnography (PSG) studies have documented several sleep abnormalities in a variety of psychiatric disorders [116]. PSG captures a wide array of data captured as a continuous recording as shown in Table 17.3, also providing the information needed to stage sleep and demonstrate other key data used to characterize disease and drug effects (Table 17.3 and Fig. 17.5). Many CNS disorders demonstrate changes in sleep parameters; however, this section will briefly discuss findings in patients with schizophrenia and major depressive disorder.

In patients with schizophrenia, differences in sleep parameters from healthy volunteers have been associated with impairment in a number of cognitive domains and diminished quality of life [118]. Among the most common sleep abnormalities observed is a decrease in slow-wave sleep (SWS), which has been correlated to deficits in memory (i.e., consolidation of procedural and declarative learning) [118]. Starting with second-generation antipsychotics (SGA), potent 5-HT2A antagonist effects have been documented for most these drugs alleviating sleep abnormalities associated with schizophrenia [119]. For example, the SGA olanzapine has been reported to increase slow-wave sleep and was positively correlated with an increase in verbal memory consolidation [120]. Thus, the use of PSG in Ph1 studies may serve as a useful biomarker for possible effects on cognition, even if the drug mechanism is novel. From an RDoC perspective, assessing changes in SWS (or ERPs as described previously) may serve to demonstrate that a IMP shares a common effect with established medications, on common neuronal pathways (or not) or specific regions of interest [121]. Conversely, drugs that are "procognitive" and/or increase

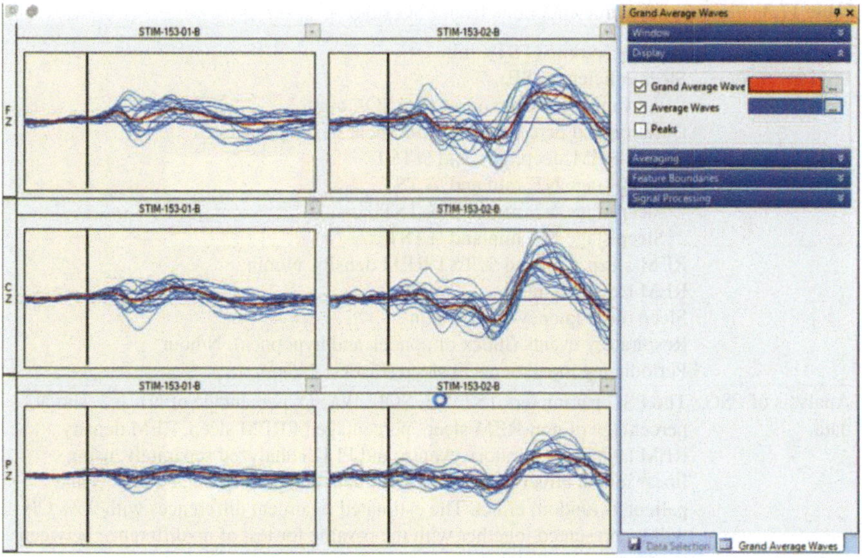

Fig. 17.4 In response to the deviant (oddball) tone during the auditory ERP paradigm, a large positive wave is produced over the 250–450 ms range with a peak around 300–350 ms range (P300). This P300 response is not observed following the standard tone, with little or no deflection observed after 200 ms. Data collected as part of the ERP Biomarker Qualification Consortium at Hassman Research Institute and Collaborative Neuroscience Research. (Data on file from Event-Related Potential (ERP) Biomarkers in Subjects With Schizophrenia and Healthy Volunteer Subjects. ERP Biomarker Qualification Consortium NCT04025502)

arousal could readily delay sleep onset and reduce sleep efficiency. Given the close inter-relationship between normal physiologic sleep and cognitive function, characterization of a drug's effects on PSG can play a role in the go/no-go decision process [122, 123].

Perhaps among the most studied psychiatric disorders where sleep disturbance is a core symptom is major depressive disorder, where disturbances of sleep continuity are common [124]. From the PSG, the sleep architecture associated with depression demonstrates a decrease in SWS (impact as described above) and disturbed rapid eye movement (REM) sleep regulation. Shortened REM latency (the time between sleep onset and the occurrence of the first REM period) is a commonplace in major depression and has been considered a biological marker of depression, including having the potential to predict the course of illness (i.e., relapse and recurrence) [125]. Additional related findings are increase in total time in REM sleep and increased REM density (frequency of rapid eye movements over the REM period) [126]. Many different antidepressant drugs, including older therapies, such as tricyclic antidepressants (TCAs), monoamine oxidase inhibitors (MAOIs), and newer serotonergic selective reuptake inhibitors (SSRIs), decrease rapid eye movement (REM) sleep [121]. It has been suggested that a reduction in REM sleep produced by many antidepressants (but not all) plays a mediating effect associated with their

Table 17.3 Pharmacodynamic endpoints from PSG studies

Derivation of PSG parameters	Total sleep time (TST), min
	Sleep efficiency (SE), %
	Wake-time after sleep onset (WASO), min
	Duration and percentage of time spent in sleep stages:
	  Non-REM sleep, min and %TST
	  Sleep stage N1, min and % TST
	  Sleep stage N2, min and % TST
	  Sleep stage N3, min and %TST
	REM-sleep, min and % TST REM density, N/min
	REM latency, min
	Sleep onset latency (SOL), min
	Respiratory events (index of apnoea and hypopnea), N/hour
	Periodic leg movements in sleep (PLDS), N/hour

Analysis of PSG data	The PSG parameters TST, SE, SOL, WASO, percentage of N1, N2, and N3, percentage of non-REM sleep, percentage of REM sleep, REM density, REM latency, respiratory events, and PLDS analyzed separately, using linear mixed effect model with treatment and period as fixed effects and patient as random effect. The estimated treatment differences with 95% CIs will be presented together with the p-value for test of no difference between treatment arms. Total sleep time
Actigraphy as a supplemental evaluation	Total sleep time (TST) – Corresponding to the total time identified as sleep, per night Wake after sleep onset (WASO) – Corresponding to the number of minutes classified as wake within the sleep period, per sleep period Sleep period efficiency (SpE) – Is defined as the ratio between the time of actual sleep on sleep period Sleep latency – Time from event marked via app to sleep as determined by the sleep detection algorithm Detection of activity (PD) vs time post dose and PK Waking hours: Effect of night's sleep and drug on activity

Def: *CI* confidence intervals, *PLDS* periodic leg movements in sleep, *PSG* polysomnography, *REM* rapid eye movement, *SE* sleep efficiency, *SpE* sleep period efficiency, *SOL* sleep onset latency, *TET* total sleep time, *WASO* wake-time after sleep onset

efficacy [127]. The effects of antidepressants is more heterogeneous when evaluating SWS; in general, antidepressants with 5-HT2A/2C receptor antagonist properties increase SWS, whereas other drugs, such as SSRIs or MAOIs, either lower SWS or with chronic use, produce no change [128]. Mirtazepine, a mechanistically differentiated drug, with no biogenic amine reuptake inhibition, demonstrates alpha-2 adrenergic (presynaptic), H1 histamine, 5HT2a, 2c, and three receptor effects [129], resulting in similar effects on sleep architecture to most antidepressants [129]. However, highly potent H1 receptor blockade leads to daytime impairment [127]. Screening new antidepressants or other psychoactive drugs using PSG is a powerful tool in drug development including extrapolation from rodent sleep studies, comparison of novel therapies to established "standards of care," and sensitivity to evaluate concentration (exposure) effects on sleep parameters (PK/PD correlations).

Quantitative PSG also enables the physiologic characterization of the sedative effects of CNS active compounds, typically evaluated during the daytime. The two

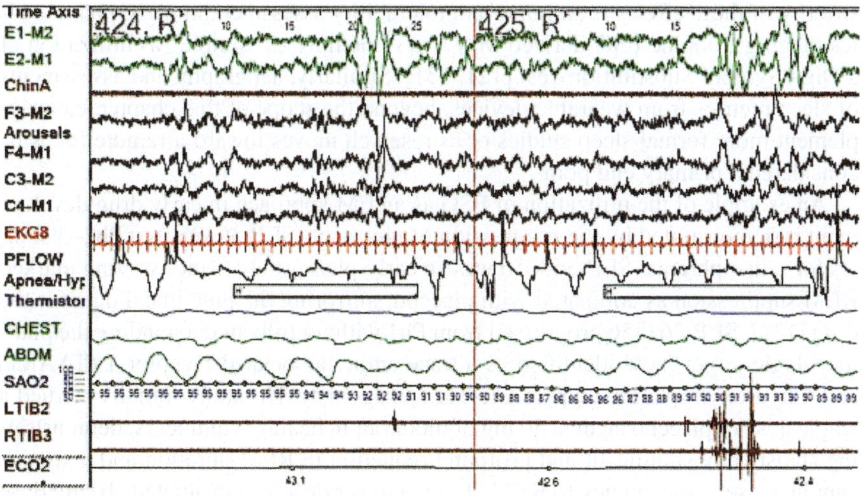

Fig. 17.5 Representative hypnogram from a 30-second epoch during a PSG study. Def: PSG polysomnography. (Adapted from: Basner [117]. Cambridge University Press, 2012)

most significant methods are the multiple sleep latency test, a technique measuring the speed at which a person falls asleep during the day (observed repeatedly over time as subjects are not permitted to remain asleep) and the Maintenance of Wakefulness Test (MWT). Both of these tests facilitate exploration of concentration-/exposure-mediated changes in arousal/sedation. While the MSLT is the traditional test evaluating excessive daytime sleepiness, the Maintenance of Wakefulness Test (MWT) is preferred when the assessment of daytime alertness is the primary goal (in contrast to diagnostic assessments). A comparison of MSLT with MWT shows that the coefficient of correlation between MSLT and MWT ($r = 0.41$) is statistically significant, however, explaining only a small amount of the observed variability (~17%) [130]. Factor analysis suggests that two factors, alertness and sleepiness, account for 91% of all variance. Their data further demonstrate that patients with diagnoses of excessive somnolence (disorders) might be discordant on the two tests, underscoring that for healthy volunteer exploratory early-phase clinical pharmacology studies that a key to success is careful screening out of patients with psychiatric and especially sleep disorders. Current consensus is the use of the MWT in subjects without sleep disorders, which may be the more suitable test in many clinical and research environments evaluating drug effects or sleep deprivation in healthy volunteer studies.

When coupled with various psychometric tests assessing attention, concentration, and psychomotor processing, a robust model for associating sleep architecture (especially SWS) with quantitative cognitive measures emerge. Additionally, pharmacologically induced sleepiness and sedative effects will have deleterious effects on cognitive function, typically correlated with the time course of drug concentration over the dosing interval (Lam). Sedation is one of the primary adverse events associated with many CNS active compounds, and early characterization of the severity of these effects is an important tolerability assessment in early-phase studies ranging from the PSG-derived MWT to something as "simple" as utilization of a Digit Symbol Substitution Test [121, 131]. Similarly, actigraphy and assessments of sleep metrics from wearable devices, beyond the scope of this chapter, can supplement more formal sleep studies or as research moves toward a remote focus be considered a primary end point.

An example of the utilization of PSG as an EM approach in early drug development is demonstrated by the progression of ulotaront (SEP-363856) to Ph2–3 studies for schizophrenia. SEP-363856 was an early-phase study demonstrating marked REM suppression as compared with placebo, mirroring the preclinical data for the drug [132]. SEP-363856 progressed from Ph1 without fully understanding the pharmacologic target (now identified as a trace amine-associated receptor 1 (TAAR1) and 5-hydroxytryptamine type 1A (5-HT1A) agonist). The study initially studied a single dose of placebo against 50 mg of ulotaront in healthy volunteers, demonstrating statistically significant and profound reduction in REM duration and a significant increase in the latency to REM. Then, a lower dose was evaluated; 10 mg dose of ulotaront demonstrated a nonsignificant reduction vs placebo for REM duration but a significant difference in latency to REM sleep. This study demonstrates the power of PSG as a tool to evaluate novel therapies and obtain dose/concentration response data (Fig. 17.6) [133].

17.3.2 Neuroimaging Biomarkers in Early-Phase CNS Drug Development

As previously described, evidence from the three-pillar model and the 5R-framework, obtaining data regarding proof of CNS exposure, target engagement, and expression of pharmacology greatly increase the likelihood of clinical success. Depending on the imaging method utilized, these studies are generally conducted in small sample sizes, can be conducted in HNVs or patients, and can include single and/or multiple doses of the IMP.

Within CNS drug development, several imaging methods have been commonly applied during early-phase development, such as PET, rsMRI and fMRI, MRS and to a lesser extent, arterial spin labeling (ASL) and diffusion tensor imaging (DTI). Generally, most imaging-based studies conducted during early clinical development are conducted in parallel with the Ph1 SAD/MAD studies, although, they could be performed sequentially after the SAD or MAD study should safety and tolerability of the dose planned for the imaging study need to be characterized. In the case of PET imaging, Ph0 "microdosing" studies may be performed in human subjects prior to the Ph1 SAD/MAD as these studies include doses that are below those exposures expected to exert a pharmacologic effect. Imaging studies may also be performed during the Ph2 POC; however, given that the purpose of the ExM imaging study is to inform CNS exposure and target engagement or modulation of known neural circuitry, the utility of these studies performed in parallel with the Ph2 POC studies is questionable.

A. Study flow chart.

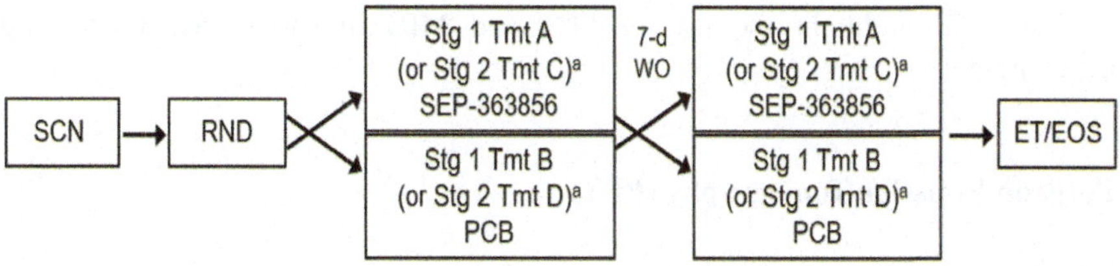

B. Time spent in REM sleep (minutes) and latency to REM between SEP-363856 (50 vs 10 mg) vs Placebo.

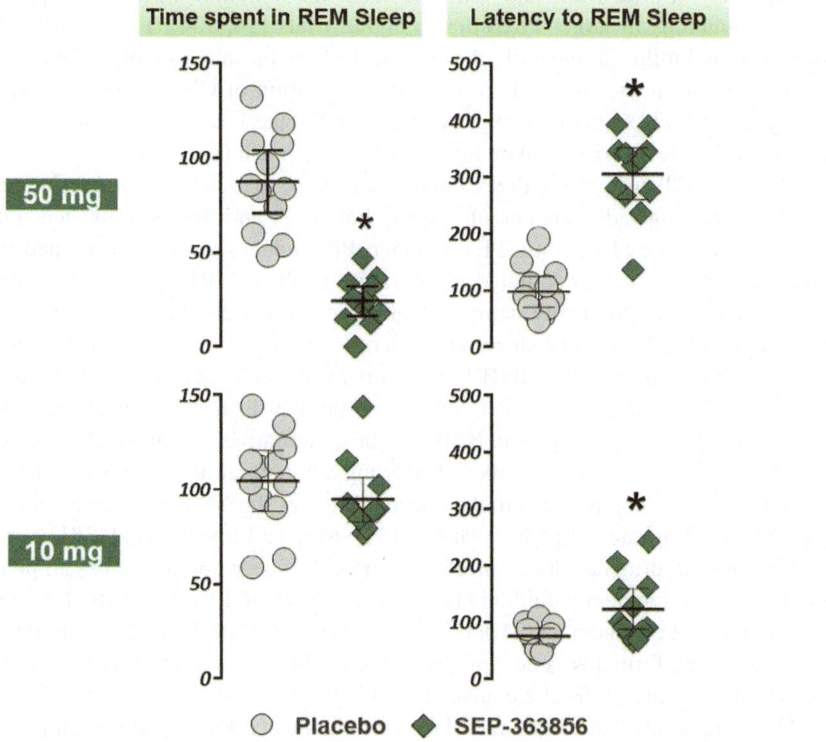

Fig. 17.6 Effects of SEP-363856 on REM sleep in healthy volunteers [133]. (**a**) Study flowchart. (**b**) Time spent in REM sleep (minutes) and latency to REM between SEP-363856 (50 vs 10 mg) vs placebo. Def: ET early term, EOS end of study, PCB placebo, SCN screening, RND randomization, STG stage, Tmt treatment, WO washout. (Adapted from: Hopkins et al. [133])

A comprehensive review of imaging and analysis methods utilized during clinical development is beyond the scope of this chapter; however, the references contained in the sections below can direct the reader to more details. The following sections will highlight the use of PET and MRI in early-phase CNS drug development.

Positron Emission Tomography (PET)

PET imaging has been utilized as both a mechanistic and functional biomarker tool in CNS drug development to demonstrate target engagement and proof of central brain penetration [24]. When available, the radiolabeling of drugs or imaging agents with positron-emitting radionucleotides (e.g., ^{11}C or ^{18}F) allows for the characterization of concentration-exposure modeling (PK/PD) with central target engagement during early clinical trials and validated from the nonclinical studies, improving dose estimations in larger Ph2b dose ranging trials [24, 134]. PET imaging has demonstrated utility in the early development of NCEs such as antipsychotics and antidepressant medications where the characterization of CNS penetration, target engagement, and extent of receptor occupancy (RO) has been useful in the selection of clinically relevant doses taken into later-phase clinical trials [135].

Briefly, PET imaging utilizes small molecules that have been labeled with positron-emitting radioisotopes of varying half-lives and then are injected into a subject, while are placed in a PET scanner. PET images are further aligned with structural images from computed tomography (CT) or MRI, whereby regions of interest can be anatomically defined. Mathematical models (kinetic tracer modeling) applied to PET data produces key outcome measures such as volume of distribution (VT); binding potential (BP), which is further defined as free-radioligand in plasma (BP_F); total plasma (BP_P); and the tissue non-displaceable measure ($BP_{ND)}$ [136]. Receptor occupancy of an NME can be quantitatively measured by comparison of baseline BP values with those following administration of the NME. The BP from PET data is compared with the concentration of the NME in plasma, whereby PK/PD modeling can define IC_{50} values [137] using an Emax model [138].

Within CNS drug development, PET imaging has been applied across all phases of clinical development including (1) characterization of drug distribution, (2) validation of target engagement (TE), (3) characterization of desired RO (using PK/PD data) that can inform dose selection decisions, (4) identification of patient subtypes, and (5) monitoring of disease course [135, 137].

For compounds that can be radiolabeled isotopically, PET imaging can provide characterization of a NCE tissue distribution. For example, during the development of BMS-181101, a drug with mixed 5-HT_{1A} and 5-HT_{1D} activity being developed for depression, a PET study in HNVs using [^{11}C]BMS-181101 demonstrated rapid cerebral clearance with limited retention [139]. The limited CNS penetration and tissue kinetics of BMS-181101 led to the subsequent discontinuation of this NME [139].

For NCEs in the development for various psychiatric disorders, PET imaging has been used extensively to demonstrate validation of target engagement and characterize RO and confirm PoP and dose selection in efficacy trials, respectively. In the case of drugs such as selective serotonin reuptake inhibitors (SSRIs) and norepinephrine inhibitors (NET), the use of novel radioligands have been used to characterize the extent of occupancy with clinical benefit [24, 135]. For example, Meyer et al. demonstrated that several SSRIs (e.g., paroxetine, sertraline, and citalopram) required 80% occupancy of the serotonin transporter (SERT) for clinical benefit [135, 140–142]. Similarly, for NCEs demonstrating antagonism of the dopamine 2 receptor (D2R) being developed as an antipsychotic, the use of radioligands such as ^{11}C-raclopride and ^{18}F-fallypride has demonstrated that NCEs that demonstrate D2R occupancy of <70% exhibit good antipsychotic efficacy against positive symptoms, while those with >80% D2R occupancy have an increased incidence of extrapyramidal symptoms [143–145] (Fig. 17.7).

In addition to providing key data related to tissue distribution and TE, PET imaging can also be used to identify selected subgroups of patients for a clinical trial and can be used to monitor the course of the disease. This strategy has been extensively applied to neurodegenerative disorders such as Alzheimer's disease (AD) and Parkinson's disease to demonstrate PoM and efficacy. For example, in patients with AD, the development of radionucleotides such as ^{11}C-Pittsburgh Compound B (PIB) or ^{18}F-florbetapir characterizes β-amyloid deposition in the brain and has been used to not only qualify subjects into clinical trials but also serve as outcome measures [146–149]. Similarly, use of PET radioligands has been utilized in other neurodegenerative diseases as an outcome measure in PoM studies [150–152].

When considering/including PET imaging into an early CDP, one must consider several factors: (1) the mechanism of action of the drug (e.g., antagonist, partial agonist, and allosteric modulator), (2) potential differences in RO due to single vs multiple dosing, (3) differences between healthy volunteers and the patient population, and (4) availability of the radionucleotide [24]. For example, although it has been demonstrated that single-dose studies can predict ROs observed from multiple dose studies, this was not the case observed with the second-generation antipsychotic ziprasidone [153]. When single oral doses of ziprasidone 40 mg was administered to HNVs, the observed D2R RO was 67%, while 60 mg doses demonstrated RO of 85%. However, results from earlier clinical trials in patients with acute schizophrenia at lower doses that were based upon the initial single-dose PET studies performed in HNVs were found to be too low to demonstrate antipsychotic efficacy [154, 155]. A multiple dose PET study performed in schizophrenic patients confirmed that the 120 mg/day dose produced target occupancy, consistent with clinical efficacy data [156].

While PET studies conducted during early phase are generally small studies (n = 20) and performed at a single site, barriers to the utilization of PET imaging during early clinical development exist. These include the unavailability of a selective radioligand, the relative short half-life of isotopically labeled radioligands such as ^{11}C (half-life around 20–90 minutes) such that onsite production using a cyclotron is required, and accessibility of a PET facility and overall cost. However, as

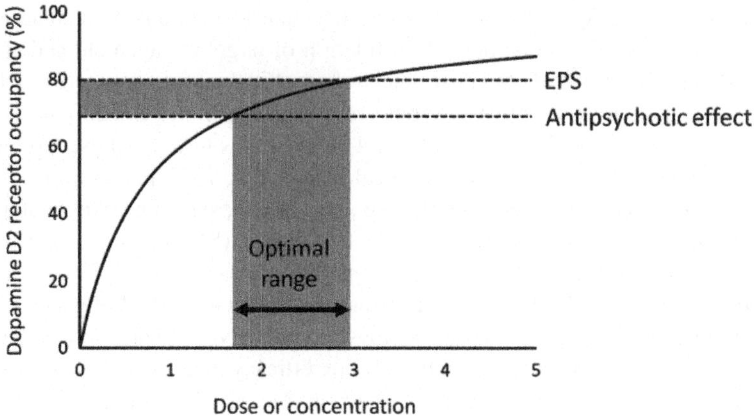

Fig. 17.7 Relationship between dopamine D2 (D2R) receptor occupancy, clinical response and motor side effects. Def BM biomarker, EM experimental medicine, MAD multiple ascending dose, SAD single ascending dose. (Adapted from: Kapur et al. [143])

discussed above, the use of PET imaging studies during early clinical development can provide information related to target engagement and dose selection that can be translated from nonclinical studies to humans and potentially avoid incorrect dose selection in later phase trials.

Magnetic Resonance Imaging (MRI)

The use of imaging techniques like MRI to support clinical trials has grown extensively over the past decade [65]. A review by Sadraee et al. (2019) of the ClinicalTrials.gov registry of 2026 entries that included fMRI, ~30% of studies involving fMRI included a drug intervention, with 33% of those trials using fMRI as the primary outcome measure [157]. Several reviews have been published outlining the advantages, disadvantages, and areas of growth for the use of fMRI techniques to support clinical studies across all phases of development [65, 158]. MRI techniques applied during clinical development have included resting state MRI (rsMRI), functional MRI (fMRI), pharmacologic MRI (phMRI), diffusion tensor imaging (DTI), and arterial spin labeling (ASL) [159–161].

While a detailed explanation of MRI physics is beyond the scope of this chapter, briefly, fMRI takes advantage of the paramagnetic differences between oxyhemoglobin (oxyHb) and deoxyhemoglobin (deoxyHb), which is reflected in the blood-oxygen-level-dependent (BOLD) signal [162, 163]. As neuronal activity increases during task activation, this results in an increase in cerebral perfusion and larger demand for oxygen by the tissues [159]. The change in the oxyHg to deoxyHg ratio produces a different BOLD signal upon the activation of a magnetic field [162]. DTI allows the assessment of white matter tracts by measurement of the random diffusion of water molecules resulting in two parameters: fractional anisotropy (FA), a measure of axonal integrity and mean diffusivity, and a measure of cellular integrity [164]. Finally, ASL uses radiofrequency pulses to magnetically label water molecules that serve as an endogenous tracer for the measurement of cerebral blood flow.

In early-phase clinical development, MRI can provide an indirect measure of target engagement by detecting functional changes within the brain in response to a drug challenge based upon the link between the molecular target and the underlying neurocircuit of interest [165]. In the absence of a PET radioligand, MRI imaging can provide evidence of central brain penetration and downstream effects of target engagement in a circuit or region of interest (ROI) that may inform dose-response relationships and guide dose selection for later phase trials [166]. Applications of MRI imaging included in many early-phase clinical study protocols have primarily included modalities such as (1) phMRI, drug-induced change in the resting state MRI; (2) rsMRI, task-free MRI that assess resting state functional connectivity between defined ROIs; and (3) task-based MRI, effects of a stimulus (e.g., finger taping and cognitive task) on the BOLD signal. Within a single study protocol, it is common to see at least two different imaging modalities (e.g., rsMRI and task-based fMRI) incorporated into a single MRI scan [167, 168]. Operational considerations for the inclusion of more than one imaging modality should include scan duration and task burden to the subject, particularly if switching between HNVs and patients as the latter group may be unable to complete the tasks, thus leading to subject fatigue that may limit data interpretation [55, 168, 169]. In addition, fMRI methods should be evaluated with respect to test-retest repeatability. Despite the increase in incorporating MRI in early clinical development, there are few publications reporting interclass correlation coefficient (ICCs) [65, 170].

During rsMRI, spontaneous brain function is measured from the BOLD signal in the absence of a task such that spatially distributed networks of temporal synchronization can be identified, referred to as resting state networks (RSNs). Upwards of 10–20 different RSNs have been identified, with common RSNs including the default mode network (DMN), sensorimotor network (SMN), and the salience network (SN) [171]. The RSN that has received the most attention is the DMN, identified by Greicius et al., using fMRI forming an intrinsic functional network that could be down-modulated during cognitive tasks [172]. Several studies have been published demonstrating abnormalities in DMN activation in a number of psychiatric disorders such as schizophrenia, anxiety, depression, autism, and AD [173]. An extension of rsMRI, pharmacologic MRI (phMRI), involves the administration of a drug where time course profiles using PK data can be compared with BOLD signal changes observed during the rsMRI compared with the predrug baseline condition. Statistical methods such as seed-based functional connectivity where correlation coefficients of one time seed and other time series data are collected can inform synchronous activity between ROIs [174]. Additionally, independent component analysis (ICA) is another statistical method to identify patterns of BOLD response related to synchronous activity between networks [175]. Several classes of drugs such as analgesics, antipsychotics, antidepressants, and pro-cognitive agents demonstrate a phMRI signal that can be based upon an a priori hypothesis based upon MOA [65, 176–180].

In addition to rsMRI and phMRI, task-based fMRI has also been applied during early clinical development to demonstrate indirect target engagement related to neurocircuitry associated with a behavioral phenotype or cognitive construct. For example, during task-based fMRI, the effects of a stimulus such as somatosensory (e.g., finger tapping or visual), emotional (e.g., faces task), or cognitive (e.g., n-back task for working memory) are measured by the BOLD response and can be compared with the subject's behavioral response (e.g., accuracy). Several studies incorporating task-based tasks during MRI imaging based upon translation of preclinical data, MOA, and underlying neurocircuitry have been performed in early-phase clinical trials in HNVs and in patients [55, 65, 168, 181]. Because of the variability in MRI data acquisition software, analysis methods, and study designs and tasks, specific guidelines to improve variability have been published [65, 182].

Similarly to rsMRI, task-based fMRI has been utilized to identify abnormalities in the underlying neurocircuitry using behavioral and cognitive tasks in patients with neuropsychiatric disorders. For example, in a meta-analysis of task-based fMRI studies in HNVs and patients with schizophrenia, patients with schizophrenia were found to have attenuation of the dorsolateral prefrontal cortex (dlPFC) and the anterior cingulate cortex (ACC) during performance on executive cognitive tasks [183]. Similarly, patients with major depression have demonstrated increased BOLD responses to negative stimuli (e.g., facial expressions) in regions such as the amygdala, putamen, and thalamus [184, 185].

Several experimental studies involving NCEs across all phases of clinical development have been published that incorporated task-based fMRI, either in HNVs and/or patients [158, 186, 187]. In the context of drug development, considerations for selecting tasks for fMRI studies should include ability of the task to elicit the known underlying circuitry, the test-retest reliability of the task, study design, decision to include HNVs or patients, and lastly the MOA of the NCE. As previously described, based upon the known MOA of PF-06412562 (D1/D5 partial agonist) and nonclinical rodent data, tasks that engaged the underlying neurocircuitry related to working memory and reward processing were used during the fMRI session and included tasks such as the n-back, the AX-Continuous performance task, and the monetary incentive delay task, respectively [55, 168]. Similarly, Krystal et al., used the MID task, a measure of reward anticipation during fMRI, evaluating a novel KOR antagonist to characterize activation of the ventral striatum [186].

The use of phMRI and task-based fMRI evaluating has been best applied during the evaluation of drugs for pain management where demonstration of modulation of neurocircuitry related to the neural processing of pain is conserved translationally from rodent to human [176, 188]. A classic example of the utilization of multimodal rsMRI and task-based fMRI imaging in early clinical development was the development of aprepitant, an NK-1 antagonist, being developed for the treatment of chronic pain [189]. Using phMRI, aprepitant-induced rsMRI BOLD changes were demonstrated in regions associated with NK-1 receptors; however, the functional connectivity (FC) between central circuitry associated with pain processing was not observed, while the μ-opioid agonist buprenorphine did exhibit analgesic-associated FC [178], and during the task-based fMRI session where subjects were subjected to a noxious heat stimulus [178]. Unlike buprenorphine, aprepitant did not attenuate the stimulus-induced fMRI BOLD response to acute pain [178]. Identification of the underlying neurocircuitry, responsible for the modulation of pain using fMRI, allows inferences to be drawn regarding how an NME modulates these circuits and thus a drug "profile" or "fingerprint" can be identified that may be translatable from animal to human [159]. This was most recently demonstrated by Duff et al., where by using machine-learning (ML) methods and data from multiple published studies of various analgesic compounds, they were able to identify "brain map fingerprints" that allowed the generation of a "go/no-go" model for analgesic drug development [176].

In addition to rsMRI and fMRI, other imaging modalities such as proton magnetic resonance spectroscopy (^{1}H-MRS) have been utilized during early drug development, particularly with NCEs that modulate glutamate or GABAergic transmission [181]. 1H-MRS can quantify the amount of central GABA glutamate, and the glutamate metabolite glutamine or glutamate+glutamine (Glx) depending on the field strength of the MRI (e.g., 1.5 vs 3 Tesla) is needed to resolve their spectral resonances. 1H-MRS can also detect other metabolites involved in central regulatory processes and metabolism [190].

^{1}H-MRS has been applied in early drug development as a stand-alone study and in combination with other imaging modalities (e.g., fMRI) [24, 191]. Similar to PET imaging, ^{1}H-MRS has been applied as a translational tool to establish TE, PoM, and patient stratification [190]. For example, several studies have evaluated the effects of NMDA antagonists such as ketamine, which increase extracellular glutamate release and can be reversed by presynaptic agonists at the mGluR 2/3 receptor or lamotrigine [170, 181, 191]. Javitt et al., 2018, used a multimodal imaging approach to characterize the effects of subanesthetic doses of ketamine in HNVs. The primary outcome measure was the change in baseline in ketamine-induced changes in phMRI, ^{1}H-MRS, and task-based fMRI. Significant increases in the BOLD signal were observed in the phMRI (Cohen $d = 5.4$; $P < 0.001$), with modest effects (Cohen $d = 0.64$; $P = 0.04$) seen with ^{1}H-MRS of Glx, immediately following ketamine infusion [191]. These data confirmed that imaging could be used as a biomarker of target engagement for glutamatergic modulating drugs.

17.4 An Example of PoM Studies Supporting the Early Clinical Development Plan

17.4.1 Development of Takeda's TAK-063, Selective Phosphodiesterase 10A (PDE10A) Inhibitor for Schizophrenia

Takeda's early-phase clinical development program (eCDP) of TAK-063, an inhibitor of phosphodiesterase 10A (PDE10A), included a comprehensive translational development strategy, which focused on translating pharmacodynamic effects observed in nonclinical models to human subjects related to its potential antipsychotic effects (Table 17.4). Inhibitors of PDE10A have been under clinical development by several different pharmaceutical companies as nonclinical studies have demonstrated a broad antipsychotic profile that included reversal of MK-801-induced hyperactivity and improvement across multiple cognitive domains [196–200]. The objectives outlined in the eCDP of TAK-063 included the establishment of the PK profile in SAD, MAD, HNVs, healthy Japanese subjects, and patients with schizophrenia and the PD effects observed in nonclinical models of antipsychotic efficacy [56, 192]. Prospective go/no-go criteria were established for each of the Ph1 studies and gated the decision to proceed to Ph2 PoC and are outlined in Table 17.3 [56].

The Ph1 program for TAK-063 included the traditional SAD and MAD performed in HNVs and healthy normal Japanese (HNJ) subjects as part of a global development strategy. Doses selected for the MAD study were based upon the safety, tolerability, and PK data from the SAD.

In nonclinical studies, the antipsychotic and procognitive effects of TAK-063 were demonstrated at dose exposures corresponding to approximately ~30% PDE10A occupancy [201]. PDE10A occupancy by TAK-063 was measured in 12 HNVs by measuring displacement of [^{11}C]T-773 during a dynamic PET scan evaluating 3, 10, 30, and 1000 mg doses [193].

In nonclinical studies, TAK-063 demonstrated reversal of MK-801 induced deficits on working memory, executive function, and attentional set-shifting [202]. TAK-063 was also found to reduce ketamine-induced increases in percent BOLD signal change in various cortical regions in a rodent phMRI study [167]. As previously mentioned, ketamine has been used as a model of hypoglutamatergic deficits producing schizophrenia-like behaviors in rodents and in humans [41, 203].

Table 17.4 Ph1 go/no-go criteria of the eCDP of TAK-063 [192–195]

Study	Objectives	Go-criteria	ClinicalTrials.gov ID
SAD/MAD in HNV and HNJ subjects	Safety, tolerability, and PK	Favorable PK profile and exposures ~1000 ng*h/ml	NCT01879722
PET occupancy in HNV	Central target engagement and distribution	Target occupancy >30%	NCT02370602
SAD/MAD in SCZ patients	Safety, tolerability, and PD effects (EEG and cognitive measures)	Favorable tolerability profile	NCT01879722
fMRI in HNVs	PoM using ketamine-induced deficits using fMRI BOLDs	Effects on rsMRI BOLD or ketamine-induced deficits	NCT01892189

Def: *AE* adverse event, *BOLD* blood oxygen level dependent, *eCDP* early clinical development plan, *EEG* electroencephalography, *fMRI* functional magnetic resonance imaging, *HNV* healthy normal volunteers, *HNJ* healthy normal Japanese, *MAD* multiple ascending dose, *PET* positron emission tomography, *PK* pharmacokinetics, *PD* pharmacodynamics, *POM* proof-of-mechanism, *SAD* single ascending dose, *SCZ* schizophrenia

Therefore, an fMRI study was conducted in HNVs during ketamine infusion with the primary end point of reduction in the percent BOLD change induced by ketamine [194]. The fMRI study included rsMRI, a spatial working memory task-induced fMRI BOLD acquisition, followed by arterial spin labeling scan [194]. Three separate doses of TAK-063 (3, 10, and 30 mg) were administered 4 hours prior to ketamine infusion. These doses were selected based upon previous Ph1 safety, tolerability and PK data, and PD data that included PET occupancy and EEG data from schizophrenic patients [56, 195]. Compared with placebo, TAK-063 reduced the ketamine-induced increases in BOLD signal during the working memory task, with the 30 mg dose level producing the most consistent response. These results satisfied the prespecified go-criteria established in the eCDP.

Takeda's TAK-063 program represents one of the most comprehensive early-phase clinical development plans to incorporate multimodal biomarkers to characterize the pharmacology of a novel PDE10A inhibitor and establish PoM. While the subsequent 6-week Ph2 trial in patients with schizophrenia failed to meet its primary end point of a change from baseline in the total PANSS score, the approach taken by Takeda represents a novel model of incorporating an RDoC-strategy with a Fast-Fail objective that may serve as a model for the development of other CNS compounds [204].

References

1. Zhu T. Challenges of psychiatry drug development and the role of human pharmacology models in early development-a drug developer's perspective. Front Psych. 2020;11:562660.
2. Rehm J, Shield KD. Global burden of disease and the impact of mental and addictive disorders. Curr Psychiatry Rep. 2019;21(2):10.
3. CNS drugs take longer to develop and have lower success rates than other drugs (press release). Tufts Center for the Study of Drug Development 2015.
4. Pangalos MN, Schechter LE, Hurko O. Drug development for CNS disorders: strategies for balancing risk and reducing attrition. Nat Rev Drug Discov. 2007;6(7):521–32.
5. Howes OD, Mehta MA. Challenges in CNS drug development and the role of imaging. Psychopharmacology. 2021;238(5):1229–30.
6. Marder SR, Laughren T, Romano SJ. Why are innovative drugs failing in phase III? Am J Psychiatry. 2017;174(9):829–31.
7. Gribkoff VK, Kaczmarek LK. The need for new approaches in CNS drug discovery: why drugs have failed, and what can be done to improve outcomes. Neuropharmacology. 2017;120:11–9.
8. Morgan P, Van Der Graaf PH, Arrowsmith J, Feltner DE, Drummond KS, Wegner CD, et al. Can the flow of medicines be improved? Fundamental pharmacokinetic and pharmacological principles toward improving Phase II survival. Drug Discov Today. 2012;17(9–10):419–24.
9. Wu SS, Fernando K, Allerton C, Jansen KU, Vincent MS, Dolsten M. Reviving an R&D pipeline: a step change in the Phase II success rate. Drug Discov Today. 2021;26(2):308–14.
10. Fernando K, Menon S, Jansen K, Naik P, Nucci G, Roberts J, et al. Achieving end-to-end success in the clinic: Pfizer's learnings on R&D productivity. Drug Discov Today. 2022;27(3):697–704.
11. Cook D, Brown D, Alexander R, March R, Morgan P, Satterthwaite G, et al. Lessons learned from the fate of AstraZeneca's drug pipeline: a five-dimensional framework. Nat Rev Drug Discov. 2014;13(6):419–31.

12. Morgan P, Brown DG, Lennard S, Anderton MJ, Barrett JC, Eriksson U, et al. Impact of a five-dimensional framework on R&D productivity at AstraZeneca. Nat Rev Drug Discov. 2018;17(3):167–81.
13. Shen J, Swift B, Mamelok R, Pine S, Sinclair J, Attar M. Design and conduct considerations for first-in-human trials. Clin Transl Sci. 2019;12(1):6–19.
14. Ivy SP, Siu LL, Garrett-Mayer E, Rubinstein L. Approaches to phase 1 clinical trial design focused on safety, efficiency, and selected patient populations: a report from the clinical trial design task force of the national cancer institute investigational drug steering committee. Clin Cancer Res. 2010;16(6):1726–36.
15. Storer BE. Design and analysis of phase I clinical trials. Biometrics. 1989;45(3):925–37.
16. Cutler NR, Sramek JJ, Greenblatt DJ, Chaikin P, Ford N, Lesko LJ, et al. Defining the maximum tolerated dose: investigator, academic, industry and regulatory perspectives. J Clin Pharmacol. 1997;37(9):767–83.
17. Sramek JJ, Murphy MF, Adcock S, Stark JG, Cutler NR. Phase 1 clinical trials of small molecules: evolution and state of the art. Rev Recent Clin Trials. 2021;16(3):232–41.
18. Stampfer HG, Gabb GM, Dimmitt SB. Why maximum tolerated dose? Br J Clin Pharmacol. 2019;85(10):2213–7.
19. Wong HH, Halford S. Dose-limiting toxicity and maximum tolerated dose: still fit for purpose? Lancet Oncol. 2015;16(13):1287–8.
20. Ting N, Chen D-G, Ho S, Cappelleri JC. Design a proof of concept trial. In: Phase II clinical development of new drugs. Singapore: Springer Singapore; 2017. p. 75–92.
21. Yan F, Thall PF, Lu KH, Gilbert MR, Yuan Y. Phase I-II clinical trial design: a state-of-the-art paradigm for dose finding. Ann Oncol. 2018;29(3):694–9.
22. Sleijfer S, Wiemer E. Dose selection in phase I studies: why we should always go for the top. J Clin Oncol. 2008;26(10):1576–8.
23. Cummings J, Feldman HH, Scheltens P. The "rights" of precision drug development for Alzheimer's disease. Alzheimers Res Ther. 2019;11(1):76.
24. Wong DF, Tauscher J, Grunder G. The role of imaging in proof of concept for CNS drug discovery and development. Neuropsychopharmacology. 2009;34(1):187–203.
25. Breithaupt-Grogler K, Hardman T, de Hoon J, Donazzolo Y, Rottey S, Sourgens H, et al. The new first-in-human EMA guideline: disruptive or constructive? Outcomes from the first EUFEMED discussion forum. Front Pharmacol. 2019;10:398.
26. Wegener G, Rujescu D. The current development of CNS drug research. Int J Neuropsychopharmacol. 2013;16(7):1687–93.
27. English BA, Thomas K, Johnstone J, Bazih A, Gertsik L, Ereshefsky L. Use of translational pharmacodynamic biomarkers in early-phase clinical studies for schizophrenia. Biomark Med. 2014;8(1):29–49.
28. Pankevich DE, Altevogt BM, Dunlop J, Gage FH, Hyman SE. Improving and accelerating drug development for nervous system disorders. Neuron. 2014;84(3):546–53.
29. Nandal S, Burt T. Integrating pharmacoproteomics into early-phase clinical development: state-of-the-art, challenges, and recommendations. Int J Mol Sci. 2017;18(2):448.
30. Brady LS, Insel TR. Translating discoveries into medicine: psychiatric drug development in 2011. Neuropsychopharmacology. 2012;37(1):281–3.
31. Wexler D, Bertelsen KM. A brief survey of first-in-human studies. J Clin Pharmacol. 2011;51(7):988–93.
32. Penel N, Kramar A. What does a modified-Fibonacci dose-escalation actually correspond to? BMC Med Res Methodol. 2012;12:103.
33. Dresser R. First-in-human trial participants: not a vulnerable population, but vulnerable nonetheless. J Law Med Ethics. 2009;37(1):38–50.
34. North B, Kocher HM, Sasieni P. A new pragmatic design for dose escalation in phase 1 clinical trials using an adaptive continual reassessment method. BMC Cancer. 2019;19(1):632.
35. O'Quigley J, Pepe M, Fisher L. Continual reassessment method: a practical design for phase 1 clinical trials in cancer. Biometrics. 1990;46(1):33–48.

36. Petroni GR, Wages NA, Paux G, Dubois F. Implementation of adaptive methods in early-phase clinical trials. Stat Med. 2017;36(2):215–24.
37. Orloff JJ, Stanski D. Innovative approaches to clinical development and trial design. Ann Ist Super Sanita. 2011;47(1):8–13.
38. Thorlund K, Haggstrom J, Park JJ, Mills EJ. Key design considerations for adaptive clinical trials: a primer for clinicians. BMJ. 2018;360:k698.
39. Sturm S, Gunther A, Jaber B, Jordan P, Al Kotbi N, Parkar N, et al. A phase 1 healthy male volunteer single escalating dose study of the pharmacokinetics and pharmacodynamics of risdiplam (RG7916, RO7034067), a SMN2 splicing modifier. Br J Clin Pharmacol. 2019;85(1):181–93.
40. Sturm S, Delporte ML, Hadi S, Schobel S, Lindemann L, Weikert R, et al. Results and evaluation of a first-in-human study of RG7342, an mGlu5 positive allosteric modulator, utilizing Bayesian adaptive methods. Br J Clin Pharmacol. 2018;84(3):445–55.
41. Moore JW, Turner DC, Corlett PR, Arana FS, Morgan HL, Absalom AR, et al. Ketamine administration in healthy volunteers reproduces aberrant agency experiences associated with schizophrenia. Cogn Neuropsychiatry. 2011;16(4):364–81.
42. Umbricht D, Schmid L, Koller R, Vollenweider FX, Hell D, Javitt DC. Ketamine-induced deficits in auditory and visual context-dependent processing in healthy volunteers: implications for models of cognitive deficits in schizophrenia. Arch Gen Psychiatry. 2000;57(12):1139–47.
43. Gilles C, Luthringer R. Pharmacological models in healthy volunteers: their use in the clinical development of psychotropic drugs. J Psychopharmacol. 2007;21(3):272–82.
44. Izzo NJ, Yuede CM, LaBarbera KM, Limegrover CS, Rehak C, Yurko R, et al. Preclinical and clinical biomarker studies of CT1812: a novel approach to Alzheimer's disease modification. Alzheimers Dement. 2021;17(8):1365–82.
45. Javitt DC, Spencer KM, Thaker GK, Winterer G, Hajos M. Neurophysiological biomarkers for drug development in schizophrenia. Nat Rev Drug Discov. 2008;7(1):68–83.
46. Ford JM, Mathalon DH, Marsh L, Faustman WO, Harris D, Hoff AL, et al. P300 amplitude is related to clinical state in severely and moderately ill patients with schizophrenia. Biol Psychiatry. 1999;46(1):94–101.
47. Baldeweg T, Klugman A, Gruzelier J, Hirsch SR. Mismatch negativity potentials and cognitive impairment in schizophrenia. Schizophr Res. 2004;69(2–3, 203):–17.
48. Luntz-Leybman V, Bickford PC, Freedman R. Cholinergic gating of response to auditory stimuli in rat hippocampus. Brain Res. 1992;587(1):130–6.
49. Preskorn SH, Gawryl M, Dgetluck N, Palfreyman M, Bauer LO, Hilt DC. Normalizing effects of EVP-6124, an alpha-7 nicotinic partial agonist, on event-related potentials and cognition: a proof of concept, randomized trial in patients with schizophrenia. J Psychiatr Pract. 2014;20(1):12–24.
50. Keefe RS, Meltzer HA, Dgetluck N, Gawryl M, Koenig G, Moebius HJ, et al. Randomized, double-blind, placebo-controlled study of encenicline, an alpha7 nicotinic acetylcholine receptor agonist, as a treatment for cognitive impairment in schizophrenia. Neuropsychopharmacology. 2015;40(13):3053–60.
51. Cutler NR, Sramek JJ, Kilborn JR. The bridging concept: optimizing dose for phase II/III in Alzheimer's disease. Neurodegeneration. 1996;5(4):511–4.
52. Nations KR, Bursi R, Dogterom P, Ereshefsky L, Gertsik L, Mant T, et al. Maximum tolerated dose evaluation of the AMPA modulator Org 26576 in healthy volunteers and depressed patients: a summary and method analysis of bridging research in support of phase II dose selection. Drugs R D. 2012;12(3):127–39.
53. Sramek JJ, Eldon MA, Posvar E, Feng MR, Jhee SS, Hourani J, et al. Initial safety, tolerability pharmacodynamics, and pharmacokinetics of CI-1007 in patients with schizophrenia. Psychopharmacol Bull. 1998;34(1):93–9.
54. Desai A, Benner L, Wu R, Gertsik L, Uz T, Marek GJ, et al. Pharmacokinetics of ASP4345 from single ascending-dose and multiple ascending-dose phase I studies. Clin Pharmacokinet. 2021;60(1):79–88.

55. Balice-Gordon R, Honey GD, Chatham C, Arce E, Duvvuri S, Naylor MG, et al. A neurofunctional domains approach to evaluate D1/D5 dopamine receptor partial agonism on cognition and motivation in healthy volunteers with low working memory capacity. Int J Neuropsychopharmacol. 2020;23(5):287–99.
56. Macek TA, Suzuki K, Asin K, Kimura H. Translational development strategies for TAK-063, a phosphodiesterase 10A inhibitor. Int J Neuropsychopharmacol. 2020;23(8):524–32.
57. Koychev I, Barkus E, Ettinger U, Killcross S, Roiser JP, Wilkinson L, et al. Evaluation of state and trait biomarkers in healthy volunteers for the development of novel drug treatments in schizophrenia. J Psychopharmacol. 2011;25(9):1207–25.
58. Cuthbert BN, Insel TR. Toward new approaches to psychotic disorders: the NIMH Research Domain Criteria project. Schizophr Bull. 2010;36(6):1061–2.
59. Kozak MJ, Cuthbert BN. The NIMH research domain criteria initiative: background, issues, and pragmatics. Psychophysiology. 2016;53(3):286–97.
60. Cools R, D'Esposito M. Inverted-U-shaped dopamine actions on human working memory and cognitive control. Biol Psychiatry. 2011;69(12):e113–25.
61. Gibbs SE, D'Esposito M. A functional magnetic resonance imaging study of the effects of pergolide, a dopamine receptor agonist, on component processes of working memory. Neuroscience. 2006;139(1):359–71.
62. van der Schaaf ME, Fallon SJ, Ter Huurne N, Buitelaar J, Cools R. Working memory capacity predicts effects of methylphenidate on reversal learning. Neuropsychopharmacology. 2013;38(10):2011–8.
63. Smith ME, Farah MJ. Are prescription stimulants "smart pills"? The epidemiology and cognitive neuroscience of prescription stimulant use by normal healthy individuals. Psychol Bull. 2011;137(5):717–41.
64. Biomarkers Definitions Working Group. Biomarkers and surrogate endpoints: Preferred definitions and conceptual framework. Clin Pharmacol Ther. 2001;69(3):89–95.
65. Carmichael O, Schwarz AJ, Chatham CH, Scott D, Turner JA, Upadhyay J, et al. The role of fMRI in drug development. Drug Discov Today. 2018;23(2):333–48.
66. Dawoud D, Naci H, Ciani O, Bujkiewicz S. Raising the bar for using surrogate endpoints in drug regulation and health technology assessment. BMJ. 2021;374:n2191.
67. Hershey LA, Tarawneh R. Clinical efficacy, drug safety, and surrogate endpoints: has aducanumab met all of its expectations? Neurology. 2021;97(11):517–8.
68. Arneric SP, Kern VD, Stephenson DT. Regulatory-accepted drug development tools are needed to accelerate innovative CNS disease treatments. Biochem Pharmacol. 2018;151:291–306.
69. Anderson DC, Kodukula K. Biomarkers in pharmacology and drug discovery. Biochem Pharmacol. 2014;87(1):172–88.
70. Davis KD, Aghaeepour N, Ahn AH, Angst MS, Borsook D, Brenton A, et al. Discovery and validation of biomarkers to aid the development of safe and effective pain therapeutics: challenges and opportunities. Nat Rev Neurol. 2020;16(7):381–400.
71. Hampel H, Lista S, Teipel SJ, Garaci F, Nistico R, Blennow K, et al. Perspective on future role of biological markers in clinical therapy trials of Alzheimer's disease: a long-range point of view beyond 2020. Biochem Pharmacol. 2014;88(4):426–49.
72. Baskaran A, Milev R, McIntyre RS. The neurobiology of the EEG biomarker as a predictor of treatment response in depression. Neuropharmacology. 2012;63(4):507–13.
73. Stampanoni Bassi M, Drulovic J, Pekmezovic T, Iezzi E, Sica F, Gilio L, et al. Cerebrospinal fluid inflammatory biomarkers predicting interferon-beta response in MS patients. Ther Adv Neurol Disord. 2020;13:1756286420970833.
74. Lesko LJ, Atkinson AJ Jr. Use of biomarkers and surrogate endpoints in drug development and regulatory decision making: criteria, validation, strategies. Annu Rev Pharmacol Toxicol. 2001;41:347–66.
75. Gromova M, Vaggelas A, Dallmann G, Seimetz D. Biomarkers: opportunities and challenges for drug development in the current regulatory landscape. Biomark Insights. 2020;15:1177271920974652.

76. Buyse M, Michiels S, Sargent DJ, Grothey A, Matheson A, de Gramont A. Integrating biomarkers in clinical trials. Expert Rev Mol Diagn. 2011;11(2):171–82.
77. Wilson FJ, Danjou P. Early decision-making in drug development: the potential role of pharmaco-EEG and pharmaco-sleep. Neuropsychobiology. 2015;72(3–4):188–94.
78. Insel T, Cuthbert B, Garvey M, Heinssen R, Pine DS, Quinn K, et al. Research domain criteria (RDoC): toward a new classification framework for research on mental disorders. Am J Psychiatry. 2010;167(7):748–51.
79. Krystal AD, Pizzagalli DA, Mathew SJ, Sanacora G, Keefe R, Song A, et al. The first implementation of the NIMH FAST-FAIL approach to psychiatric drug development. Nat Rev Drug Discov. 2018;18(1):82–4.
80. Grabb MC, Hillefors M, Potter WZ. The NIMH 'Fast-Fail Trials' (FAST) initiative: rationale, promise, and Progress. Pharmaceut Med. 2020;34(4):233–45.
81. Grabb MC, Cross AJ, Potter WZ, McCracken JT. Derisking psychiatric drug development: the NIMH's fast fail program, a novel precompetitive model. J Clin Psychopharmacol. 2016;36(5):419–21.
82. Smit DJA, Andreassen OA, Boomsma DI, Burwell SJ, Chorlian DB, de Geus EJC, et al. Large-scale collaboration in ENIGMA-EEG: a perspective on the meta-analytic approach to link neurological and psychiatric liability genes to electrophysiological brain activity. Brain Behav. 2021;11(8):e02188.
83. Glasser MF, Smith SM, Marcus DS, Andersson JL, Auerbach EJ, Behrens TE, et al. The human connectome project's neuroimaging approach. Nat Neurosci. 2016;19(9):1175–87.
84. Brady LS, Potter WZ. Public-private partnerships to revitalize psychiatric drug discovery. Expert Opin Drug Discov. 2014;9(1):1–8.
85. Brady LS, Winsky L, Goodman W, Oliveri ME, Stover E. NIMH initiatives to facilitate collaborations among industry, academia, and government for the discovery and clinical testing of novel models and drugs for psychiatric disorders. Neuropsychopharmacology. 2009;34(1):229–43.
86. Insel TR. The NIMH Research Domain Criteria (RDoC) Project: precision medicine for psychiatry. Am J Psychiatry. 2014;171(4):395–7.
87. Cuthbert BN, Insel TR. Toward the future of psychiatric diagnosis: the seven pillars of RDoC. BMC Med. 2013;11:126.
88. Harrison LA, Kats A, Williams ME, Aziz-Zadeh L. The importance of sensory processing in mental health: a proposed addition to the research domain criteria (RDoC) and suggestions for RDoC 2.0. Front Psychol. 2019;10:103.
89. Ross CA, Margolis RL. Research domain criteria: strengths, weaknesses, and potential alternatives for future psychiatric research. Mol Neuropsychiatry. 2019;5(4):218–36.
90. Patrick CJ, Hajcak G. Reshaping clinical science: introduction to the special issue on psychophysiology and the NIMH research domain criteria (RDoC) initiative. Psychophysiology. 2016;53(3):281–5.
91. Paul SM, Mytelka DS, Dunwiddie CT, Persinger CC, Munos BH, Lindborg SR, et al. How to improve R&D productivity: the pharmaceutical industry's grand challenge. Nat Rev Drug Discov. 2010;9(3):203–14.
92. Szabo ST, Kinon BJ, Brannan SK, Krystal AK, van Gerven JM, Mahableshwarkar A, et al. Lessons learned and potentials for improvement in CNS drug development: ISCTM section on designing the right series of experiments. Innov Clin Neurosci. 2015;12(3–4):11S–25S.
93. Der-Avakian A, Barnes SA, Markou A, Pizzagalli DA. Translational assessment of reward and motivational deficits in psychiatric disorders. Curr Top Behav Neurosci. 2016;28:231–62.
94. Bruijnzeel AW. Kappa-opioid receptor signaling and brain reward function. Brain Res Rev. 2009;62(1):127–46.
95. Carlezon WA Jr, Beguin C, Knoll AT, Cohen BM. Kappa-opioid ligands in the study and treatment of mood disorders. Pharmacol Ther. 2009;123(3):334–43.
96. Carlezon WA Jr, Krystal AD. Kappa-opioid antagonists for psychiatric disorders: from bench to clinical trials. Depress Anxiety. 2016;33(10):895–906.

97. Knutson B, Westdorp A, Kaiser E, Hommer D. FMRI visualization of brain activity during a monetary incentive delay task. NeuroImage. 2000;12(1):20–7.
98. Knutson B, Bhanji JP, Cooney RE, Atlas LY, Gotlib IH. Neural responses to monetary incentives in major depression. Biol Psychiatry. 2008;63(7):686–92.
99. Stoy M, Schlagenhauf F, Sterzer P, Bermpohl F, Hagele C, Suchotzki K, et al. Hyporeactivity of ventral striatum towards incentive stimuli in unmedicated depressed patients normalizes after treatment with escitalopram. J Psychopharmacol. 2012;26(5):677–88.
100. O'Donnell P, Rosen L, Alexander R, Murthy V, Davies CH, Ratti E. Strategies to address challenges in neuroscience drug discovery and development. Int J Neuropsychopharmacol. 2019;22(7):445–8.
101. Hu S, Lai Y, Valdes-Sosa PA, Bringas-Vega ML, Yao D. How do reference montage and electrodes setup affect the measured scalp EEG potentials? J Neural Eng. 2018;15(2):026013.
102. Javitt DC, Siegel SJ, Spencer KM, Mathalon DH, Hong LE, Martinez A, et al. A roadmap for development of neuro-oscillations as translational biomarkers for treatment development in neuropsychopharmacology. Neuropsychopharmacology. 2020;45(9):1411–22.
103. Kantrowitz JT, Epstein ML, Beggel O, Rohrig S, Lehrfeld JM, Revheim N, et al. Neurophysiological mechanisms of cortical plasticity impairments in schizophrenia and modulation by the NMDA receptor agonist D-serine. Brain. 2016;139(Pt 12):3281–95.
104. O'Donnell BF, Vohs JL, Krishnan GP, Rass O, Hetrick WP, Morzorati SL. The auditory steady-state response (ASSR): a translational biomarker for schizophrenia. Suppl Clin Neurophysiol. 2013;62:101–12.
105. Luck SJ, Gaspelin N. How to get statistically significant effects in any ERP experiment (and why you shouldn't). Psychophysiology. 2017;54(1):146–57.
106. Clementz BA, Trotti RL, Pearlson GD, Keshavan MS, Gershon ES, Keedy SK, et al. Testing psychosis phenotypes from bipolar-schizophrenia network for intermediate phenotypes for clinical application: biotype characteristics and targets. Biol Psychiatry Cogn Neurosci Neuroimaging. 2020;5(8):808–18.
107. Wilson FJ, Leiser SC, Ivarsson M, Christensen SR, Bastlund JF. Can pharmaco-electroencephalography help improve survival of central nervous system drugs in early clinical development? Drug Discov Today. 2014;19(3):282–8.
108. de la Salle S, Choueiry J, Shah D, Bowers H, McIntosh J, Ilivitsky V, et al. Effects of ketamine on resting-state EEG activity and their relationship to perceptual/dissociative symptoms in healthy humans. Front Pharmacol. 2016;7:348.
109. Light GA, Swerdlow NR, Rissling AJ, Radant A, Sugar CA, Sprock J, et al. Characterization of neurophysiologic and neurocognitive biomarkers for use in genomic and clinical outcome studies of schizophrenia. PLoS One. 2012;7(7):e39434.
110. Light GA, Swerdlow NR, Thomas ML, Calkins ME, Green MF, Greenwood TA, et al. Validation of mismatch negativity and P3a for use in multi-site studies of schizophrenia: characterization of demographic, clinical, cognitive, and functional correlates in COGS-2. Schizophr Res. 2015;163(1–3):63–72.
111. Olichney JM, Yang JC, Taylor J, Kutas M. Cognitive event-related potentials: biomarkers of synaptic dysfunction across the stages of Alzheimer's disease. J Alzheimers Dis. 2011;26(Suppl 3):215–28.
112. Katada E, Sato K, Sawaki A, Dohi Y, Ueda R, Ojika K. Long-term effects of donepezil on P300 auditory event-related potentials in patients with Alzheimer's disease. J Geriatr Psychiatry Neurol. 2003;16(1):39–43.
113. Onofrj M, Thomas A, Luciano AL, Iacono D, Di Rollo A, D'Andreamatteo G, et al. Donepezil versus vitamin E in Alzheimer's disease: part 2: mild versus moderate-severe Alzheimer's disease. Clin Neuropharmacol. 2002;25(4):207–15.
114. Papadaniil CD, Kosmidou VE, Tsolaki A, Tsolaki M, Kompatsiaris IY, Hadjileontiadis LJ. Cognitive MMN and P300 in mild cognitive impairment and Alzheimer's disease: a high density EEG-3D vector field tomography approach. Brain Res. 2016;1648(Pt A):425–33.

115. Cecchi M, Moore DK, Sadowsky CH, Solomon PR, Doraiswamy PM, Smith CD, et al. A clinical trial to validate event-related potential markers of Alzheimer's disease in outpatient settings. Alzheimers Dement (Amst). 2015;1(4):387–94.
116. Krystal AD. Psychiatric disorders and sleep. Neurol Clin. 2012;30(4):1389–413.
117. Basner RC, editor. Case studies in polysomnography interpretation. Cambridge University Press.
118. Manoach DS, Stickgold R. Does abnormal sleep impair memory consolidation in schizophrenia? Front Hum Neurosci. 2009;3:21.
119. Krystal AD, Goforth HW, Roth T. Effects of antipsychotic medications on sleep in schizophrenia. Int Clin Psychopharmacol. 2008;23(3):150–60.
120. Goder R, Fritzer G, Gottwald B, Lippmann B, Seeck-Hirschner M, Serafin I, et al. Effects of olanzapine on slow wave sleep, sleep spindles and sleep-related memory consolidation in schizophrenia. Pharmacopsychiatry. 2008;41(3):92–9.
121. DeMartinis NA, Winokur A. Effects of psychiatric medications on sleep and sleep disorders. CNS Neurol Disord Drug Targets. 2007;6(1):17–29.
122. Killgore WD. Effects of sleep deprivation on cognition. Prog Brain Res. 2010;185:105–29.
123. Krystal AD, Durrence HH, Scharf M, Jochelson P, Rogowski R, Ludington E, et al. Efficacy and safety of doxepin 1 mg and 3 mg in a 12-week sleep laboratory and outpatient trial of elderly subjects with chronic primary insomnia. Sleep. 2010;33(11):1553–61.
124. Trivedi MH, Rush AJ, Armitage R, Gullion CM, Grannemann BD, Orsulak PJ, et al. Effects of fluoxetine on the polysomnogram in outpatients with major depression. Neuropsychopharmacology. 1999;20(5):447–59.
125. Sculthorpe LD, Douglass AB. Sleep pathologies in depression and the clinical utility of polysomnography. Can J Psychiatr. 2010;55(7):413–21.
126. Palagini L, Baglioni C, Ciapparelli A, Gemignani A, Riemann D. REM sleep dysregulation in depression: state of the art. Sleep Med Rev. 2013;17(5):377–90.
127. Dolder CR, Nelson MH, Iler CA. The effects of mirtazapine on sleep in patients with major depressive disorder. Ann Clin Psychiatry. 2012;24(3):215–24.
128. Sharpley AL, Cowen PJ. Effect of pharmacologic treatments on the sleep of depressed patients. Biol Psychiatry. 1995;37(2):85–98.
129. Croom KF, Perry CM, Plosker GL. Mirtazapine: a review of its use in major depression and other psychiatric disorders. CNS Drugs. 2009;23(5):427–52.
130. Sangal RB, Thomas L, Mitler MM. Maintenance of wakefulness test and multiple sleep latency test. Measurement of different abilities in patients with sleep disorders. Chest. 1992;101(4):898–902.
131. Jaeger J. Digit symbol substitution test: the case for sensitivity over specificity in neuropsychological testing. J Clin Psychopharmacol. 2018;38(5):513–9.
132. Dedic N, Dworak H, Zeni C, Rutigliano G, Howes OD. Therapeutic potential of TAAR1 agonists in schizophrenia: evidence from preclinical models and clinical studies. Int J Mol Sci. 2021;22(24):13185.
133. Hopkins SC, Dedic N, Koblan KS. Effect of TAAR1/5-HT1A agonist SEP-363856 on REM sleep in humans. Transl Psychiatry. 2021;11(1):228.
134. George N, Gean EG, Nandi A, Frolov B, Zaidi E, Lee H, et al. Advances in CNS imaging agents: focus on PET and SPECT tracers in experimental and clinical use. CNS Drugs. 2015;29(4):313–30.
135. Arakawa R, Takano A, Halldin C. PET technology for drug development in psychiatry. Neuropsychopharmacol Rep. 2020;40(2):114–21.
136. Innis RB, Cunningham VJ, Delforge J, Fujita M, Gjedde A, Gunn RN, et al. Consensus nomenclature for in vivo imaging of reversibly binding radioligands. J Cereb Blood Flow Metab. 2007;27(9):1533–9.
137. Hargreaves RJ, Rabiner EA. Translational PET imaging research. Neurobiol Dis. 2014;61:32–8.

138. Laruelle M, Slifstein M, Huang Y. Positron emission tomography: imaging and quantification of neurotransporter availability. Methods. 2002;27(3):287–99.
139. Christian BT, Livni E, Babich JW, Alpert NM, Dischino DD, Ruediger E, et al. Evaluation of cerebral pharmacokinetics of the novel antidepressant drug, BMS-181101, by positron emission tomography. J Pharmacol Exp Ther. 1996;279(1):325–31.
140. Meyer JH, Wilson AA, Ginovart N, Goulding V, Hussey D, Hood K, et al. Occupancy of serotonin transporters by paroxetine and citalopram during treatment of depression: a [(11) C]DASB PET imaging study. Am J Psychiatry. 2001;158(11):1843–9.
141. Meyer JH, Wilson AA, Sagrati S, Hussey D, Carella A, Potter WZ, et al. Serotonin transporter occupancy of five selective serotonin reuptake inhibitors at different doses: an [11C]DASB positron emission tomography study. Am J Psychiatry. 2004;161(5):826–35.
142. Arakawa R, Stenkrona P, Takano A, Svensson J, Andersson M, Nag S, et al. Venlafaxine ER blocks the norepinephrine transporter in the brain of patients with major depressive disorder: a PET study using [18F]FMeNER-D2. Int J Neuropsychopharmacol. 2019;22(4):278–85.
143. Kapur S, Zipursky R, Jones C, Remington G, Houle S. Relationship between dopamine D(2) occupancy, clinical response, and side effects: a double-blind PET study of first-episode schizophrenia. Am J Psychiatry. 2000;157(4):514–20.
144. Farde L, Nordstrom AL. PET examination of central D2 dopamine receptor occupancy in relation to extrapyramidal syndromes in patients being treated with neuroleptic drugs. Psychopharmacol Ser. 1993;10:94–100.
145. Nordstrom AL, Farde L, Wiesel FA, Forslund K, Pauli S, Halldin C, et al. Central D2-dopamine receptor occupancy in relation to antipsychotic drug effects: a double-blind PET study of schizophrenic patients. Biol Psychiatry. 1993;33(4):227–35.
146. Cummings J, Lee G, Zhong K, Fonseca J, Taghva K. Alzheimer's disease drug development pipeline: 2021. Alzheimers Dement (N Y). 2021;7(1):e12179.
147. Swanson CJ, Zhang Y, Dhadda S, Wang J, Kaplow J, Lai RYK, et al. A randomized, double-blind, phase 2b proof-of-concept clinical trial in early Alzheimer's disease with lecanemab, an anti-Abeta protofibril antibody. Alzheimers Res Ther. 2021;13(1):80.
148. van Dyck CH, Nygaard HB, Chen K, Donohue MC, Raman R, Rissman RA, et al. Effect of AZD0530 on cerebral metabolic decline in Alzheimer disease: a randomized clinical trial. JAMA Neurol. 2019;76(10):1219–29.
149. Mega MS, Dinov ID, Porter V, Chow G, Reback E, Davoodi P, et al. Metabolic patterns associated with the clinical response to galantamine therapy: a fludeoxyglucose f 18 positron emission tomographic study. Arch Neurol. 2005;62(5):721–8.
150. Merchant KM, Cedarbaum JM, Brundin P, Dave KD, Eberling J, Espay AJ, et al. A proposed roadmap for Parkinson's disease proof of concept clinical trials investigating compounds targeting alpha-synuclein. J Parkinsons Dis. 2019;9(1):31–61.
151. Squitieri F, Ciammola A, Colonnese C, Ciarmiello A. Neuroprotective effects of riluzole in Huntington's disease. Eur J Nucl Med Mol Imaging. 2008;35(1):221–2.
152. Dupont AC, Largeau B, Guilloteau D, Santiago Ribeiro MJ, Arlicot N. The place of PET to assess new therapeutic effectiveness in neurodegenerative diseases. Contrast Media Mol Imaging. 2018;2018:7043578.
153. Vernaleken I, Fellows C, Janouschek H, Brocheler A, Veselinovic T, Landvogt C, et al. Striatal and extrastriatal D2/D3-receptor-binding properties of ziprasidone: a positron emission tomography study with [18F]Fallypride and [11C]raclopride (D2/D3-receptor occupancy of ziprasidone). J Clin Psychopharmacol. 2008;28(6):608–17.
154. Daniel DG, Zimbroff DL, Potkin SG, Reeves KR, Harrigan EP, Lakshminarayanan M. Ziprasidone 80 mg/day and 160 mg/day in the acute exacerbation of schizophrenia and schizoaffective disorder: a 6-week placebo-controlled trial. Ziprasidone Study Group. Neuropsychopharmacology. 1999;20(5):491–505.
155. Keck P Jr, Buffenstein A, Ferguson J, Feighner J, Jaffe W, Harrigan EP, et al. Ziprasidone 40 and 120 mg/day in the acute exacerbation of schizophrenia and schizoaffective disorder: a 4-week placebo-controlled trial. Psychopharmacology. 1998;140(2):173–84.

156. Mamo D, Kapur S, Shammi CM, Papatheodorou G, Mann S, Therrien F, et al. A PET study of dopamine D2 and serotonin 5-HT2 receptor occupancy in patients with schizophrenia treated with therapeutic doses of ziprasidone. Am J Psychiatry. 2004;161(5):818–25.
157. Sadraee A, Paulus M, Ekhtiari H. fMRI as an outcome measure in clinical trials: a systematic review in clinicaltrials.gov. Brain Behav. 2021;11(5):e02089.
158. Suhara T, Chaki S, Kimura H, Furusawa M, Matsumoto M, Ogura H, et al. Strategies for utilizing neuroimaging biomarkers in CNS drug discovery and development: CINP/JSNP working group report. Int J Neuropsychopharmacol. 2017;20(4):285–94.
159. Borsook D, Becerra L, Hargreaves R. A role for fMRI in optimizing CNS drug development. Nat Rev Drug Discov. 2006;5(5):411–24.
160. Wang DJ, Chen Y, Fernandez-Seara MA, Detre JA. Potentials and challenges for arterial spin labeling in pharmacological magnetic resonance imaging. J Pharmacol Exp Ther. 2011;337(2):359–66.
161. Jenkins BG. Pharmacologic magnetic resonance imaging (phMRI): imaging drug action in the brain. NeuroImage. 2012;62(2):1072–85.
162. Gore JC. Principles and practice of functional MRI of the human brain. J Clin Invest. 2003;112(1):4–9.
163. Khanna N, Altmeyer W, Zhuo J, Steven A. Functional neuroimaging: fundamental principles and clinical applications. Neuroradiol J. 2015;28(2):87–96.
164. Moseley M. Diffusion tensor imaging and aging - a review. NMR Biomed. 2002;15(7–8):553–60.
165. Hargreaves RJ, Hoppin J, Sevigny J, Patel S, Chiao P, Klimas M, et al. Optimizing central nervous system drug development using molecular imaging. Clin Pharmacol Ther. 2015;98(1):47–60.
166. Nathan PJ, Phan KL, Harmer CJ, Mehta MA, Bullmore ET. Increasing pharmacological knowledge about human neurological and psychiatric disorders through functional neuroimaging and its application in drug discovery. Curr Opin Pharmacol. 2014;14:54–61.
167. Tomimatsu Y, Cash D, Suzuki M, Suzuki K, Bernanos M, Simmons C, et al. TAK-063, a phosphodiesterase 10A inhibitor, modulates neuronal activity in various brain regions in phMRI and EEG studies with and without ketamine challenge. Neuroscience. 2016;339:180–90.
168. Arce E, Balice-Gordon R, Duvvuri S, Naylor M, Xie Z, Harel B, et al. A novel approach to evaluate the pharmacodynamics of a selective dopamine D1/D5 receptor partial agonist (PF-06412562) in patients with stable schizophrenia. J Psychopharmacol. 2019;33(10):1237–47.
169. Esposito F, Otto T, Zijlstra FR, Goebel R. Spatially distributed effects of mental exhaustion on resting-state FMRI networks. PLoS One. 2014;9(4):e94222.
170. De Simoni S, Schwarz AJ, O'Daly OG, Marquand AF, Brittain C, Gonzales C, et al. Test-retest reliability of the BOLD pharmacological MRI response to ketamine in healthy volunteers. NeuroImage. 2013;64:75–90.
171. Lee MH, Smyser CD, Shimony JS. Resting-state fMRI: a review of methods and clinical applications. AJNR Am J Neuroradiol. 2013;34(10):1866–72.
172. Greicius MD, Krasnow B, Reiss AL, Menon V. Functional connectivity in the resting brain: a network analysis of the default mode hypothesis. Proc Natl Acad Sci U S A. 2003;100(1):253–8.
173. Simon R, Engstrom M. The default mode network as a biomarker for monitoring the therapeutic effects of meditation. Front Psychol. 2015;6:776.
174. Cole DM, Smith SM, Beckmann CF. Advances and pitfalls in the analysis and interpretation of resting-state FMRI data. Front Syst Neurosci. 2010;4:8.
175. Smucny J, Wylie KP, Tregellas JR. Functional magnetic resonance imaging of intrinsic brain networks for translational drug discovery. Trends Pharmacol Sci. 2014;35(8):397–403.
176. Duff EP, Vennart W, Wise RG, Howard MA, Harris RE, Lee M, et al. Learning to identify CNS drug action and efficacy using multistudy fMRI data. Sci Transl Med. 2015;7(274):274ra16.

177. Wager TD, Atlas LY, Lindquist MA, Roy M, Woo CW, Kross E. An fMRI-based neurologic signature of physical pain. N Engl J Med. 2013;368(15):1388–97.
178. Upadhyay J, Anderson J, Schwarz AJ, Coimbra A, Baumgartner R, Pendse G, et al. Imaging drugs with and without clinical analgesic efficacy. Neuropsychopharmacology. 2011;36(13):2659–73.
179. Sarpal DK, Robinson DG, Lencz T, Argyelan M, Ikuta T, Karlsgodt K, et al. Antipsychotic treatment and functional connectivity of the striatum in first-episode schizophrenia. JAMA Psychiat. 2015;72(1):5–13.
180. Risacher SL, Wang Y, Wishart HA, Rabin LA, Flashman LA, McDonald BC, et al. Cholinergic enhancement of brain activation in mild cognitive impairment during episodic memory encoding. Front Psych. 2013;4:105.
181. Kantrowitz JT, Grinband J, Goff DC, Lahti AC, Marder SR, Kegeles LS, et al. Proof of mechanism and target engagement of glutamatergic drugs for the treatment of schizophrenia: RCTs of pomaglumetad and TS-134 on ketamine-induced psychotic symptoms and pharmacoBOLD in healthy volunteers. Neuropsychopharmacology. 2020;45(11):1842–50.
182. Botvinik-Nezer R, Holzmeister F, Camerer CF, Dreber A, Huber J, Johannesson M, et al. Variability in the analysis of a single neuroimaging dataset by many teams. Nature. 2020;582(7810):84–8.
183. Minzenberg MJ, Laird AR, Thelen S, Carter CS, Glahn DC. Meta-analysis of 41 functional neuroimaging studies of executive function in schizophrenia. Arch Gen Psychiatry. 2009;66(8):811–22.
184. Harmer CJ, Cowen PJ, Goodwin GM. Efficacy markers in depression. J Psychopharmacol. 2011;25(9):1148–58.
185. Wessa M, Lois G. Brain functional effects of psychopharmacological treatment in major depression: a focus on neural circuitry of affective processing. Curr Neuropharmacol. 2015;13(4):466–79.
186. Krystal AD, Pizzagalli DA, Smoski M, Mathew SJ, Nurnberger J Jr, Lisanby SH, et al. A randomized proof-of-mechanism trial applying the 'fast-fail' approach to evaluating kappa-opioid antagonism as a treatment for anhedonia. Nat Med. 2020;26(5):760–8.
187. Borsook D, Becerra L, Fava M. Use of functional imaging across clinical phases in CNS drug development. Transl Psychiatry. 2013;3:e282.
188. Upadhyay J, Geber C, Hargreaves R, Birklein F, Borsook D. A critical evaluation of validity and utility of translational imaging in pain and analgesia: utilizing functional imaging to enhance the process. Neurosci Biobehav Rev. 2018;84:407–23.
189. Borsook D, Upadhyay J, Klimas M, Schwarz AJ, Coimbra A, Baumgartner R, et al. Decision-making using fMRI in clinical drug development: revisiting NK-1 receptor antagonists for pain. Drug Discov Today. 2012;17(17–18):964–73.
190. Egerton A. The potential of (1)H-MRS in CNS drug development. Psychopharmacology. 2021;238(5):1241–54.
191. Javitt DC, Carter CS, Krystal JH, Kantrowitz JT, Girgis RR, Kegeles LS, et al. Utility of imaging-based biomarkers for glutamate-targeted drug development in psychotic disorders: a randomized clinical trial. JAMA Psychiat. 2018;75(1):11–9.
192. Goldsmith P, Affinito J, McCue M, Tsai M, Roepcke S, Xie J, et al. A randomized multiple dose pharmacokinetic study of a novel PDE10A inhibitor TAK-063 in subjects with stable schizophrenia and Japanese subjects and modeling of exposure relationships to adverse events. Drugs R D. 2017;17(4):631–43.
193. Takano A, Stenkrona P, Stepanov V, Amini N, Martinsson S, Tsai M, et al. A human [(11)C]T-773 PET study of PDE10A binding after oral administration of TAK-063, a PDE10A inhibitor. NeuroImage. 2016;141:10–7.
194. Yurgelun-Todd DA, Renshaw PF, Goldsmith P, Uz T, Macek TA. A randomized, placebo-controlled, phase 1 study to evaluate the effects of TAK-063 on ketamine-induced changes in fMRI BOLD signal in healthy subjects. Psychopharmacology. 2020;237(2):317–28.

195. Tsai M, Chrones L, Xie J, Gevorkyan H, Macek TA. A phase 1 study of the safety, tolerability, pharmacokinetics, and pharmacodynamics of TAK-063, a selective PDE10A inhibitor. Psychopharmacology. 2016;233(21–22):3787–95.
196. Zagorska A, Partyka A, Bucki A, Gawalskax A, Czopek A, Pawlowski M. Phosphodiesterase 10 inhibitors - novel perspectives for psychiatric and neurodegenerative drug discovery. Curr Med Chem. 2018;25(29):3455–81.
197. Menniti FS, Chappie TA, Schmidt CJ. PDE10A inhibitors-clinical failure or window into antipsychotic drug action? Front Neurosci. 2020;14:600178.
198. DeMartinis N 3rd, Lopez RN, Pickering EH, Schmidt CJ, Gertsik L, Walling DP, et al. A proof-of-concept study evaluating the phosphodiesterase 10A inhibitor PF-02545920 in the adjunctive treatment of suboptimally controlled symptoms of schizophrenia. J Clin Psychopharmacol. 2019;39(4):318–28.
199. Walling DP, Banerjee A, Dawra V, Boyer S, Schmidt CJ, DeMartinis N. Phosphodiesterase 10A inhibitor monotherapy is not an effective treatment of acute schizophrenia. J Clin Psychopharmacol. 2019;39(6):575–82.
200. Umbricht D, Abt M, Tamburri P, Chatham C, Holiga S, Frank MJ, Collins AGE, Walling DP, Mofsen R, Gruener D, Gertsik L, Sevigny J, Sanjay K, Dukart J. Proof-of-mechanism study of the phosphodiesterase 10 inhibitor RG7203 in patients with schizophrenia and negative symptoms. Biological psychiatry global open. Science. 2021;1(1):70–7.
201. Takano A, Stepanov V, Nakao R, Amini N, Gulyas B, Kimura H, et al. Brain PET measurement of PDE10A occupancy by TAK-063, a new PDE10A inhibitor, using [11 C]T-773 in nonhuman primates. Synapse. 2016;70(6):253–63.
202. Shiraishi E, Suzuki K, Harada A, Suzuki N, Kimura H. The phosphodiesterase 10A selective inhibitor TAK-063 improves cognitive functions associated with schizophrenia in rodent models. J Pharmacol Exp Ther. 2016;356(3):587–95.
203. Kapur S, Seeman P. NMDA receptor antagonists ketamine and PCP have direct effects on the dopamine D(2) and serotonin 5-HT(2)receptors-implications for models of schizophrenia. Mol Psychiatry. 2002;7(8):837–44.
204. Macek TA, McCue M, Dong X, Hanson E, Goldsmith P, Affinito J, et al. A phase 2, randomized, placebo-controlled study of the efficacy and safety of TAK-063 in subjects with an acute exacerbation of schizophrenia. Schizophr Res. 2019;204:289–94.

Index

M. Macaluso et al. (eds.), *Drug Development in Psychiatry*, Advances in Neurobiology 30, https://doi.org/10.1007/978-3-031-21054-9

The manufacturer's authorised representative in the EU is Springer Nature Customer Service Centre GmbH, Europaplatz 3, 69115 Heidelberg, Germany. If you have any concerns regarding our products, please contact ProductSafety@springernature.com

Printed and bound by CPI Group (UK) Ltd, Croydon, CR0 4YY
28/09/2026
02224056-0010